Climate Change and Cities
First Assessment Report of the
Urban Climate Change Research Network

Urban areas are home to over half of the world's people and are at the forefront of the climate change issue. Climate change exerts added stress on urban areas through increased numbers of heat waves threatening the health of the elderly, the infirm, and the very young; more frequent and intense droughts and inland floods compromising water supplies; and for coastal cities, enhanced sea level rise and storm surges affecting essential infrastructure, property, ecosystems, and inhabitants. At the same time, cities are responsible for no less than 40% of global greenhouse gas emissions, and given current demographic trends, this level will likely only increase over time. These challenges highlight the need for cities to rethink how assets and people are deployed and protected, how infrastructure investments are prioritized, and how climate will affect long-term growth and development plans.

Work on the *First Assessment Report on Climate Change and Cities* (ARC3) was launched by the Urban Climate Change Research Network (UCCRN) in November 2008 with the goal of building the scientific basis for city action on climate change. The authors include experts from cities in both the developing and developed world, representing a wide range of disciplines. The book focuses on how to use climate science and socio-economic research to map a city's vulnerability to climate hazards, and how cities can enhance their adaptive and mitigative capacity to deal with climate change over different timescales.

The volume is structured to communicate to a range of groups important for urban decision-making:

- The *Executive Summary* is invaluable for mayors, city officials, and policymakers;

- The *Urban Climate, Land Use*, and *Governance* chapters are of great interest to urban sustainability officers and urban planners;

- The *Sector* chapters are important for mid-level urban stakeholders in agencies charged with developing climate change mitigation and adaptation programs;

- The entire volume, including the framing *Urban Climate Change in Context* and the *Cities, Disasters, and Climate Risk* chapters, provides a broad spectrum of climate change knowledge to researchers, professors, and advanced students.

Cynthia Rosenzweig is a Senior Research Scientist at the NASA Goddard Institute for Space Studies where she heads the Climate Impacts Group. She recently co-chaired the New York City Panel on Climate Change, a body of experts convened by the Mayor to advise the city on adaptation for its critical infrastructure. She co-led the Metropolitan East Coast Regional Assessment of the U.S. National Assessment of the Potential Consequences of Climate Variability and Change, sponsored by the U.S. Global Change Research Program. She was a Coordinating Lead Author of the IPCC Working Group II Fourth Assessment Report, and served on the IPCC Task Group on Data and Scenario Support for Impact and Climate Analysis. A recipient of a Guggenheim Fellowship, she joins impact models with climate models to project future outcomes of both land-based and urban systems under altered climate conditions. She is a Professor at Barnard College and a Senior Research Scientist at the Earth Institute at Columbia University.

William D. Solecki is a Professor in the Department of Geography, Hunter College, City University of New York. He has led or co-led numerous projects on the process of urban environmental change and transformation. As Director of the CUNY Institute for Sustainable Cities, he has worked extensively on connecting cutting-edge urban environmental science to everyday practice and action in cities. He most recently served as Co-Chair of the New York City Panel on Climate Change, as Co-Principal Investigator of the Integrated Assessment for Effective Climate Change Adaptation Strategies in New York State (ClimAID), and as Co-Leader of the Metropolitan East Coast Assessment of the US National Assessment of the Potential Consequences of Climate Variability and Change. He is a Lead Author of the IPCC Working Group II Fifth Assessment Report. He is also a member of the Scientific Steering Committee of the Urban and Global Environmental Change core project of the International Human Dimensions Programme.

Stephen A. Hammer is the President of Mesacosa LLC, a consultancy that conducts research on urban energy and climate issues in cities around the globe. He was formerly Executive Director of the Energy Smart Cities Initiative, a project of the Joint U.S.-China Collaboration on Clean Energy (JUCCCE), running energy and climate policy training and technical assistance programs for local governments in China. He is also the past director of the Urban Energy Program at Columbia University's Center for Energy, Marine Transportation and Public Policy. He has authored or co-authored dozens of policy studies and journal articles on urban sustainability planning, urban energy systems, distributed generation technology and the impacts of climate change on local and regional energy networks. He is an Adjunct Professor at Columbia University's School of International and Public Affairs, a member of New York City Mayor Bloomberg's Energy Policy Task Force, and a consultant to the OECD and the World Bank.

Shagun Mehrotra is Managing Director of Climate and Cities, an international policy advisory facility at Columbia University's Center for Climate Systems Research. As a Columbia University Faculty Fellow, he provides research and policy advice focusing on infrastructure economics and finance, development economics and poverty reduction in slums. He has developed a comprehensive framework for city climate risk assessment that combines hazards, vulnerabilities and agency. Previously, he was on the staff of the World Bank, leading infrastructure reform of state-owned utilities in Africa. Over the last decade, his advice has been sought by national and local governments in East Africa, South-East Asia, China and India, as well as the United Nations Human Development Report, the Bill and Melinda Gates Foundation, and the Columbia Earth Institute. His co-authored book, *Bankruptcy to Billions: How the Indian Railways Transformed*, was recently launched by the President of India.

Climate Change and Cities
First Assessment Report of the
Urban Climate Change Research Network

Edited by

Cynthia Rosenzweig

NASA Goddard Institute
for Space Studies and
Columbia University, New York

William D. Solecki

Hunter College,
City University of New York

Stephen A. Hammer

Mesacosa, LLC and
Columbia University, New York

Shagun Mehrotra

Columbia University,
New York

Urban Climate Change Research Network
Center for Climate Systems Research
Earth Institute, Columbia University

CAMBRIDGE UNIVERSITY PRESS
Cambridge, New York, Melbourne, Madrid, Cape Town,
Singapore, São Paulo, Delhi, Tokyo, Mexico City

Cambridge University Press
The Edinburgh Building, Cambridge CB2 8RU, UK

Published in the United States of America by Cambridge University Press, New York

www.cambridge.org
Information on this title: www.cambridge.org/9781107004207

First published 2011

Printed in the United Kingdom at the University Press, Cambridge

A catalog record for this publication is available from the British Library

ISBN 978-1-107-00420-7 Paperback

Contents

Foreword - *Anna Tibaijuka*

The world rapidly urbanizing, and a majority of the global population will experience climate change in cities. Climate change will exacerbate the existing urban environmental management challenges in cities – in most cases making existing problems much worse. Additionally, it is the urban poor, who often are forced to live in flood- and landslide-prone areas and who face other vulnerabilities, who will bear a disproportionate share of the effects of climate change. Though cities are vulnerable to the effects of climate change, they are also uniquely positioned to take a global leadership role in both mitigating and adapting to it.

As cities begin to develop climate change action plans there is great need for a mechanism by which research and expert knowledge may contribute to the development and implementation of effective urban climate change policies and programs. Since responding to the complex challenges of climate change mitigation and adaptation requires a knowledge-based approach, the First UCCRN Assessment Report on Climate Change and Cities (ARC3) provides a tool for policymakers as they "mainstream" responses to climate change in urban areas.

The ARC3, a project of the Urban Climate Change Research Network (UCCRN), is innovative and important. It supports the work of local government officials and local researchers and complements the work of the already-existing body of knowledge developed by the Intergovernmental Panel on Climate Change (IPCC) by addressing the needs of cities. ARC3 provides the scientific base needed for sound mitigation and adaptation decision-making on a sector-by-sector basis, mirroring the administrative structure of a city.

The UCCRN's goal of turning the ARC3 process into an on-going initiative is also critical. Climate science is ever-evolving and cities are constantly reacting to and proactively addressing their unique climate change challenges. With the support of the research community, cities around the world will now have access to the latest information and the most robust understanding of climate change available.

I therefore applaud the work of groups such as UCCRN and the many researchers from both developing and developed cities contributing to this important research initiative and creating a mechanism to help cities further empower themselves. We will promote the use of the information compiled in ARC3 through our Cities and Climate Change Initiative and through our collaboration frameworks with other organizations, including the Joint Work Programme between the World Bank, UN-HABITAT and UNEP, supported by the Cities Alliance.

I am convinced that this body of knowledge will be of direct benefit and inspiration to the cities which we are supporting to develop climate action plans. It will help cities make more informed decisions about how climate change will affect public health, local infrastructure, and in turn, their own economic vitality in the coming decades.

Anna Tibaijuka
Former Under-Secretary General of the United Nations
Former Executive Director, UN-HABITAT

Foreword - *Rajendra Kumar Pachauri*

Clearly, cities are playing an increasing role in responding to climate challenges and are therefore in need of knowledge to aid in their policy development. The First Assessment Report on Climate Change in Cities (ARC3), a project of the Urban Climate Change Research Network (UCCRN), is particularly useful in this regard. The ARC3 provides a scientific assessment of climate change in cities, presenting the information necessary for sound mitigation and adaptation decision-making on a sector-by-sector basis. By specifically addressing climate change in cities, the ARC3 supports the work of local governments, officials and researchers, and complements the work of the Intergovernmental Panel on Climate Change (IPCC).

The Fourth Assessment Report of the Intergovernmental Panel on Climate Change (IPCC) provided the global community with up-to-date knowledge about the impacts of climate change. It projected that climate change will lead to a number of consequences for urban areas, including declining air quality, an increased number and severity of heat waves in cities in which heat waves already occur, increased pressure on infrastructure, and augmented stress on water resources. Furthermore, the Fourth Assessment Report noted that residents of some cities in the world, including some in Europe and the USA, have experienced high levels of mortality due to the impacts of extreme climate events. The 2003 European heat-related deaths and the deaths of over 1,000 people in New Orleans due to Hurricane Katrina are two examples of this. The IPCC Fifth Assessment, now underway, will continue to address these important issues, building on the work of the ARC3.

Due to the evolving nature of climate science, developing the ARC3 process into one that issues reports for cities on a regular basis is important. The UCCRN researchers will thus provide crucial information to urban decision-makers in a timely way as scientific understanding progresses.

I am highly appreciative of the work of the members of the UCCRN from developing and developed cities who are participating in the ARC3 activity. Policymakers, administrators, and researchers from cities around the world will benefit from the information provided in ARC3, enabling them to develop effective programs for mitigating and adapting to climate change.

Rajendra Kumar Pachauri
Chair, Intergovernmental Panel on Climate Change

Foreword - *Michael R. Bloomberg*

Cities are the vanguard in the battle against climate change. We are the source of approximately 80 percent of global greenhouse gas emissions. And densely populated urban areas, particularly coastal cities, will disproportionately feel the impacts of climate change. Those of us in local government recognize the importance of national and international leadership on climate change. But we also are not waiting for others to act first.

Through *PlaNYC*, New York City's comprehensive sustainability plan that we released in April 2007, we are working to create a greener, greater New York. Many of the 127 initiatives in the plan focus on reducing our greenhouse gas emissions. These initiatives, including the Greener, Greater Buildings Plan, which will increase the energy efficiency of existing buildings, will help us meet our goal of reducing the city's carbon emissions by 30 percent by 2030.

Four years after launching *PlaNYC*, we are seeing the benefits of our efforts. Our greenhouse gas emissions are down by over 12 percent from 2005 levels, and we've developed the City's first official climate change projections. We are now in the process of updating *PlaNYC*, and also doing more to draw on the creativity of New Yorkers in every borough. At the same time, we are learning from our colleagues across the world who are undertaking ambitious climate change mitigation and adaptation programs.

Five years ago, 18 of the world's great cities came together, to share best practices and make common cause in the effort to reduce green-house gas emissions. This group of 18 eventually grew into what is now the C40 Climate Leadership Group, a network of 40 of the largest cities in the world.

The Urban Climate Change Research Network recognizes the potential and responsibility of cities to enact change, and highlights the strategies employed by cities across the globe who are leading the way towards a sustainable future. The strategies explored in this text will not only help to guide individual local government efforts, but also help to tell the story of the critical importance of local action. The best scientific data tell us that it is long past time to address that challenge. And the best demographic data tell us that cities must lead the way.

Cities have demonstrated that we are prepared to boldly confront climate change. As mayors, we know that we don't have the luxury of simply talking about change without delivering it. The ARC3 project will help ensure that we not only create a greener, greater New York for future generations, that we continue to learn from the lessons of our counterparts across the world, and that we share our progress and our story with our partners throughout government, academia, and the private sector.

Michael R. Bloomberg
Mayor, New York City

Foreword - *Jeffrey D. Sachs*

The twenty-first century will be the age of sustainable development - or the age of ruin. Worldwide economic growth over the past two centuries has brought remarkable progress but also remarkable risk. By mobilizing fossil fuels, humanity lifted itself from the ancient scourges of hunger, disease, and early death. Living standards and income levels in many parts of the world soared beyond the wildest of expectations. Yet these gains are now bringing new and grave threats as well. Humanity has inadvertently pushed against the planet's safe boundaries regarding greenhouse gas emissions, land use changes, pollution, and human-induced threats to biodiversity and public health. In the coming decades, the core challenge of societies around the world will be to refashion our ways of life – living patterns, technologies, and economic systems – so that we can combine the benefits of economic development with sustainable management of the Earth's ecosystems.

Cities will be at the center of this unique and unprecedented challenge. During the past decade, humanity reached the decisive halfway point on the path to urbanization. From time immemorial until the Industrial Revolution two centuries ago, almost all of humanity lived as subsistence farmers in the rural areas. Starting two centuries ago, with the mobilization of new energy resources and technologies, including in food production, humanity began a long-term transition to urban living. As farmers became more productive, a declining share of the population could feed the rest. For generations now, the children of farm families have been heading to the cities for a new urban life. Today, the UN tells us that a little more than 50 percent of the world now lives in cities, and that by 2050 the proportion is likely to rise to nearly 70 percent.

How cities are structured – in the patterns of residential settlements, commercial and industrial land use, energy systems, transport networks, water and sewerage infrastructure, public health management, and more – will not only determine the quality of life of the majority of the world's population, but also whether humanity, at long last, is able to live sustainably with nature. To learn to do so is vital. Our livelihoods and very lives will depend on it. But it will not be easy by any means. The scale, scope, and complexity of the challenge will rival any that humanity has faced in recent centuries.

The new volume produced by the scientists of the Urban Climate Change Research Network (UCCRN) is a lifeline to sustainability. We should be grateful that leading scientists from around the world have taken up the challenge of sustainable urbanization, with a specific focus on the interrelationship of city life and human-induced climate change. The authors of this remarkable report, the First Assessment Report of the UCCRN, are at the cutting edge of global science and policy. Every essay emphasizes the complexity of the challenges ahead, and how we are just at the start of reshaping our cities for sustainability.

As this report makes amply clear, climate change will be a vital entry point for achieving sustainable development in the world's cities. While climate change is just one of several environmental challenges facing the world, it is the largest, most complex, and most urgent. There can be no answers to other challenges of sustainable development – safe water, clean and abundant energy, and urban public health – unless they are also answers to the climate-change conundrum.

As this volume explains, there are two interrelated aspects of the climate change puzzle. The first is adaptation. Human-induced climate change is already underway and will intensify in the coming decades.

The cities will be threatened in several major ways, and every city must plan ahead to confront, manage, and where possible, fully head off the growing risks. Heat waves will threaten lives of vulnerable populations such as the elderly. Droughts, floods, and other natural hazards will become more frequent, though the vulnerability of specific cities will vary widely depending on their physical geography, climatology, level of economic development, the quality of governance, social cohesion, and the financial capacity to adjust. Rising sea levels may play havoc with coastal cities, submerging some areas, and making others far more vulnerable to storm surges, or adversely impacting key infrastructure.

The other major challenge is climate change mitigation: reducing humanity's greenhouse gas emissions in order to slow and eventually to stop or even reverse the human impacts on the climate. Mitigation is every bit as complex as adaptation, and often the two are closely intertwined. Green buildings can both reduce energy use and also increase resilience to heat waves and other climate hazards. Mitigation will require major long-term changes to energy systems, the design of buildings, transport networks, and urban spatial patterns and zoning. Changing these fundamental attributes of cities will often involve making deep changes in the fabric of city life and its underlying economics. Yet the task of mitigation, essentially moving to a low-carbon society, will have to be carried out in thousands of cities around the world. The process will require decades of persistent and creative policymaking to achieve. There is no better place to start charting that transition than with this pioneering report.

Humanity is in uncharted territory. We must steer future technologies and urban development in a directed and coherent manner, consistent with the best science, social fairness, and economic efficiency. This book is a remarkable, cutting-edge, how-to manual at the start of a decades-long process. The authors don't claim to have all of the answers. Indeed, they constantly emphasize the uncertainties around climate forecasts, technological options, and social best practices. Yet the tools described here are the best around for getting started.

The volume is extraordinary on several counts. First, it is comprehensive, in that it considers every major dimension of adaptation and mitigation that cities will confront. Second, it is remarkably broad ranging in its case studies of dozens of cities around the world. These cases are enormously interesting and enormously instructive. Third, it draws on the very best current knowledge by recognized leaders in their respective fields. Fourth, and impressively, it is very clearly written. This is not a theoretical tome. This is a volume that can guide policymakers in cities and national governments around the world to launch their own climate assessments, and to begin developing meaningful climate solutions for their cities. By complementing the work of the already existing body of knowledge developed by the Intergovernmental Panel on Climate Change (IPCC), this First Assessment Report on Climate Change and Cities (ARC3) provides a rigorous set of analytical tools for effective mitigation and adaptation decision-making, and in a sector-by-sector approach that is likely to be of practical benefit for city planners, managers, businesses, and non-governmental organizations.

Over one hundred scholars around the world, representing a diverse group of developing and developed country cities, have collaborated on the ARC3. The work is a triumph, a must-read study for city planners, mayors, and managers around the world. The lead editors, Cynthia Rosenzweig, William D. Solecki, Stephen A. Hammer, and

Shagun Mehrotra, merit our special thanks and admiration for taking on a challenge of such global significance, and for bringing the best of the world's scientific knowledge together in such a useful and comprehensive manner.

Jeffrey D. Sachs
Director of the Earth Institute at Columbia University
Special Advisor to UN Secretary General Ban Ki-Moon on the
Millennium Development Goals

Preface

This volume is the Urban Climate Change Research Network's First Assessment Report on Climate Change and Cities (ARC3). It contains an Executive Summary and the four sections of the report.

This report would not be possible without the tremendous support of the Cities Alliance, UN-HABITAT, United Nations Environment Programme (UNEP), and the World Bank. We especially thank William Cobbett and his team at Cities Alliance, Jean Christophe, Ricardo Jimenez, Sid Henderson, Neelam Tutej, Kevin Milroy, Viorica Revutch, Phyllis Kibui, and Madhavan Balachandra.

At UN-HABITAT, we are thankful to Anna Tibaijuka and Joan Clos, the outgoing and incoming Executive Directors, and their team led by Rafael Tuts with Robert Kehew and Bernhard Barth, as well as others who provided useful reviews of ARC3.

At the World Bank, we thank Inger Andersen, Vice President, Sustainable Development Department, and her team at the Urban Anchor led by Abha Joshi Ghani; Dan Hoornweg and Anthony Bigio of the Urban Anchor have been unfailingly supportive. At UNEP we would like to thank Soraya Smaoun.

We would also like to thank the Sector Managers and Directors at the World Bank and the leaders of the UNFCCC and the IPCC who have supported the need for ARC3. The Global Facility for Disaster Reduction and Recovery (GFDRR) and the U.S. Geological Survey also provided much-appreciated support for the ARC3 initiating workshop, through the enthusiastic leadership of Saroj Jha and DeWayne Cecil, respectively. They are all exemplary international public servants committed to the development of effective ways for cities to confront climate change challenges and to identify opportunities in resolving them.

We appreciate the advice provided by the members of the UCCRN Steering Group – Albert Bressand, Richenda Connell, Peter Droege, Alice Grimm, Saleemul Huq, Eva Ligeti, Claudia Natenzon, Ademola Omojola, Roberto Sanchez, and Niels Schulz – whose wisdom has guided the establishment of the network and the development of the ARC3 process.

We gratefully acknowledge the discussions and feedback during sessions with Mayors, their advisors, leaders of major institutions, urban policymakers, and scholars. In particular, we thank everyone who participated in ARC3 consultations: scholars at NCCARF 2010 Climate Adaptation Futures Conference in Australia; Konrad Otto-Zimmermann, Monica Zimmermann, Yunus Arikan, and participants at ICLEI's Resilient Cities Adaptation Summit in Bonn; scholars and practitioners at UGEC's Global Summit in Phoenix, Arizona; Mayors and city leaders at the C40 Large City Climate Change Summits in New York, Seoul, and Hong Kong; and at the Mayors Summit held during the COP15 in Copenhagen and the World Council of Mayors Summit held in Mexico City before COP16 in Cancun. At the UN-HABITAT's World Urban Forum in Rio de Janeiro we benefited from the interaction with a broad array of urban stakeholders who shared their thoughts on how to maximize the effectiveness of the ARC3 process. We extend special gratitude to the urban leaders who represent a diverse group of cities, who have commended UCCRN and ARC3. We also give a special thanks to the many students at Columbia University (New York), The Daly College (Indore), and Tec de Monterry (Mexico City) for their keen interest in the emerging field of urban climate change, which helped push the ideas for this volume forward.

This report is the product of the work of the over 100 dedicated members of the UCCRN ARC3 writing team representing more than 50 cities in developing and developed countries. We express our sincere thanks to each of them for their sustained and sustaining contributions, and to their institutions for supporting their participation. We especially thank Shobhakar Dhakal (Tsukuba), Toshiaki Ichinose (Tokyo), Haluk Gerçek (Istanbul), Claudia Natenzon (Buenos Aires), Martha Barata (Rio de Janeiro), and Ademola Omojola (Lagos) for their efforts on behalf of the UCCRN in Asia, the Middle East, Latin America, and Africa regions.

We profoundly appreciate Joseph Gilbride and Somayya Ali for their tremendous work as the UCCRN ARC3 Project Managers, without whom the ARC3 could not have been completed in such a comprehensive and timely way. We also acknowledge the exceptional commitment of the ARC3 research assistants and interns, Jeanene Mitchell, Shailly Kedia, Young-Jin Kang, Masahiko Haraguchi, Steve Solecki, Casey Jung, Irune Echevarria, Lumari Pardo-Rodriguez and Kimberly Peng. At the Goddard Institute for Space Studies, we thank Daniel Bader, José Mendoza, Richard Goldberg and Adam Greeley for their technical expertise, and George Ropes, of www.climateyou.org, for his superb editing.

We recognize with great esteem the expert reviewers of the ARC3 without whom the independent provision of sound science for climate change mitigation and adaptation in cities cannot proceed.

It is a great honor that the ARC3 is being published by Cambridge University Press. We would especially like to thank Matt Lloyd, Editorial Director, Science, Technology and Medicine, Americas; Laura Clark, Assistant Editor; Abigail Jones, Production Editor; and their staff for their expert partnership in the publication of this volume.

Finally, we are deeply grateful to the Columbia University Earth Institute and its Director Jeffrey Sachs for their support for the UCCRN ARC3 process from its inception.

Cynthia Rosenzweig, William D. Solecki, Stephen A. Hammer, and Shagun Mehrotra,
Editors
First UCCRN Assessment Report on Climate Change and Cities

Climate Change and Cities

First Assessment Report of the Urban Climate Change Research Network

Executive Summary

Executive Summary

Cities[1] are home to over half of the world's people and are at the forefront of the climate change issue. Climate change exerts added stress on urban areas through increased numbers of heat waves threatening the health of the elderly, the infirm, and the very young; more frequent and intense droughts and inland floods compromising water supplies; and for coastal cities, enhanced sea level rise and storm surges affecting inhabitants and essential infrastructure, property, and ecosystems. At the same time, cities are responsible for no less than 40% of global greenhouse gas emissions, and given current demographic trends, this level will likely only increase over time. These challenges highlight the need for cities to rethink how assets are deployed and people protected, how infrastructure investments are prioritized, and how climate will affect long-term growth and development plans.

Work on the *First Assessment Report on Climate Change and Cities* (ARC3) was launched by the Urban Climate Change Research Network (UCCRN) in November 2008 at a major workshop in New York City with the goal of building the scientific basis for city action on climate change. Eventually more than 100 lead and contributing authors from over 50 cities around the world contributed to the report, including experts from cities in both the developing and developed world, representing a wide range of disciplines. The book focuses on how to use climate science and socio-economic research to map a city's vulnerability to climate hazards, and how cities can enhance their adaptive and mitigative capacity to deal with climate change over different timescales.

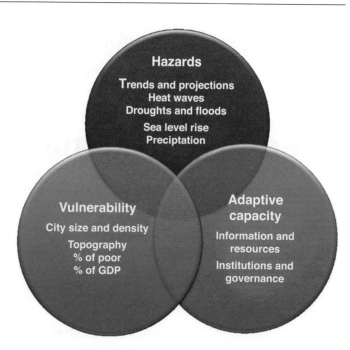

Figure 1: *Urban climate change vulnerability and risk assessment framework.*

Source: Mehrotra et al. (2009).

Key findings

Defining the risk framework

A new vulnerability and risk management paradigm is emerging as a useful framework for city decision-makers to analyze how their city should seek to adapt to the anticipated impacts of climate change. The UCCRN climate change vulnerability and risk assessment framework (Figure 1) is composed of three sets of indicators:

- *Climate hazards* facing the city, such as more frequent and longer duration heat waves, greater incidence of heavy downpours, and increased and expanded coastal or riverine flooding;

- *Vulnerabilities* due to a city's social, economic, or physical attributes such as its population size and density, topography, the percentage of its population in poverty, and the percentage of national GDP that it generates;

- *Adaptive capacity aspects*, factors that relate to the ability of a city to act, such as availability of climate change information, resources to apply to mitigation and adaptation efforts, and the presence of effective institutions, governance, and change agents.

In most cities, readily available data exist about climate hazards (trends and projections), population and geographic features, and insti-tutional capacity that can serve as a foundation for adaptation planning efforts. In other cities that are still in the early stages of efforts to assess local vulnerabilities and climate risks, work can nonetheless begin by using generalized climate risks and information from similar urban areas as a starting point for local climate planning efforts.

For example, in Sorsogon City in the Philippines, the city government developed its local vulnerability assumptions using climate change projections and risk assessments from national government agencies and private research institutions.

Urban climate: processes, trends, and projections

Cities already face special climatic conditions that must be accounted for when preparing long-term climate change adaptation plans. These include:

- *Urban heat island.* Cities already tend to be hotter than surrounding suburban and rural areas due to the absorption of heat by concrete and other building materials and the removal of vegetation and loss of permeable surfaces, both of which provide evaporative cooling.

- *Air pollution.* The concentration of residential, commercial, industrial, electricity-generating, and transportation activities (including automobiles, railroads, etc.) contributes to air pollution, leading to acute and chronic health hazards for urban residents.

- *Climate extremes.* Major variability systems such as the El Niño-Southern Oscillation, the North Atlantic Oscillation, and

---

1   Cities are defined here in the broad sense to be urban areas, including metropolitan and suburban regions.

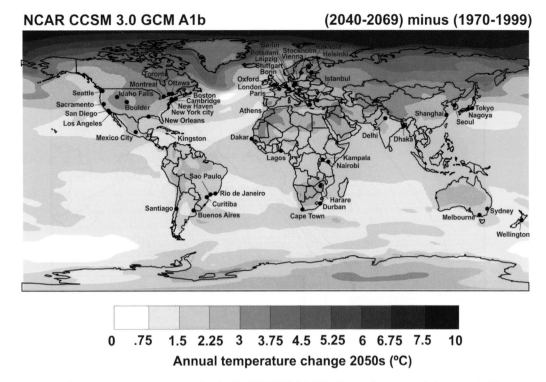

Figure 2: *Cities represented in ARC3 and 2050s temperature projections for the NCAR CCSM 3.0 GCM with greenhouse gas emissions scenario A1b.*

Source: NCAR CCSM 3.0 – Collins et al. (2006); Emissions Scenario A1b – Nakicenovic et al. (2000).

oceanic cyclonic storms (e.g., hurricanes and typhoons) affect climate extremes in cities. How these systems will interact with anthropogenic climate change is uncertain, but awareness of their effects can help urban areas to improve climate resilience.

Existing city-specific climate data and downscaled projections from global climate models can provide the scientific foundation for planning efforts by city decision-makers and other stakeholder groups (Figure 2). In twelve cities analyzed in depth in this report (Athens, Dakar, Delhi, Harare, Kingston, London, Melbourne, New York, São Paulo, Shanghai, Tokyo, and Toronto), average temperatures are projected to increase by between 1 °C and 4 °C by the 2050s. Most cities can expect more frequent, longer, and hotter heat waves than they have experienced in the past. Additionally, variations in precipitation are projected to cause more floods as the intensity of rainfall is expected to increase. In many cities, droughts are expected to become more frequent, more severe, and of longer duration.

Coastal cities should expect to experience more frequent and more damaging flooding related to storm events in the future due to sea level rise. In Buenos Aires, for example, damage to real estate from flooding is projected to total US$80 million per year by 2030, and US$300 million per year by 2050. This figure does not account for lost productivity by those displaced or injured by the flooding, meaning total economic losses could be significantly higher.

Sector-specific impacts, adaptation, and mitigation

Climate change is expected to have significant impacts on four sectors in most cities – the local energy system; water supply, demand, and wastewater treatment; transportation; and public health. It is critical that policymakers focus their attention on understanding the nature and scale of the impacts on each sector, developing adaptation and mitigation strategies, and determining policy alternatives.

Climate change and urban energy systems

Cities around the world have prioritized efforts to reduce energy consumption and the associated carbon emissions. This has been done both for localized efficiency reasons – to reduce the effects of high energy costs on household budgets, for example – as well as to respond to concerns that activities in cities are responsible for a large share of global greenhouse gas emissions. Emphasis is now being placed on urban energy system adaptation, as well, because climate change impacts such as the loss of key supply sources or transmission and distribution assets can jeopardize public health and the economic vitality of a city. For example, in New York City, power plants were historically sited on the waterfront to facilitate fuel supply delivery and to provide access to cooling waters. The majority of these facilities are at an elevation of less than 5m, making them susceptible to increased coastal flooding due to sea level rise (Figure 3).

Increases in the incidence or duration of summertime heat waves may result in higher rates of power system breakdown or failure, particularly if sustained high demand – driven by high rates of air conditioning use – stresses transmission and distribution assets beyond their rated design capacity. In Chinese cities, the number of households with air conditioners has increased dramatically in the past 15 years (Figure 4), although the extent to which usage is nearing a point where system vulnerabilities are heightened is still unclear. In cities heavily reliant on hydropower, changing precipitation patterns resulting from climate change may be problematic, if availability is reduced during summertime periods when demand is greatest.

Power Plants along the East River, New York City

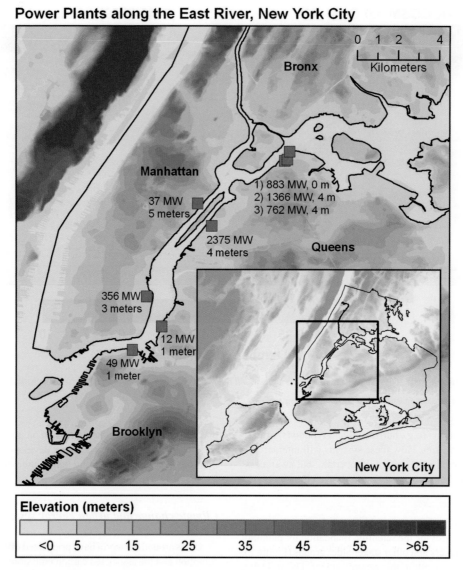

Figure 3: *Location and elevation of power plants along the East River in New York City.*

Source: Power plant data for 2000 from eGRID (US EPA, 2002) to reflect with recently retired plants deleted. New York City digital elevation model is from the USGS (1999), which has a vertical error of approximately +/–4 feet.

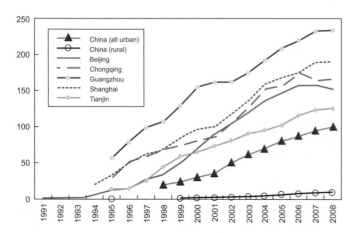

Figure 4: *Number of air conditioners per 100 households in selected Chinese cities.*

Source: CEIC (2010).

For any given city, local analyses are necessary to determine the overall impact of climate change on energy demand, as it may increase *or* decrease depending on which of the seasonal effects of climate change (i.e., reduction in energy demand in cooler seasons and increased demand in warmer seasons) are most significant.

Cities can take robust steps to reduce their energy demand and thus their carbon emissions, and it is increasingly clear that many of these steps also provide significant adaptation benefits. These steps include:

- Develop demand management programs to cut peak load, reducing carbon emission levels and simultaneously lessening stress on the system during times of heightened vulnerability.

- Capitalize on the natural replacement cycle to update power plants and energy networks to reduce their carbon intensity and simultaneously increase their resilience to flooding, storm, and temperature-related risks.

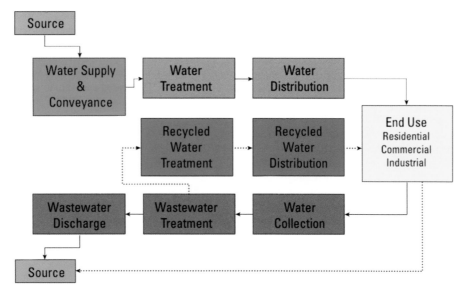

Figure 5: *Typical water-use cycle for cities and other developed supplies; dotted arrows indicate pathways that sometimes occur.*

Source: Modified from Klein et al. (2005).

• Diversify local power supply sources to increase the share of renewables, thereby enhancing system resiliency and reducing carbon emissions.

Climate change, water, and wastewater in cities

Urban water systems include water supply sources, conveyance, distribution, reuse, treatment, and disposal elements, all of which may be vulnerable to a changing climate (Figure 5).

Within cities, impervious surfaces and increased precipitation intensity can overwhelm current drainage systems. In Mexico City, the city's 27 treatment facilities currently handle only a fraction of the total sewage generated citywide, and as the local population increases, the ability of the system to accommodate runoff has become compromised, raising the risk of flooding around the city.

In many cities, the quantity and quality of the water supply will be significantly affected by the projected increases in both flooding and droughts, amplifying the need for cities to focus on upgrading their supply networks to maximize the availability of existing supplies. For example, in developed country cities, leakage from the supply distribution system can be severe, resulting in system losses of between approximately 5% and more than 30%. In developing country cities, the supply problem is often different, as significant numbers of people rely on informal water supply systems. In Lagos, for instance, 60% of the population uses informal distribution systems (Figure 6), which are far more vulnerable to drought-induced stoppages.

A range of adaptation measures will be required to ensure water supplies of adequate quantity and quality, especially in coastal regions where water sources and infrastructure are subject to the impacts of rising sea level, higher storm surge, salt-water intrusion, and land subsidence. Cities are pursuing a range of strategies to address these water and wastewater challenges, including:

• Reduce non-revenue water, which constitutes a significant fraction of supply in many urban areas, through leak detection and repair and reduction in unauthorized withdrawals;

• Review and modify surface water and groundwater sources, storage facilities, and intakes where appropriate to make supplies less vulnerable to climate-induced risks such as floods and droughts;

• Implement innovative local supply augmentations where feasible through techniques such as rainwater harvesting and water reuse, as well as through improved water accounting from better observation networks and holistic modeling;

• Practice demand management through appropriate pricing (including social, environmental and economic objectives), public education on water use and conservation, improved toilet and shower codes, updated drought management plans, and targeted land-use strategies; and

• Encourage the use of water-efficient processes in domestic, industrial, and agricultural uses.

Figure 6: *Informal urban water supply: a water vendor's cart in Lagos.*

Photo by Ademola Omojola.

Climate change and urban transportation systems

Globally, according to the IPCC 2007 report, the transport sector accounted for 23% of the world's greenhouse gas emissions related to energy in 2004, although in some cities, the percentage is much higher, a reflection of local land use and mobility patterns. Cities are adopting a range of strategies to reduce transport-related emissions, including promoting transit-oriented development, reclaiming roadways to provide more space for bicycles and pedestrian walkways, and increasing the amount of mass transit systems available around the city.

Regulatory and pricing instruments are also increasingly being deployed to reduce the volume, timing, or location of private vehicle use, often with significant impact. In London, a congestion pricing program resulted in a 12% decrease in traffic levels in the congestion pricing zone, while in Stockholm, there was a 22% reduction in vehicle passages in the congestion zone. Beijing, Bogota, and Mexico City have all pursued limits on the number of days vehicles can be driven, but this approach may penalize households in locations where public transportation is inadequate. Other cities have focused on promoting more efficient fuels and technology as a means of reducing transport-related carbon emissions. In Delhi, for instance, all public transport buses were converted to compressed natural gas (CNG)-operated systems, in response to public action and right-to-clean-air campaigns that brought the issue to the attention of the Supreme Court of India. The Court subsequently issued a series of judgments regulating public transport and air quality. A key lesson from this experience is that leadership for change in cities can arise from diverse stakeholders – be it citizen groups, the private sector, or the judiciary – as well as from city government itself (Figure. 7).

Some of these mitigation strategies will bring climate change adaptation co-benefits, such as new energy-efficient fuel technologies that provide better temperature control for passengers, but others are being undertaken specifically to maintain the integrity of essential transportation infrastructure assets under changing climate conditions, such as improved engineering and management. Maputo is one of four cities benefiting from a UN HABITAT-supported initiative focused on climate planning, with a specific goal of identifying the hard approaches (sea walls, engineered levees, pump stations) and soft, ecosystem-based approaches (wetlands, parks, and planted levees) designed to protect local transportation system assets from coastal flooding. Mitigation and adaptation strategies for city transportation systems include:

- Integrate land use and transportation planning to increase the density of the urbanized portion of developed land, plan for mixed-use development, and enhance the proximity of travelers to transit and/or to their destinations to reduce vehicle miles of travel;

- Construct transport systems with materials that are more resilient to higher temperatures and the potentially corrosive effects of increased exposure to sea water due to sea level rise and coastal flooding;

- Consider the appropriateness of rezoning as an adaptation solution, retreating from the shoreline, and building new transportation facilities farther inland on higher ground;

- Protect transport systems from increased precipitation and flash flooding through the use of larger culverts and catch basins, and replacement of impermeable road surfaces with permeable material and impermeable roof surfaces with green roofs; and

- Introduce operational measures, including traffic closures during extreme weather events; moving rolling stock to flood-protected and/or wind-protected locations; closing traffic on tall bridges during high winds; and using media to indicate hazardous road conditions and safer alternate transportation routes and modes.

Climate change and human health in cities

Cities are subject to unique health risks since larger populations and higher population density amplify the potential for negative outcomes. Climate change is likely to exacerbate existing health risks in cities and to create new ones. Specific impacts include:

- Direct physical injuries and deaths from extreme weather events such as tropical cyclones, storm surges, intense rainfall that leads to flooding, or ice storms that damage trees and overhead structures and produce dangerous transport conditions;

- Illnesses resulting from the aftermath of extreme weather events that destroy housing, disrupt access to clean water and food, and increase exposure to biological and chemical contaminants;

- Water-borne diseases following extended or intense periods of rainfall, ground saturation and floods, and saline intrusion due to sea level rise; all of which compound existing deficiencies in local water services (Figure 8);

- Food-borne diseases resulting from bacterial growth in foods exposed to higher temperatures;

- Illnesses and deaths from an expanded range of vector-borne infectious diseases;

- Respiratory illnesses due to worsening air quality related to changes in temperature;

- Morbidity and mortality, especially among the elderly, small children, and people whose health is already compromised, as a result of stress from hotter and longer heat waves.

City health agencies can contribute to improvement of knowledge of the health effects of climate change on urban populations and work together with other responsible agencies to reduce the vulnerability of city dwellers to climate variability. Adaptation strategies, many linked to other sectors, land use planning, and governance, include:

Figure 7: *Compressed natural gas public bus, Delhi. Photo by Shagun Mehrotra.*

Figure 8: *Potential health hazards in Kibera related to sanitation and water systems. Photo by Shagun Mehrotra.*

- Expand health surveillance and early warning systems utilizing both technology and social networks, especially for the elderly, very young, and the poor;

- Reduce the size of the urban heat island effect through passive approaches such as tree planting, green and reflective roofs, and permeable pavements, thereby minimizing heat stress on all citizens.

- Emphasize water and energy system climate resilience strategies, because of the key role they play in protecting the public during and after extreme weather events; and

- Regulate settlement in flood plains to minimize exposure to coastal storms and inland flooding.

Cross-cutting issues

A city's land use and governance practices are integrally bound up in the climate change issue. Past zoning and land use decisions are key factors because they create the essential circumstances from which climate-related vulnerabilities may arise. Local powers and the larger governance environment will influence what can actually be done, and at what pace. Progress in addressing climate change requires strategic management, science-based policies, efficient financing, jurisdictional coordination, and citizen participation.

The role of urban land in climate change

The built environment or structural aspects of cities, streets, buildings, and infrastructure systems contribute significantly to the emission of greenhouse gases, and can also amplify climate change impacts. The structure, orientation, and conditions of buildings and streetscapes can increase the need for cooling and heating buildings, which are associated with the level of energy use and greenhouse gas emissions in a city. Swaths of impervious surfaces can intensify flooding and are direct determinants of the heat island effect. The presence or lack of street trees and parks, and the extent of wastewater and drainage systems can either impede or enhance the natural processes of evapotranspiration, in addition to amplifying flooding and drought effects.

A city's natural setting, its urban form and built environments are relatively static factors, but they are subject to future modification through urban planning and management. For example, Shanghai has sought to increase the level of vegetation around the urban core to mitigate the urban heat island; since 1990, urban greenery per capita has increased from 1.0 m^2 to 12.5 m^2, resulting in decreasing temperatures. In Tokyo, the municipal government has similarly expanded its expenditures on tree planting, park development, and the use of paved surfaces that block heat and absorb moisture.

Stockholm is engaged in a long-term planning initiative to both mitigate and adapt to climate change. The Stockholm Royal Seaport is a new development district with strict environmental requirements on buildings. All buildings will be placed 2.5 m above the average sea level; building materials will be required to resist high humidity; and other requirements call for greenery on roofs, walls, and yards.

These examples represent a starting point for initiatives that local authorities can use to respond to climate change. These initiatives can be pursued through legal and political systems, planning departments, zoning regulations, infrastructure and urban services, real estate markets, and fiscal arrangements. Other specific adaptation and mitigation initiatives related to urban land use include:

- Reduce sprawl by increasing population and building densities, mixing land uses to reduce automobile traffic, and more frequent use of public transit;

- Change building codes to reduce energy use for heating and cooling;

- Restrict land use in areas subject to climate change impacts such as sea level rise and riverine flooding;

- Change building codes and land regulations to reduce damage from climate change hazards, e.g., elevating buildings in flood-prone areas;

- Increase urban tree coverage and vegetation to reduce the heat island effect;

Cities and climate change: The challenges for governance

Local governments face many challenges in their efforts to mitigate and adapt to climate change. For any city, climate is but one of many issues on the local agenda. Governments are also faced with the trade-offs between current priorities and long-term risks, a situation compounded by the uncertainties that may surround the timing and severity of climate-related impacts in a city.

Most cities undertaking climate plans find themselves constrained by fiscal and policymaking limitations. Jurisdictional conflicts over who can or must take action on a specific mitigation or adaptation initiative can make progress challenging. For example, in Mexico City, administrative boundaries do not align with the city's geographic boundaries and carbon-relevant functioning. Similar issues exist in Paris, where the *Plan Climat de Paris* is focused on the 105 km^2 area under the direct control of the *Mairie de Paris*, a fraction of the Paris metropolitan region which totals 700 km^2 and is under the jurisdiction of three other *départements*. In Durban, local officials are seeking to ensure that climate change does not get pigeonholed as simply an environmental issue, but instead is more appropriately seen as a development-related challenge.

Despite these difficulties, cities around the world are committing to action on climate change, entering into dialogues with state, provincial, and national governments to discuss their climate policy agendas. Cities are also increasingly focused on data gathering, both to improve internal management practices and to allow for comparison with other cities around the world.

In examining how cities are delivering effective action on climate change adaptation and mitigation, four key factors emerge:

- *Effective leadership* is critical for overcoming fragmentation across neighborhoods and sectors when building consensus on the climate change agenda in cities;

- *Efficient financing* is a core requirement for empowered governance in cities; success to date with efforts to confront climate change challenges has been hampered due to deficient financing;

- *Jurisdictional coordination* across city, state, and national governments is one of the most pressing challenges common to cities worldwide; and,

- *Citizen participation* can help in development of inclusive local government decision-making on climate change.

Cities act

Cities around the world are highly vulnerable to climate change, but have great potential to lead on both adaptation and mitigation efforts. Despite the economic and political constraints that many cities face, they are serving as important laboratories for climate change action.

These efforts have produced much helpful climate risk and response information. In order to effectively address the challenges presented by climate change, cities need to incorporate climate science, adaptation strategies, and mitigation actions into daily decision-making and long-term plans and investments. Many cities in both developing and developed country cities are also centers for research and house extensive research communities that are able and willing to help develop plans for assessing and acting on climate change.

Many cities are developing both near- and long-term climate action plans–but many more need to bring climate adaptation and mitigation into their everyday operations as well as their longer-term planning process. The *First Assessment Report on Climate Change and Cities* (ARC3) of the Urban Climate Change Research Network (UCCRN) provides knowledge to urban policy-makers for science-based city climate actions through an on-going information collection, review, and sharing process.

About the *First Assessment Report on Climate Change and Cities* (ARC3)

The First Assessment Report on Climate Change and Cities (ARC3) presents a comprehensive assessment of the most significant issues for cities as they face the climate change challenge. It was launched by the Urban Climate Change Research Network (UCCRN) in November 2008, with the goal of providing the scientific basis for city action in the mitigation of and adaptation to climate change. The ARC3 seeks to synthesize our current state of knowledge about how cities will be affected by climate change and the steps being taken to address climate change at the local level. It is specifically intended both to identify and to fill data gaps in the existing climate change literature, the majority of which has been compiled to analyze the information at a global, national, or regional scale.

To ensure that the information provided would be of use to urban decision-makers, UCCRN first conducted a needs assessment via a survey of city leaders in both developed and developing countries around the world. The content and structure of ARC3 reflects feedback received from respondents to this survey.

The report encompasses nine chapters which are divided into four sections: **Introduction** (Urban climate change in context), **Defining the risk framework** (Cities, disasters, and climate risk; and Urban climate: processes, trends, and projections), **Urban sectors** (Climate change and urban energy systems; Climate change, water, and wastewater in cities; Climate change and urban transportation systems; and Climate change and human health in cities), and **Cross-cutting issues** (The role of urban land in climate change; and Cities and climate change: The challenges for governance). The report represents the work of more than 100 lead and contributing authors from over 50 cities around the world. ARC3 authors are experts in climate change adaptation and mitigation, and include physical scientists, geographers, planners, engineers, social scientists, and policy experts. Each chapter of ARC3 has gone through a multi-stage expert review process.

Contact: www.uccrn.org

References

CEIC (2010). *Webceic Data manager* (on-line database), China Premium Database, New York, USA: ISI Emerging Markets.

Collins, W. D., *et al.* (2006). The Community Climate System Model Version 3 (CCSM3). *Journal of Climate*, **19**, 2122-2143, doi: 10.1175/JCLI3761.1

Klein, G., M. Krebs, V. Hall, T. O'Brien, and B.B. Blevins (2005). *California's Water-Energy Relationship*. California Energy Commission Final Staff Report CEC-700-2005-011-SF.

Mehotra, S., C. E. Natenzon, A. Omojola, *et al.* (2009). *Framework for City Climate Risk Assessment*. Commissioned research, World Bank Fifth Urban Research Symposium, Marseille, France.

Nakicenovic, N., *et al.* (2000). *Special Report on Emissions Scenarios: A Special Report of Working Group III of the Intergovernmental Panel on Climate Change.* Cambridge, UK: Cambridge University Press.

US EPA (2002). eGRID 2002 Archive. Available at www.epa.gov/cleanenergy/energy-resources/egrid/archive.html, accessed September 2008.

USGS (1999) *New York City Area Digital Elevation Model, 1/3 Arc Second*, US Geological Survey, EROS Data Center.

Part I

Introduction

1

Urban climate change in context

Authors:
Cynthia Rosenzweig (New York City), William D. Solecki (New York City), Stephen A. Hammer (New York City), Shagun Mehrotra (New York City, Delhi)

This chapter should be cited as:
Rosenzweig, C., W. D. Solecki, S. A. Hammer, S. Mehrotra, 2011: Urban Climate Change in Context. *Climate Change and Cities: First Assessment Report of the Urban Climate Change Research Network*, C. Rosenzweig, W. D. Solecki, S. A. Hammer, S. Mehrotra, Eds., Cambridge University Press, Cambridge, UK, 3–11.

1.1 Introduction

Cities,[1] as home to over half the world's people, are at the forefront of the challenge of climate change. Climate change exerts added stress on urban environments through increased numbers of heat waves threatening the health of the elderly, the ill, and the very young; more frequent and intense droughts and inland floods threatening water supplies; and for coastal cities, enhanced sea level rise and storm surges affecting people and infrastructure (Figure 1.1) (IPCC, 2007). At the same time, cities are responsible for a considerable portion of greenhouse gas emissions and are therefore crucial to global mitigation efforts (Stern, 2007; IEA, 2008). Though cities are clearly vulnerable to the effects of climate change, they are also uniquely positioned to take a leadership role in both mitigating and adapting to it because they are pragmatic and action-oriented; play key roles as centers of economic activity regionally, nationally, and internationally; and are often first in societal trends. There are also special features of cities related to climate change. These include the presence of the urban heat island and exacerbated air pollution, vulnerability caused by growing urban populations along coastlines, and high population density and diversity. Further attributes of cities specifically relevant to climate change relate to the presence of concentrated, highly complex, interactive sectors and systems, and multi-layered governance structures.

The Urban Climate Change Research Network (UCCRN) is a group of researchers dedicated to providing science-based information to decision-makers in cities around the world as they respond to climate change. The goal is to help cities develop effective and efficient climate change mitigation and adaptation policies and programs. By so doing, the UCCRN is developing a model of within- and across-city interactions that is multidimensional, i.e., with multiple interactions of horizontal knowledge-sharing from the developing to developed cities and vice versa. The UCCRN works simultaneously by knowledge-sharing among small to mid-sized to large to megacities as well. Free-flowing multidimensional interactions are essential for optimally enhancing science-based climate change response capacities. The temporal dimension is also critical – the need to act in the near term on climate change in cities is urgent both in terms of mitigation of greenhouse gas emissions and in terms of climate change adaptation. The UCCRN is thus developing an efficient and cost-effective method for reducing climate risk by providing state-of-the-art knowledge for policymakers in cities across the world in order to inform ongoing and planned private and public investments as well as to retrofit existing assets and management practices.

1.1.1 Contributions to climate change and cities

There are several ongoing efforts that focus on climate change and cities. On the institutional side, ICLEI – Local Governments

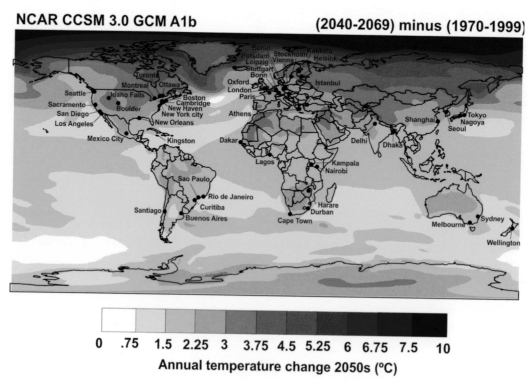

Figure 1.1: *Cities represented in ARC3 and 2050s temperature projections for the NCAR CCSM 3.0 GCM with greenhouse gas emissions scenario A1b.*

Source: NCAR CCSM 3.0 – Collins et al. (2006); Emissions Scenario A1b – Nakicenovic et al. (2000).

---

1 We define "cities" here in the broad sense to be urban areas, including metropolitan and surrounding suburban regions.

for Sustainability has played a major role in encouraging mitigation efforts by local governments around the world. The UN-HABITAT has started a cities and climate change program, as has the World Bank and Organization for Economic Co-operation and Development (OECD). On the research side, the International Institute for Environment and Development (IIED) has focused on community aspects of climate change vulnerability in developing cities, and the World Bank has sponsored a set of research studies for the Fifth Urban Research Symposium held in Marseille in June 2009. The Global Report on Human Settlements 2011 of UN-HABITAT is focused on cities and climate change.

1.1.2 Focus on the urban poor

One of several major foci of these efforts is the developmental needs of the urban poor. In urban areas, inequities among socio-economic groups are projected to become more pronounced as climate change progresses (Mehrotra et al., 2009). For example, the urban poor are less able to move from highly vulnerable locations in coastal and riverine areas at risk of enhanced flooding. This will lead to changes in the spatial distribution and density of both formal and informal urban settlements. Factors that affect social vulnerability to climate change include age, lack of material resources, access to lifelines such as transportation and communication, and lack of information and knowledge. As warmer temperatures extend into higher latitudes and hydrological regimes shift, some vector-borne diseases may extend their ranges, either re-emerging or becoming new problems. Water-borne diseases may shift ranges as well due to changes in water temperature, quantity, and quality in future climates. Such changes can cause serious consequences, especially in densely populated informal settlements in cities in developing countries.

1.1.3 Intersection of climate change and disaster risk reduction and recovery

A related focus is on the intersection of climate change and disaster risk reduction and recovery. Climate change and its effects have the potential to increase exposure to a range of urban risks, and in turn influence how disaster reduction strategies need to be conceived and executed by urban decision-makers. Current efforts by both the climate change and the disaster risk reduction communities are focused on identifying pathways and opportunities for integrating climate change adaptation strategies and disaster risk reduction activities into the daily programmatic activities of a broad range of stakeholder agencies in cities. Disaster- and hazard-related organizations as well as stakeholders who manage risk in other agencies will need to take on these challenges.

Many impacts of climate change in cities, especially in the short and medium term, will be felt in the form of enhanced variability and changing frequency or intensity of extreme events. These will often be considered "disasters." As disaster risk reduction and climate change adaptation are beginning to

converge, there is the potential for them to be largely managed as one integrated agenda. There is not, however, complete overlap between the two fields. One difference is that disaster risk reduction covers geophysical disasters, such as earthquakes, which do not overlap with climate risk.

While climate risk is certainly connected with the increased likelihood of damaging extreme events, climate change is also conceptualized as an emerging public policy issue relating to slow transitions in many urban sectors, ranging over water supply and sanitation, public health, energy, transportation, and migration, among others, which demands response within existing management cycles and planning activities. A central challenge for cities is not only to define the links between climate change and adaptation and disaster risk reduction with respect to extreme events but also with respect to these other climate change policies, which if mismanaged could aggravate vulnerability.

In order for cities to develop effective and resource-efficient integrated strategies for climate change adaptation and disaster risk reduction, the two efforts need to be connected wherever possible with ongoing policies and actions that link the two from both directions. The climate change adaptation community of researchers and stakeholders emerged in relative isolation from long-standing disaster management policy and practices. Cross-fertilization is now being encouraged between these two groups to take advantage of already-recognized best management practices and to reduce redundancy of efforts, which cities cannot afford (IPCC, 2011). Instead, disaster risk reduction strategies and/or adaptation strategies can contribute to reduction of poverty levels and vulnerability, and promote economic development and resilience in an era of increasing climate change.

1.2 ARC3 structure and process

The ARC3 represents a new "science-into-action" model for integrating climate change into highly complex urban areas that is of great interest to both academic researchers and urban decision-makers as they seek to bridge theory and practice. The first ARC3 provides an in-depth review of research on climate science, mitigation, and adaptation addressed from an urban perspective for key city sectors, including energy and buildings, water supply and wastewater treatment, transportation, and human health as well as for land use and governance. For each of these the urban sector or system is described, specific urban climate risks are identified, adaptation strategies for both climate extremes (including disasters) and mean changes are discussed, as well as potential mitigation actions. Where information is available, the economic aspects of adaptation and mitigation are considered. Policy options are brought forward in each of the chapters, and the important role of communities in cities, both as vulnerable populations and as participants in formulating responses to climate change, is also highlighted. Knowledge and applications gaps are identified.

1.2.1 Developing and developed cities

The ARC3 recognizes that there are both similarities and differences between developed and developing city responses to climate change. For example, there is a great deal of fundamental information on climate change projections, vulnerabilities, and risk assessment methods that has a common base in both types of cities. At the same time, there are great differences in the circumstances in developing country cities. These are discussed throughout the chapters, with key points brought forward as city case studies. The city case studies, which illustrate challenges, "best practices," and available tools to facilitate actions in developing and developed cities, are presented throughout the text. The case studies cover the status and activities related to climate change on a city-by-city basis. There are several types of case studies included throughout ARC3: those developed by the chapter authors; those invited from others that apply entirely to the chapter topic; and a third category, "cross-cutting case studies" that touch on many different urban climate change topics that a particular city or organization is addressing. The case studies have been developed by authors drawn from both the research and practitioner communities; such teams are helping to build a cadre of knowledge-providers to aid in implementation of climate change actions in cities around the world.

1.2.2 Multiple stresses and risk management

The chapters characterize urban-specific issues integrating social, economic, and physical aspects. There is explicit recognition that cities face multiple stresses – including population pressure, urban poverty, and pollution – and that climate change and these multiple stresses will likely be manifested in intertwined ways.

Another topic covered is the framing of climate change as a risk-management issue. This is evolving into a new paradigm for both mitigation and adaptation (NPCC, 2010). As described in the previous section, disaster risk reduction is an important set of activities challenging many cities, especially in the developing world, and approaches to fostering the intersection of the risk reduction community with the climate change community are put forward.

1.3 What urban decision-makers need to know about climate change

The process of creating ARC3 began with a scoping activity to solicit the key questions that urban policy makers need answered and to survey current topics in urban climate change research (UCCRN, 2008). Five questions were developed and sent to mayors, their staffs, and/or UCCRN research partners in approximately 45 C40 cities around the world and decision-makers from 15 cities (from both developing and developed countries) responded. Informal discussions about ARC3 also took place with city leaders at the C40 Large Cities Climate Summits held in New York in 2007 and in Seoul in 2009.

The survey questions focused on the climate-related challenges that cities face and information needs for adaptation and mitigation policies. The survey questions included:

What climate-related challenges does your city face?

Where are adaptation policies and actions most urgently needed?

In what fields do you see potential for strong mitigation efforts in your city?

What policy mechanisms is your city potentially or actually implementing?

What other special issues would you like the ARC3 to address?

The main themes that emerged as a result of the city-stakeholder responses and that have been incorporated into the content of the ARC3 are vulnerability, science–policy links, urbanization and land-use issues, and equity and environmental justice. These themes were then incorporated into the sector chapter templates. The volume thus responds to expressed stakeholder needs as well as it draws from and provides broader perspectives on urban mitigation and adaptation to climate change.

1.3.1 Vulnerability

Regarding vulnerability, the stakeholder concerns deal with the impacts on their inhabitants of the predicted increase in hotter days, likelihood of more intense rain events at potentially more frequent intervals, flooding that will likely result from these rain events, and sea level rise and storm surge. One inland city's authorities pointed out that they also need to be concerned with sea level rise since they are a likely destination of people seeking refuge from higher seas and the accompanying damages from coastal flooding.

1.3.2 Science–policy linkages

The second theme that emerged from the stakeholder survey is the challenge of linking science and policy. City decision-makers are unsure how much mitigation and adaptation they should undertake, when, and at what cost. Methods for risk assessment are needed since there is a large amount of uncertainty in climate change science. Related questions brought forward by stakeholders include: Who should be assessing the risks a city faces, how can risks be assessed effectively and efficiently, and how often should risks be reassessed?

1.3.3 Urbanization

The third theme is the challenge of dealing with ongoing urbanization in conjunction with the climate-related challenges that cities face. A major concern is the potential for increasing flooding to disrupt urban development along coasts and rivers. For developing cities, a key issue is how to plan for new infrastructure, taking climate change adaptation and mitigation into account. For developed cities, the challenge is often related to retrofitting existing infrastructure to make it more energy efficient and climate change resilient, but siting and protecting new developments along coasts and rivers are challenges in developed cities as well. Another question raised by city stakeholders relates to the urban heat island effect: What will be the impacts on their city's population and the surrounding areas of the combined effects of the urban heat island and climate change?

1.3.4 Equity and environmental justice

The fourth theme to emerge from the urban stakeholder responses relates to equity and environmental justice. Many of the same populations are vulnerable to the effects of climate change regardless of whether they find themselves in a developed or developing city. These populations tend to be the elderly, the very young, and the poor in cities everywhere.

1.3.5 Climate risks, adaptation and mitigation, and governance

Some specific climate change topics that respondents wanted ARC3 to address include urban climate risks, adaptation and mitigation and their interactions, and strategies for effective policy development.

City leaders wanted to understand key climate processes that pertain specifically to urban areas, such as how urban areas are simulated in climate models and the critical interactive processes related to the urban heat island effect, climate change, and their effects on urban populations.

Urban policymakers are particularly interested in knowing how to decide when and how much to adapt to climate change. They want to ensure that adaptation is flexible enough to deal with the uncertainties in climate projections. Further, they asked for help in identifying the point at which adaptation needs to go beyond simply making incremental improvements and in undertaking explicit revision of existing standards and practices.

In regard to mitigation, city decision-makers require information on determining the role of renewable energy in urban areas and its relation to emerging technologies at various scales. They also see the need to link climate change to the broader energy agenda – including access to energy, poverty and equity issues, fuel choice, and energy network infrastructure – rather than considering it in isolation.

Urban policymakers are well aware that adaptation and mitigation responses are interactive, and that their interactions can be positive or negative. The goal is to enhance the synergies between mitigation and adaptation, while minimizing the constraints. Key policy areas for explicitly addressing these challenges are: retrofitting existing urban residential development and infrastructure to reduce greenhouse gas emissions as well as heat stress of residents; and devising strategies for seaports and airports, since they are both vulnerable to climate change and central to reductions of greenhouse gas emissions.

At the governance level, a key issue that the ARC3 addresses is the need to delineate the role of the city authority in regard to climate change compared to regional, national, and international bodies. This is germane to another need expressed by city stakeholders, which is determining climate change risks and levels of acceptability.

1.4 Urban climate change issues covered in ARC3

The ARC3 is divided into four parts: Introduction; Defining the risk framework; Urban sectors; and Cross-cutting issues. The chapters within these sections relate to assessment of urban vulnerability and key climate hazards, mitigation and adaptation responses in urban sectors, and the roles of land use planning and governance in responding to climate change challenges.

1.4.1 Vulnerability and risk assessment

Estimation of spatially and temporally disaggregated risks is a critical prerequisite for the assessment of effective and efficient adaptation and mitigation climate change strategies and policies in complex urban areas (Chapter 2). Risk may be considered as the intersection of three vectors – hazards, vulnerabilities, and adaptive capacity. These vectors consist of a combination of physical science, geographical, and socio-economic elements that can be used by municipal governments to create and carry out climate change action plans. Some of these elements include climate indicators, global climate change scenarios, downscaled regional scenarios, changes anticipated in extreme events (including qualitative assessment of high-impact, low-probability phenomena), qualitative assessment of high-impact and low-probability events, associated vulnerabilities, and the ability and willingness to respond. The focus is on articulating differential impacts on poor and non-poor urban residents as well as sectorally disaggregating implications for infrastructure and social well-being, including health.

1.4.2 Urban climate hazards

Cities already experience special climate conditions in regard to the urban heat island and poor air quality (Chapter 3). In addition to these, climate change is projected to bring more frequent, intense, and longer heat waves in cities, and most cities are expected to experience an increase in the percentage of their precipitation in the form of intense rainfall events. In many cities, droughts are expected to become more frequent, severe, and of longer duration. Additionally, rising sea levels are extremely likely in coastal cities, and are projected to lead to more frequent and damaging flooding related to coastal storm events in the future.

In regard to critical urban infrastructure, degradation of building and infrastructure materials is projected to occur, especially affecting the energy and transportation sectors. The gap between water supply and demand will likely increase as drought-affected areas expand, particularly for cities located in the lower latitudes, and as floods intensify. While precipitation is expected to increase in some areas, particularly in the mid and high latitudes, water availability is projected to eventually decrease in many regions, including cities whose water is supplied primarily by meltwater from mountain snow and glaciers.

Overall, climate change and increased climate variability will alter the environmental baselines of urban locales, shifting temperature regimes and precipitation patterns (Mehrotra *et al.*, 2009). Changes in mean climate conditions and frequency of extreme events will have direct impacts on water availability, flooding and drought periodicity, and water demand. These dynamic changes will affect system processes within multiple sectors in cities interactively, increasing the uncertainty under which urban managers and decision-makers operate.

1.4.3 Energy and buildings

Climate change will affect urban energy through the complex regulatory, technical, resource, market, and policy factors that influence the design and operation of local energy systems. A key attribute for effective climate change response is the ease with which changes can be made to address climate change mitigation and adaptation (Chapter 4). The International Energy Agency (IEA) estimates that 67 percent of global primary energy demand – or 7,903 Mtoe* – is associated with urban areas (IEA, 2008). While literature on the impacts of climate change on this sector is still limited, urban energy systems can be dramatically affected by climate change at all parts of the process including supply, demand, operations, and assets (Figure 1.2).

In developed countries, climate change concerns are leading cities to explore ways to reduce greenhouse gas emissions associated with fossil fuel combustion and to increase the resiliency of urban energy systems. In developing countries, cities often lack access to adequate, reliable energy services, a significant issue. In these cities, scaling up access to modern energy services to reduce poverty, promote economic development, and improve social institutions often takes precedence over climate-related concerns.

However, if adoption of mitigation measures brings greater reliance on renewable sources of energy (including biomass-based cooking and heating fuels), some cities may become even more vulnerable to climate change, since production of biomass-based fuel is itself subject to changing climate regimes.

1.4.4 Urban water supply and wastewater treatment

Long-term planning for the impacts of climate change on the formal and informal water supply and wastewater treatment sectors in cities is required, with plans monitored, reassessed, and

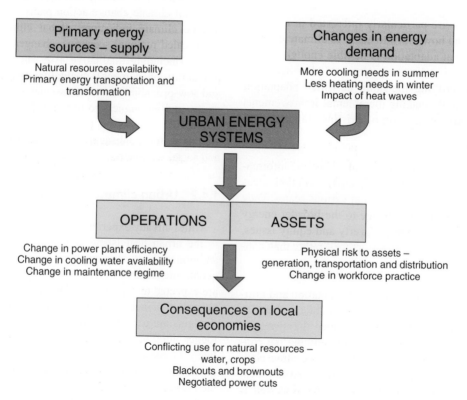

Figure 1.2: *Impacts of climate change on urban energy systems.*

---

*Million tons of oil equivalent

revised every 5–10 years as climate science progresses and data improve (Chapter 5). What is needed as well is the development of a new culture of water value, use, and consumption, based on balanced perspectives of its economic, physical, ecological, social, political, and technical dimensions.

Supply, demand, and quality in informal water supply systems in poor cities need to be better understood, with the purpose of improving these systems and their resilience in the context of climate change. More information on comparative performance among cities, as well as on city and regional hydrologic budgets, is required to guide efficient resource allocation and climate change responses in the urban water sector. Integrated water management includes supply, quality, and wastewater treatment both in cities and in their surrounding regions, and effective planning links beyond the water sector to other sectors, such as energy and disaster risk reduction.

The roles of institutions managing formal and informal water resources in urban areas should be analyzed and reassessed, to ensure that institutions are appropriate to changing challenges, including climate change impacts. This may include collaboration between informal and formal sectors where possible. Urban governance issues regarding water supply and demand in both the formal and informal sectors are likely to become increasingly important and contested, and may require changes in water law and management practices.

In regard to immediate adaptation strategies, programs for effective leak detection and repair and the implementation of stronger water conservation/demand management actions – beginning with low-flow toilets, shower heads, and other fixtures – should be undertaken in formal and, to the extent relevant, informal water supply systems. As higher temperatures bring higher evaporative demand, water reuse also can play a key role in enhancing water-use efficiency, especially for landscape irrigation in urban open spaces. Urban-scale water marketing through the informal private sector can be a mechanism by which to increase efficiency, improve system robustness, and facilitate integration of multisector use in some urban circumstances. Water banking (in which water in wet years is saved in, for example, aquifers for use in dry years) by urban water system managers is a way of hedging against uncertainties and improving system robustness. Rainwater capture can also be undertaken as a conservation adaptation to reduce pumped groundwater and related energy use.

1.4.5 City transportation systems

Urban transportation comprises the facilities and services to move people and materials throughout the city and its surrounding region. Cities encompass many modes of transport, including personal vehicles traveling on surface roads and public transport via bus, rail, and airplanes (Chapter 6). Rail transit systems are often critically important in urban areas, with very large extents and high rates of passenger service. For instance, the Metro-North railway in New York City serves 1.5 billion passengers annually. In coastal cities, rail transit systems contain many points of climate change vulnerability to enhanced flooding from sea level rise, such as public entrances and exits, ventilation facilities, and manholes. These facilities are vulnerable to inland flooding as well. Most importantly, large portions of transit networks are of a hub-and-spoke design and converge on single points giving relatively little flexibility if any one area is disabled during extreme events, which are projected to increase in the future.

Surface transportation refers to both road-based transit (e.g., buses) and vehicular travel, much of which has high-volume traffic and key infrastructure located near coasts and rivers in many cities and thus vulnerable to sea level rise and inland flooding. Tunnels, vent shafts, and ramps are clearly at risk. Flooding necessitates the use of large and numerous pumps throughout these systems, as well as removal of debris and the repair or replacement of key infrastructure, such as motors, relays, resistors, and transformers.

Besides sea level rise and storm surge vulnerability, steel rail and overhead electrical wire associated with transportation systems are particularly vulnerable to excessive heat. Overheating can deform transit equipment, for example, causing steel rail lines to buckle, throwing them out of alignment, which potentially can cause train derailments. Heat can also reduce the expected life of train wheels and automobile tires. Roadways made of concrete can buckle, and roads of asphalt can melt. This is especially dangerous under congested conditions where heavy vehicles sit on hot surfaces for long periods of time, adding to the stress on materials.

Urban transportation adaptation strategies can focus effectively on both usage and technology. Usage strategies involve the ability to provide alternative means of transport during the periods in which acute climate impacts occur. These include being able to substitute roadways and rail lines for similar facilities in other areas, if possible. Examples of types of adaptation strategies for specific impacts are: changing to heat-resistant materials; sheltering critical equipment from extreme rainfall and wind; raising rail and road lines; increasing the deployment and use of pumps; installing drainage systems to convey water from facilities rapidly; and installing barriers such as seawalls at vulnerable locations.

Urban transportation systems also play an important role in mitigation of greenhouse gas emissions. Such mitigation actions can be implemented via transport and land use policies; transport demand management; and supply of energy-efficient transport infrastructure and services.

1.4.6 Climate change and human health in cities

Climate change can best be conceptualized as an amplifier of existing human health problems, attenuating or aggravating multiple stresses and, in some cases, potentially pushing a highly stressed human health system across a threshold of sustainability (Figure 1.3) (Chapter 7). Protection of the health of the

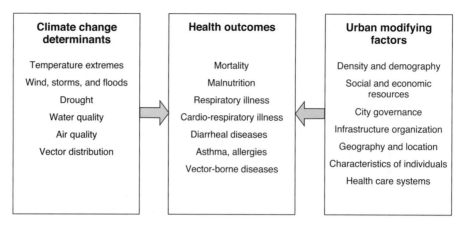

Climate change determinants	Health outcomes	Urban modifying factors
Temperature extremes	Mortality	Density and demography
Wind, storms, and floods	Malnutrition	Social and economic resources
Drought	Respiratory illness	City governance
Water quality	Cardio-respiratory illness	Infrastructure organization
Air quality	Diarrheal diseases	Geography and location
Vector distribution	Asthma, allergies	Characteristics of individuals
	Vector-borne diseases	Health care systems

Figure 1.3: *Climate change determinants and urban modifying factors on health outcomes in cities.*

world's urban populations requires the involvement of all groups (government, business, academia, and communities), levels of government (international, national, regional, and local) and diverse disciplines (health, planning, engineering, meteorology, ecology, etc.).

Since the infrastructure for health protection is already swamped in many developing country cities, climate change adaptation strategies should focus on the most vulnerable urban residents. Such strategies need to promote "co-benefits" such that they ameliorate the existing and usually unequally-distributed urban health hazards, as well as helping to reduce vulnerability to climate change impacts. This involves health programs developed in partnership with public and private organizations and agencies to guide investments and technology choices that benefit the current health of urban residents at the same time as preparing for and responding to climate change.

1.4.7 Urban land use and climate change

Urban land factors that affect climate change risk include the natural features of a city's geography, e.g., coasts and flood plains; its urban form, e.g., is the city compact or characterized by "sprawl"; and the nature of the built environment, e.g., what

is the extent of impervious surfaces that can exacerbate runoff (Chapter 8). A city's urban planning and management structure also affects its ability to respond to climate change, since planning and management agencies and organizations can contribute greatly to the development of efficient and effective processes for both mitigation and adaptation (Table 1.1). Through urban planning and management, cities determine their land use, neighborhood densities, character of the built environment, parks and open spaces, as well as public infrastructure and facilities. Planning and management departments administer public services and regulate and provide incentives for private infrastructure providers and land markets. A key climate change response mechanism relates to property rights and land tenure. For example, how property rights and land tenure are structured in a city will play a key role in responding to the threat of sea level rise in regard to its coastal development.

1.4.8 City governance

Climate change presents city governments with several challenges, including the need for political and fiscal empowerment at the local level to deal with local impacts and specific mitigation measures; the presence of multiple jurisdictions among cities, metropolitan regions, states, and nations; and often weak planning and management structures (Chapter 9). These challenges highlight the need for science and evidence-based policy formulation in regard to climate change. However, data and measurement capability are often lacking, especially in cities in developing countries. Beyond the need to develop specific near-term adaptation and mitigation measures, city governments face the challenge of addressing deeper and enduring risks and long-term vulnerabilities. Since city administrations tend to be rather short-lived, long-term risks are often ignored. A final governance challenge involves the need to be inclusive of all communities in an urban area, especially since vulnerability to climate change varies widely among socio-economic groups. To answer these climate change challenges successfully, city governments need to enhance the potential for science-based policymaking, effective leadership, efficient financing, jurisdictional coordination, land use planning, and citizen participation.

Table 1.1: *Urban planning strategies for climate change hazards.*

Climate hazards	Major urban planning strategies
Temperature, heat waves	Change building codes to withstand greater energy loads.
Precipitation, floods, and droughts	Restrict development in areas prone to floods, landslides, fires; change building codes to encompass more drainage.
Sea level rise and storm surge	Restrict development in coastal areas; change building codes to reduce impact, e.g., elevate buildings, build protective works.
	For existing urban areas subject to coastal areas where protection efforts not feasible, develop plans for retreat and new settlements.

1.5 Conclusion

The form and function of the ARC3 is designed to be multi-dimensional. While the core of the effort reflects a cutting-edge climate change assessment focused on cities, it has emerged out of a process that explicitly aims to link on-the-ground scientific expertise in the service of the needs and requirements of local city decision-makers. The presentation and organization of the assessment are designed to bridge the science–policy divide in a range of urban contexts. Key dimensions are the development of risk assessment and management frameworks that take urban climate hazards, sensitivity, adaptive capacity, and agency into account, interactive consideration of mitigation and adaptation in critical urban sectors – energy, water, transportation, and human health – and the inclusion of overarching integrating mechanisms of urban land use and governance. Throughout, the goal is to contribute to effective, ongoing, and beneficial processes in the diverse cities of the world to respond to the risks of current climate extremes and future climate changes. These responses include effective planning to safeguard all urban inhabitants from climate risks equitably, while mitigating greenhouse gas emissions and thus contributing to reduction of the magnitude and impact of future changes.

REFERENCES

Collins, W. D., *et al.* (2006). The Community Climate System Model Version 3 (CCSM3). *Journal of Climate*, **19**, 2122-2143, doi: 10.1175/JCLI3761.1

IEA (2008). *World Energy Outlook 2008*. International Energy Agency, OECD Paris.

IPCC (2007). *Climate Change 2007: Impact, Adaptation and Vulnerability. Contribution of Working Group II to the Fourth Assessment Report of the Intergovernmental Panel on Climate Change*, M. L. Parry, O. F. Canziani, J. P. Palutikof, P. J. van der Linden and C. E. Hanson, Eds., Cambridge, UK: Cambridge University Press.

IPCC (2011). *Managing the Risks of Extreme Events and Disasters to Advance Climate Change Adaptation*. Special Report (in preparation, due late 2011).

Mehrotra, S., C. E. Natenzon, A. Omojola, *et al.* (2009). *Framework for City Climate Risk Assessment*. Commissioned research, World Bank Fifth Urban Research Symposium, Marseille, France.

Nakicenovic, N., *et al.* (2000). *Special Report on Emissions Scenarios: A Special Report of Working Group III of the Intergovernmental Panel on Climate Change*. Cambridge, UK: Cambridge University Press.

New York City Panel on Climate Change (2010). *Climate Change Adaptation in New York City: Building a Risk Management Response*. eds. C. Rosenzweig and W.D. Solecki. Prepared for use by the New York City Climate Change Adaptation Task Force. In *Annals of the New York Academy of Sciences*, 2010. New York, NY. 354 pp. http:www.nyas.org.

Stern, N. (2007). *The Economics of Climate Change: The Stern Review*. Cambridge, UK: Cambridge University Press.

Urban Climate Change Research Network (2008). *ARC3 Progress Overview and Stakeholder Input*. Unpublished manuscript.

Part II

Defining the risk framework

2

Cities, disasters, and climate risk

Coordinating Lead Authors:

Shagun Mehrotra (New York City, Delhi), Cynthia Rosenzweig (New York City), William D. Solecki (New York City)

Lead Authors:

Claudia E. Natenzon (Buenos Aires), Ademola Omojola (Lagos), Regina Folorunsho (Lagos), Joseph Gilbride (New York City)

This chapter should be cited as:

Mehrotra, S., C. Rosenzweig, W. D. Solecki, C. E. Natenzon, A. Omojola, R. Folorunsho, J. Gilbride, 2011a: Cities, disasters and climate risk. *Climate Change and Cities: First Assessment Report of the Urban Climate Change Research Network*, C. Rosenzweig, W. D. Solecki, S. A. Hammer, S. Mehrotra, Eds., Cambridge University Press, Cambridge, UK, 15–42.

2.1 Introduction[1]

Cities are central to the climate change challenge, and their position is ever more important as the world's population is becoming increasingly urban. City governments can play an active role in attempting to mitigate climate change, as well as in sheltering their residents from the negative consequences of climate change. In this chapter, we examine the connections between cities and the management of these negative consequences of climate change.

Climate change affects hazard, vulnerability, and risk exposure in cities through a variety of direct and indirect relationships. Cities in many ways were first created as a means to more efficiently protect populations from hazards, whether they be physical (e.g., storms, droughts) or social (e.g., war, civil unrest) in origin. The very fact that cities are population centers illustrates the tension that city managers face with respect to hazards. They can be expected to help protect the populations that live within their cities' borders; while, at the same time, the concentration of population in cities means that when disaster strikes a large number of people could be adversely impacted.

City governments are beginning to put a greater focus on adapting their cities to the inevitable effects of climate change. In its 2007 Fourth Assessment Report (AR4), the Intergovernmental Panel on Climate Change (IPCC, 2007a) concluded that there is a greater than 90 percent chance that the average global temperature increase over the last century was primarily caused by human activity. However, in the context of cities, several climate-induced challenges, such as increased flooding potential and its impacts on water supply, are still largely understudied.

Climate change and increased climate variability will alter the environmental baselines of urban locales. Shifts in climate and increased frequency of extreme events have direct impacts on water availability and quality, flooding and drought periodicity, and water demand. These dynamic changes will affect system processes within multiple sectors in cities interactively, increasing the uncertainty under which urban managers and decision-makers must operate.

In turn, a central objective of this chapter is to review how this new information and uncertainty about climate risk is being integrated into effective and efficient adaptation planning at the city level. To address this issue, the chapter first focuses broadly on the connection between climate change adaptation and more established disaster risk reduction strategies. Derived from these connections, we then present a climate risk assessment-based analysis of adaptive capacity in a diverse set of four city-focused case studies.

2.2 Urban disasters and hazard dynamics in the context of climate change

At the global level, the IPCC Working Group I identified four major aspects of climate change relevant to cities (IPCC, 2007a). First, heat waves are very likely to increase in frequency over most land areas. Second, heavy precipitation events are very likely to increase in frequency over most areas; available data suggest that a significant current increase in heavy rainfall events is already occurring in many urbanized regions. The resulting risk poses challenges to urban society, physical infrastructure, and water quantity and quality. Third, the area affected by drought is likely to increase. There is high confidence that many semi-arid areas will suffer a decrease in water resources due to climate change. Drought-affected areas are projected to increase in extent, with the potential for adverse impacts on multiple sectors, including food production, water supply, energy supply, and health. Fourth, it is likely that intense tropical cyclone activity will increase. It is also likely that there will be increased incidence of extreme high sea levels (excluding tsunamis).

Further, the IPCC (2007b) Working Group II lays emphasis on conceptual issues regarding urban climate change with a focus on economic and social sectors in its Chapter 7 on Industry, Settlements, and Society. The review identifies four key findings with *very high* or *high* confidence. First, climate change effects can amplify the risks that cities face from non-climate stresses. These non-climate stresses include: large slum populations that live in low-quality housing lacking access to basic social services; city-wide lack of access to effective and efficient physical infrastructure; often poor quality of urban air, water, and waste disposal systems; lack of land use planning and other urban governance systems among others. Further, the climate change associated risks for cities stem primarily from extreme events – implying that cities need to assess risk for droughts, floods, storms, and heat waves, in order to plan and implement adaptation strategies. However, gradual changes such as a rise in mean temperature do affect cities in at least two significant ways: by increasing the frequency and intensity of extreme events and burdening existing infrastructure.

2.3 Climate risk and key urban sector impacts

Climate change and disasters affect critical socio-economic sectors in settlements (both formal and informal), water and energy supply and demand, and public health. Climate change impacts on these key sectors and the implications for disaster risk reduction effectiveness are presented in the following section. The impacts include (i) increased system variability and

---

1 The basic outline of the chapter was presented at the World Bank Fifth Urban Research Symposium, June 2009, Marseille, France. Case studies were adapted from an unpublished manuscript. Mehrotra *et al*. (2009) *Framework for City Climate Risk Assessment*.

potential instability, (ii) increased potential for declines in systems productivity, (iii) challenges to current system efficiency, and (iv) exacerbation of existing inequities.

In urban areas, inequities will become more apparent as populations will be differentially able to relocate away from highly vulnerable locations, leading to changes in the spatial distribution and density of both formal and informal settlements. Degradation of building materials is also projected to occur. As warmer temperatures extend into higher latitudes, diseases never before seen might appear, or diseases that had long been considered eradicated may re-emerge. The health ramifications in some areas could be disastrous, especially in densely populated informal settlements. Water supply and demand will be affected as areas under drought will expand. While precipitation is expected to increase in some areas, water availability is projected to eventually decrease in many regions, including those where water is supplied by meltwater from mountains. For energy supply and demand, climate change will put increased pressure on energy infrastructure via rising cooling demand, as well as greater likelihood of supply disruption from extreme event impacts.

Increased variability and instability of sector operation could in some cases lead to beneficial conditions. For example, future climate change could result in more favorable winter conditions in many northern-latitude cities. In most situations, however, the impacts will be more likely to be detrimental to sector operation, especially in poor countries that will be affected the most by climate change, and result in declines in system efficiency and production, e.g., decline in local agricultural productivity, more frequent drinking water shortages in cities, and loss of housing to flooding and storm surge. Even if these impacts are not evident in the short or medium term, enhanced climate change increases the potential for these types of losses, and potential secondary impacts such as increased rural-to-urban migration if rural agricultural systems become less viable. Local and national management regimes need to be ready to respond to these new possibilities. For example, the drought of record or similar baseline can no longer be seen as the most extreme and as the benchmark for disaster and hazard planning.

Sectoral impacts from climate change will reflect the underlying social and environmental conditions within cities. People and places that are most vulnerable and least resilient will be those most affected by climate change. Urban residents of already drought-prone and water-limited areas of cities will be more likely exposed to increased climate variability. Other urban poor already living in higher-risk sites in cities, such as floodplains or hill slopes, will be affected by more frequent and more severe floods or more massive landslides and mass wasting of hillsides.

In summary, climate change-related, sectoral impacts present a variety of challenges and opportunities for disaster risk reduction strategies. The challenges include a need for continuing reevaluation of existing plans and efforts to determine if they are to be responsive to the increased dynamism and changing baseline of the local environment; to the presence of increased numbers of more intensively marginalized and less-resilient populations and places; and to increased competition for limited funding and resources for climate change-related impacts (i.e., government funding and attention could be drawn away from disaster planning if the focus turns to address increasingly difficult economic and social stresses present in many cities).

The question of how to develop effective climate change adaptation strategies within the context of disaster risk reduction management approaches and a multitude of other demanding city-scale public policy pressures is discussed in the next section.

2.4 Climate change adaptation and disaster risk reduction: comparison, contrasts, and emerging synergies

The basic assumptions and institutional conditions under which climate change adaptation strategies and disaster risk reduction activities have emerged are reviewed here, as well as areas where additional potential opportunities for synergy within the urban context are starting to emerge.

Urban climate change adaptation strategy efforts have responded to increased awareness of the potential threat of climate change and enhanced climate variability. In the past decade, government and international organizations within cities have begun to assess how climate change could have a wide variety of primary and secondary impacts. The foundation for emerging climate change and adaptation policy has been science-based studies and assessments. Premier examples of such efforts are the four major assessment reports of the Intergovernmental Panel on Climate Change, which have been produced since 1990. These reports, along with a host of regional and national scientific reports as well as city-based assessments, serve as the basis for city-scale action. The administrative response to the emerging pressure of climate change impacts and associated scientific assessment, however, remains largely diffuse and uneven, often-times driven by a single agency or concerned officials within several agencies, without firm legislative mandates, and/or without financial or human resources to implement action plans. In many least developed country cities, if a climate change adaptation agenda exists it is often buried in environmental ministries or agencies that have very little voice and often even less influence.

In other managerial contexts, urban theorists and practitioners, with their emphasis on rational comprehensive planning as the primary tool to regulate city development, have evolved into a more bottom-up community-oriented approach of advocacy planning widely used in urban governance in the past decade (Campbell and Fainstein, 1996). Here as well, most city managers do not yet address climate change in their policy planning and strategies largely because city-specific risks remain undefined and more short-term problems, such as lack of basic services or overextended and aging infrastructure, take precedence.

Where climate change concerns are actually being recognized by policymakers and managers at the local level, climate risk literature looks predominantly at hazards – temperature, precipitation, and sea level trends and projections. This emphasis can be explained by the fact that in the near future most climate change impacts are likely to be in the form of enhanced climate variability, i.e., increased frequency and intensity of extreme events. While the observed and projected trends in climate parameters are a prerequisite to any assessment of climate risk and associated management strategies, in the context of cities two additional vectors are critical and often neglected – namely vulnerability and adaptive capacity. Vulnerability of a city is determined by a host of internal characteristics of the city set within a larger socio-environmental context. Adaptive capacity is a function of the ability and willingness of the city stakeholders to respond to and prepare for future climate-induced stresses.

In contrast to recent developments in climate change adaptation strategy action, disaster risk reduction planning developed out of much longer term efforts over the past half century in cities to provide emergency disaster response and recovery services for affected populations and economies (Blaikie *et al.* 1994; Helmer and Hilhorst, 2006; O'Brien *et al.*, 2006). Previous to this, indigenous disaster resilience techniques were used. Disaster risk reduction planning and efforts in most cities emerged only after a major catastrophe occurred and where the potential for coherent local and national response efforts was possible. Prior to the past several decades, deficient communication and transportation infrastructure, political instability, and/or lack of governance structure made the response to disasters in cities quite limited. These challenges are still present in many less-developed countries' cities.

In further contrast to climate change adaptation, disaster risk reduction efforts typically grew out of public safety and civil defense mandates in the wake of a major event, instead of scientific study as with climate change and IPCC-style assessments. In its origin, urban disaster management focused largely on response and recovery to address the immediate pain and suffering of disaster aftermath rather than the underlying socio-economic foundations of vulnerability and adaptive capacity.

In the past two decades, as it became clear that disasters cannot be managed only as one-time events to be dealt with through humanitarian response and economic reconstruction, a significant amount of focus in disasters work has shifted toward a new approach that addresses the root causes of disaster vulnerability through either structural or non-structural adaptations (Blaikie *et al.*, 1994). The new perspective has engendered considerable focus on risk assessment, institutional capacity building, risk mitigation investments, emergency preparedness, and catastrophe risk financing. Examples of significant improvements include development and implementation of building codes, early warning systems, drainage systems, hazard mapping, agricultural insurance, and financial risk pools.

Another central difference between climate change adaptation and disaster risk reduction as they are currently implemented in cities is the type of events they address and the implication of the events. Coupled with these distinctions are differing definitions and terminologies (for terms like hazard, risk, and sensitivity), which further divide the communities. For example, disaster risk reduction focuses on extreme events, the damage they cause, and short-term response. These extreme events often are conceptualized and defined as aberrant, isolated moments outside the norm, and the appropriate government response should to be bring the environment and social life in the affected zone "back to normal" as soon as possible. Climate change can exacerbate existing disaster risks and increase the potential for new risk levels, but in many cases causes gradual changes that are not typically associated with disasters – e.g., shifts in ecosystem zones that can affect local and regional hydrology, and spread of public health disease vectors, and changes in heating and cooling energy demand. Under climate change, extreme events and gradual shifts are seen as part of an increasingly variable and dynamic physical environment and that new "normal" conditions are constantly evolving.

While the origins and basic precepts are different, there is a large amount of common ground between climate change adaptation and disaster risk reduction planning in cities (Huq *et al.*, 2004). They share a common goal: managing hazards, building resilience and adaptive capacity in vulnerable communities. Adaptation strategies and disaster risk reduction increasingly are connected by their dual focus on the significance and impact of extreme climate events; analysis of the root cause of exposure and vulnerability and integration of these concepts into planning and action; and the significant role of management. Both communities seek to mainstream their activities (i) through the development of management plans aiming to incorporate these into the local and regional development plans and strategies, and (ii) by having existing agencies and departments and local governments integrate key sectoral guidelines and issues into their planning and implementation activities. Both communities support local capacity building and have an increasing recognition of the importance of both expert knowledge and local knowledge in addressing their policy concerns. We further explore and assess this blending of the two sets of activities in the next section.

2.5 Convergence of climate change adaptation and disaster risk reduction in the urban context

Climate change and potential adaptation strategies are connected with existing disaster response and hazard management strategies in several ways (ICS, 2008; International Strategy For Disaster Reduction, 2010; Mercer, 2010). Disaster risk reduction management includes several phases such as risk assessment and preparedness planning; response, relief, and recovery; and structural hazard mitigation activities and non-structural hazard mitigation activities. Conditions of climate change and increased climate variability can affect the success and effectiveness of critical phases of disaster and hazard management. Such interactions are used as opportunities to more closely link climate risk adaptation and disaster risk reduction strategies and avoid the potential of the two remaining separate, become duplicative, fragmented, and overlapping risk assessments that do not well serve local decision-makers.

Climate change provides disaster risk reduction in cities several important opportunities for advancement, and vice versa (Pelling 2003, Pelling et al., 2009; Prabhakar et at., 2009). The most important opportunity is that climate change raises the profile and importance of disaster risk reduction efforts. Climate change can help to connect disaster management to cutting-edge national and international policy issues, which are garnering the attention of governments at all levels and the NGO community as well. Conversely, since the disaster risk reduction agenda has often been built up over the past decade in the finance and planning agencies in many cities, this presence can help the climate change adaptation agenda to evolve from its current often-constrained situation in environment departments. Furthermore, disaster risk reduction planning frequently has well-developed platforms and coordination mechanisms, such as interagency task forces and mutual aid agreements, at the regional and city level to which climate change adaptation planning can be linked. Governments typically have plans, institutions, and policies already in place for disasters, while climate change adaptation planning is still very much under development.

The disaster management community also provides the main entry points to climate change adaptation for the general public. The public in cities often connect greenhouse gas reduction when hearing of climate change, but the questions "How do we deal with disaster impacts here and now?" and "What are the solutions to reduce the disaster impacts/vulnerabilities?" are questions that disaster management officials deal with on a regular basis. Thus, to incentivize officials at all levels to engage in the climate change agenda, disaster reduction can provide a key policy entry point.

City governments have begun to recognize the importance of disaster risk reduction at times other than during and just after a disaster event (Van Aalst, 2006; Van Aalst et al., 2007). Climate change heightens the awareness and concern for the potential of climate-related perturbations, such as cyclones, droughts, and floods. In many cities, the specter of increased intensity and frequency of extreme events is the primary policy concern associated with climate change. As this concern is raised, so is the focus on disaster risk reduction strategies and planning. Resilience initiatives emerge as one of the best "no regrets" actions that can be taken in the short term to reduce disaster vulnerability, from both current threats and those emerging from climate change. This increased interest not only presents opportunities for those promoting in disaster response and recovery, but also more importantly for long-term disaster risk exposure reduction and non-structural hazard mitigation strategies (e.g., promoting

resettlement away from low-lying coastal sites that will affected by sea level rise, increased storm surge, and inundation).

Disaster preparedness planning is predicated on understanding the hazard and risk potential within the locale through analysis of historical hazard events and ongoing socio-physical shifts – e.g., increased urbanization and changing flood potential. Climate change increases the need for reassessment of disaster planning assumptions because of the potential for change of the environmental baselines (e.g., coastal evacuation plans need to be reevaluated because of heightened storm surge and flooding potential). Similarly, disaster recovery and reconstruction efforts will need to be reexamined, because in the future more-numerous climate-related disasters could have greater threat to life and property and be associated with more widespread and higher numbers of displaced people, homelessness, and property damage.

Structural and non-structural hazard mitigation policies will need to be reevaluated in the context of climate change (Thomalla et al., 2006). Additionally, the efficacy of large-scale infrastructure and capital improvement projects will need to be reevaluated. Projects possibly deemed as not cost-effective or overly ambitious, such as storm barriers or other flood-control devices in cities, might become feasible and needed as climate change impacts increase the possibility of more frequent disasters and greater losses per event. Non-structural approaches, which currently might seem as not appropriate or necessary, might need to be considered. Moving critical infrastructure out of highly vulnerable floodplains or coastal zones is just one example.

The interactions between climate change and disaster management provide opportunities to promote more structured coordination, planning, and communication, and eventually the effectiveness of both. The blending of initially diverse policy and research arenas is not without precedent within disaster management. During the 1980s, the previously disparate streams of natural hazards and technological hazards (including hazardous chemical releases and exposures) were effectively blended in an all-hazards management approach in some governments, such as the in USA through its Federal Emergency Management Agency (FEMA). Lessons learned from this experience can contribute to understanding how best to link climate change adaptation and disaster risk reduction.

The climate change community is now providing increasingly sophisticated climate models and scenarios that are enabling the disaster risk reduction managers in cities to better evaluate in terms of risks and vulnerabilities (World Bank, 2008). In turn, the disaster risk reduction community has brought to climate change planning a long history of experience and lessons learned and well-developed tools, such as probabilistic risk assessment techniques and frameworks for decision-making. For example, the disaster risk reduction community has developed the Hyogo Framework for Action 2005–2015 (HFA), an international agreement to build the resilience of nations and communities to disasters. The HFA provides the foundation for the implementation of disaster risk reduction. Agreed at the World Conference on Disaster Reduction in January 2005, in Kobe, Japan, by 168 governments, its intended outcome is the substantial reduction of losses from disasters in the next 10 years.

Critical components in this process of convergence include the development of common protocols for data gathering and reporting, and for assessment. Within the past half decade, climate change and climate variability studies increasingly have been able to utilize conceptual assessment frames present within the disaster risk reduction community. Critical concepts being explored include hazard, vulnerability, and adaptive capacity. In the next section, we detail these concepts and their emerging connections to climate change adaptation strategy development and risk assessment.

2.6 Climate risk in cities: assessing the nexus of disaster reduction and adaptation

A critical need for sector-specific analyses of climate hazards and vulnerabilities, and location-specific adaptive capacities and mitigation strategies, has been highlighted in the most recent IPCC Working Group III AR4 (IPCC, 2007c), and the World Resource Institute (Baumert *et al.*, 2005; Bradley *et al.*, 2007), and OECD (Hunt and Watkiss, 2007; Hallegatte *et al.*, 2008). Several examples in the literature exist which highlight these omissions and deficiencies. For example, most analyses on climate impact assessments and cities have neglected non-coastal cities, and uncertainties in assessing the economic impacts need to be incorporated, particularly for developing-country cities where variances are likely to be higher (Hunt and Watkiss, 2007, and Hallegatte *et al.*, 2008). Overall, there is a pressing call for understanding risks associated with climate change as they pertain to different types of cities – coastal versus non-coastal and developed versus developing; different sectors – physical infrastructure such as energy, transport, water supply, as well as social services such as health and environmental management (and the complex interactions among these sectors); and differential impacts on the poor or the young and old, who are more vulnerable than the rest of the urban population.

Furthermore, city-specific efforts should graduate from awareness raising to impact assessments – including costing impacts and identifying co-benefits-and-costs – and adaptation analysis so that "no-regret adaptation options" can be adopted to increase resilience of cities to climate change. Economic costs of climate change in cities should "bracket" for uncertainty and assess both intra- and inter-sectoral and systemic risks to address direct and indirect economic impacts.

Understanding how cities craft institutional mechanisms to respond to climate change is another important element of the urban risk assessment process. This has been briefly explored, where relevant, for detailed case studies on Quito and Durban (see Carmin and Roberts, 2009) and the eight-stage Risk,

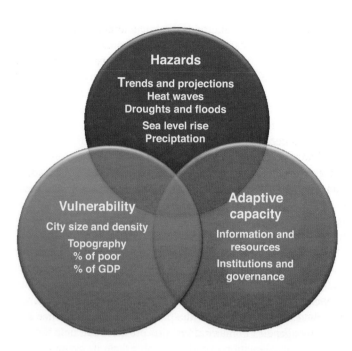

Figure 2.1: *Urban climate change vulnerability and risk assessment framework.*
Source: Mehrotra et al. (2009).

Uncertainty and Decision-Making Framework developed by the UK Climate Impacts Programme (2009), which aims "to help decision-makers identify and manage their climate risks in the face of uncertainty."

To further explore current practices and potential of urban climate risk assessment, we define and examine the characteristics and interplay of three basic risk elements in framework format: *hazards*, *vulnerability*, and *adaptive capacity* (see Figure 2.1). (Note: framework approach adapted from Mehrotra, 2003; Rosenzweig and Hillel, 2008; Mehrotra *et al.*, 2009; also see Rawls, 1971; Shook, 1997; and Sen, 1999.) The challenge in the research community is to translate information on each and from climate science into knowledge that triggers a realistic assessment of the vulnerability of the city and its systems so as to facilitate the development of pragmatic adaptation strategies. In the remainder of this chapter we detail this challenge through the articulation of four city-level case studies around the three following objectives:

1. Characterize the hazards associated with climate change at the city-level;
2. Identify the most vulnerable segments (people, locations, sectors) of the city, and
3. Assess the city's ability to adapt to anticipated changes in climate.

2.6.1 Hazards

Hazards are defined in the framework as the climate-induced stresses and climate-related extremes and are identified through observed trends and projections derived from global climate models (GCMs) and regional downscaling (see Table 2.1). Extreme events affected by climate change include heat waves, droughts, inland floods, accelerated sea level rise, and floods for coastal cities. The variables examined to track these climate-related hazards are temperature, precipitation, and sea level (see Figures 2.2–2.22). In essence, the hazard element presents an array of climate change information and insights into the key stresses that potentially have the greatest consequence for any specific city. In this regard, it is critical to draw attention to both the variation in climate means and the change in frequency and intensity of extreme events. The latter offers opportunities for linkages with disaster risk reduction programs and has received perhaps more attention, while the former has critical long-term implications for city infrastructure and development, and tends to receive less attention because the mean changes are gradual.

Analysis of hazards specific to a particular city should include observed and projected data on key climate parameters – temperature, precipitation, sea level rise, among others. Further, each hazard needs to be analyzed for variance in climate parameters over the short and long term and, where relevant, for frequency as well as intensity of extreme events. Climate change scenarios provide a reasonable understanding of potential future climate conditions (Parsons *et al.*, 2007). It is not expected that a single climate model will project exactly what will happen in the future, but by using a range of climate model simulations along with scenarios incorporating different atmospheric concentrations of greenhouse gases, a range of possible climate outcomes are produced and can be presented as projections that demonstrate the current expert knowledge.

Local climate change information for cities can be derived from the scenarios of greenhouse gas emissions and global climate model simulations described above. For the case-study cities that are examined below, quantitative projections are made for key climate variables such as the change in mean temperature that reflect a model-based range of values for the city-specific model grid boxes (see Table 2.5). Further, there is a need for a nuanced understanding of the complex interactions between hazards and the city. This is because the city can be both a producer as well as a receiver of these hazards. For instance, New York City alone contributes about 0.25 percent of global greenhouse gas emissions (The City of New York, 2007). On the other hand, increase in sea level also increases the city's susceptibility to flooding. In addition, while both urban heat island and global warming increase the ambient temperature of the city, one is internally generated while the other is externally induced.

2.6.2 Vulnerability

Vulnerability is defined as the physical attributes of the city and its socio-economic composition that determine the degree of its susceptibility. The variables affecting vulnerability include flood proneness (proximity to coast or river), land area, elevation, population density, percentage of poor, and quality of infrastructure. The OECD's work on city vulnerability in the context of climate change points to such variables as location, economy, and size as well (Hunt and Watkiss, 2007). More detailed indica-

tors such as composition of the poor population – age, gender, labor force composition, and the like – need to be taken into consideration when in-depth city vulnerability analysis is conducted. For our evaluation here a more restricted set of variables that is readily available for most cities was utilized. These variables illustrate that such physical and socio-economic characteristics affect a city's risk.

Vulnerability is defined as the extent to which a city is predisposed to "adverse effects of *climate change*, including *climate variability* and extremes" (IPCC, 2007d). However, unlike the IPCC definition of vulnerability that includes adaptive capacity, we decouple the two here and address them individually, considering vulnerability to be determined by the physical and underlying social conditions of the city while adaptive capacity is determined predominantly by the change agents. In turn, vulnerability is a function of a host of city characteristics, including but not limited to the location of a city, particularly its proximity to a salt water coastline or other large water body, topography or any other physical attributes of the landscape or physical geography that make the city susceptible to climate variations.

Social factors that determine the degree of vulnerability of a city include its population size and composition, density, size of city, quality of infrastructure, type and quality of its built environment and its regulation, land use, governance structure and the like. A critical factor determining the vulnerability of the poor as opposed to the non-poor population of the city is the percentage of the population living in slums. These are households that lack access to one or more of the following: improved water supply, improved sanitation, sufficient living space, structurally sound dwellings, and security of tenure (UN-HABITAT, 2003). The contrast between the formally planned part of the city and the slums is stark and is a key determinant of the differential vulnerability of the poor as opposed to the non-poor (UN-HABITAT, 2008a).

2.6.3 Adaptive capacity

Adaptive capacity includes institutional attributes of the city and its actors that determine the degree of its capability to respond to potential climate change impacts. Thus they provide measures of the ability (institutional structure, caliber, resources, information, analysis), and willingness of actors (local governments, their constituent departments, private sector, civil society, NGOs, academics) to adapt to climate change. Variables that can determine the extent of a city's ability to adapt include the structure and capacity of institutions, presence of adaptation and mitigation programs, and motivation of change agents. Here it is critical to draw a distinction with the term "resilience" that the IPCC (2001b) Working Group II assessment defined as "amount of change a system can undergo without changing state." In contrast, adaptive capacity does not assume a steady state of a city and its integrated systems; rather it measures the ability and willingness to not only cope but to respond positively to the stresses that climate change imposes.

Adaptive capacity is the ability and willingness of the city's key stakeholders to cope with the adverse impacts of climate change and depends on the awareness, capacity, and willingness of the change agents. A quick measure of institutional awareness is the presence of a comprehensive analysis of climate risks for the city and corresponding adaptation and mitigation initiatives. Capacity here refers to the quality of institutions at various levels of governments – local, regional, and national – and within local government, across various departments. Further, the capacity of the private sector, non-governmental organizations, and community groups to respond also matters. Finally, the willingness to act is of the essence. In this regard, identifying in substantial detail the leading actors for climate response – government, private sector, and civil society – and mapping their initiatives is essential in estimating adaptive capacity of a city.

2.7 Climate risk assessment in selected case studies

We attempt to illustrate and assess the relative role of each of the three elements through a presentation of four major case studies for the metropolitan areas of Buenos Aires, Delhi, Lagos, and New York. These cities are located in four different global sub-regions and have a range of socio-economic conditions and vulnerability to climate hazards. A primary criterion for selecting these cities is that all four have leadership that is committed to addressing the issue of climate change and thus are exemplars for other cities within their respective region, and globally. Further, as these are all megacities and important national urban centers in their respective countries, not only do they constitute a significant share of the national GDP but also help to shape the direction of national urban development policies. See Table 2.1 for demographic parameters for the case study cities.

Most aspects of the risk case studies articulated in this chapter are equally applicable to small and medium sized cities, as in many cases time-series data on climate parameters are available. Smaller cities may have fewer resources to apply to the development of climate risk responses and thus may have additional needs for national and international guidance and support. However, the diverse urban conditions in the case study cities allow for some generalized lessons to be drawn regarding effective and efficient urban responses to climate change. The combination of city cases allows for a comparison among developing countries as well as contrasts between developed and developing country cities, their challenges and responses.

For each city, available knowledge is analyzed for various aspects of climate risks (including uncertainty). Background information from the case study cities has been evaluated and selected variables have been assigned to the framework components. The case studies allow for the understanding of the transferability of the climate risk framework to a variety of cities and what "climate services" of data analysis, access, and processing need to be provided at the international level.

To provide concrete examples of how climate risk information can be communicated, current trends in key climate variables (including temperature, precipitation, and the incidence of extreme events) for each of the case study cities have been determined, and recent IPCC 2007 projections (up to 16 models and three emissions scenarios) have been used to create city-focused downscaled climate model projections. The degree to which these models are able to replicate observed climate and climate trends in the past several decades is described. We also explore discrepancies, if any, between the identified risks, vulnerabilities, social capacities, and the current responses of cities to climate-related hazards. This addresses the important question of real-versus-perceived needs.

2.7.1 Buenos Aires, Argentina

Buenos Aires is the third largest city in Latin America, and is the political and financial capital of Argentina. The city is composed of several sub-jurisdictions that were added as the city expanded since its inception in the fifteenth century as a Spanish port. The Greater Buenos Aires Agglomeration (AGBA) has over 12 million inhabitants (National Population Census, 2001), with 77 percent of the population living in the surrounding provincial boroughs, and 23 percent in the central urban core of Buenos Aires City (Instituto Nacional de Estadística y Censos (INDEC), 2003). Buenos Aires City (CABA) is administered by an autonomous government elected directly by its citizens.

2.7.1.1 Hazards and vulnerability

Increases in sea and river levels, rising temperature and precipitation, along with increased frequency of extreme events such

Table 2.1: *Demographics for the case study cities (metropolitan area).*

Metropolitan area	Population	Area	Population density	Slum population as a percentage of national urban population
Buenos Aires	12.0 million	3,833 km^2	3,131 people per km^2	26.2 percent
Lagos	7.9 million	1,000 km^2	7,941 people per km^2	65.8 percent
Delhi	12.9 million	9,745 km^2	1,324 people per km^2	34.8 percent
New York	8.2 million	790 km^2	10,380 people per km^2	N/A

Sources: Authors' compilation from city, state, and national statistics and census bureaus of Argentina, Nigeria, India, and USA; slum data from UN HABITAT, 2008a,b.

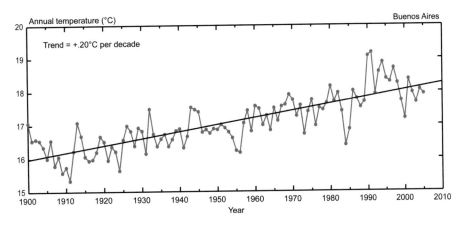

Figure 2.2: *Observed temperatures, Buenos Aires.*

as flooding caused by heavy (convective) rains and storm surges, as well as droughts, are the primary climate-induced hazards for Buenos Aires. The city has a humid subtropical climate with long hot summers, and winters with low precipitation caused by the central semi-permanent high-pressure center in the South Atlantic. This pressure system can cause strong south-southeast winds in the autumn and summer causing floods along the shores (Camillioni and Barros, 2008). For details on observed and projected temperature and precipitation trends for Buenos Aires see Figures 2.2, 2.3, 2.4, and 2.5. In regard to extreme events, there is an increase in frequency of extreme precipitation and associated city floods, see Table 2.2.

Further, occurrence of precipitation events of more than 100 millimeters within 24 hours has increased from 19 times between 1911 and 1970 to 32 times between 1980 and 2000. Such observed increases in the quantity and frequency of extreme precipitation not only adversely affect urban infrastructure, but also damage private property and disrupt the economic and social functioning of the city. With respect to sea level rise and drainage, the city is

located along the shores of the La Plata River and spreads over the *pampa*, a wide fertile plain, and adjoining the Paraná river delta. As a result, the entire metropolitan area is less than 30 meters above mean sea level. As the city grew, several rivulets that formed the natural drainage system were replaced with a system of underground stormwater drains (Falczuk, 2008).

Spatial distribution of poor versus non-poor

Over the 1990s the city experienced sprawl, with developers building gated communities on the periphery of the metropolitan area, extending the city over an area one-and-a-half times the size of the CABA (Pírez, 2002). With disparity on the rise and migration of the non-poor from the city center to the periphery, the city has been further spatially segregated by income groups. This condition was further intensified with the economic crisis of 2001, which created the "new poor" consisting of the newly unemployed middle class.

The precise distribution and enumeration of the slums is complicated by two additional factors. First is the process of "urban invasions" whereby squatter settlements crop up sporadically across the city. Second, like all other urban data for AGBA, information on the poor is parsed into 30 administrative units. For this research, data for slums and other dilapidated housing in the CABA were derived from an Ombudsman survey in 2006 (see Table 2.3 for quantification of low-income housing), which found that about 20 percent of all households in the urban core of the AGBA live in poor housing conditions.

Additionally, the survey identified 24 new settlements with 13,000 inhabitants located under bridges or simply "under the sky" (Defensoría del Pueblo de la Ciudad de Buenos Aires, 2006). However, unlike developing-country slums, most households have land tenure and property rights related to their homes due to a well-established public housing program in Argentina. Mapping the spatial distribution of differential vulnerability of the poor and non-poor to floods and other hazards is critical to crafting a climate-risk assessment of Buenos Aires (see Figure 2.6).

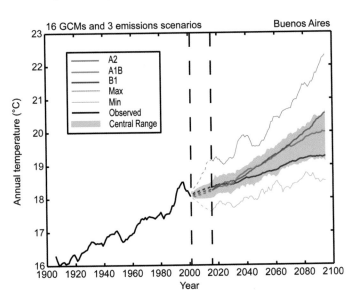

Figure 2.3: *Projected temperatures, Buenos Aires.*

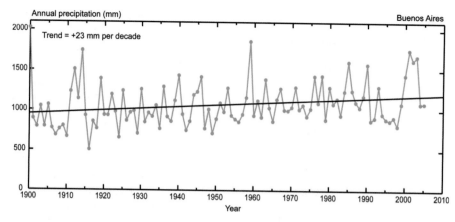

Figure 2.4: *Observed precipitation, Buenos Aires.*

Low elevation urban areas

In its present configuration, a quarter of the metropolitan area is susceptible to floods (Clichevsky, 2002; Menéndez, 2005). Urban expansion continues over the basins of the Matanza-Riachuelo, Reconquista, and Luján rivers, as well as the estuary of the La Plata River (see Figure 2.6). These areas consist of a combination of new gated communities, real estate speculation sites, as well as illegal plots in the flood plain targeted toward the housing needs of the poor. With a lack of regulation governing such urban development and the creation of unprotected infrastructure in the flood plain, the vulnerability of this part of the city is increasing (Ríos and González, 2005).

To assess the vulnerability of the low-elevation areas of the city, a review of past urban floods was undertaken. As reported in newspapers and official assessments, floods impaired all modes of public and private transportation, including domestic flights, road, and rail; disrupted energy supplies, telephone lines, and traffic lights; flooded buildings; and created an overall disruption of city life. Streets and cellars were waterlogged and people

living in low-elevation neighborhoods in the suburbs were evacuated (González, 2005). In sum, the economic costs were high. Unlike urban disasters in other developing countries, the death toll in Buenos Aires related to flooding disasters tends to be low. The primary costs are the disruption of the economic activity of the city and damage to public and private property.

As the metropolitan area has been expanding into the flood plains, a simulation to quantify the population vulnerable to sea level rise was conducted. Barros *et al.* (2008) observed: "Assuming little change in population density and distribution, under the scenario of maximum sea-level rise during the 2070 decade […] the number of people living in areas at flood risk with a return period of 100 years is expected to be about 900,000, almost double the present at-risk population." The potential damage to public and private assets can be assessed from a recent survey that estimated that 125 public offices, 17 social security offices, 205 health centers, 928 educational buildings, 306 recreational areas, and 1,046 private industrial complexes are currently at risk of floods.

A conservative estimate by Barros *et al.* (2008) is that at present the damage to real estate from floods is about US$30 million per year. Assuming a business-as-usual scenario, which includes a 1.5 percent annual growth in infrastructure and construction and no adoption of flood-protection measures, the projected annual cost of damages is US$80 and US$300 million by 2030 and 2070 respectively. These estimates do not include the losses to gated communities of the non-poor being built in the coastal area, largely located less than 4.4 meters above mean sea

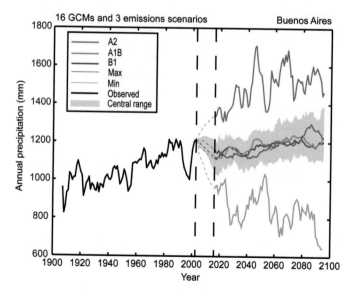

Figure 2.5: *Projected precipitation, Buenos Aires.*

Table 2.2: *Extreme events in Buenos Aires.*

City	Extreme temperatures	Extreme precipitation
Buenos Aires	January & February 1987; December & January 1988; December & January 1996–97; January & February 2001; January 2002; January 2003; January 2004; December 2004	March 1988; May 2000; February 2003

Table 2.3: *Slum population in Buenos Aires City (CABA).*

Housing types by building quality	Number of units
Slums (*villas miseria*)	< 120,000
Properties of other people (*inmuebles tomados*)	200,000
Tenement house (*casas de inquilinato*)	70,000
Lodges	70,000
Rooms in relatives' houses, rental rooms, or overcrowded houses	120,000

Source: Office of the National Ombudsman, Buenos Aires, 2006.

level. Nor does this account for the loss in productivity of the labor force, which can be significant given the size of the population likely to be affected. Thus, the costs of not responding to climate change in the course of urban development are projected to be significant and disruptive.

2.7.1.2 Adaptive capacity: current and emerging issues

The Argentinean government's response to global climate change has been dominated by mitigation efforts related to poli-cies and programs to reduce greenhouse gas emissions (Pochat *et al.*, 2006), with relatively little attention to adaptation. The lead national agency to address climate issues is the Secretariat for Environmental and Sustainable Development.

In 1993 Argentina became a signatory to the United Nations Framework Convention on Climate Change. In response, the federal government established the office for Joint Implementation, but in 1998 this was renamed the Office for Clean Development Mechanisms. Further, in 1999 Argentina adopted the objectives of the Greenhouse Gases Reduction Programme, and in 2001 signed on to the Kyoto Protocol. To institutionalize the response to climate change, in 2003 a Climate Change Unit was established within the Secretariat for Environmental and Sustainable Development. In 2007 this evolved into the Climate Change Office. In addition, the government has been supporting a range of research programs, such as the National Program on Climate Scenarios, which was initated in 2005. Through these institutional arrangements, first and second national reports were prepared in 1997 and 2006 respectively. The third version is under preparation (Pochat *et al.*, 2006).

However, the roles and responsibilities of governmental agencies in regard to climate change remain fragmented, while adaptation responses, specifically at the city level, remain to be addressed. In addition, four ministries with a dozen departments

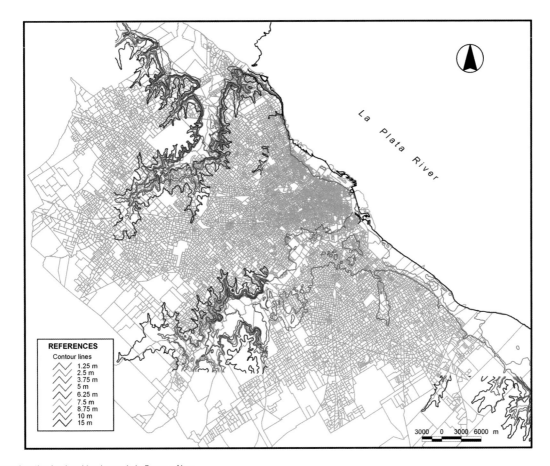

Figure 2.6: *Low-elevation land and land parcels in Buenos Aires.*

Source: Elaborated by S. Gonzalez. Based on AIACC Project, LA 26.

and institutions are involved in flood monitoring and broader disaster management systems (Natenzon and Viand, 2008). Gradually, lower levels of government such as the states and local authorities are taking an interest in addressing climate change mitigation and adaptation, and a range of stakeholders such as NGOs, the media, and citizen groups are participating.

Conflicting plans and multiple jurisdictions reduce the efficacy of climate change response plans at the city level as well (Murgida and González, 2005). For example, in 2007 an Office for Climatic Protection and Energy Efficiency was established within the Ministry for Environment of Buenos Aires City. With the arrival of a new administration in December 2007 this ministry was restructured into the Ministry of Environment and Public Space, with a new Environmental Protection Agency. The Office for Climatic Protection and Energy Efficiency was dismantled despite the fact that previously initiated programs and projects such as "Clean Production" and "Air Quality" continue to be implemented (Murgida, 2007).

The primary obstacles to institutional action at the metropolitan level are lack of actionable climate information, as well as vertical and horizontal fragmentation of jurisdictions with divergent interests and responses. Administrative units within the AGBA address flood management but lack an integrated strategy. For example, within Buenos Aires City two different plans – the Urban Environmental Plan and the Buenos Aires 2015 Strategic Plan – are being implemented simultaneously but with a lack of effective coordination. Further, in practice there are two critical legislative instruments to regulate urban development in the city – namely, the Building Code enacted in 1944 and the Urban Planning Code enacted in 1977. These are complemented by additional measures such as the Flood Control Plan, and post-2001 flood tax rebates for affected communities. However, these plans, codes and norms are inconsistent. For instance, the Urban Planning Code incentivizes the occupation of vulnerable low-lying areas within the city contradicting the flood prevention plans (González, 2005).

Moreover, constantly changing organizational roles and responsibilities of government agencies tasked to address climate change pose a challenge. For instance, in 2005 the Buenos Aires State Government created a unit to address climate change within the provincial Ministry for Environmental Policy. This office continues to be operational under the new local government that was elected in 2007, but the unit has been moved to the Ministry of Social Development and has a reduced mandate. This lack of action orientation is compounded by a general lack of public awareness of the risks associated with climate change (Assessment of Impacts and Adaptation to Climate Change, 2005).

Additionally, there is a mismatch in terms and scales. While the climate adaptation strategies such as flood prevention and management need to take a long-term view and plan for the metropolitan region as a whole, most planning interventions address short-term needs and do not take a citywide view (Murgida and

Natenzon, 2007). "By analyzing who participated in the planning process and in which areas they did so, it becomes evident that the vast majority of interventions were partial, some were very specific, and a few encompass different areas and spheres" (Pírez, 2008). These issues become further complicated for the metropolitan region due to the overlap and aggregation of administrative units that lack a central governing authority.

The community of scientists and researchers has taken on an unusual task of coordinating climate-related programs and policies. A leading example of this effort was the launch of the Global Climate Change Research Program at Buenos Aires University (PIUBACC) in May 2007. The objective of this program is to map and link all research as well as city development projects within the metropolitan area so as to provide the government, civil society, and more-specifically interested groups directly involved in climate change programs with a holistic and scientific assessment of climate change risks. Additionally, the scientists are drawing transferable lessons from community knowledge on flood management along the La Plata River coast with a dual focus on the vulnerability of the poor and on adaptation to storm-surge floods (Barros et al., 2005).

2.7.2 Delhi, India

Metropolitan Delhi has a population of 16 million, and is rapidly urbanizing with a 3.85 percent annual growth rate over the 1990s amounting to half a million additional inhabitants each year. In 1901 Delhi had 400,000 inhabitants. Furthermore, rising per capita incomes are increasing energy consumption and over-stretching its infrastructure. Delhi is a city of wide income contrasts – in 2000, 1.15 million people were living below the National Poverty line. On the other hand, Delhi's Gross State Domestic Product at current prices was about US$27 billion during 2007 (Department of Planning, 2008).

Delhi has three distinct seasons – summer, winter, and monsoons with extreme temperatures and concentrated precipitation. Summers begin in mid March, lasting for three months, and are dry and hot with temperatures peaking at about 40 °C in the months of May and June. Monsoons are between mid June and September, during which period Delhi receives most of its 600 millimeters of annual rainfall, with July and August getting as much as 225 millimeters each, see Figures 2.7–2.10 for seasonal variation in temperature and precipitation. Winters are dry and last from November to mid March, with December and January being the coldest months with temperatures as low as 7 °C (Delhi, 2009).

2.7.2.1 Hazards and vulnerability

The National Action Plan for Climate Change and related analysis provides an overview of climate change issues confronting India as recognized by the federal government (see Government of India, 2002, 2008). Through a review of research on climate science, policy papers, and practitioner notes, five hazards are

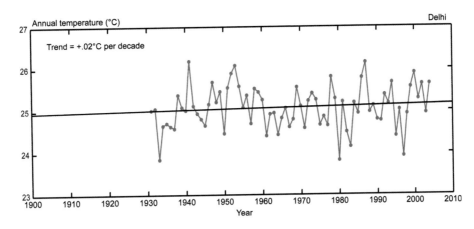

Figure 2.7: *Observed temperatures, Delhi.*

identified (Revi, 2007). First, although there are uncertainties with scaling down global models such that they reflect regional climate conditions such as the Indian monsoon, temperature, precipitation, and sea-level are likely to rise. For a summary of observed and projected temperatures for Delhi see Figures 2.7 and 2.8. Mean extreme temperatures, as well as maxima and minima, are expected to increase by 2 to 4 °C, likely to result in an average surface warming of 3.5 to 5 °C within this century.

Second, average mean rainfall is projected to increase by 7 to 20 percent due to the increase in mean temperature and its impact on the Indian monsoon cycles within the latter half of this century (see Figures 2.9 and 2.10). However, some drought-prone areas are expected to become drier and flood-prone areas will very likely experience more intense periods of precipitation. Third, 0.8 meters is the projected centennial rise in mean sea level. Fourth, extreme events like the Mumbai flood of 2005 are expected to be more frequent in western and central India. A combination of these hazards exposes the cities in this region to a range of other climate-induced extreme events such as droughts,

temporary and permanent flooding, both inland and in coastal areas, and cyclones.

Delhi's physical infrastructure, social services, and slum populations make the city highly vulnerable. Demand for basic infrastructure services such as water, electricity, and public transport far exceeds supply (Delhi Development Authority, 2005). To add to the existing conditions, climate change-induced variability in rains could worsen the severe shortage of drinking water in summers and aggravate the floods in the monsoon season, thus making the existing energy shortage more challenging to address. With regard to transportation, Delhi has the highest per capita vehicular population in India – 5.4 million automobiles for 15 million people. This poses a challenge for a city with mixed land use and varying urban densities within the metropolitan region to introduce effective modes of public transport. Carbon emissions from vehicles, traffic congestion, and increasing particulate matter all pose challenges. These and other challenges create widespread public health risks to the inhabitants of Delhi. For example, lack of adequate sanitation facilities for the poor poses a problem for a rapidly growing city where a large proportion of the population lives in slums.

The hyper-dense nature of the slums, despite Delhi's relatively low population density and the centrality of the poor in provision of services – from household help to a range of labor-intensive and low-wage tasks – poses an enormous challenge. About 45 percent of the city's population live in a combination of unregulated settlements, including unauthorized colonies, villages, and slums. Further, three million people live along the Yamuna River, which is prone to flooding, where 600,000 dwellings are classified as slums.

Moreover, increasing competition for scarce basic services caused by the rapidly growing population of Delhi poses public health as well as quality-of-life challenges. For example, some poor settlements lack basic amenities resulting in open defecation. Although the extent of the impacts remains to be assessed, potential climate change impacts added to current local environmental stresses are likely to intensify this crisis. Moreover, the low quality of housing in slums and their proximity to

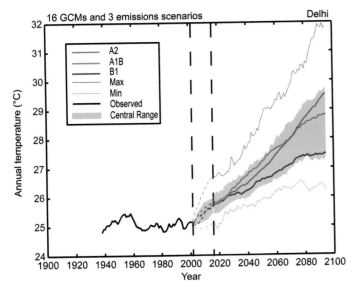

Figure 2.8: *Projected temperatures, Delhi.*

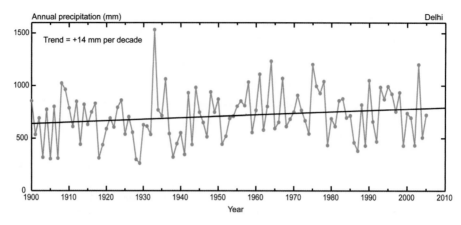

Figure 2.9: *Observed precipitation, Delhi.*

environmentally degraded land and flood-prone areas further exacerbate the vulnerability of the poor. Within the slums, climate-induced stress is likely to affect certain social groups more than others, particularly the elderly, women, and children.

2.7.2.2 *Adaptive capacity: current and emerging issues*

The government of Delhi has made many efforts towards climate change mitigation, but there is less emphasis on adaptation. In addition to the issues of energy, water, and transportation, mitigation projects also encompass public health and other social and economic development efforts. Climate change mitigation efforts by the government of Delhi were introduced first in the government departments and are being gradually expanded to include other stakeholders – schools, households, and firms. Most initiatives remain project-oriented (Department of Environment, 2008).

Some projects, such as the Bhagidari program, seek participation from neighborhood groups, private-sector associations, schools, and non-governmental organizations to enhance civil

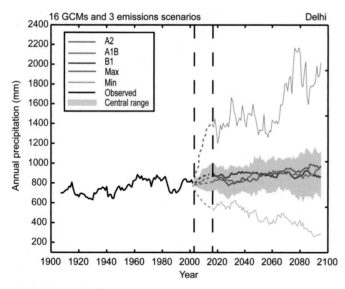

Figure 2.10: *Projected precipitation, Delhi.*

society engagement in environmental management, creating an expanded policy space for addressing climate change. Such collaboration holds the potential to address broader issues of climate adaptation by building awareness as well as capacity of stakeholders to respond. However, the most striking of all climate mitigation initiatives in Delhi so far is the establishment of the world's largest fleet of compressed natural gas (CNG) fueled public transport in response to a Supreme Court order. This has resulted in 130,000 CNG-powered vehicles, 145 CNG fuel stations, as well as improved vehicular emission standards like those adopted by the European Union. The greatest lesson from this initiative is in recognizing the diverse set of triggers and actors that can start adaptation and mitigation programs.

Some mitigation measures have co-benefits for adaptation. For instance, adoption of green building technology that is mandatory for the Public Works Department and the Airport Authority was introduced to address mitigation, but has adaptation benefits as well. Expected greenhouse gas emissions reductions are 35 to 50 percent in general energy consumption and up to 100 percent in energy for water heating. Moreover, the New Delhi Municipal Council aims to reduce other demand for energy and has set time-bound efficiency targets. The Municipal Corporation of Delhi is making efforts to install compact fluorescent lamps and capacitor banks to increase energy efficiency. Further, the government has a program that subsidizes electric vehicles and is encouraging the introduction of the Reva car, as well as battery-operated two- and three-wheelers.

Delhi also has expanded its forest cover over the past ten years. The cities greening program is considered to be one of the largest in the world. The forest cover has grown from 3 percent in 1998 to 19 percent in 2005. The city planted 1.7 million trees in 2007 and the forest cover grew to 300 square kilometers. To maintain the momentum, the city planned to plant 1.8 million saplings in the fiscal year 2009, increasing the greenery cover to a total of 326 square kilometers. The city also has a policy to plant ten trees for every tree chopped down. This project is done in collaboration with several stakeholders including school children, female homemakers, and neighborhood associations. The saplings are distributed *gratis* through a host of distributors. This afforestation effort is part of a Clean Development Mechanism (CDM) project proposal. To scale up mitigation efforts, the Delhi government has established a program with the aim to raise awareness about carbon credits and clean development mechanisms among various departments. The objective is to develop a holistic approach towards reducing greenhouse gas emissions and enable projects that can redeem carbon credits. In essence, these mitigation projects can prove vital for adaptive capacity as well. For example, green roofs and walls, and tree planting help to cool the urban environment and reduce heat island effects, as do many of the energy-efficiency projects related to buildings.

While the neglect of adaptation remains a concern, another co-benefit to mitigation efforts in Delhi is the climate change awareness and administrative capacity being built as a result of mitigation projects that may help as adaptation projects and policy measures are introduced. Not only is the government developing financial incentives to introduce programs and adopting a multisectoral approach that involves various departments within the city jusrisdiction, they are also learning to utilize mechanisms such as UNFCCC's CDM funds that are likely be equally relevant for adaptation. Illustrations of such efforts are the CDM projects and certified emission reductions (CERs) in the water sector.

While Delhi is making major efforts towards mitigation of climate change through carbon emission reductions and other environmental improvements, there is a significant lack of awareness about the need for adaptation to climate change. Therefore, the city has not yet planned for adaptation. Further, Delhi's response to climate change is often less than effective as well as piecemeal because its efforts are primarily project-oriented. In the experience of the Delhi government, incentives – subsidies and grants – have been effective for initiating projects, but operation and management frequently remain neglected. For instance, subsidies to install rainwater-harvesting systems have created demand, but subsequent maintenance is too often ignored and many systems fall into disrepair. Such experiences hold the potential to inform adaptation efforts as well.

Gradually the city is developing a programmatic approach, but there is a need to coordinate between departments and among levels of government. For example, while the Prime Minister of India has recently released the National Action Plan for Climate Change, Delhi's local efforts will need to be reconciled with regional and national priorities.

2.7.3 Lagos, Nigeria

Lagos is Africa's second most populous city and has grown explosively, from 300,000 in 1950 to an expected 18 million by 2010, ranking it as one of the world's ten largest cities. The metropolitan area, an estimated 1,000 square kilometers, is a group of islands surrounded by creeks and lagoons and bordered by the Atlantic Ocean. Lagos is the commercial and industrial hub of Nigeria, a country with a GDP three times larger than any other country in West Africa. Lagos is home to a large amount of commercial infrastructure, and has greatly benefited from Nigeria's natural resources of oil, natural gas, coal, fuel wood, and water. For an overview of the state of Nigerian cities see UN-HABITAT (2004).

The climate of Lagos is affected by Atlantic Ocean and atmospheric interactions both within and outside its environment, in which the Inter-Tropical Convergence Zone (ITCZ) plays a controlling factor. The movement of the ITCZ is associated with the warm humid maritime tropical air mass with its southwestern winds and the hot and dry continental air mass with its dry northeasterly winds. Maximum temperatures recorded during the dry season are high and range from 28 to 33 °C when the region is dominated by the dry northeast trade winds. Minimum temperature of about 26 °C is experienced during the wet season of May to September.

The city of Lagos experiences relatively high to very high temperatures throughout the year. The mean annual temperature is about 28 °C and the maximum and minimum temperatures are 33 °C and 26 °C respectively. High to very high monthly rainfall is also experienced between May and November, although significant variations in monthly rainfall peak values are experienced. For example, between 1950 and 2006, more than ten instances were recorded with a maximum rainfall of over 700 millimeters. Minimum monthly rainfall of less than 50 millimeters is experienced between December and March. Occasionally, extreme precipitation events are experienced in June. On June 17, 2004, for example, 243 millimeters of rain was experienced in Victoria Island and the Lagos environs. This resulted in flooding of streets and homes, collapse of bridges, and massive erosion of the main road linking Lekki to Lagos Island. About 78 percent of the total rainfall amount for the month was experienced in one day in June. The city was ill-prepared for that amount of rainfall.

2.7.3.1 Hazards and vulnerability

Ekanade *et al.* (2008) describe the nature and magnitude of the climate change hazards for the city level using different greenhouse gas emission scenario models. The IPCC (2001a) Special Report on Emission Scenarios A2 and B1 climate change scenarios were utilized to project 30-year timeslices for temperature and rainfall values for the City of Lagos and Port Harcourt and the coastal areas of Nigeria. This study did not, however, project sea level rise.

Records from the two stations (Ikeja and Lagos) used in this analysis show that monthly maximum temperature was increasing at about 0.1 °C per decade from 1952 to 2006, while monthly minimum was decreasing at about 0.5 °C per decade: since the 1900s average temperature has increased 0.07 °C per decade (see Figure 2.11). At the extremes, monthly maximum temperatures for Lagos have reached above 34 °C during seven of the last twenty years. The number of heat waves in Lagos has also increased since the 1980s (see Table 2.4). There have been very few incidences of unusually cold months of less than 20 °C

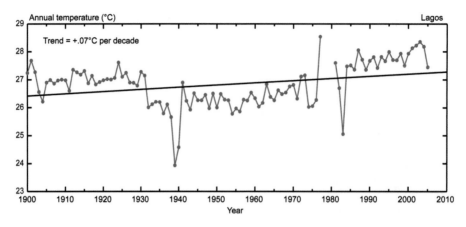

Figure 2.11: *Observed temperatures, Lagos.*

since 1995. Projected temperature for Lagos for the 2050s anticipates a 1 to 2 °C warming (see Figure 2.12).

According to historical records, the total annual precipitation in Lagos has decreased by 8 millimeters per decade since 1900 (see Figure 2.13). In keeping with the overall precipitation trends, most of Lagos has experienced decreases in rainfall amounts during the rainy season. For example, between 1950 and 1989 more than 20 months experienced rainfall amounts of over 400 millimeters. In the recent period between 1990 and 2006, however, very few (four) rainy months recorded over 400 millimeters of rain. In the twenty-first century, precipitation in Lagos is expected to be less frequent but more intense; projected precipitation for Lagos for the 2050s anticipates an uncertain 5 percent change in mean precipitation (see Figure 2.14; Table 2.4).

Coastal storm surge also is a concern. Lagos, as well as the entire Nigerian coast, is projected to experience more storm surges in the months of April to June and September to October annually. This increase in storminess is projected to be accompanied by greater extreme wave heights along the coasts. According to Folorunsho and Awosika (1995) the months of April and August are usually associated with the development of low-pressure systems far out in the Atlantic Ocean (in the region known as the "roaring forties"). Normal wave heights along the Victoria Beach range from 0.9 m to 2 m. However, during these swells, wave heights can exceed 4 m. The average high high-water (HHW) level for Victoria Island is about 0.9 m above the zero tide gauge with tidal range of about 1 m. However, high water that occurs as surges during these swells has been observed to reach well over 2 m above the zero tide gauge. These oceanographic conditions are aggravated when the swells coincide with high tides and spring tide.

An extreme event, which can be considered a case study for future threats, was observed between August 16 and 17, 1995, when a series of violent swells in the form of surges were unleashed on the whole of Victoria Beach in Lagos. The most devastating of these swells occurred on August 17, 1995, and coincided with high tide, thus producing waves over 4 m high causing flooding. Large volumes of water topped the beach and the Kuramo waters. A small lagoon separated from the ocean by a narrow – 50 m wide – strip of beach was virtually joined to the Atlantic Ocean. Many of the streets and drainage channels were flooded, resulting in an abrupt dislocation of socio-economic activities in Victoria and Ikoyi Islands for the period of the flood.

In addition, coastal erosion is very prevalent along the Lagos coast. Bar Beach in Lagos has an annual erosion rate of 25 to 30 m. Earlier IPCC scenarios have been used to estimate the effects of 0.2, 0.5, 1, and 2.0 m sea level rise for Lagos. Along with coastal flooding and erosion, another adverse effect of sea level rise on the Lagos coastal zone as earlier assessed by Awosika *et al.* (1993a, b) is increased salinization of both ground and surface water. The intrusion of saline water into groundwater supplies is likely to adversely affect water quality, which could impose enormous costs on water treatment infrastructure.

Lagos has an extremely dense slum population, many of whom live in floating slums. These are neighborhoods that extend out

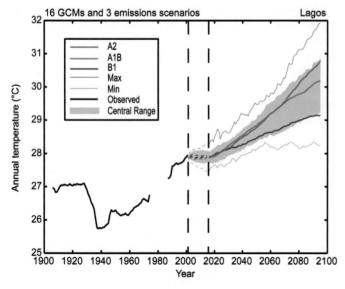

Figure 2.12: *Projected temperatures, Lagos.*

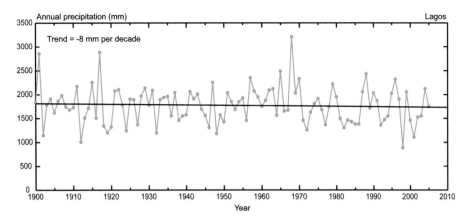

Figure 2.13: *Observed precipitation, Lagos.*

into the lagoons scattered throughout the city. The barrier lagoon system in Lagos, which comprises Lagos, Ikoyi, Victoria, and Lekki, will be adversely affected through the estimated displacement of between 0.6 and 6 million people for sea level rise of between 0.2 and 2 m (Awosika *et al.*, 1993a) (see Table 2.5).

In their study of the impacts and consequences of sea level rise in Nigeria, French *et al.* (1995) recommended that buffer zones be created between the shoreline and the new coastal developments. A more generalized multisectoral survey of Nigeria's vulnerability and adaptation to climate change was funded by the Canadian International Development Agency (CIDA) through its Climate Change Capacity Development Fund (CCCDF). This study has served to create awareness of climate change issues and of the need for manpower development.

Even more worrisome is the general sensitivity of Lagos to climate change due to its flat topography and low-elevation location, high population, widespread poverty, and weak institutional structures. Many more vulnerabilities stem from these characteristics including the high potential for backing up of water in

drainage channels, inundation of roadways, and severe erosion. The barrier lagoon coastline in the western extremity, including the high-value real estate at Victoria Island and Lekki in Lagos, could lose well over 584 and 602 square kilometers of land respectively from erosion, while inundation could completely submerge the entire Lekki barrier system (Awosika *et al.*, 1993a, b). Moreover, flooding poses greater threats to the urban poor in several African cities (Douglas and Alam, 2006). See Figure 2.15 for topography identifying low-lying areas that overlap with built-up areas and are prone to flooding.

Intense episodes of heat waves will likely severely strain urban systems in Lagos, by inflicting environmental health hazards on the more vulnerable segments of the population, imposing extraordinary consumption of energy for heating and air conditioning where available, and disrupting ordinary urban activities.

It is very likely that heat-related morbidity and mortality will increase over the coming decades; however, net changes in mortality are difficult to estimate because, in part, much depends on complexities in the relationships among mortality, heat, and other stresses. High temperatures tend to exacerbate chronic health conditions. An increased frequency and severity of heat waves is expected, leading to more illness and death, particularly among the young, elderly, frail, and poor. In many cases, the urban heat island effect may increase heat-related mortality. High temperatures and exacerbated air pollution can interact to result in additional health impacts.

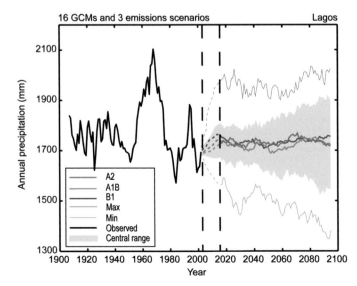

Figure 2.14: *Projected precipitation, Lagos.*

Table 2.4: *Extreme events in Lagos.*

City	Extreme temperatures	Extreme precipitation
Lagos	March 1988; March 1990; February 1998; March 1988; March 2001; March 2002; February 2003; March 2003; August 2004	May 1958; June 1962; July 1968; August 1998; November 1998

Source: Center for Climate Systems Research, Columbia University.

Table 2.5: *Estimation of internally displaced people by sea level rise scenarios in Lagos.*

Sea level rise scenarios	0.2 m	0.5 m	1.0 m	2.0 m
By shoreline types, number of people displaced (in millions)				
Barrier	0.6	1.5	3.0	6.0
Mud	0.032	0.071	0.140	0.180
Delta	0.10	0.25	0.47	0.21
Strand	0.014	0.034	0.069	0.610
Total	0.75	1.86	3.68	7.00

Source: Awosika et al. (1992).

Impacts are projected to be widespread, as urban economic activities will probably be affected by the physical damage caused to infrastructure, services, and businesses, with repercussions on overall productivity, trade, tourism, and on the provision of public services.

The degraded state of the urban form and poverty is indicative of the expected low resilience of most of the inhabitants of the city to external hazard stressors such as those often associated with climate extreme events. Most of the city's slums are located on marginal lands that are mainly flood-prone with virtually no physical and social infrastructure. Furthermore, some of the planned and affluent neighborhoods in many parts of the city still experience flooding during "normal" rainfall. This may be attributed to the little-to-no attention often given to the provision and maintenance of sewer and storm drains in these supposedly "planned" affluent neighborhoods. For instance, Ikoyi, one of the most highly priced neighborhoods in the city, was actually developed from an area originally covered by about 60 percent wetland.

2.7.3.2 Adaptive capacity: current and emerging issues

Even with active membership in the C40 Large Cities Climate Leadership network, Lagos megacity still does not have a comprehensive analysis of the possible climate risks facing it, especially with respect to inundation due to sea level rise. The implication is that there is an urgent need to address the obvious lack of awareness of the vulnerability of Lagos to climate change and the need to begin to plan adaptation strategies. Recently, tackling the problem of flooding and coastal erosion has been given more attention by the Lagos State government in the form of a sea wall along Bar Beach in Victoria Island. This activity, however, is evidence that local awareness appears to be lacking the full scope of the city's vulnerability to climate change.

Although the attention of the city managers is more focused on filling its long physical infrastructural gap due to years of neglect, the lack of concern or awareness of likely sea level rise in Lagos is worrisome. There continues to be sand-filling of both the Lagos Lagoon and the Ogun River flood plain in the Kosofe

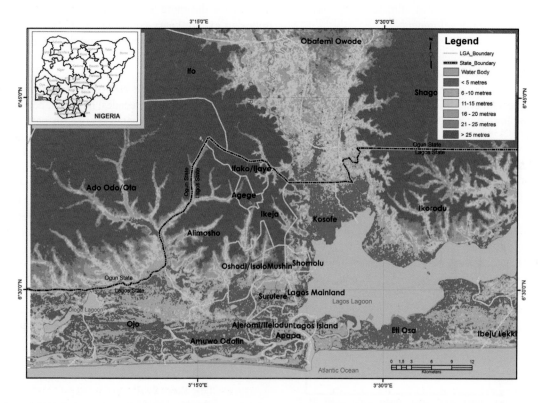

Figure 2.15: *Lagos topography.*

Source: Ademola Omojola – Author's analysis, derived from the Shuttle Radar Topography Mission (SRTM) Digital Terrain Model (DTM) data. The DTM data were color-coded, clipped for the study area, and the administrative map superimposed.

local government area to about 2 m above sea level for housing developments. Such activities need to be done with projections of sea level rise due to climate change as part of the planning process.

Currently, the leading actor on climate change issues in the city is the Lagos State government, which has been influenced by its membership of the C40 Large Cities Climate Summit. Some of the mitigation actions being pursued by the Lagos State government in the city include:

1. Improvement of the solid waste dump sites that are notable point sources of methane – a greenhouse gas – emissions in the city;
2. The new bus rapid transport (BRT) mass transit system is already shopping for green technology to power vehicles in its fleet;
3. Commencement of tree planting and city greening projects around the city; and
4. Proposed provision of three air-quality monitoring sites for the city.

The full picture of the nature of climate change and variability, its magnitude, and how it will affect the city is yet to be analyzed to support any informed adaptation actions. Thus, the climate risk reduction adaptation actions presently taken in the city are primarily spin-offs from the renewed interest of the city's management in reducing other risks and taking care of developmental and infrastructural lapses, rather than being climate change driven. Some of these adaptation activities include:

1. The sea wall protection at Bar Beach on Victoria Island to protect the coastal flooding and erosion due to storm surges;
2. Primary and secondary drainage channel construction and improvement to alleviate flooding in many parts of the city;
3. Cleaning of open drains and gutters to permit easy flow of water and reduce flooding by the Lagos State Ministry of Environment Task Force, locally referred to as "Drain Ducks";
4. Slum upgrade projects by the LMDGP project; and
5. Awareness and education campaigns such as the formation of Climate Clubs in primary and secondary schools in Lagos, and organization of training sessions and workshops on climate change issues.

Due to the increasing activity in the Ogun State sector of the city, a regional master plan for the years 2005–2025 (Ogun State Government, 2005) has been developed for its management. However, the issue of climate change risks to infrastructure and the different sectors such as water and wastewater (Iwugo *et al.*, 2003), health, and energy, is not yet reflected in the report.

"Normal" rainfall is known to generate extensive flooding in the city largely because of inadequacies in the provision of sewers, drains, and wastewater management, even in government-approved developed areas. Consequently, an increase in the intensity of storms and storm surges is likely to worsen the city's flood risks. Since the local governments work closely with the people and communities threatened by climate risks, there is a need to create awareness at the local government level. There is an urgent need to empower them intellectually, technically, and financially to identify, formulate, and manage the climate-related emergencies and disasters, as well as longer-term risks, more proactively.

2.7.4 New York City, United States

With 8.2 million people, a $1.1 trillion GDP (Bureau of Economic Analysis, 2008) and an operating budget of over $40 billion, New York City is the largest city in the USA both in population and in economic productivity (The City of New York, 2007, 2009). The distribution of wealth within the city, however, has been described as an "hourglass economy," where there is a shrinking of the middle class and growth in both the upper- and lower-income populations (Rosenzweig and Solecki, 2001).

New York City is an archipelago with five boroughs spread out over three islands – Long Island, Manhattan, and Staten Island – and the mainland of the USA. New York is one of the world's most important international financial hubs. As a coastal city, most of New York City sits at a relatively low elevation with approximately 1 percent of the city lower than 3 meters (10 feet) (Rosenzweig and Solecki, 2001). Much of Manhattan's very low-lying land is home to some of the most important economic infrastructure in the world.

New York City has a temperate, continental climate characterized by hot and humid summers as well as cold winters and consistent precipitation year round. Using a baseline period of 1971–2000, records show an average temperature of 12.7 °C with precipitation averaging 109 to 127 centimeters per year. Recent climate trends show an increase in average temperature of 1.4 °C since 1900 and a slight increase in mean annual precipitation (New York City Panel on Climate Change (NPCC), 2009) (see Figures 2.16 and 2.20).

2.7.4.1 *Hazards and vulnerabilities*

New York City is susceptible to mid-latitude cyclones and nor'easters, which peak from November to April. These storms contribute greatly to coastal erosion of vital wetlands that help defend areas of the city from coastal flooding. Tropical cyclones (hurricanes) also have the potential to reach New York City, usually during the months of August and September. There is some indication that intense hurricanes will occur more frequently in the future, but this is an area of active scientific research.

Based on climate model projections and local conditions, sea level is expected to rise by 4 to 12 centimeters by the 2020s and 30 to 56 centimeters by the 2080s (see Figures 2.18 and 2.19); when the potential for rapid polar ice-melt is taken into account based on current trends and paleoclimate studies, sea level rise projections increase to between 104 and 140 centimeters (NPCC, 2009). The possibility of inundation during coastal storms is greatly enhanced with the projected effects of sea level rise.

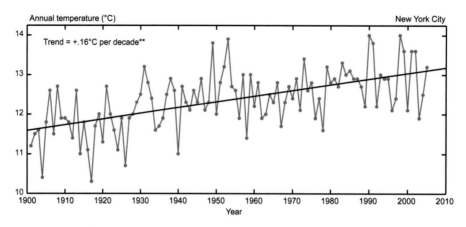

Figure 2.16: *Observed temperatures, New York City.*

Another hazard to impact New York City as a result of climate change is rising mean temperature, along with the associated increase in heat waves. The annual mean temperature in New York City has increased nearly 2 °C since 1900 (NPCC, 2009). Climate models predict that the average temperature will increase by between 1 and 1.5 °C by 2020 and 2 to 4 °C by the 2080s as seen in Figure 2.17 (NPCC, 2009). As defined by the New York Climate Change Task Force, a heat wave is any period of three straight days with a temperature over 32 °C. The frequency of heat waves is projected to increase as the number of days over 32 °C increases. These higher temperatures will also intensify the urban heat island in New York City, since urban materials absorb radiation throughout the daytime and release it during the night, causing minimum temperatures to rise (Rosenzweig and Solecki, 2001; Kinney *et al.*, 2008). These sustained, higher temperatures exacerbate the effects of heat on humans (Basu and Samet, 2002).

Inland floods and droughts are two more hazards that confront New York City. Climate models indicate that precipitation

in New York City is likely to increase by up to 5 percent by the 2020s and between 5 and 10 percent by the 2080s, as seen in Figure 2.21 (NPCC, 2009). These increases are projected to come in the form of more intense rain events. This means more days without precipitation between larger and more intense storms. As extreme rain events are expected to increase in intensity while decreasing in frequency, many of the rivers and tributaries that flow through New York City and feed into the bodies of water that surround the city may breech their banks more frequently, as they will likely be unable to handle the volume of water flowing into them as runoff.

Droughts may also prove to be a hazard as a result of climate change if the period between rain events increases. A major concern is the New York City water supply, which is drawn from up to 100 miles north of the city. The higher levels of precipitation associated with climate change are expected to be offset by the greater rates of evaporation associated with temperature increase, thus increasing the likelihood of drought (NPCC, 2009).

The impacts of these climate hazards are interconnected and affect many systems in New York City differently but simultaneously. Roadways and subways, as well as ferry ports, industries located along the coast, and wastewater treatment facilities are susceptible to inundation. More hot days will increase electricity demand to run cooling systems, thereby increasing CO_2 emissions. The erosion of natural defenses such as coastal wetlands increases the likelihood of flooding of nearby neighborhoods and industries.

Some populations are more vulnerable than others and these vulnerabilities are frequently differentiated along the lines of inland versus proximity to coast, young versus old, and rich versus poor. One key climate change vulnerability is related to air quality and human health, since degradation of air quality is linked with warmer temperatures. The production of ozone (O_3) and particulate matter with diameters below 2.5 micrometers ($PM_{2.5}$) in the atmosphere is highly dependent on temperature (Rosenzweig and Solecki, 2001). Therefore, increased temperatures are likely to make managing these pollutants more difficult. Both of these pollutants affect lung functioning,

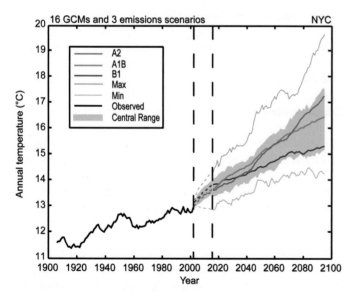

Figure 2.17: *Projected temperatures, New York City.*

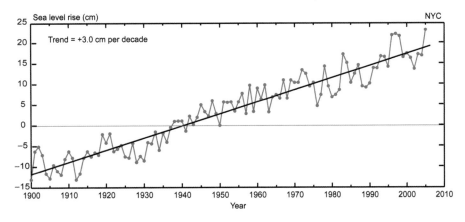

Figure 2.18: *Observed sea level rise, New York City.*

with higher ozone levels being associated with increased hospital admissions for asthma. Further, the elderly and those suffering from heart- and lung-related diseases have been shown to be more susceptible to the effects of heat, often resulting in death from heat stroke and heat-related causes (Knowlton *et al.*, 2007).

New York City is vulnerable to heat waves and, as an archipelago, is particularly vulnerable to the effects of storm surge as a result of sea level rise. Projected sea level rise of 30 to 58 centimeters – or 104 to 140 centimeters, if rapid polar ice-melt is considered – is not expected to inundate the city extensively; rather, the problem emerges when larger storms such as the 1-in-100 year storm, which are expected to become more frequent, produce a greater storm surge that will likely cause damaging floods (NPCC, 2009) (see Figure 2.22).

Certain populations are more vulnerable to the effects of heat and higher sea levels. Approximately 967,022 people in New York City are aged 65 or older and of those it is estimated

that 43 percent are living with some sort of disability (US Census Bureau, 2008). These two factors contribute to the extreme vulnerability to heat of the elderly (Basu and Samet, 2002). According to the Department of Health for the City of New York, during the heat wave of 2006 over half of those who died in New York City were over age 65 and all but five people were known to have suffered from some type of medical condition (Department of Health and Mental Hygiene (DOHMH), 2006).

New York City is a densely populated city with approximately 10,380 people in each of its 305 square miles or 790 square kilometers (Department of City Planning, 2009). Within this area there are clear pockets of wealth and poverty. The areas of low per capita income are in northern Manhattan, above Central Park, the borough of the Bronx and parts of Brooklyn. Sea level rise and coastal flooding are concerns for certain parts of these areas including Coney Island, Brighton Beach, and Jamaica Bay. One of the more recurring vulnerabilities for these populations is extreme heat and the diminished air quality that accompanies the heating trend that New York City has seen over the last 100 years and that is projected to continue. The US Census Bureau has estimated that for the period 2005–2007 about 20 percent of the people in New York City were living below the poverty line as established by the US Government (US Census Bureau, 2008). During the heat wave of 2006, 38 of those who died of heat stroke did not have functioning air conditioning in their apartment (DOHMH, 2006).

2.7.4.2 *Adaptive capacity: current and emerging issues*

The environment in which New York City makes climate change adaptation and mitigation decisions is highly complex. Due to shared regional transportation, water, and energy systems, the stakeholders in any decision include numerous local governments, multiple state governments, businesses, and public authorities.

The foundation for tackling the challenges of climate change in New York City began in the mid 1990s when the New York Academies of Science published *The Baked Apple? Metropolitan New York in the Greenhouse* in 1996. Shortly thereafter, the

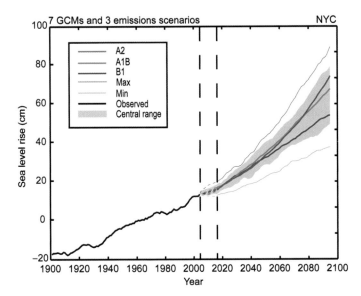

Figure 2.19: *Projected sea level rise, New York City.*

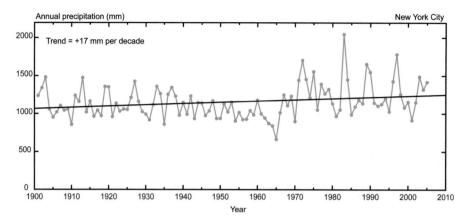

Figure 2.20: *Observed precipitation, New York City.*

Earth Institute at Columbia University, through the Center for Climate Systems Research (CCSR) released *Climate Change and a Global City: The Potential Consequences of Climate Variability and Change* (Rosenzweig and Solecki, 2001). This report covered the Metro–East Coast Region and served as the first assessment of climate change and cities in the USA. In 2008, CCSR worked with the New York City Department of Environmental Protection to develop a sector-specific climate assessment and action plan for New York City's water system (New York City Department of Environmental Protection, 2008).

The New York City administration through its Office of Long-Term Planning and Sustainability created the NYC Climate Change Adaptation Task Force in 2008, which is now working with local experts, city departments, and stakeholders to develop a comprehensive, integrated climate change risk assessment and adaptation plan for the critical infrastructure of the metropolitan region. The NYC Climate Change Adaptation Task Force is made up of representatives from over 40 city and regional departments and industries. The city administration also convened the New York Panel on Climate Change (NPCC) to provide expert infor-

mation about climate change risks and adaptation. The NPCC is made up of climate change scientists and experts from the legal field, insurance, telecommunications, and transportation, and has provided the climate risk information needed to create actionable guidelines and plans for adapting the city's critical infrastructure for the projected effects of climate change (NPCC, 2009). The NPCC has also worked with the NYC Climate Change Adaptation Task Force to develop a common set of definitions for adaptation assessment.

The next step is to begin planning and making specific adaptation investments across the city. In the past in New York City, this has tended to be on a project basis and so has been less coordinated across sectors. Having brought decision-makers from all key departments in the city and from numerous sectors, the New York City climate change adaptation process is helping to facilitate more open avenues of communication and coordination within and among departments.

2.8 Cities, disasters risk reduction and climate change adaptation: critical observations and conclusions

The case study risk assessment process through the lens of the preceding disaster risk reduction and climate change adaptation strategy discussion produced several critical observations particularly important for understanding the challenges and opportunities for cities facing climate change. These include the following:

First, a multidimensional approach to risk assessment is a prerequisite to effective urban development programs that incorporate climate change responses. At present most climate risk assessment is dominated by an overemphasis on hazards. The application of the climate risk framework developed in this chapter provides more nuanced and more actionable insights into the differential risks. These insights reflect the exposure to hazards and the spectrum of vulnerabilities faced by urban households, neighborhoods, and firms. However, a critical issue

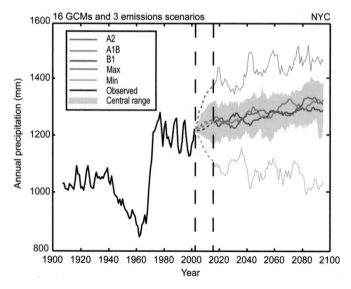

Figure 2.21: *Projected precipitation, New York City.*

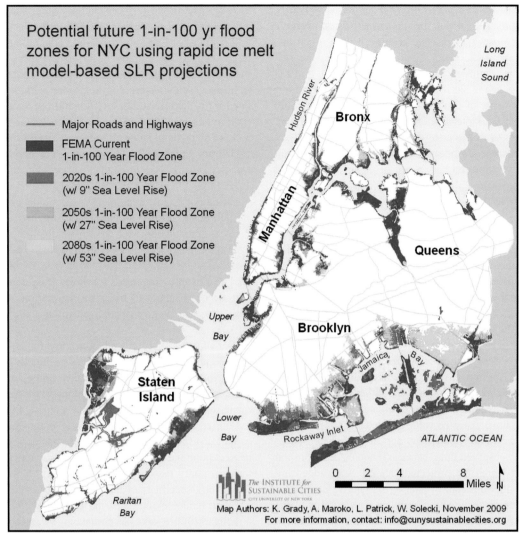

Potential future 1-in-100 yr flood zones for NYC using rapid ice melt model-based SLR projections

— Major Roads and Highways

FEMA Current
1-in-100 Year Flood Zone

2020s 1-in-100 Year Flood Zone
(w/ 9" Sea Level Rise)

2050s 1-in-100 Year Flood Zone
(w/ 27" Sea Level Rise)

2080s 1-in-100 Year Flood Zone
(w/ 53" Sea Level Rise)

Long Island Sound

Hudson River

Bronx

Manhattan

Queens

Upper Bay

Brooklyn

Jamaica Bay

Staten Island

Lower Bay

Rockaway Inlet

ATLANTIC OCEAN

Raritan Bay

The INSTITUTE for SUSTAINABLE CITIES
CITY UNIVERSITY OF NEW YORK

0   2   4   8
Miles   N

Map Authors: K. Grady, A. Maroko, L. Patrick, W. Solecki, November 2009
For more information, contact: info@cunysustainablecities.org

Note. This map is subject to limitations in accuracy as a result of the quantitative models, datasets, and methodology used in its development. The map and data should not be used to assess actual coastal hazards, insurance requirements or property values or be used in lieu of Flood Insurance Rate Maps issued by FEMA.

Interpretation. The floodplains delineated above in no way represent precise flood boundaries but rather illustrate three distinct areas of interest: 1) areas currently subject to the 1-in-100 year flood that will continue to be subject to flooding in the future, 2) areas that do not currently flood but are expected to potentially experience the 1-in-100 year flood in the future, and 3) areas that do not currently flood and are unlikely to do so in the timeline of the climate projection scenarios used in this research (end of the current century).

Figure 2.22: *Sea level rise projections.*

Source: The Institute of Sustainable Cities (2009).

that requires further research is identifying when strategic retreat may be more cost effective than adaptation and under what socio-economic conditions is it desirable and feasible.

Second, mismatches between needs and responses are occurring in regard to who should mitigate, how much to adapt, and why. Cities need climate change risk assessment, in addition to analysis of mitigation options, in order to decide for themselves what is the appropriate balance of mitigation and adaptation. Climate change risk frameworks, such as those described in this chapter, can help cities to address the issue of mismatches; that is, the difference between the city's response to climate change as opposed to the actual needs. For example, it appears that some developing countries may be over-focusing on mitigation when they could be addressing

adaptation more due to the presence of critical climate risks in the near-term as well as in future decades. The 17 largest economies account for most of the greenhouse gas emissions, the root cause of climate change (US Department of State, 2009). And while many cities within these major economies have a significant role in mitigation, it may be prudent for cities in low-income countries with large populations of poor households to incorporate climate risk into ongoing and planned investments as a first step (Mehrotra, 2009). However, since cities play an important role in greenhouse gas emissions in both developed and developing countries, there is also motivation for cities to lead on mitigation activities as well. Emissions from cities everywhere burden the environment, which is a global public good, and thus can be regulated through a combination of market and non-market incentives at the urban scale.

Third, the vertically and horizontally fragmented structure of urban governance is as much an opportunity as an obstacle to introducing responses to climate change. While much has been researched about the need for an integrated and coordinated approach, the fragmented governance structure of cities is unlikely to change in the short term and offers the opportunity to have multiple agents of change. Examples in the case study cities show that early adopters on climate change solutions play an important role. As in the case of New York City, the city's early action has become a model for other cities within the USA and internationally. The broad spectrum of governmental, civil society, and private sector actors in cities encourages a broader ownership of climate change adaptation programs.

Further, gaps and future research for scaled-down regional and local climate models were identified. In addition to the difficulties global climate models have with simulating the climate at regional scales, especially for locations with distinct elevational or land–sea contrasts, they also continue to have difficulty simulating monsoonal climates. Such is the case for climate projections for some of the case study cities in this chapter, especially related to projected changes in precipitation. This is because simulation of seasonal periods of precipitation is challenging in terms of both timing and amount; in some cases the baseline values used for the projection of future changes are extreme – either too high or too low. Therefore, the percentage changes calculated vary greatly and can, on occasions, have distorted values. Especially for precipitation projections, the future trends may appear to be inconsistent when compared to observed data, because the averages from the baseline period to which the projected changes were added are inaccurate, either due to a lack of data or extreme values within the time period that are skewing the averages. The inability of the global models to simulate the climate of individual cities raises the need for further research on regional climate modeling.

However, what is important to focus on in these future climate projections is the general trends of the projected changes and their ranges of uncertainty. These refer to attributes such as increasing, decreasing, or stable trends, and information about the uncertainty of projections in particular due to climate sensitivity or greenhouse gas emission pathways through time. Information on climate model projections regarding the extreme values and the central ranges both provide useful information to city decision-makers.

Annex 2.1 Global institutional structure for risk assessment and adaptation planning[2])

Moving forward, there is a need for a programmatic science-based approach to addressing climate risk in cities of developing countries that are home to a billion slum dwellers, and quickly growing. These cities are most unprepared to tackle climate change-induced stresses that are likely to exacerbate the existing lack of basic urban services such as water supply, energy, security, health, and education, as well as of disaster preparedness and response. Most climate change adaptation efforts until now have focused on lengthy descriptive papers or small experimental projects, each valuable in their own right, but insufficient to inform policymakers. Further, there is little recognition of the need for a flexible strategy to adaptation; instead there are sporadic medium- and long-term project-oriented responses, often lacking analysis of basic climate parameters such as temperature variability, precipitation shifts, and sea level rise data. Three central elements of a comprehensive approach could include:

1. Establishment of a scientific body, which can verify as well as advise on technical matters of climate change science as it pertains to mitigation and impacts on cities;
2. A systematic approach to climate change adaptation;
3. Ongoing assessment of climate change knowledge for cities.

The most pressing needs for developing-country cities in low-income countries is to focus on adaptation, outlined here, but similar measures are essential for mitigation efforts as well.

For adaptation, there is a requirement for assessing risk, evaluating response options, making some politically complex decisions on implementation choices and implementing projects, monitoring process and outcomes, and continuously reassessing for improvements as the science and practice evolve. Further, in order to maximize impact, there is a need to leverage ongoing and planned capital investments to reduce climate-risk exposure, rather than to neglect potential climate impacts. Practitioners and scholars agree that a primary reason why cities neglect climate change risks is lack of city-specific relevant and accessible scientific assessments. Thus, to reduce climate risk in developing-country cities there is a demand for city-specific climate risk assessment as well as the crafting of flexible adaptation and mitigation strategies to leverage existing and planned public investments.

To inform action, the experience of the Climate Impacts Group at NASA's Goddard Institute for Space Studies and scholars at Columbia University and the City University of New York points to a need for at least a four-track approach.

Track 1: Addressing the need for a global progress assessment. This is an across-city global assessment that captures the state-of-knowledge on climate and cities. The First UCCRN Assessment Report on Climate Change and Cities (ARC3) is one of the only ongoing efforts that addresses risk, adaptation, and mitigation, and derives policy implications for the key city sectors – urban climate risks, health, water and sanitation, energy, transportation, land use, and governance. This assessment is an effort by more than 50 scholars located all over the world and offers sector-specific recommendations for cities to inform action. The aim is to continue assessments (on the order of every four years) and offer Technical Support after the UN Framework Convention on Climate Change COP16 has been held in December, 2010 in Mexico City.

---

2 Annex 2.1 is an excerpt from a White Paper authored by Shagun Mehrotra (April, 2009).

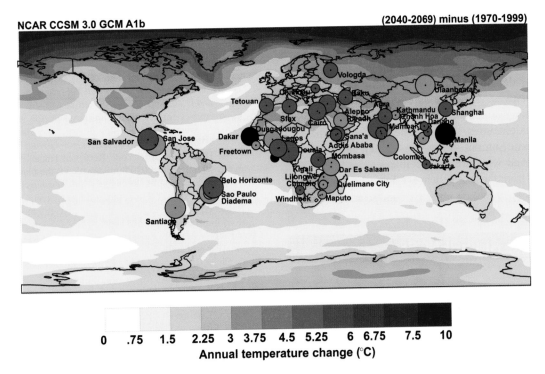

Figure 2.23: *Cities involved in Cities Alliance activities, 2008, and temperature projections for the 2050s. Green circles represent national slum upgrading plans: blue, city development strategies; and black, both.*

Source: This volume, chapter 1; data from Cities Alliance Annual Report 2008.

Track 2: Across-city rapid risk assessment. Mainstreaming climate risk assessment into City Development Strategies as well as pro-poor programs like Cities Alliance's citywide and nationwide slum upgrading. This effort is critical to inform ongoing large-scale capital investments in cities that often lack basic climate risk considerations (see Figure 2.23). For most cities illustrated in Figure 2.23, between 50 and 100 years of observed climate data are available but remain to be analyzed.

Track 3: City-specific in-depth sectoral assessment. General assessments are insufficient as many cities lack in-house expertise for technical analysis of city-specific climate impacts. To fill this gap, there is a requirement to craft city-specific risk and adaptation assessments for city departments (sector by sector) to redirect existing and planned investments. Cities like New York, London, and Mexico City have initiated this demanding, yet essential task. Such risk analysis needs to disaggregate risk into hazards (external climate-induced forcing), vulnerability (city-specific characteristics, such as location, and percentage of slum population), and agency (ability and willingness of the city to respond). The process will engage in-city experts and stakeholders in each city in the assessment process in order to develop local adaptive capacity.

Track 4: Learning from experience. This task involves deriving adaptation lessons from the early climate change

Table 2.6: *Four tracks with objectives, and related outputs.*

Track	Objective	Output
1. Global progress assessment	To provide state-of-knowledge	Providing a global First UCCRN Assessment Report on Climate Change and Cities (ARC3), a comprehensive assessment of risks, adaptation, mitigation options, and policy implications
2. Across-city rapid risk assessment	To inform ongoing urban investments that lack climate risk considerations	Integrate climate risk assessment into City Development Strategies and pro-poor programs
3. City-specific in-depth sectoral assessment	To redirect existing and planned investments	Crafting city-specific risk and adaptation assessments for each city department (sector by sector)
4. Learning from experience	To derive adaptation lessons from the early adopters	Detailed case-studies of implementation mechanisms from London, Mexico City, New York City, etc.

adopters like London, Mexico City, and New York City. It focuses on answering such questions as: How are London, Mexico City, and New York crafting a response to climate change? What are the institutional arrangements? What are the roles of mayoral leadership, public demand, offices of long-term planning, and civil society initiatives? How are assessments financed – for example foundations, scientist-volunteered, tax dollars? What are the positive externalities – such as establishment of the C40 Large Cities Climate Summit – of scaling-up both nationally and internationally? And what are transferable lessons? What can other cities learn from these experiences? Table 2.6 summarizes the four tracks along with their objectives and related outputs.

REFERENCES

Abosede, F. (2006). Housing in Lagos mega city: improving livability, inclusion and governance. Paper presented at the Social and Economic Rights Action Center's (SERAC) international conference: *Building Nigeria's Capacity to Implement Economic, Social and Cultural Rights: Lessons Learned, Challenges and the Way Forward*, Abuja, Nigeria, September 27–28, 2006.

Abosede, F. A. (2008). The challenges of slum upgrading: the Lagos experience. Paper presented at the International Policy Dialogue: *The Challenges of Slum Upgrade: Sharing Sao Paulo's Experience*, Sao Paulo, Brazil, March 10–14, 2008.

Aggarwal, D. and M. Lal (2001). *Vulnerability of Indian Coastline to Sea Level Rise*, New Delhi, India: Centre for Atmospheric Sciences, Indian Institute of Technology.

Assessments of Impacts and Adaptations to Climate Change (2005). *Informe final del proyecto LA 26* [LA 26 project, final report assessment]. Retrieved January 25, 2009, from www.atmo.at.fcen.uba.ar/~lcr/libros/Cambio_Climatico-Texto.pdf.

Awosika, L. F., G. T. French, R. J. Nicholls, and C. E. Ibe. (1993a). Impacts of sea level rise on Nigeria. In *Proceedings of IPCC Symposium: The Rising Challenge of the Sea*, Margarita, Venezuela, March 14–19, 1992.

Awosika, L. F., O. Ojo, T. A. Ajayi, *et al.* (1993b). *Implications of Climate Changes and Sea Level Rise on the Niger Delta, Nigeria, Phase 1*. Report, Nairobi, Kenya: UNEP.

Barros, V., A. Menéndez, C. Natenzon, *et al.* (2005). Climate change vulnerability to floods in the metropolitan region of Buenos Aires City. Paper presented at the *Assessments of Impacts and Adaptations to Climate Change (AIACC) Workshop: Climate Change*, Bellagio, Italy, May 7–12. Retrieved January 25, 2009, from www.aiaccproject.org/working_papers/working_papers.html.

Barros, V., A. Menéndez, C. Natenzon, *et al.* (2008). Storm surges, rising seas and flood risks in metropolitan Buenos Aires. In N. Leary *et al.* (Eds.), *Climate Change and Vulnerability*, London, UK: Earthscan, pp. 117–133.

Basu, R. and J. M. Samet (2002). Relation between elevated ambient temperature and mortality: a review of the epidemiologic evidence. *Epidemiologic Reviews*, **24**, 190–202.

Baumert, K., T. Herzog, and J. Pershing (2005). *Navigating the Numbers: Greenhouse Gas Data and International Climate Policy*, Washington, DC, USA: World Resource Institute.

Blaikie, P., T. Cannon, I. Davies, and B. Wisner (1994). *At Risk: Natural Hazards, People's Vulnerability and Disaster*. London, UK: Routledge.

Bradley, R., B. Staley, T. Herzog, J. Pershing, and K. Baumert (2007). *Slicing the Pie: Sector-based Approaches to International Climate Agreements*, Washington, DC, USA: World Resource Institute.

Bureau of Economic Analysis (2008). *News Release: GDP by Metropolitan for 2006 and Revised 2004–2005*. Retrieved January 25, 2009, from www.bea.gov/newsreleases/regional/gdp_metro/gdp_metro_newsrelease.htm.

Camillioni, I. and V. Barros (2008). Climate. In *Environmental Atlas of Buenos Aires*. Retrieved January 24, 2009, from www.atlasdebuenosaires.gov.ar/aaba/index.php?option=com_content&task=view&id=339&Itemid=188&lang=en.

Campbell, S. and S. S. Fainstein (Eds.) (1996). *Readings in Planning Theory*. Cambridge, UK: Blackwell Publishers.

Carmin, J. and D. Roberts (2009). Government institutions and innovations in governance for achieving climate adaptation in cities. *Fifth Urban Research Symposium*, Marselles, France. June, 2009.

City of New York (2007). *plaNYC: A Greener Greater New York*. New York, USA: The City of New York. Retrieved January 5, 2009, from www.nyc.gov/html/planyc2030/html/downloads/the-plan.shtml.

City of New York (2009). *Financial Plan, Summary: Fiscal Years 2009–2013*. New York, USA: Office of Management and Budget, May 1. Retrieved June 9, 2009, from www.nyc.gov/html/omb/html/publications/finplan05_09.shtml.

Clichevsky, N. (2002). *Pobreza y políticas urbano-ambientales en Argentina* [Poverty and urban-environmental policies in Argentina], Santiago de Chile: CEPAL – ECLAC, División de Medio Ambiente y Asentamientos Humanos.

Defensoría del Pueblo de la Ciudad de Buenos Aires (2006). *Informe de situación. Desalojos de "Nuevos Asentamientos Urbanos"* [Situation report. Eviction of "New urban settlements"]. Retrieved January 25, 2009, from www.defensoria.org.ar/institucional/doc/asentamientos.doc.

Delhi (2009). In *Encyclopædia Britannica*. Retrieved June 2, 2009, from Encyclopædia Britannica Online: www.britannica.com/EBchecked/topic/156501/Delhi.

Department of City Planning (2009). New York City: Department of City Planning. Retrieved January 15, 2009, from www.nyc.gov/html/dcp/.

Department of Environment (2008). *Delhi, Amongst the Greenest Capitals of the World*, New Delhi, India: Department of Environment, Government of National Capital Territory of Delhi.

Department of Health and Mental Hygiene (2006). *DOHMH Vital Signs: 2006 Heat Wave Investigation Report*. Retrieved from www.nyc.gov/html/doh/downloads/pdf/survey/survey-2006heatdeaths.pdf – 2006–11–15.

Department of Planning (2008). *Economic Survey of Delhi 2007–08*, New Delhi: The Government of the National Capital Territory of Delhi, Government of India. Retrieved January 05, 2009, from www.delhi.gov.in/wps/wcm/connect/doit_planning/Planning/Economic±Survey±of±Delhi/.

Delhi Development Authority (2005). *Master Plan for Delhi, 2021*, New Delhi, India: Development Authority, Government of National Capital Territory of Delhi, Government of India. Retrieved January 5, 2009, from www.dda.org.in/planning/draft_master_plans.htm.

Directorate for Statistics and Census (2007). *Anuario Estadístico 2006* [Statistical Yearbook 2006]. Gobierno de la Ciudad de Buenos Aires, Ministerio de Hacienda. Retrieved January 25, 2009, from http://buenosaires.gov.ar/areas/hacienda/sis_estadistico/anuario_2006/ cap_11.htm.

Douglas, I. and K. Alam (2006). *Climate Change, Urban Flooding and the Rights of the Urban Poor in Africa: Key Findings from Six African Cities*, London, UK: ActionAid International. Retrieved from www.actionaid.org.uk/doc_lib/urban_flooding_africa_report.pdf.

Ekanade O., A. Ayanlade, and I. O. O. Orimogunje (2008). Geospatial analysis of potential impacts of climate change on coastal urban settlements in Nigeria for the 21st century. *Journal of Geography and Regional Planning*, **1**(3), 49–57. Retrieved from www.academicjournals.org/JGRP.

Falczuk, B. (2008). Surface water. In *Environmental Atlas of Buenos Aires*. Retrieved January 24, 2009, from /www.atlasdebuenosaires.gov.ar/aaba/index.php?option=com_content&task=view&id=339&Itemid=188&lang=en.

Federal Republic of Nigeria (2006). *Report of the Presidential Committee on Redevelopment of Lagos Mega-city Region*, Lagos: Lagos Megacity Project.

Folorunsho, R. and L. F. Awosika (1995). Nigerian meteorological induced changes along the Nigerian coastal zone and implications for integrated coastal zone management plan. In *Proceedings of BORDOMER '95, International Convention: Rational Use of Coastal Zone*, Bordeaux, France, February 6–10, 1995.

French, G. T., L. F. Awosika, and C. E. Ibe (1995). Sea level rise and Nigeria: Potential impacts and consequences. *Journal of Coastal Research*, **14**, 224–242.

González, S. (2005). Ciudad visible vs. ciudad invisible: la gestión del riesgo por inundaciones en la Ciudad de Buenos Aires. *Territorios, 13*, 53–67.

Government of India (2002). *Climate Change and India*, New Delhi, India: Government of India. Retrieved January 12, 2009 from http://envfor. nic.in/cc/index.htm.

Government of India (2008). *National Action Plan for Climate Change*, New Delhi, India: Government of India.

Hallegatte, S., F. Henriet, and J. Corfee-Morlot (2008). *The Economics of Climate Change Impacts and Policy Benefits at City Scale: A Conceptual Framework* (Environment Working Papers No. 4), Paris, France: Environment Directorate, Organisation for Economic Co-operation and Development.

Helmer, M., and D. Hilhorst (2006). Natural disasters and climate change. *Disasters* **30**(1) 1–4.

Hunt, A. and P. Watkiss (2007). *Literature Review on Climate Change Impacts on Urban City Centres: Initial Findings* (ENV/EPOC/GSP(2007)10), Paris: Environment Directorate, Organisation for Economic Co-operation and Development.

Huq, S., H. Reid, M. Konate, *et al.* (2004). Mainstreaming adaptation to climate change in Least Developed Countries (LDCs). *Climate Policy*, 4(1) 25–43.

ICS (International Council for Science), Planning Group on Natural and Human-induced Environmental Hazards and Disasters (2008). *A Science Plan for Integrated Research on Disaster: Addressing the Challenge of Natural and Human-Induced Environmental Hazards*. ICSU.

Instituto Nacional de Estadística y Censos (2003). *¿Qué es el Gran Buenos Aires?* [What is Greater Buenos Aires?]. Retrieved January 25, 2009, from www.indec.gov.ar/glosario/folletoGBA.pdf.

International Strategy for Disaster Reduction (2010). Hyogo Framework for Action 2005–2015: Building the Resilience of Nations and Communities to Disasters. **http://www.unisdr.org/eng/hfa/hfa.htm**.

IPCC (2001a). *Special Report on Emissions Scenarios: A Special Report of Working Group III of the Intergovernmental Panel on Climate Change*. N. Nakicenovic, J. Alcamo, G. Davis, *et al.* (Eds.). Cambridge, UK: Cambridge University Press.

IPCC (2001b). *Climate Change 2001: Working Group II: Impacts, Adaptation and Vulnerability*. Assessment Report, Intergovernmental Panel on Climate Change.

IPCC (2007a). *Climate Change 2007: The Physical Science Basis. Contribution of Working Group I to the Fourth Assessment Report of the Intergovernmental Panel on Climate Change*. S. Solomon, D. Qin, M. Manning, *et al.* (Eds.). Cambridge, UK and New York, USA: Cambridge University Press.

IPCC (2007b). *Climate Change 2007: Impacts, Adaptation And Vulnerability. Contribution of Working Group II to the Fourth Assessment Report of the Intergovernmental Panel on Climate Change*. M. L. Parry, O. F. Canziani, J. P. Palutikof, P. J. van der Linden, and C. E. Hanson (Eds.). Cambridge, UK: Cambridge University Press.

IPCC (2007c). *Climate Change 2007: Mitigation. Contribution of Working Group III to the Fourth Assessment Report of the Intergovernmental Panel on Climate Change*. B. Metz, O. R. Davidson, P. R. Bosch, R. Dave, and L. A. Meyer (Eds.). Cambridge, UK and New York, USA: Cambridge University Press.

IPCC (2007d). *Climate Change 2007: Synthesis Report*. Contribution of Working Groups I, II and III to the Fourth Assessment Report of the Intergovernmental Panel on Climate Change [Core Writing Team, R. K. Pachauri A. and Reisinger (Eds.)]. Geneva, Switzerland: IPCC.

Iwugo, O. K., B. D'Arcy, and R. Anoch (2003). Aspects of land based pollution of an African coastal megacity of Lagos. Paper presented at the *International Water Association International Conference*: *Diffuse Pollution and Basin Management*, Dublin, Ireland. August 17–22, 2003.

Kasperson, J. X. and R. E. Kasperson (Eds.) (2001). *Global Environmental Risk*, Tokyo, Japan: United Nations University Press.

Kinney, P. L., M. S. O'Neill, M. L. Bell, and J. Schwartz (2008). Approaches for estimating effects of climate change on heat-related deaths: challenges and opportunities. *Environmental Science and Policy*, **11**, 87–96.

Knowlton, K., L. Barry, A. G. Richard, *et al.* (2007). Projecting heat-related mortality impacts under a changing climate in the New York City region. *American Journal of Public Health*, **97**(11), 2028–2034. Retrieved January 31, 2009, from ABI/INFORM Global database.

Mehrotra, S. (2003). *Unfinished Promise: Institutionalising Local Knowledge for Risk Reduction in Earthquake-Prone Settlements of the Poor: Porbandar District, Gujarat, India*, ProVention Consortium, Applied Research Grant proposal, Washington, DC, USA: World Bank.

Mehrotra, S. (2009). *Climate and Cities: Crafting a Global Institutional Structure for Risk Assessment and Adaptation Planning*, White paper, New York, USA: Columbia University, and NASA Goddard Institute of Space Studies.

Mehrotra, S., C. E. Natenzon, A. Omojola, *et al.* (2009). *Framework for City Climate Risk Assessment, Buenos Aires, Delhi, Lagos, and New York*. Washington, DC, USA: World Bank.

Menéndez, A. (Ed.) (2005). Componente B1, Vulnerabilidad de la zona costera [B1 Component, Vulnerability of the coastal zone]. In *Cambio Climático: Segunda Comunicación Nacional Gobierno de la República Argentina Informe final*, Fundación Di Tella, Fundación Bariloche, Global Environment Facility, World Bank.

Mercer, J.A. (2010). Disaster risk reduction or climate change adaptation: Are we reinventing the wheel? *Journal of International Development* **22** 247–264.

Murgida, A. (2007). *Informe Final: Amenazas-Vulnerabilidad-Adaptación y Mitigación, Ciudad Autónoma de Buenos Aires* [Final Report Assessment: Threats-Vulnerability-Adaptation and Mitigation, Autonomous City of Buenos Aires], Buenos Aires, Argentina: Oficina de Protección Climática y Eficiencia Energética, GCABA, mimeo.

Murgida, A. M. and S. G. González (2005). Social risk, climate change and human security: an introductory case study in Metropolitan Area of Buenos Aires (Argentina). In *Proceedings of International Workshop: Human Security and Climate Change*, Oslo, Norway, June 21–23, 2005.

Murgida, A. M. and C. E. Natenzon (2007). Social 'downscaling': a few reflections on adaptation in urban environments. In *Proceedings of III Regional Conference on Global Changes, Round table 4: Urban aspects*. São Paulo, Brazil. November 4–8, 2007.

Natenzon, C. and J. M. Viand (2008). Gestión de los desastres en Argentina: instituciones nacionales involucradas en la problemática de las inundaciones [Disaster management in Argentina: national institutions involved in the flooding problem]. In A. I. Geraiges de Lemos, J. L. Sanches Ross, and A. Luchiari (Org.), *América Latina: sociedade e meio ambiente* (pp. 171–185), São Paulo, Brazil: Expressão Popular.

New York City Department of Environmental Protection (2008). *The Assessment and Action Plan: A Report Based on the Ongoing Work of the DEP Climate Change Task Force*, New York, USA: The New York City Department of Environmental Protection.

New York City Panel on Climate Change (2009). *Adaptation Assessment Guidebook*. Prepared for use by the New York City Climate Change Adaptation Task Force. Retrieved from www.nyc.gov/html/om/pdf/2009/NPCC_CRI.pdf.

NYSE Euronext (2009). *NYSE Euronext*. Retrieved June 9, 2009, from www.nyse.com/about/listed/1170350259411.html.

O'Brien, G., P. O'Keefe, J. Rose, and B. Wisner (2006). Climate change and disaster management. *Disasters, 30*(1), 64–80.

Ogun State Government (2005). *Ogun State Regional Plan 2005–2025: Development Pressure Area, DPA report*.

Parsons, E. A., V. R. Burkett, K. Fisher-Vanden, *et al.* (2007). *Global-change Scenarios: Their Development and Use, Sub-report 2.1B of synthesis and assessment product 2.1 by the US climate change science program and the subcommittee on global change research*, Washington, DC, USA: Office of Biological & Environmental Research, Department of Energy.

Pelling, M. (2003). *The Vulnerability of Cities*. London, UK: Earthscan.

Pelling, M., A. Maskrey, P. Ruiz, L. Hall (Eds.) (2004). *A Global Report: Reducing Disaster Risk a Challenge for Development*. New York: United Nations Development Programme, Bureau for Crisis Prevention and Recovery.

Pírez, P. (2002). Buenos Aires: fragmentation and privatization of the metropolitan city. *Environment and Urbanization*, **14**(1), 58–76.

Pírez, P. (2008). Planning. In *Environmental Atlas of Buenos Aires*. Retrieved January 28, 2009, from www.atlasdebuenosaires.gov.ar/aaba/index.php?option=com_content&task=view&id=34&Itemid=30&lang=en.

Pochat, V., C. E. Natenzon, and A. Murgida (2006). Domestic policy frameworks on adaptation to climate change in water resources. Argentina country case study. In UNFCCC/OECD Global Forum on Sustainable Development: *Working Together to Respond to Climate Change: Annex I Expert Group Seminar*. Paris, France, March 27–28. Retrieved from www.oecd.org/dataoecd/29/2/36448827.pdf.

Prabhakar, S. V. R. K, A. Srinivasan, and R. Shaw (2009). Climate change and local level disaster risk reduction planning: need, opportunities and challenges. *Mitigation and Adaptation Strategies to Global Change*, **14**, 7–13.

Rawls, J. (1971). *A Theory of Justice*. London, UK: Oxford University Press.

Revi, A. (2007). *Climate Change Risk: An Adaptation and Mitigation Agenda for Indian Cities*, India Background Paper for Global Urban Summit, Bellagio, Italy, July 2007.

Ríos, D. and González, S. (2005). La aglomeración gran Buenos Aires [The greater Buenos Aires agglomeration]. In *Cambio Climático, Segunda Comunicación Nacional. Gobierno de la República Argentina. Informe final*, Fundación Di Tella, Fundación Bariloche, Global Environment Facility, World Bank.

Rosenzweig, C. and Hillel, D. (2008). *Climate Variability and the Global Harvest: Impact of El Niño and Other Oscillations on Agro-Ecosystems*. New York, USA: Oxford University Press.

Rosenzweig, C. and Solecki W. D. (Eds.) (2001). *Climate Change and a Global City: The Potential Consequences of Climate Variability and Change – Metro East Coast*. Report for the US Global Change Research Program, *National Assessment of the Potential Consequences of Climate Variability and Change for the United States*. New York, USA: Earth Institute at Columbia University.

Satterthwaite, D., Huq, S., Pelling, M., Reid H., and Lankao-Romero, P. (2007). *Adapting to Climate Change in Urban Areas: The Possibilities and Constraints in Low- and Middle-income Nations* (Climate Change and Cities Series, Discussion Paper, No. 1), London, UK: International Institute for Environment and Development (IIED).

Sen, A. (1999). *Development as Freedom*, New York, USA: Anchor Books.

Servicio Meteorológico Nacional (2008). *Informes estacionales* [Climate Data]. Retrieved January 24, 2009, from www.smn.gov.ar/?mod=clima&id=5

Shook, G. (1997). An assessment of disaster risk and its management in Thailand. *Disasters*, **21**(1), 77–88.

Thomalla, F., T. Downing, E. Spanger-Siegfried, G. Han, and J. Rockstrom (2006). Reducing hazard vulnerability: towards a common approach between disaster risk reduction and climate adaptation. *Disasters*, **30**(1), 39–48.

UK Climate Impacts Programme (2009). Retrieved from www.ukcip.org.uk/index.php.

United Nations Development Programme (2004). *Human Development Report*. Retrieved from www.hdr.undp.org.

UN-HABITAT (2003). *Slums of the World: The Face of Urban Poverty in the New Millennium?* Nairobi, Kenya: Global Urban Observatory, UN-HABITAT.

UN-HABITAT (2004). *State of the Lagos Mega City and other Nigerian Cities Report*. Lagos, Nigeria: UN-HABITAT and Lagos State Government.

UN-HABITAT (2006). A tale of two cities. In *World Urban Forum III, An International UN-Habitat Event: Urban Sustainability*, Vancouver, Canada, June 19–23, 2006.

UN-HABITAT (2008a). *The State of World's Cities 2008/2009: Harmonious Cities*, London, UK: Earthscan.

UN-HABITAT (2008b). *The State of African Cities 2008: Framework for Addressing Urban Challenges in Africa*, Nairobi, Kenya: UN-HABITAT.

US Census Bureau (2008). *State and County QuickFacts*. Retrieved January 25, 2009, from http://quickfacts.census.gov/qfd/states/36/3651000.html.

US Department of State (2009). *Major Economies Forum on Energy Security and Climate Change*. Retrieved from www.state.gov/g/oes/climate/mem/#.

Van Aalst, M.K. (2006). The impacts of climate change on risk of natural disasters. *Disasters*, **30**(1), 5–18.

Van Aalst, M. K., T. Cannon, and I. Burton (2007). Community level adaptation to climate change: The potential role of participatory community risk assessment. *Global Environmental Change*, **18**, 165–179.

World Bank, Global Facility for Disaster Reduction and Recovery (2008). *Climate Resilient Cities, 2008 Primer: Reducing Vulnerabilities to Climate Change Impacts and Strengthening Disaster Risk Management in East Asian Cities*. Washington, DC, USA: World Bank/International Bank for Reconstruction and Development.

World Bank (2009). *Glossary*. Retrieved June 9, 2009, from http://youthink.worldbank.org/glossary.php#ppp.

3

Urban climate

Processes, trends, and projections

Coordinating Lead Authors:

Reginald Blake (New York City, Kingston), Alice Grimm (Curitiba), Toshiaki Ichinose (Tokyo), Radley Horton (New York City)

Lead Authors:

Stuart Gaffin (New York City), Shu Jiong (Shanghai), Daniel Bader (New York City), L. DeWayne Cecil (Idaho Falls)

This chapter should be cited as:

Blake, R., A. Grimm, T. Ichinose, R. Horton, S. Gaffin, S. Jiong, D. Bader, L. D. Cecil, 2011: Urban climate: Processes, trends, and projections. *Climate Change and Cities: First Assessment Report of the Urban Climate Change Research Network*, C. Rosenzweig, W. D. Solecki, S. A. Hammer, S. Mehrotra, Eds., Cambridge University Press, Cambridge, UK, 43–81.

3.1 Introduction

Cities play a multidimensional role in the climate change story. Urban climate effects, in particular the urban heat island effect, comprise some of the oldest observations in climatology, dating from the early nineteenth century work of meteorologist Luke Howard (Howard, 1820). This substantially predates the earliest scientific thought about human fossil fuel combustion and global warming by chemist Svante Arrhenius (Arrhenius, 1896). As areas of high population density and economic activity, cities may be responsible for upwards of 40 percent of total worldwide greenhouse gas emissions (Satterthwaite, 2008), although various sources have claimed percentages as high as 80 percent (reviewed in Satterthwaite, 2008). Figure 3.1 shows a remotely sensed map of nocturnal lighting from urban areas that is visible from space and vividly illustrates one prodigious source of energy use in cities. Megacities, often located on the coasts and often containing vulnerable populations, are also highly susceptible to climate change impacts, in particular sea level rise. At the same time, as centers of economic growth, information, and technological innovation, cities will play a positive role in both climate change adaptation and mitigation strategies.

Urban population recently surpassed non-urban population worldwide and is projected to grow from 50 percent currently to 70 percent by 2050 (UNFPA, 2007). The urban population growth rate will be even more rapid in developing countries. In terms of absolute numbers, urban population will grow from ~3.33 billion today to ~6.4 billion in 2050, about a 90 percent increase. These numbers underscore the fact that urban climate is becoming the dominant environment for most of humanity.

This chapter presents information on four interrelated components of urban climate: (i) the urban heat island effect and air pollution, (ii) the current climate and historical climate trends, (iii) the role of natural climate variability, and (iv) climate change projections due to worldwide greenhouse gas increases. Figure 3.2 provides a schematic of key interactions within the urban climate system.

Teasing out the relative influences of these components on urban climate is challenging. Natural variability can occur at multidecadal timescales, comparable to the timescales used for historical analysis and to long-term greenhouse gas forcing. Another challenge is that different climate factors may not be independent. For example, climate change may influence the amplitude and periodicity of natural variability, such as the intensity and frequency of the El Niño–Southern Oscillation (ENSO).

To survey these issues, twelve cities are selected as examples. The twelve focus cities in this chapter – Athens (Greece), Dakar (Senegal), Delhi (India), Harare (Zimbabwe), Kingston (Jamaica), London (UK), Melbourne (Australia), New York City (USA), São Paulo (Brazil), Shanghai (China), Tokyo (Japan), and Toronto (Canada) (Figure 3.3) – share a range of characteristics but also differ in key respects. They are all large, dynamic, and vibrant

Figure 3.1: *View of Earth at night: Areas of light show densely populated, urban areas.*

Source: NASA

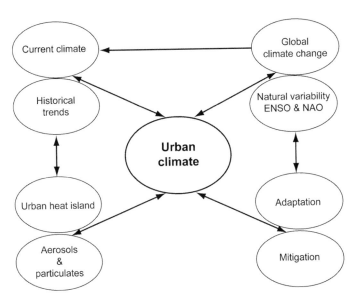

Figure 3.2: *Conceptual framework of this chapter showing major components impacting urban climate.*

urban areas, and act as hubs of social and economic activity. They also feature long-term twentieth-century climate records that allow trend detection, projections, and impact analysis. All the cities selected are likely to experience significant climate change this century, and several illustrate the influence of climate variability systems such as ENSO. Some of the city examples demonstrate unique vulnerabilities to extreme climate events.

With regard to differences, the selection includes a range for geography and economic development levels. The various geographic locations allow examination of climate change impacts in multiple climate zones. Covering multiple climate zones also allows for the examination of the influence of the major natural climate variability systems such as ENSO.

A range of economic development examples is important for vulnerability studies. Adaptation planning for climate change

tends to be more difficult in cities with limited resources; typically but not always the case in developing countries. Our focus cities include examples from both the developed and developing world, with some having already taken steps to prepare for climate change and others that have yet to start. Table 3.1 gives some summary statistics on socio-economic and geographic data, and mean climate for the focus cities.

3.1.1 Effects of cities on temperature: urban heat islands

Urban areas are among the most profoundly altered landscapes away from natural ecosystems and processes. Figure 3.4 illustrates a recent reconstruction of the verdant "Mannahatta" island circa 1609 as compared to the current landscape 400 years later (Sanderson and Boyer, 2009). Pondering this visually arresting reconstruction, it is not surprising that cities have altered microclimates with, among other effects, significantly elevated surface and air temperatures.

The elevation in temperatures is most generally explained in terms of the basic surface energy balance processes of shortwave and longwave radiation exchange, latent, sensible, and conductive heat flows (Oke, 1987). With respect to shortwave, or solar, radiation, surface albedo refers to the reflectivity of a surface to visible light and is measured from 0 to 100 percent reflectivity. The regional albedo of cities is significantly lower than natural surfaces due to the preponderance of dark asphalt roadways, rooftops, and urban canyon light trapping. These urban features have typical albedo values below 15 percent (Table 2-4 in Rosenzweig *et al.*, 2006). This leads to efficient shortwave radiation absorption. The urban skyline, with deep urban canyons, results in a greatly reduced skyview at street level and this impedes longwave radiative cooling processes. This urban vertical geometry further impacts winds, generally reducing ventilation and sensible heat cooling. The replacement of natural soil and vegetation with impervious surfaces leads to greatly reduced evapotranspiration and latent heat cooling. The dense

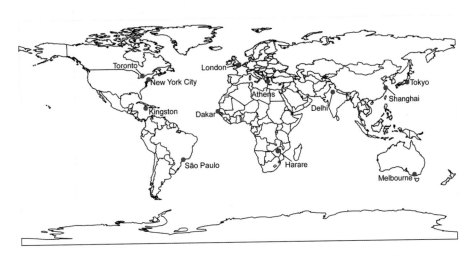

Figure 3.3: *Map of cities highlighted in this chapter. The cities were selected based on socio-economic factors and the availability of long-term climate data.*

Table 3.1: *City statistics and mean temperature and precipitation for 1971–2000.**

City	Latitude	Longitude	Population	Mean annual temperature	Annual precipitation
Athens	37.9 N	23.7 E	789,166 (2001)	17.8 °C	381 mm
Dakar	14.7 N	17.5 W	1,075,582 (2007)	24.0 °C	357 mm
Delhi	28.6 N	77.2 E	9,879,172 (2001)	25.1 °C	781 mm
Harare	17.8 S	31.0 E	1,435,784 (2002)	18.1 °C	830 mm
Kingston	17.9 N	76.8 W	579,137 (2001)	27.5 °C	691 mm
London	51.3 N	0.4 W	7,556,900 (2007)	9.7 °C	643 mm
Melbourne	37.8 S	145.0 E	3,806,092 (2007)	15.7 °C	652 mm
New York City	40.8 N	74.0 W	8,274,527 (2007)	12.8 °C	1181 mm
São Paulo	23.5 S	46.4 W	11,016,703 (2005)	19.5 °C	1566 mm
Shanghai	31.5 N	121.4 E	14,348,535 (2000)	16.4 °C	1155 mm
Tokyo	35.7 N	139.8 E	8,489,653 (2005)	16.2 °C	1464 mm
Toronto	43.7 N	79.0 W	2,503,281 (2006)	7.5 °C	793 mm

*Population data for all cities are from the United Nations Statistics Division Demographic Yearbook, United Kingdom National Statistics Office, and Census of Canada. Annual temperature and precipitation statistics are computed for all cities using data from the National Climatic Data Center Global Historical Climatology Network (NCDC GHCNv2), UK Met Office and Hadley Centre, Australian Bureau of Meteorology, and Environment Canada.

Figure 3.4: *Manhattan-Mannahatta: on right is a reconstruction of Manhattan Island circa 1609 (called "Mannahatta" by the Lenape native Americans), as compared to today, based on historical landscape ecology and map data.*

Sources: Markley Boyer / The Mannahatta Project / Wildlife Conservation Society and the aerial view of modern Manhattan, Amiaga Photographers

impervious surfaces with high heat capacity create significant changes in heat storage and release times as compared to natural soil and vegetated surfaces.

There are additional atmospheric and heat source processes in cities that interact with these energy balances. Aerosols tend to reduce the amount of incoming solar radiation reaching the surface (a net cooling effect), while elevated ambient urban carbon dioxide levels may further reduce net radiative cooling.

The high density of population and economic activity in urban areas leads to intense anthropogenic heat releases within small spatial scales. These include building heating and cooling systems, mass transportation systems and vehicular traffic, and commercial and residential energy use. Anthropogenic heat emission has been well documented and researched in developed countries as a major factor causing the heat island phenomenon (Ohashi *et al.*, 2007). As economic development, urbanization, and population growth continue in the developing countries, anthropogenic heat has increased there as well (Ichinose and Bai, 2000). Growth in urbanization increases energy demand in general and electricity demand in particular.

In analogy with the well-established urban heat island, it is tempting to define additional atmospheric urban "islands" such as rainfall islands, and relative humidity islands, which refer to potential urban alterations of precipitation and reductions in urban soil moisture availability due to impervious surfaces. In some cities, there may be good evidence for their existence and effects, such as in Shanghai (Section 3.1.3.1). In general though, the case for additional urban atmospheric islands, such as rainfall islands, is not as straightforward as the heat island and needs further research and characterization.

There can be little question, however, about the broad array of quality-of-life issues that are generally negatively impacted by excess urban heat. These include extreme peak energy demands, heat wave stress and mortality risk, air quality deterioration, seasonal ecological impacts including thermal shocks to waterways following rain events and impacts on urban precipitation.

3.1.2 Effects of cities on local precipitation

There is a longstanding interest in the question of urban impacts on precipitation both locally and regionally. Although there is evidence that urbanization and precipitation are positively correlated, a consensus on the relationship has not yet been reached. Early studies by Horton (1921) and Kratzer (1937, 1956) provided indications that urban centers do play a role in strengthening rainfall activity. Studies by Landsberg (1956), Stout (1962), and Changnon (1968) discussed the extent to which urbanization may induce and strengthen precipitation. The strongest argument used in those studies to substantiate the role of urban centers on rainfall was the enhancement of downwind rainfall. Landsberg (1970) and Huff and Changnon (1972a,b) used observed data to support this hypothesis.

Balling and Brazel (1987), Bornstein and LeRoy (1990), Jauregui and Romales (1996), Selover (1997), Changnon and Westcott (2002), and Shepherd *et al.* (2002) have shown evidence that corroborates the earlier findings of enhanced precipitation due to urbanization. However, the hypothesis has been disputed, and even challenged, by other data studies that show no local effect on precipitation (Tayanc and Toros, 1997) or even deficits in precipitation that accompany urbanization (Kaufmann *et al.*, 2007).

Recent studies by Burian and Shepherd (2005) and Shepherd (2006) point specifically to increases in downwind rainfall due to urbanization. Although topological effects may be partially responsible for this finding in the Shepherd (2006) case, Chen *et al.* (2007) supports the Shepherd (2006) study. Shepherd *et al.* (2002) and Simpson (2006) have defined an "urban rainfall effect," which is defined as the impact of urban centers on enhancing downwind and peripheral rainfall. However, there is again no consensus for a unifying theory for the urban rainfall effect. Various explanations of why urbanization positively impacts convection have evolved over the years (Shepherd, 2005). The arguments include sensible heat flux enhancement, urban heat island-induced convection, the availability of more cloud condensation nuclei in urban areas, urban canopy alteration or disruption of precipitation systems, and increased surface roughness convergence.

3.1.3 Climate change impacts on the urban heat island processes

Climate change may well modify the urban heat island and rainfall effects, but the quantitative extents are unknown at this time. This is a ripe area for potential future climate research. For example, warmer winter temperatures may decrease energy combustion for heating, but also increase summertime air conditioning needs, thus affecting these anthropogenic heat sources. Higher temperatures in summer are likely to lead to higher levels of pollutants such as ozone. Changes in temperature, precipitation and ambient levels of carbon dioxide will all impact local vegetation and ecosystems with effects on urban parks and vegetation restoration, which is an important adaptation strategy. Climate-induced changes in winds could also impact urban climate. It should also be noted that urbanization often increases the impact of a given climate hazard. For example, a precipitation event is more likely to produce flooding when natural vegetation is replaced with concrete, and temperatures above a certain level will cause greater mortality when air pollution levels are high.

3.1.3.1 *Shanghai: many urban environmental islands*

Due to its land use patterns (Figure 3.5) and high densities of population and buildings, Shanghai experiences perhaps five interrelated microclimatic impacts, which we frame as metaphorical "islands" in analogy with the heat island: the heat island, dry island, moisture island, air pollution island, and rain island.

The Shanghai heat island appears from afternoon to midnight under certain weather patterns. According to ground observation records from the past 40 years, there is an apparent mean annual temperature difference of 0.7 °C between downtown and the suburban areas (16.1 °C and 15.4 °C, respectively). The corresponding mean annual extreme maximum temperatures are 38.8 °C and 37.3 °C, respectively. This urban warmth tends to appear from afternoon to midnight in mid autumn and in early winter, summer, and spring under clear conditions with low winds (Figure 3.5).

The second and third Shanghai islands can be referred to as the dry and moisture island effects. The elevated inner city temperatures and greater soil moisture availability in rural areas should result in lower urban relative humidity. Annual average data from Shanghai confirm this effect (Figure 3.5). However, the daily cycle of urban–rural humidity can be more complex due to differences in dewfall, atmospheric stability, and freezing, which can dry out rural air overnight and lead to greater urban humidity at night (Oke, 1987). This can lead to a cycle of relative dry and moisture islands that alternate during the day and night.

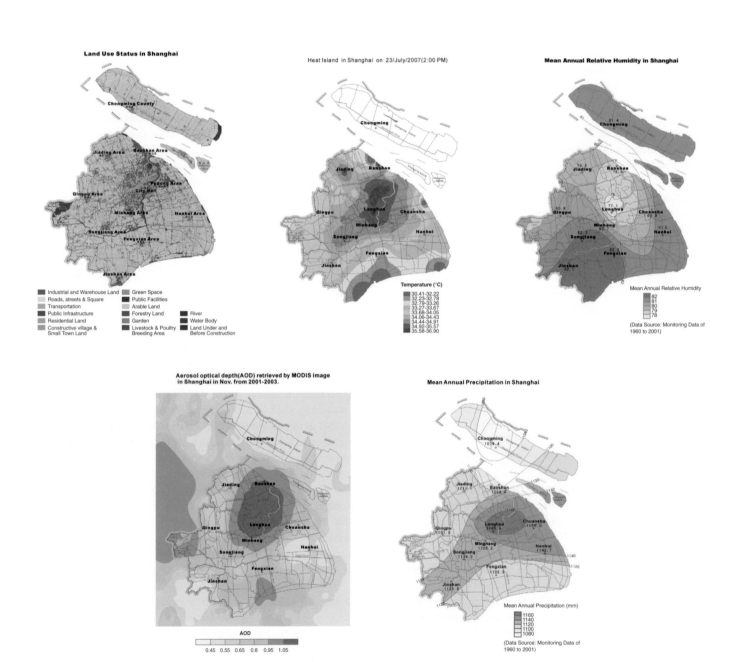

Figure 3.5: *Shanghai urban islands effects. From top left to bottom right are Shanghai's land use pattern, temperature, relative humidity, optical depth, and precipitation. For all the climate variables, a distinct pattern emerges in the area of greatest development.*

Source: Jiong (2004).

These differences in relative humidity may have implications for human comfort indices during heat waves and cold periods.

The fourth island is an air pollution island, since urban air quality is poorer than that of suburban and rural areas. The inversion layer over the urban heat island holds back the diffusion of atmospheric pollutants, increasing pollution levels locally. This in turn leads to acid rain. In 2003, the average pH value of precipitation in Shanghai was 5.21, with a percentage of acid rain of 16.7 percent. In the downtown area, where industry, commerce, traffic, and residents interact closely, air pollution is severe, as shown in Figure 3.5.

The fifth Shanghai urban island is termed the urban rainfall effect. According to ground observation records in the flooding season (May–September) and non-flooding season (October–April in the following year) over the period 1960–2002, the central city experiences greater precipitation than the outlying regions (urban rainfall effect), with an average precipitation that is 5–9 percent higher than in the surrounding regions. There are a number of hypothetical scenarios that may produce an urban rainfall effect: (1) the urban heat island effect may contribute to the rising of local air currents, which help to develop convective precipitation; (2) reduced urban windspeeds may slow the movement of storm systems over urban areas and could lengthen the duration of rain events; (3) aerosol pollutants may provide potential rainfall "nucleation sites" in low-forming clouds over the city. Thus, the urban heat island, the urban wind regime, and the urban pollutant island may all be partially responsible for the higher precipitation amount in the central city and the leeward area (Figure 3.5).

3.2 Natural variability of the climate system

Cities are vulnerable to modes of natural climate variability (that is, not caused by climate change), or preferred oscillations in atmospheric circulation. These preferred oscillations tend to recur at seasonal to multi-year timescales, have somewhat predictable impacts on temperature and precipitation in specific regions, and impact large portions of the planet.

Modes of climate variability centered in one location are able to impact temperature and precipitation in distant regions through a process called teleconnections, whereby wave motions transport energy outward from climate variability source regions along preferred paths in the atmosphere (Hurrel et al., 2003). Natural climate variability can cause significant impacts, for example by influencing the frequency and intensity of extremes of temperature and precipitation. It is, therefore, important for stakeholders to recognize these risks, which may be overlooked when assessing climate hazards associated with long-term global change.

Modes of natural climate variability have some level of forecast predictability, which can help cities prepare for climate

extremes associated with the modes. However, many cities, especially in the developing world, lack capacity to make forecasts on their own. It must also be kept in mind that these modes of variability are only responsible for a portion of the temperature and precipitation variability experienced by cities. It is therefore not uncommon for cities to experience climate anomalies of opposite sign to what the particular phase of the mode would suggest. For example, in Zimbabwe during the El Niño year of 1997–1998, drought conditions were predicted, and policy decisions were made based on the expectation of drought and reduced maize yields. However, rainfall ended up being above normal, which reduced perceived credibility of climate forecasts (Dilley, 2000). Conveying subtleties such as probability, uncertainty, and risk management to stakeholders around the globe can be a major challenge.

One lesson is that intra-regional rainfall patterns in these El Niño (and other modes) impacted areas can be very complex (Ropelewski, 1999). Another is that, while climate models of projected impacts are improving, achieving accurate seasonal climate predictions months in advance is still a research goal. In addition, with global climate change, patterns of natural climate variability themselves may change (in terms of strength, frequency, and duration of the modes). Even if the modes remain the same, climate change may alter the teleconnection patterns that drive regional climate impacts.

We highlight two specific modes that have large impacts on temperature and precipitation in many cities around the world: the El Niño–Southern Oscillation and the North Atlantic Oscillation. Additional modes are briefly described, followed by a discussion of interactions between climate variability and climate change.

3.2.1 El Niño–Southern Oscillation

Among climate oscillations, the coupled atmosphere–ocean phenomenon, known as the El Niño–Southern Oscillation (ENSO), impacts the most cities worldwide. The oceanic manifestations of this mode, known as El Niño episodes, are mainly characterized by the warming of the tropical central and eastern Pacific. La Niña tends to exhibit the opposite cooling effects. The main atmospheric manifestation, known as the Southern Oscillation, is a seesaw of the global-scale tropical and subtropical surface pressure pattern, which also involves changes in the winds, tropical circulation patterns, and precipitation (Trenberth and Caron, 2000).

ENSO is measured by several indices, including the surface pressure difference between Darwin and Tahiti (Southern Oscillation Index, SOI) and the sea surface temperatures in some selected regions in the central and eastern equatorial Pacific (Figure 3.6). El Niño episodes occur about every three to seven years, but their frequency and intensity vary on inter-decadal timescales.

There are tropical–extra-tropical teleconnections (Figure 3.7) caused by energy transfer from the tropics, which also cause

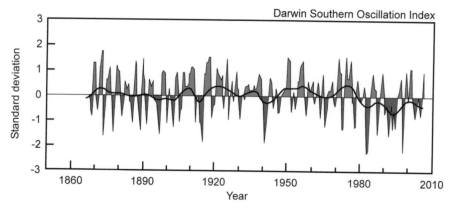

Figure 3.6: *Southern Oscillation Index. The SOI is one measure of the El Niño–Southern Oscillation, a pattern of natural climate variability. Red indicates El Niño episodes, and blue La Niña episodes.*

Source: Trenberth (1989), with data from UCAR/NCAR.

circulation, precipitation, and temperature variations in sub-tropical and higher latitudes, such as the Pacific–North American (PNA) and Pacific–South American (PSA) patterns. These teleconnections are associated with precipitation anomalies (departures from long-term averages) in cities in South and Southeast Asia, Australia, southern Africa, the southern United States, and tropical and subtropical South America.

The impacts of ENSO vary by season in different regions of the world, and therefore in different cities. For example, in

southeastern South America, impacts are strongest during the austral spring (Grimm *et al.*, 2000). Opposite effects may occur in different periods of an ENSO episode, and thus annual anomalies can be weak even if strong anomalies happen during particular months or seasons (Grimm, 2003, 2004). For example, in Jakarta, Indonesia, reduced precipitation in the fall of El Niño years is often partially offset by enhanced precipitation the following spring (Horton, 2007).

Box 3.1 describes the impacts of ENSO in Rio de Janeiro.

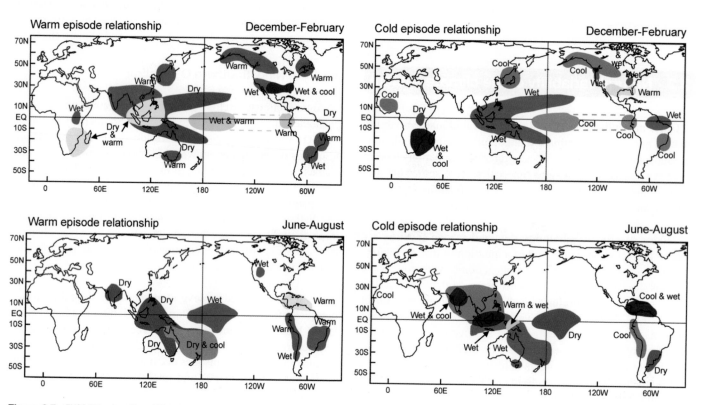

Figure 3.7: *ENSO Teleconnections. Teleconnections for both warm (El Niño) and cold (La Niña) episodes are shown. The variation of the teleconnections by season for each episode is also included.*

Source: NOAA NCEP Climate Prediction Center

[ADAPTATION] Box 3.1   Climate-proofing Rio de Janeiro, Brazil[i]

Alex de Sherbinin

CIESIN, Earth Institute at Columbia University

Daniel J. Hogan

State University of Campinas, Brazil

As flooding and landslides in April 2010 and January 2011 have demonstrated, Rio de Janeiro and the surrounding region continue to be at high risk of climate impacts. Approximately 200 deaths were attributed to the April 2010 floods and landslides, and several thousand people were made homeless, while 450 died in mudslides in the state of Rio de Janeiro in early 2011.[ii] Efforts will need to be made to "climate proof" the metropolitan region, increasing its resilience to floods induced by climate change and variability.

The city and metropolitan region of Rio de Janeiro have populations of 5 million and 11 million, respectively. Guanabara Bay, where Rio is situated, faces almost due south. Rio's dramatic topography has made it prone to landslides and flooding in low-lying areas (Box Figure 1). With the Atlantic rainforest having been stripped away from many hillsides, the thin soils have become prone to landslides, and the granite and gneiss bedrock has been left exposed to weathering, making it more prone to decomposition and erosion.

The coastline in this area was characterized by lagoons, estuaries, and low-lying coastal marshes, many of which have been filled in. The flat topography of low-lying areas, combined with a lack of drainage, has continued to result in flooding during the summer rainy season (January–March). The few remaining lagoons, mangroves, and marshes have been affected by land fill and sedimentation, reducing their absorptive capacity during extreme rainfall events. Low-lying areas around Lagoa de Tijuca and Lagoa de Jacarepaguá will largely be submerged with sea level rise of 0.5–1m.[iii]

The city receives higher than normal precipitation during the summer months of some El Niño events, but the connection with ENSO is not consistent. In the summer of 1998, in the end of an El Niño episode, the city was affected by severe floods as a result of two intense periods of rainfall in early February that produced a total of 480 mm of rain, which constitutes one-third the annual average rainfall of 1,200–1,500 mm (depending on location). The flooding in early April 2010 was precipitated by 288 mm of rainfall in a 24-hour period. These floods were associated with a weak El Niño event, but the floods in early 2011 were actually associated with a La Niña event, so the linkage is not consistent.[iv]

Rio's peculiar physical setting, and the circumscribed nature of suitable building sites, has spawned two kinds of response. One is the construction of high-rise apartments close to the coastline (e.g., Copacabana, Ipanema and Leblon) and in flood-prone areas further inland; the other is unregulated construction on steep slopes, particularly on the Tijuca mountain range. The unregulated construction of *favelas* (shanty towns) has a long history, and stems from the invasion of both private and public urban lands by poor urban squatters who become de facto (and in some cases de jure) owners of plots of land.[v]

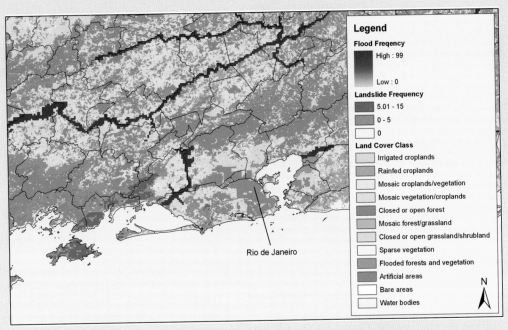

Box Figure 3.1: *Rio de Janeiro: flood and landside risk (Global Risk Data Platform Preview). Flood frequency is the expected average number of events per 100 years. Landslide frequency is the expected annual probability and percentage of pixel of occurrence of a potentially destructive landslide event ×1,000,000.*

Sources: UNEP/DEWA/GRID-Europe. 2009. Flood frequency. In: ISDR (2009) Global Assessment Report on Disaster Risk Reduction. United Nations, Geneva, Switzerland; International Centre for Geohazards (NGI). 2009. Frequency of landslides triggered by precipitations. In: ISDR (2009) Global Assessment Report on Disaster Risk Reduction. United Nations, Geneva, Switzerland; European Space Agency. 2009. Globcover for South America (global legend).

In terms of human vulnerability, Rio's income distribution is highly skewed. The existence of pockets of extreme poverty side-by-side with affluent neighborhoods is characteristic of the city. The 2000 census reported that 1.1 million people live in *favelas*, or 20 percent of the municipality's population. The city is also densely settled: the municipality has an average density of 4,640 persons per sq km but densities in the smaller administrative units of the metro area are between 8,000 and 12,000 persons per sq km.

Rio has a large migrant population from the poorest parts of Brazil's arid northeast region. The fact that many of them do not have personal experience with mudslides or mass wasting may account for their building practices. Migrants move up hillsides in search of new land, consistently eating away at the vegetation cover on the slopes above the *favelas*, despite government efforts to cordon off such areas to prevent further development. New regulations have been put in place that restrict building in hazard-prone areas. Efforts to "regularize" *favelas* have also been underway for several years, with various government programs to undertake cadastral surveys, grant deeds to de facto owners and provide basic infrastructure. These same plans limit the further expansion of *favelas* in flood-prone or steeply sloped areas. Rio de Janeiro has invested more than US$600 million in its *Programa Favela Bairro* to improve access to basic infrastructure, health and education for half a million of its poorest residents.[vi] In terms of social cohesion, the *favelas* do have some rudimentary organization, including neighborhood watches and self-improvement societies. These can be important for self-help and early warning systems. The drug traffic entrenched in many *favelas* (and the violence it spawns) is a major counterforce to these elements – a problem that has no end in sight.

Although *favelas* have always suffered during rainy seasons, the development up slope and paving of walkways has had the effect of increasing runoff to the low-lying areas. Runoff is channeled down cemented and quasi-natural watercourses to the narrow coastal lowlands, where they join canals whose limited flow capacity causes frequent flooding. By contrast, the Baixada Fluminense, a large marshy lowland somewhat removed from the steeper parts of the city, has had reasonably adequate drainage since the 1930s.[vii]

Generally, precipitation extremes are expected to increase in severity with climatic change, and these will have adverse impacts on Rio, given that the city already experiences extreme flooding and landslides on a roughly 20 year basis. Poor neighborhoods are particularly vulnerable to this extreme precipitation: roughly 300 people died and more than 20,000 people were made homeless during a 1967 flood, and in 2010 the Morro do Bumba slum built on a former garbage dump in Niteroi collapsed and slid downhill, burying homes and killing more than 200 people. Extreme and unpredictable rainfalls and floods converge with projected sea-level rise to increase stresses that will be difficult for Rio to handle owing to the city's topography (narrow coastal shelf backed by steep mountains subject to mass erosion), poor building conditions, the lack of secure land tenure for a notable portion of the city's population, poverty coupled with large income inequalities, and large problems with sanitation systems and sewage disposal.

Little in the way of concrete flood protection infrastructure has been set up in the wake of the 1988 floods. It is possible to speak of highly vulnerable sub-populations living in *favelas* and near waterways, and relatively less vulnerable upper classes living in high-rise apartments in locations less susceptible to inundation. This speaks to the need to upgrade slums, limit settlement on steep slopes and unstable locations, relocate some settlers on some slopes, revegetate hillsides, create more green verges near waterways for water absorption, and improve drainage systems in low-lying areas. Part of this will entail cleaning and maintenance of existing waterways and canals. A number of more specific suggestions for climate proofing are provided in the volume by Gusmão *et al.* (2008).

---

[i]This is dedicated to Daniel J. Hogan, who passed away in 2010. Dan was a leader in the field of population-environment studies and did much work on urbanization impacts on the environment in Brazil.

[ii]"April 2010 Rio de Janeiro floods and mudslides", Wikipedia, accessed on January 19, 2011 at http://en.wikipedia.org/wiki/April_2010_Rio_de_Janeiro_floods_and_ mudslides; "Brazil mudslide death toll passes 450", CBC News, accessed on January 20, 2011 at http://www.cbc.ca/world/story/2011/01/13/brazil-flood-deaths.html.

[iii]Gusmão, P. P., P. Serrano do Carmo, and S. B. Vianna (2008). *Rio Proximos 100 Anos*. Rio De Janiero: Instituto Municipal de Urbanismo Pereira Passos.

[iv]National Weather Service's Climate Prediction Center for a list of El Niño and La Niña years, accessed on January 20, 2011 at http://www.cpc.ncep.noaa.gov/ products/analysis_monitoring/ensostuff/ensoyears.shtml.

[v]Fernandes, E. (2000). The legalisation of *favelas* in Brazil: problems and prospects, *Third World Planning Review*, **22** (2), 167–188.

[vi]UN HABITAT (2006). *State of the World's Cities*, London, UK: Earthscan.

[vii]Cunha, L. R. and M. Miller Santos (1993) The Rio reconstruction project: the first two years. In *Towards A Sustainable Urban Environment: The Rio de Janeiro Study*, World Bank Discussion Paper 195, Washington, DC, USA: World Bank.

3.2.1.1 São Paulo and the El Niño–Southern Oscillation

The ENSO can have a strong effect on rainfall in São Paulo, Brazil, although this city is situated between regions of opposite effects of ENSO in the rainy season, Central-East and South Brazil. In Central-East, ENSO affects the frequency of extreme daily precipitation events in opposite wasy in the early and peak summer rainy season (Grimm and Tedeschi, 2009). In São Paulo, the behavior is different: there is significant impact in the early rainy season (October–November) but no consistent impact in the peak rainy season (January–February). Enhanced precipitation in the early rainy season, which is more frequent during El Niño episodes, can have a significant impact, since in this case urban floods and landslides may become more frequent during the rainy season. Flooding and landslides are disruptive, costly, and deadly.

3.2.2 Effects of the North Atlantic Oscillation on cities

The North Atlantic Oscillation (NAO) is the dominant pattern of atmospheric circulation variability in the North Atlantic region

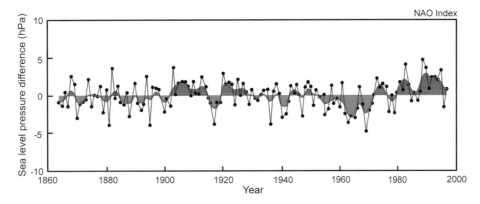

Figure 3.8: *NAO Index. One measure of the NAO Index is the difference in sea level pressure between the polar low near Iceland and the subtropical high near the Azores. Red indicates the positive phase of the NAO and blue indicates the negative phase of the NAO.*

Source: Hurrell (1995), with data from NCAR.

(Hurrell *et al.*, 2003). Although concentrated there, the impacts on climate extend over cities in a much larger region, from central North America to Europe, Asia, and Africa. When the NAO index is positive, there is a higher than usual subtropical high pressure (located in the general area spanning Portugal and the Azores) and a deeper than normal Iceland low pressure (Visbeck *et al.*, 2001; Figure 3.8) This larger pressure difference produces stronger-than-average westerly winds at middle latitudes, resulting in warm and wet winters in eastern USA and most of Europe and cold and dry winters in Canada and Greenland. Drier and colder conditions also prevail over southern Europe, the Mediterranean, and northern Africa. The oscillation and its aforementioned impacts are strongest in winter, but are present throughout the year. The NAO index varies from month to month and from year to year; however, the long-term average tends to remain in one phase for multi-year periods.

The negative NAO index refers to a weaker and eastward-displaced subtropical high, and a weaker Iceland low, leading to anomalies that are largely opposite of those associated with the positive NAO index. The mid-latitude winds are weaker and in a more west–east direction during negative NAO events. Warm and moist air is brought into southern Europe and cold temperatures prevail in northern Europe and in eastern USA. Canada and Greenland experience higher winter temperatures.

The impacts of NAO on precipitation and temperature in North America and Europe have urban consequences for hydropower generation (through water availability), energy consumption (through temperatures and humidity), water supply, and ecosystems (drought and forest fires, for example) among other sectors. Cities highlighted in this chapter that are impacted by the NAO include New York City, Toronto, London, and Athens. A London example is presented below.

3.2.2.1 *London and the North Atlantic Oscillation*

One city featured in this chapter that is affected by the NAO is London. Changes in the phase of the NAO have an impact on temperature and precipitation in the city, especially during the wintertime. The general relationship between the phase of the

NAO and climate in London is that the positive phase brings wetter and warmer conditions, while the negative phase brings drier and colder conditions (Wilby *et al.*, 1997). In early 2009, London experienced a rare heavy snowfall, which crippled the transportation system and forced many businesses to close. At that time, the NAO was in the negative phase, which helped transport cold air to the city. The snowfall in 2009 is an example of how expected relationships of natural climate variability do not always hold. While the colder temperatures were typical for the negative NAO, the increased precipitation was not.

3.2.3 Other modes of natural climate variability

In addition to ENSO and the NAO, there are other modes of natural climate variability that impact cities. These include the Pacific Decadal Oscillation (PDO), the Southern Annular Mode (SAM), the Atlantic Multidecadal Oscillation (AMO), the Indian Ocean Dipole (IOD), and the Madden–Julian Oscillation (MJO). The PDO is a longer timescale (decadal compared to inter-annual) pattern of variability that affects the climate in the Pacific Basin. The SAM is a mode of extra-tropical variability in the Southern Hemisphere that affects the climate in South America and Australia. The AMO is a mode of natural variability of North Atlantic Ocean sea-surface temperatures that influences Atlantic Basin hurricane activity. The Indian Ocean Dipole (IOD) is a coupled ocean–atmospheric mode centered on the tropical Indian Ocean Basin. The Madden–Julian Oscillation is an intra-seasonal mode of variability that primarily influences precipitation in the tropics.

3.2.4 Natural variability and global climate change

The frequency and intensity of the modes of natural climate variability and their associated teleconnections may change with global climate change. Because of the non-linearity of the climate system, it is not possible to simply add the teleconnections of a particular mode to a changing background climate (Hoerling *et al.*, 1997). It is important to analyze long-term observations of the modes to learn how they have varied in the past and if they are exhibiting any trends in the current climate. Climate change and climate variability may interact in complex and unpredictable ways. For example, shifts in mean wind patterns (such as the jet stream) associated with climate change may modify the regional teleconnections of the modes. It is possible that the accuracy of seasonal forecasts may change in some regions as climate change accelerates.

There remains much uncertainty as to how El Niño/La Niña conditions may change with a warmer climate. While some studies suggest that with increased greenhouse gas concentrations, the El Niño pattern may become more dominant, there is great uncertainty surrounding this issue (Collins, 2005).

For the NAO, some scientific evidence including modeling studies suggests that, with a warmer atmosphere, ocean temperatures will rise in a pattern that reinforces the positive phase of the NAO. As was the case with projections of ENSO with global warming, future patterns of the NAO also remain extremely uncertain (Visbeck *et al.*, 2001).

3.3 Global climate change and its impact on urban areas

Along with all other planetary surfaces, urban climates are subject to global changes due to radiative forcing. The impact of global change on cities is the subject of this section.

3.3.1 The climate system and global climate change

The elements of the global climate system include the atmosphere, biosphere, hydrosphere, cryosphere, and lithosphere (Figure 3.9). The climate system is coupled, in the sense that the components interact at many spatial and temporal scales.

The Earth's climate is determined by the long-term balance between incoming solar radiation and outgoing terrestrial radiation. Incoming solar radiation is partly absorbed, partly scattered, and partly reflected by gases in the atmosphere, by aerosols, and by clouds. Under equilibrium conditions, there is an energy balance between the outgoing terrestrial or "longwave" radiation and the incoming solar or "shortwave" radiation. Greenhouse gases are responsible for an approximately 30 °C elevation of global average surface temperature. Since the Industrial Revolution, increasing greenhouse gas concentrations due to fossil fuel combustion, cement-making, and land use changes has increased the mean surface temperature of the Earth by approximately an additional 1 °C.

These and other climate changes and impacts have been documented by an international panel of leading climate scientists, the Intergovernmental Panel on Climate Change (IPCC), formed in 1988 to provide objective and up-to-date information regarding the changing climate. Key findings of the 2007 Fourth

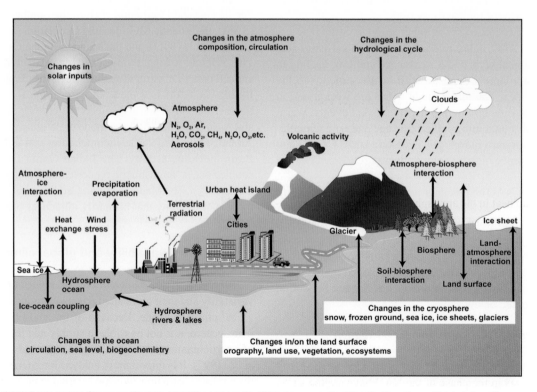

Figure 3.9: *The global climate system. Shown are the many interactions among the different components of the climate system, which includes cities.*

Source: IPCC, WGI (2007).

Assessment Report (AR4; IPCC, 2007) included the following (as summarized in Horton *et al.*, 2010):

- there is a greater than 90 percent chance that warming temperatures are *primarily* due to human activities (IPCC, 2007);
- atmospheric concentrations of carbon dioxide (CO_2) are now more than one-third higher than pre-industrial levels;
- concentrations of other important greenhouse gases methane (CH_4) and nitrous oxide (N_2O) have increased by more than 100 percent and close to 20 percent respectively over the same time period;
- further increases in greenhouse gas concentrations are projected to lead to further temperature increases and associated changes in the climate system;
- over the twenty-first century, global average temperature is expected to increase by between 1.8 and 4.0 °C.

Warming is expected to be largest over land and in the high-latitude North, where some Arctic cities may experience warming exceeding 8 °C by 2100. Outside the tropics and subtropics, the largest warming is generally expected in winter. Generally speaking, precipitation is expected to increase in high-latitude cities and decrease in subtropical cities. Hot extremes and cold extremes in cities are generally expected to increase and decrease respectively.

As CO_2 continues to be absorbed by the oceans, ocean acidification will accelerate, with potentially large implications on marine ecosystems. While the implications may be largest for those coastal cities where marine ecosystems are a source of economic livelihood and sustenance, in a global economy all the world's cities would be indirectly impacted by large-scale ocean acidification.

3.3.2 Drivers of global climate change

The global changes described above can largely be attributed to three drivers, or causes, of climate change: greenhouse gases, aerosols, and land use changes. These three drivers, with emphasis on urban contributions, are described below.

3.3.2.1 *Greenhouse gases*

Gases that trap heat in the atmosphere are referred to as greenhouse gases. Three primary greenhouse gases that are directly linked to anthropogenic activities are carbon dioxide (CO_2), methane (CH_4), and nitrous oxide (N_2O). As centers of population, economic activity, and energy use, cities are responsible for a large portion of greenhouse gas emissions. On a per capita basis, however, urban residents in developed nations probably have lower emission rates than ex-urban residents due to the inherent energy efficiencies of mass transit and multiple-resident buildings.

Some greenhouse gases, such as carbon dioxide, occur in the atmosphere through natural processes, while others, such as the hydrofluorocarbons, are created solely from human activities.

Carbon dioxide, which accounts for over 75 percent of all greenhouse gas emissions, is emitted into the atmosphere through the burning of fossil fuels and the clearing of land for agriculture. Methane is emitted from natural gas production, livestock, and agricultural production. Application of nitrogen fertilizer for agricultural production also leads to nitrous oxide being emitted into the atmosphere. Most hydrofluorocarbons, another important greenhouse gas, come from industrial activity.

3.3.2.2 *Aerosols*

Aerosols are atmospheric particles of both natural and anthropogenic origin. Natural aerosol sources include volcanoes and sea salt, while key anthropogenic sources include fossil fuel combustion and biomass burning. Aerosol concentrations tend to be higher in urban than rural areas, although during times of extensive biomass burning rural areas can have comparable concentrations. Climatic, human health, and visibility effects of aerosols have been documented. (Shu *et al.*, 2000, 2001; Pawan *et al.*, 2006; Eri *et al.*, 2009). Aerosols modify the earth's energy budget by scattering and absorbing short- and longwave radiation. As a prime radiative forcer (Charlson *et al.*, 1992), aerosol particles also influence cloud optical properties, cloud water content, and lifetime. That is referred to as the indirect effect of aerosols, or indirect climate forcing (Harshvardhan, 2002). Volcanic aerosols can lead to brief periods of global cooling, and biomass burning of agricultural regions has been shown to affect regional weather. While greenhouse gas effects are primarily global and regional in scale, aerosol effects span from the global to the urban scale.

3.3.2.3 *Land use change*

Land use change and urbanization influence the climate through changes in surface albedo, land roughness, hydrological and thermal features. Across the globe, human activities have changed the face of the planet. Deforestation has modified the climate by changing solar absorption and moisture transfer rates, as well as increasing carbon dioxide levels. Agricultural expansion has modified these processes as well. Urbanization and development have led to decreased groundwater absorption by the land and more heat absorption as the built environment expands (Zheng *et al.*, 2002; Gao *et al.*, 2003; Pielke, 2005; Lian and Shu, 2007). Like aerosol effects, the climatic effects of land use changes range from large regional scales to urban scales.

3.4 Observed climate change in cities

We present observed long-term climate data for the 12 cities to evaluate whether there are long-term trends. The urban heat island effect that may exist in each case should in principle be captured by the urban weather stations used. The selected cities include a wide range of climate zones. Tropical cities such as Dakar are characterized by warm temperatures throughout the year. The mid-latitude cities, in contrast, all experience a continental climate, with large temperature differences between summer and winter seasons. Some cities such as Delhi have a monsoonal climate, with the majority of the precipitation occurring during one part of the year, while other cities such as New York experience substantial precipitation during all months. These differences in background climate can be thought of as the foundation for the unique blend of climate change hazards each city faces, since climate changes are superimposed on a city's baseline climate.

Over the past century, there have been significant observed trends in climate hazards including annual mean temperature, annual precipitation, and extreme events, such as heat waves and intense precipitation, at global scales (IPCC, 2007). Trend analyses are a first step towards attributing changes to factors ranging from variability to greenhouse gases, to changes in the urban environment.

The observed annual temperature and precipitation values for each city were computed using a monthly data set.[1] The months in the data set that were missing were replaced by the climatological average for that month over the full time series for the city. The period over which the observed trends were analyzed was 1900 to 2005 unless otherwise noted. For the temperature trends, the following cities did not have data sets running the entire length of this period; Toronto (1938–2005), Delhi (1931–2005), Kingston (1943–2005), and Harare (1900–2002). For the precipitation trends, Toronto (1938–2005) and Harare (1937–2002). A brief city-by-city summary for observed trends in temperature and precipitation is shown in Table 3.2. The statistical significance of the trends is also included. Graphs of the observed climate data for each of the 12 cities are shown in Figure 3.10.

3.4.1 Temperature

Observations of annual mean temperature are used to determine whether or not a city is experiencing warming or cooling. In addition to mean temperature, maximum temperature and minimum temperature can also be used to identify changes in the climate. For these variables, finer temporal scales (daily data) are useful for analysis of observed climate trends. However, because of the limited data records for some cities, obtaining climate data on these small timescales can be difficult. These additional temperature variables can also be used in the analysis of extreme climate events, including hot days and heat waves.

For the 12 cities that were selected for the chapter, observed trends in annual mean temperature showed warming in 10 of the cities. Over the past century the most rapid rate of warming has occurred in São Paulo. The city had a trend of +0.27 °C per decade. The two cities that showed cooling (temperatures decreasing over the time period) were in Africa. Dakar and Harare had observed trends in annual mean temperature of –0.09 °C and –0.06 °C respectively. However, for these two cities, while the overall trend is negative, both have seen warming in the past two decades.

Table 3.2: *Observed climate trends in cities.*

City	Years	Number of missing months	Temperature trend (°C per decade)
Athens	1900–2005	15	0.03
Dakar	1900–2005	57	–0.09**
Delhi	1931–2005	10	0.03
Harare	1900–2002	44	–0.06**
Kingston	1943–2005	59	0.19**
London	1900–2005	0	0.08**
Melbourne	1900–2005	0	0.14**
New York	1900–2005	0	0.16**
São Paulo	1900–2005	30	0.27**
Shanghai	1900–2005	33	0.05**
Tokyo	1900–2005	2	0.26**
Toronto	1938–2005	4	0.10

City	Years	Number of missing months	Precipitation trend (mm per decade)
Athens	1900–2005	11	2
Dakar	1900–2005	24	–20**
Delhi	1900–2005	18	16*
Harare	1937–2002	26	–21
Kingston	1900–2005	19	–15
London	1900–2005	0	2
Melbourne	1900–2005	0	–4
New York	1900–2005	0	18
São Paulo	1900–2005	16	29**
Shanghai	1900–2005	24	–3
Tokyo	1900–2005	2	–17*
Toronto	1938–2005	4	7

Annual temperature and precipitation statistics are computed for all cities using data from the National Climatic Data Center Global Historical Climatology Network (NCDC GHCN v2), UK Met Office and Hadley Centre, Australian Bureau of Meteorology, and Environment Canada.

The single star (*) indicates that the trend is significant at the 95 percent level and the double star (**) indicates that trend is significant at the 99 percent level.

Although most of the cities did show a warming trend, the trends in each individual city vary by their rate of change. For example, one city may have had a slow increase for a large part of the twentieth century with a rapid warm up at the end of the period, while another may have seen a moderate increase over the entire time period. Further investigation of the trend for each individual city is necessary to understand the possible causes and potential impacts of warming temperatures. The rate at

---

1 For data sources, see Table 3.2

which the urban heat island is increasing in each city will affect observed temperature trends. For example, approximately one-third of the warming trend in New York City has been attributed to urban heat island intensification (Gaffin *et al.*, 2008). In developing countries, the rate is probably higher (Ren *et al.*, 2007).

3.4.1.1   Tokyo's heat island

As temperatures rise in Tokyo, residents are experiencing more health problems, including heatstroke and sleeping difficulties. Both are associated with higher nighttime temperatures (Figure 3.11).

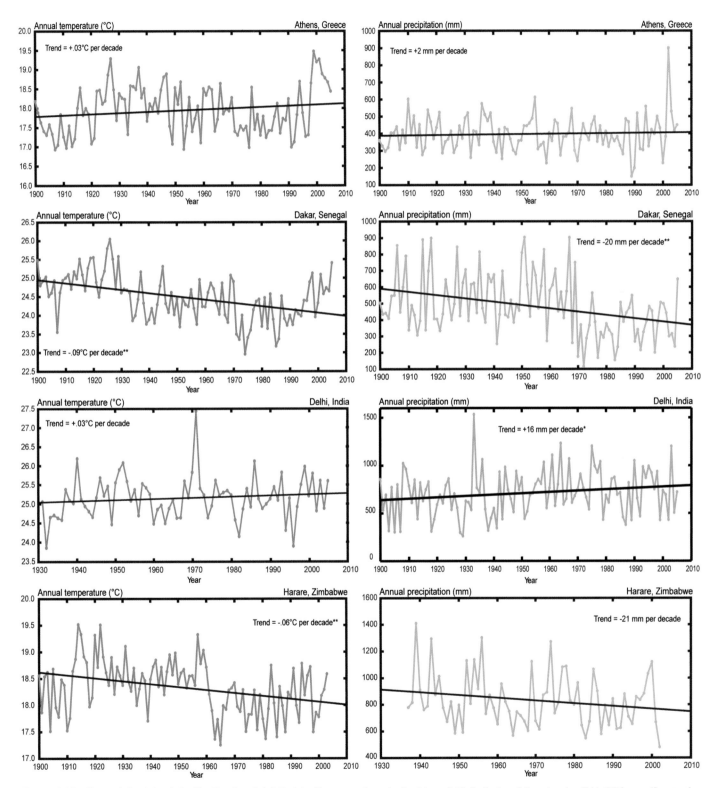

Figure 3.10: *Observed climate trends in cities. Trends and statistical significance are shown for the data available for the twentieth century (see Table 3.2 for specific years for each city). Note differences in temperature and precipitation scales.*

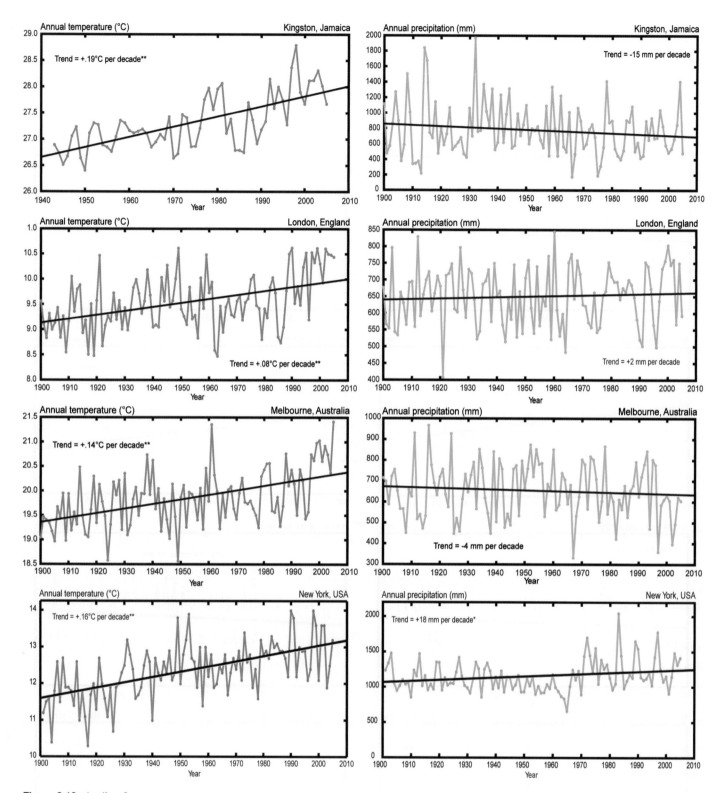

Figure 3.10: *(continued)*

Additional impacts include ecosystem changes such as earlier flowering dates and changes in insect and amphibian populations.

3.4.1.2 Observed temperature trend in São Paulo, Brazil

For the cities chosen for the chapter, São Paulo had the most rapid rate of increase in mean annual temperature. Over the period from 1900 to 2005, mean temperature in the city rose at a rate of 0.27 °C per decade. Much of the warming is occurring at night, with minimum temperature increasing at a faster rate than maximum temperature. The average of these two variables, maximum and minimum temperatures, is the mean temperature. São Paulo's warming was greater in winter than in summer. Some of the warming in São Paulo may be

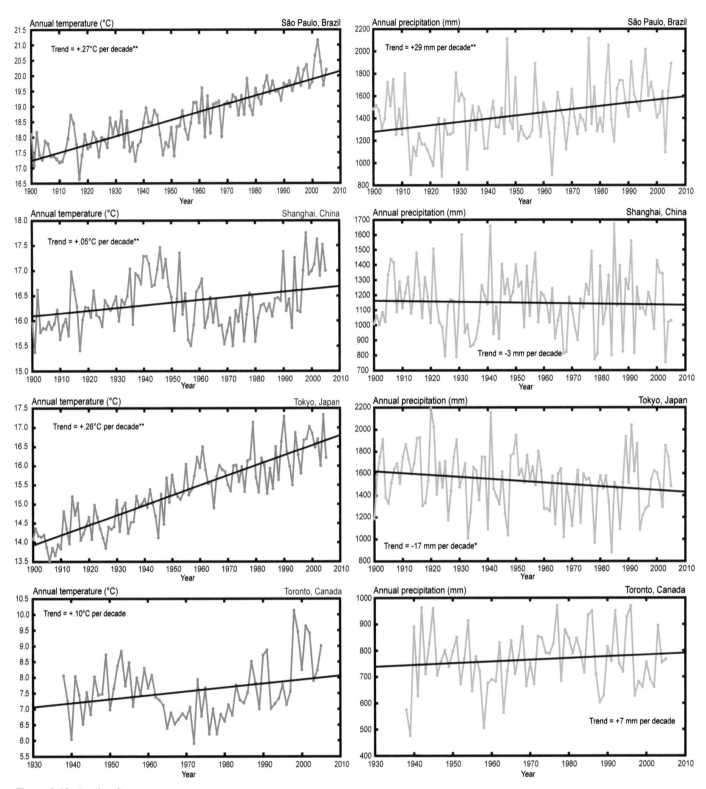

Figure 3.10: *(continued)*

caused by land use changes due to urbanization (Marengo and Camargo, 2008).

The observed warming in mean annual temperature appears to be greatest over approximately the past 20 years, as compared to the 20 years prior to that. Some of this variation between time slices can be explained by changes in the frequency of modes of natural climate variability. ENSO events in recent years coincide with the more rapid rates of warming. In addition, warming temperatures in the South Atlantic Ocean also may influence the warming mean temperatures in São Paulo (Marengo and Camargo, 2008).

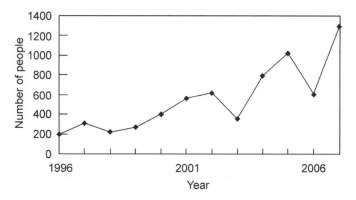

Figure 3.11: *Urban heat island impacts in Tokyo, Japan. The number of people sent to the hospital with heatstroke has been increasing over time.*

Source: Tokyo Metropolitan Government

3.4.2 Precipitation

Observations of annual total precipitation reveal precipitation changes in cities on decadal timescales. Natural climate variability can greatly influence annual precipitation (more so than temperature). Changes in the inter-annual variability of precipitation can also provide insight into how the climate is changing. For example, in New York City, while annual precipitation has only increased slightly in the past century, inter-annual variability has become more pronounced. In addition, analyzing trends in monthly rainfall can also be useful, as many cities experience two distinct seasons, such as the monsoons in Delhi, India. Data on shorter timescales, such as daily and weekly, can be used in extreme events analysis of droughts, floods, and intense precipitation rates. Again, with limited data records for some cities, obtaining climate data on these small timescales can be difficult.

Precipitation trends for the 12 cities reflect the regional nature of precipitation changes with climate change, relative to the largely homogenous changes expected in temperature. Of the 12 cities looked at in the chapter, half saw increasing trends in annual precipitation. Over the past century the most rapid rate of precipitation increase has occurred in São Paulo (+29 mm per decade), and the largest decrease has occurred in Harare (−21 mm per decade).

For the cities where observed trends in annual precipitation showed increasing (decreasing) precipitation, the trends do not reveal whether the wetter (drier) conditions are caused by more (less) frequent heavy rainfall events or more (less) persistent lighter rainfall. Also, the direction of trends for extreme precipitation events, such as days where precipitation is greater than 50 mm, will not necessarily correspond to the trend in annual precipitation. For example, a city could experience a decreasing trend in annual precipitation, but have an increase in short-duration intense precipitation events.

As was the case with temperature, there are multidecadal fluctuations in observed precipitation trends specific to each city. This is especially true for this climate variable given the large

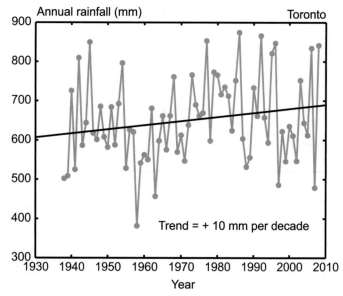

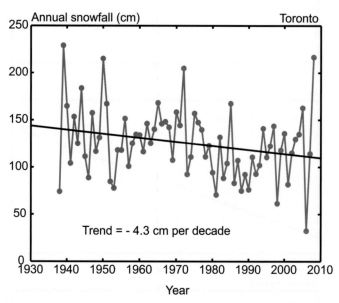

Figure 3.12: *Observed precipitation trends from 1938 to 2005 in Toronto, Canada: rain above; snow below. Showfall trend is significant at the 95% level.*

Source: Data are from Environment Canada.

inter-annual variability of precipitation. Cities with increasing annual precipitation over the time period still may have pronounced periods of drought, while those experiencing a drying trend may still have years with higher than average rainfall.

3.4.2.1 Observed precipitation trend in Toronto, Canada

Toronto is one city selected for this chapter that receives precipitation in the form of snow. While total precipitation has been increasing over the past century in the city, most of the increase has been in the form of more rainfall. Snowfall in Toronto has been on the decline, leading to a decrease in the percentage of total precipitation falling as snow. The trends for observed annual rainfall and snowfall in Toronto are presented in Figure 3.12, with rainfall and snowfall shown in green and blue respectively. Rainfall has increased at a rate of 10 mm per decade while snowfall has decreased at a rate of 4.3 cm per decade between 1938 and 2008.

To some extent, trends in precipitation for a given city depend on annual mean temperatures. In a warmer climate, the atmosphere can hold more moisture, allowing for more precipitation. But higher temperatures also can alter the type of precipitation that will fall. For Toronto, warmer temperatures have allowed for more precipitation; in the spring, precipitation that once fell as snow is now falling as rain. This trend is not evident in many of the colder parts of Canada. In colder areas the increases in total annual precipitation (like those observed in Toronto) have

included increases in snowfall, especially during the winter months. Temperatures in these areas of the country have warmed enough to allow for more precipitation but have not crossed the threshold for snow to become rain (Zhang *et al.*, 2000).

3.4.3 Sea level rise

Sea level rise analysis was not performed for all of the coastal cities due to limited data sets. However, coastal cities are extremely vulnerable to rising sea levels, since approximately 35 percent of world population lives within 100 km of the coast (Hachadoorian *et al.*, 2011; L. Hachadoorian, pers. comm.).

[ADAPTATION]   Box 3.2   Sorsogon City, Philippines, responding to climate change

Bernhard Barth and Laids Mamonong

UN-HABITAT

Sorsogon City is one of 120 cities in the Philippines in the Asia Pacific region. It has a land area of 313 square kilometers with a population of 151,454 (as of 2007) growing at a rate of 1.78 percent annually. Its economy is based mainly on agriculture, fishing, trade, and services. It is the capital and the administrative, commercial, and educational center of Sorsogon Province.

In August 2008 the city launched a Climate Change Initiative, championed by the new mayor. Until then the popular perception was that climate change is a global and national issue requiring limited action from the local government. A series of briefings for decision-makers and local leaders was conducted to enhance basic understanding of climate change and the important role of local government. This resulted in an expressed commitment from decision-makers in developing their city's climate change profile and defining responsive local actions.

Various city stakeholders worked together with the local government in the conduct of a participatory vulnerability and adaptation (V&A) assessment. Using climate change projections and risk assessments from national government agencies and private research institutions, the city government developed its local vulnerability assumptions. To assess local impacts, the city gathered and analyzed its own recorded observations. These were further substantiated by local people's accounts of their personal experiences. During city consultations, residents recounted how typhoons and storm surges over the past decade had become stronger and more destructive. These records and personal accounts were recorded as evidence of climate change impacts through community risk mapping. Using hand-drawn maps, local people graphically described the changes in the reach of tidal flooding and identified the areas gradually lost due to sea level rise and erosion. This participatory exercise promoted ownership by the locals of the assessment process and results, and increased their awareness of climate change impacts. Moreover, the process empowered the people to work together with the local government in finding practical solutions that they can personally act on.

As noted in the city's climate change profile, the city was badly hit by two super-typhoons in 2006, causing widespread

devastation within a two-month interval and leaving in their wake a total of 27,101 families affected and 10,070 totally damaged houses (Box Figure 3.2). The first typhoon, in just 5 hours, caused damage to public infrastructure estimated at 208 million pesos or US$4.3 million. The city is projected to experience more cases of prolonged monsoon rains resulting in total rainfall exceeding 2,800 to 3,500 mm per year.

The V&A assessment revealed that the city's geographical location and previous stresses make it sensitive to changes in extremes – such as tropical cyclones, storm surges, and extreme rainfall/flooding – and changes in means – such as increased temperature, increased precipitation, and sea level rise. With sea level rise projected to accelerate, the city's built-up areas situated near the coast present the highest vulnerability to climate change impacts because they have the highest concentration of people, especially informal settlers, living in inadequate structures in danger zones. These areas are also the hub for economic activities (accounting for 60 percent of the economy) and the location of basic lifelines such as water, electricity, and basic service facilities. Around 36 percent of the total population, or 55,000 people, are vulnerable to flooding. Over 35,000 people from nine coastal villages are threatened by sea level rise and storm surge, of whom 22,000 are women.

Box Figure 3.2: *Section of seawall in Beacon District, Sorsogon, partially destroyed by a 2006 storm. Many of Sorsogon's informal settlements are just behind the seawall.*

Source: UN-HABITAT/Bernhard Barth

Knowing these climate change vulnerabilities (areas, population, economic activities, policy gaps), the city government is now engaging local communities and the private sector in climate change adaptation planning. Using tools from UN-HABITAT's Sustainable Cities Programme, the local government conducted multisector city consultations that resulted in the identification of four priority "quick-win" responses to increase people's resilience to climate change, namely: (i) improving settlements and basic infrastructure, (ii) enhancing livelihoods, (iii) developing climate and disaster risk management systems, and (iv) improving environmental management and climate change mitigation actions. Issue working groups composed of representatives from people's organizations, NGOs, private sector, and LGU were organized to develop the action agenda per "quick-win" area and ensure its implementation.

So far the following important lessons have been learned:

1. There is a need to promote and advocate awareness on climate change among the general public and stakeholders through various media and community activities. This would broaden/establish partnerships among the private, public, academic, civil society, and neighbourhood associations for convergence of efforts on adaptation and mitigation.
2. The city government's capacity must be developed to make it more responsive to increase its resilience to climate change impacts. A framework must be developed to help and guide the city in integrating climate change considerations in the land use and development plans. A stronger link with national climate change programs is critical especially in enhancing building code and land use planning parameters.
3. The city needs to learn from good practices by other cities. It should also share its own experience in engaging various stakeholders in defining a collective climate change action.
4. It is crucial for the business sector to play a vital role in providing green building technology development and in promoting risk-resilient communities through the use of appropriate and innovative technologies in housing and infrastructure development.

The above lessons have become major considerations as the city works on mainstreaming climate change risk management into its local governance processes and implementing local climate change adaptation actions.

Source: UN-HABITAT (2010). Sorsogon City Climate Change Vulnerability and Adaptation Assessment. Available at: http:www.unhabitat.org.ph/climate-change/knowledge-productsoulputs.

Cities in this chapter at risk of sea level rise include New York, São Paulo, Tokyo, London, Shanghai, Melbourne, and Dakar. For all of these coastal cities, the sea level rise they are experiencing is caused by a combination of global and local factors. While the rate of sea level rise for each city due to global thermal expansion and meltwater from glaciers and ice sheets is the same, city-specific factors include land subsidence and local ocean height. These city-specific terms are necessary not only to determine the local rate of sea level and compare it to other cities and the global trend, but also for sea level projections.

3.4.3.1   Observed sea level rise in Dakar, Senegal

One city that is at risk of and has been experiencing rising sea levels over the past century is Dakar. Unfortunately, due to limited data availability, a common phenomenon in developing country cities, very little can be said about sea level rise trends. While sea level has been increasing at a rate of 1.5 cm per decade over the 11-year data record, for such short timescales natural variations and cycles can dominate any climate change signal. This example highlights the need for expanded data collection and quality control in many cities.

3.4.4   Extreme events

Extreme events can be defined as climate variables experienced in a limited duration. Temperature extremes include hot days where temperatures exceed a specified threshold, and heat waves–consecutive hot days. Precipitation extremes, which cover varying timescales, include intense/heavy precipitation events and droughts. Coastal storms and tropical cyclones are also types of climate extreme events.

Limited data availability at short timescales constrains analysis of urban extreme events. Extreme events also differ by city; for many inland locations, for example, coastal storm surges can be ignored. Some extreme events occur more frequently in association with certain phases of climate variability patterns such as ENSO. Because some variability patterns are somewhat predictable, there is an opportunity for seasonally forecasting these events, allowing cities to prepare for them in advance.

3.4.4.1   Hot days in Melbourne, Australia

One city that experiences hot days and heat waves is Melbourne. As defined by the World Meteorological Organization (WMO), hot days have maximum temperature exceeding 35 °C. For Melbourne, the trend from 1900 to 2008 shows no significant increase in the number of hot days, shown in Figure 3.13. The trend in hot days does not reveal a significant increase, even though annual average temperatures have risen significantly over the past century.

Combined with prolonged periods of dry weather, consecutive hot days (heat waves) have the potential to greatly impact cities and their surrounding areas. Specifically for Melbourne, years with above-average hot days, combined with other meteorological conditions, yield an increased threat of wildfires. Although the fires themselves are often in outlying areas away from the city, infrastructure, agriculture, ecosystems, water, and human resources critical to the city's survival may be impacted. Fires can also directly impact the city by reducing air quality. The most recent extreme warmth in early 2009, the heat wave of 2006, and in the summer of 1983 are all examples of years with increased fire activity and high numbers of hot days.

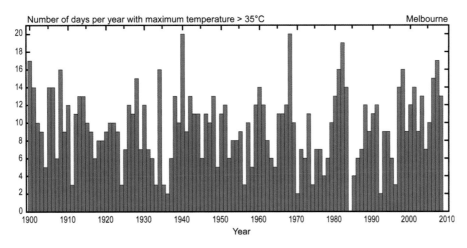

Figure 3.13: *Observed temperature extremes, hot days with temperatures above 35 °C, in Melbourne, Australia.*

Source: Data are from Bureau of Meteorology, Australia.

3.4.4.2 *Drought in Harare, Zimbabwe*

Drought is a precipitation extreme event that occurs over longer timescales, ranging from months to years. Unlike other extreme events, droughts lack a formal definition or index that is applicable globally, which makes assessments of their severity, trends in their frequency, and future projections difficult. Because of the varying indices and definitions, a qualitative assessment of drought based on precipitation is appropriate for multi-city analyses.

One city that experiences frequent and prolonged periods of drought is Harare, Zimbabwe. Over the past century, droughts have occurred several times, including 1991/1992, 1994/1995, and 1997/1998. Analysis of precipitation data reveals that precipitation in Harare has been declining over the past century at a rate of –21 mm per decade. In addition, of the ten driest July–June periods between 1938 and 2002, five have been since 1980. These results suggest that drier conditions may be becoming more frequent.

The droughts that occur in Harare have a strong connection to ENSO events. While ENSO is not the only factor that affects droughts in Zimbabwe, using this as one forecast predictor can help the city prepare and issue drought warnings with ample lead-time. However, reliance on just this predictor can be dangerous, as was the case in 1997/1998. Substantial preparations for a drought were made that year with the prediction of a strong El Niño, yet conditions did not become dry as expected (Dilley, 2000).

3.5 Future climate projections

This section describes climate change projection methods and results. The projections can be used by urban stakeholders to identify key sector-specific impacts of climate change, as a

first step for developing adaptation strategies. The projection component is derived from the global climate models that focus on broad geographical scales, rather than urban climate processes. However, historical data are used for projection baselines, and these baselines reflect the effects of urban climate processes. Some of the challenges of and potential solutions to incorporating urban processes in city-specific climate projections are discussed throughout the section.

3.5.1 Global climate change projection methods

3.5.1.1 *Global climate models*

The theoretical impact of atmospheric greenhouse gases on the planet's energy budget has been recognized since the nineteenth century (Arrhenius, 1896). Global climate models (GCMs) are mathematical representations of climate system interactions that help quantify the impact of greenhouse gases and other system interactions. Climate models also include other climate drivers, including aerosols, land-cover changes, and solar variability. Figure 3.14 depicts physical processes that a climate model simulates.

The simulations conducted for the IPCC 2007 report were run at higher spatial resolution than prior simulations and more accurately included complex physical processes such as turbulent fluxes. Current-generation climate models capture many aspects of climate, including twentieth-century warming (when simulated using historical greenhouse gas concentrations and volcanic aerosols) (Figure 3.15). They also capture key climate characteristics of paleoclimates such as the cool Last Glacial Maximum (~21,000 years ago) and warm mid-Holocene (~6,000 years ago). Model successes over a range of historical climate conditions increase confidence that future simulations will be realistic as CO_2 concentrations continue to increase.

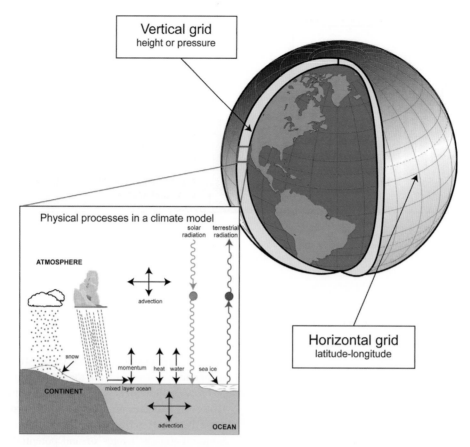

Figure 3.14: *Global Climate Model. Global climate models break the surface down into a series of gridboxes. Within each gridbox, equations are used to simulate the physical processes in the climate system.*
Source: NOAA.

3.5.1.2 Sources of uncertainty: emissions scenarios and climate sensitivity

Despite climate model advances, projections continue to be characterized by large uncertainty (for more information, see Horton *et al.*, 2010). The critical uncertainties concern:

- *Future concentrations of greenhouse gases* and other climate drivers that alter the global energy balance including aerosols and black carbon;

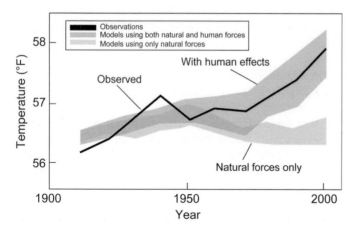

Figure 3.15: *Twentieth-century observations and GCM results. The observed temperature series follows the model simulation that includes both human effects and natural forces.*
Source: United States Global Change Research Program (2009).

- *Sensitivity of the climate system* to changing concentrations of climate drivers;
- *Climate variability*, which is largely unpredictable beyond one year in advance, and can mask the climate change signal at urban scales;
- *Local physical processes* such as the urban heat island, and coastal breezes that occur at smaller spatial scales than GCMs can resolve.

These uncertainties are partially addressed here by using a suite of GCMs and greenhouse gas emissions scenarios, averaged over 30-year time periods to cancel out a portion of the unpredictable natural variability. By presenting the projections as changes through time, uncertainties associated with local physical processes are minimized, although they cannot be eliminated entirely.

3.5.1.3 Greenhouse gas emissions scenarios

Greenhouse gas emissions scenarios, described in the IPCC Special Report on Emissions Scenarios (SRES; Nakicenovic *et al.*, 2000), are plausible descriptors of possible future socio-economic, technological, and governance conditions that drive energy demand and fuel choices (Parson *et al.*, 2007). The greenhouse gas concentrations associated with each scenario were used as inputs for global climate model simulations. Three greenhouse

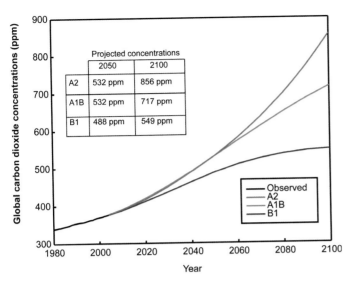

Figure 3.16: *Atmospheric CO$_2$ concentrations resulting from three emissions scenarios. Observed CO$_2$ concentrations through 2003, and future CO$_2$ for 2004 to 2100. The table gives the value of the CO$_2$ concentration for each scenario in 2050 and 2100.*

Source: IPCC (2007).

Table 3.3: *IPCC AR4 climate models.*

Institution	Model	Atmospheric resolution (latitude × longitude)
Bjerknes Centre for Climate Research, Norway	BCCR	2.8 × 2.8
National Weather Research Centre, METEO-FRANCE, France	CNRM	2.8 × 2.8
Canadian Center for Climate Modeling and Analysis, Canada	CCCMA	3.75 × 3.75
CSIRO Atmospheric Research, Australia	CSIRO	1.9 × 1.9
Geophysical Fluid Dynamics Laboratory, USA	GFDL1	2.0 × 2.5
Geophysical Fluid Dynamics Laboratory, USA	GFDL2	2.0 × 2.5
NASA Goddard Institute for Space Studies	GISS	4.0 × 5.0
Institute for Numerical Mathematics, Russia	INMCM	4.0 × 5.0
Pierre Simon Laplace Institute, France	IPSL	2.5 × 3.75
Frontier Research Center for Global Change, Japan	MIROC	2.8 × 2.8
Meteorological Institute of the University of Bonn, Germany	MIUB	3.75 × 3.75
Max Planck Institute for Meteorology, Germany	MPI	1.9 × 1.9
Meteorological Research Institute, Japan	MRI	2.8 × 2.8
National Center for Atmospheric Research, USA	CCSM	1.4 × 1.4
National Center for Atmospheric Research, USA	PCM	2.8 × 2.8
Hadley Centre for Climate Prediction, Met Office, UK	HadCM3	2.5 × 3.75

gas emissions scenarios (A2, A1B, and B1), were selected for use in this chapter. Figure 3.16 shows the CO$_2$ concentration associated with each of these three SRES scenarios. The three scenarios span a reasonably wide range of concentrations and allow for nominal high-, medium- and low-path CO$_2$ futures.

Greenhouse gas concentrations could exceed the levels in these SRES scenarios, due to increases in human emissions (for example, more rapid economic and population growth) or due to carbon and methane cycle feedbacks. As an example of the latter, increasing Arctic temperatures are leading to accelerated permafrost melting, which could lead to the release of stored methane to the atmosphere. As greenhouse gas emissions and concentrations are monitored in the coming decades, it will be possible to better assess the likelihood of high-end emissions scenarios.

3.5.1.4 Global climate model selection

From the IPCC climate model database, we selected 16 prominent models that had available output for each of the three emissions scenarios. The selected models and their space resolution are listed in Table 3.3. The outputs are collected by the World Climate Research Program (WCRP) and the Program for Climate Model Diagnosis and Intercomparison (PCMDI) (www-pcmdi.llnl.gov/ipcc/about_ ipcc.php), at the Lawrence-Livermore Laboratory in Livermore, California.

While it is unlikely that the future will closely follow a single emissions scenario, or that a GCM projection will capture all aspects of the climate system's response, using a suite of emissions scenarios and climate model simulations increases the probability that the projection range will span the actual future climate outcome. More information about the rationale for this approach, used by the IPCC, can be found in IPCC, 2007.

3.5.1.5 Model-based probability

Our combination of 16 global climate models and three emissions scenarios produces a distribution of 48 outputs for temperature and precipitation – in essence a "model-based" probability distribution. As described in Horton *et al.* (2010), future time periods are compared to a 1971–2000 model baseline. Temperature change is expressed as a difference, and precipitation as a percentage change. It should be noted that the model-based probability distribution will differ from the true probability distribution, for a range of reasons described in Horton and Rosenzweig (2010).

3.5.1.6 Downscaling

As described in Horton and Rosenzweig (2010), the projections use GCM output from a single model gridbox. Depending on the model, resolution can be as fine as ~75 – ~100 miles or as coarse as ~250 – ~275 miles. Changes from the relevant gridbox are applied to observed station data, a procedure known as the Delta Method that corrects for biases over the *baseline period* caused by factors such as the spatial mismatch between a gridbox and a point location. More sophisticated statistical and dynamical downscaling techniques can apply more localized *change projections*. While such techniques should be pursued in the future, a host of issues may hinder their utility. These issues include: the computational cost of dynamical downscaling and the fact that statistical downscaling may be less valid as climate statistics change (Christensen *et al.*, 2007).

3.5.2 Temperature, precipitation, sea level rise, and extreme events climate projections

The climate projections for temperature and precipitation for the 12 cities are presented first. For sea level rise and extreme event projections, more detailed methodologies are provided in a case study format focusing on New York City.

It is important to recognize the uncertainties associated with climate models. Therefore, the direction of change suggested here is as important as the specific numbers, which are not precise. Table 3.4 summarizes the temperature and precipitation projections for each of the cities for three future timeslices. Shown is the middle 67 percent (central range) of the model projections. Graphs showing the projections for each of the cities are shown in Figure 3.17.

3.5.2.1 Temperature

All cities are projected to see warming, with temperature increases of between 1.5 and 5.0 °C possible by the 2080s. The city expected to see the greatest warming is Toronto, Canada, with projected increase in temperature by the 2080s ranging from 2.5 to 5.0 °C.

Warming is expected to increase with distance from the equator; Toronto, for example, is expected to warm approximately two times as much as Kingston by 2100 under the A2 scenario. Warming is also generally expected to be greater in interior continental regions than in coastal regions, due to the moderating inertial influence of the oceans. This helps explain why more warming is expected in Toronto than London, even though they are at comparable latitudes. In the extra-tropical regions, warming will generally be greatest in winter.

Temperature projections for Delhi, India

Delhi, India, is one city where annual mean temperature is projected to increase over the next century. A slight upward, although not statistically significant, trend has been observed in the city from 1931 to 2005. This trend is consistent with trends from the region around Delhi in India. The region as a whole has experienced a warming trend over the past century, with the greatest warming occurring in the past three decades. The long-term warming trend appears to be driven by warmer maximum temperatures, while the more recent trend is a result of warmer maximum and minimum temperatures (Kothawale and Rupa Kumar, 2005).

Looking at Table 3.4, Delhi is projected to see an increase in annual mean temperature of between 2.5 and 4.5 °C by the

Table 3.4: *GCM projections for temperature and precipitation.*

City Name	Temperature[a]			Precipitation[a]		
	2020s	2050s	2080s	2020s	2050s	2080s
Athens	0.5 to 1.5 °C	1.0 to 4.0 °C	2.0 to 4.0 °C	−10 to +0%	−20 to −5%	−25 to −10%
Dakar	0.5 to 1.0 °C	1.0 to 2.0 °C	1.5 to 3.0 °C	−10 to +10%	−15 to +10%	−20 to +15%
Delhi	0.5 to 1.5 °C	1.5 to 2.5 °C	2.5 to 4.5 °C	−10 to +20%	−15 to +35%	−15 to +35%
Harare	0.5 to 1.5 °C	1.5 to 2.5 °C	2.0 to 4.0 °C	−10 to +5%	−10 to +5%	−15 to +5%
Kingston	0.5 to 1.0 °C	1.0 to 1.5 °C	1.5 to 2.5 °C	−10 to +5%	−25 to +0%	−30 to −5%
London	0.5 to 1.0 °C	1.0 to 2.0 °C	1.5 to 3.0 °C	−5 to +5%	−5 to +5%	−5 to +5%
Melbourne	0.5 to 1.0 °C	1.0 to 2.0 °C	1.5 to 3.0 °C	−10 to +5%	−15 to +0%	−20 to +0%
New York	1.0 to 1.5 °C	1.5 to 3.0 °C	2.0 to 4.0 °C	+0 to +5 %	+0 to +10%	+5 to +10%
Sao Paulo	0.5 to 1.0 °C	1.0 to 2.0 °C	1.5 to 3.5 °C	−5 to +5%	−5 to +5%	−10 to +5%
Shanghai	0.5 to 1.0 °C	1.5 to 2.5 °C	2.0 to 4.0 °C	−5 to +5%	+0 to +10%	+0 to +15%
Tokyo	0.5 to 1.0 °C	1.5 to 2.5 °C	2.0 to 4.0 °C	−5 to +5%	+0 to +5%	+0 to +10%
Toronto	1.0 to 1.5 °C	2.0 to 3.0 °C	2.5 to 5.0 °C	+0 to +5%	+0 to +10%	+0 to +15%

Shown is the central range (middle 67 percent) of values from model-based probabilities; temperatures ranges are rounded to the nearest half-degree and precipitation to the nearest 5 percent.
[a] Based on 16 GCMs and three emissions scenarios.

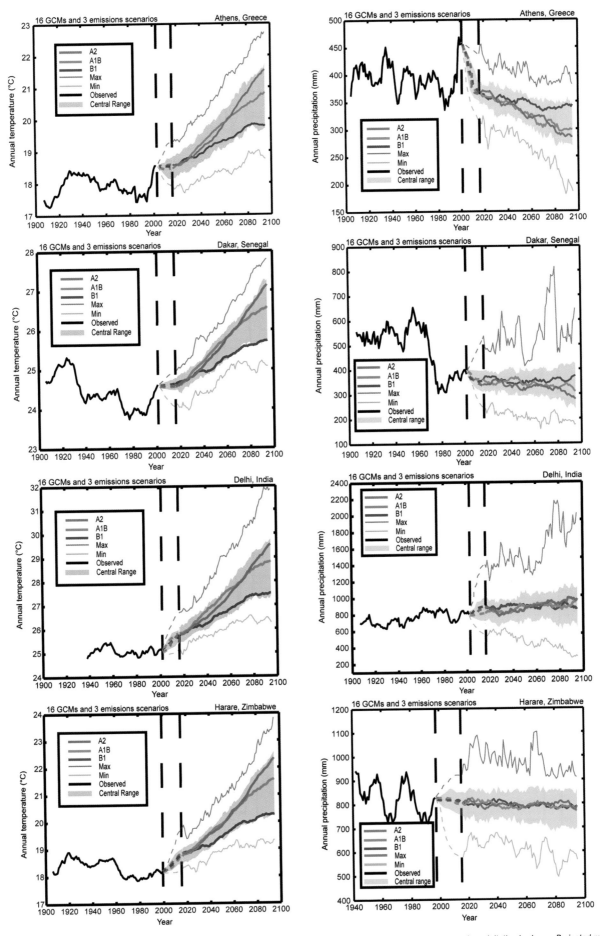

Figure 3.17: *Observed and projected temperature and precipitation. Combined observed (black line) and projected temperature and precipitation is shown. Projected model changes through time are applied to the observed historical data. The three thick lines (green, red, and blue) show the average for each emissions scenario across the 16 GCMs. Shading shows the central range. The bottom and top lines, respectively, show each year's minimum and maximum projections across the suite of simulations. A 10-year filter has been applied to the observed data and model output. The dotted area between represents the period that is not covered due to the smoothing procedure. WCRP, PCMDI, and observed data sources found in Table 3. 2. Note differences in temperature and precipitation scales.*

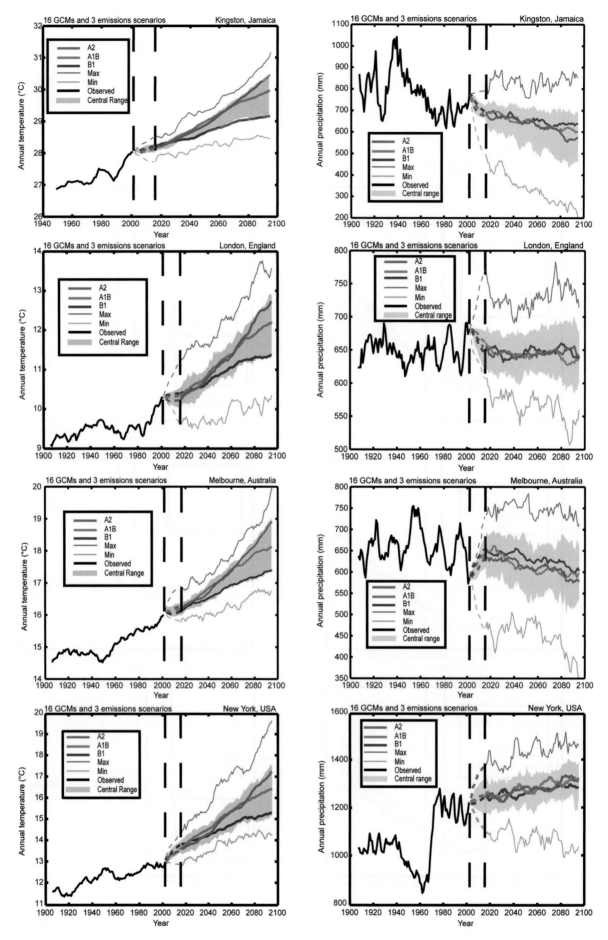

Figure 3.17: *(continued)*

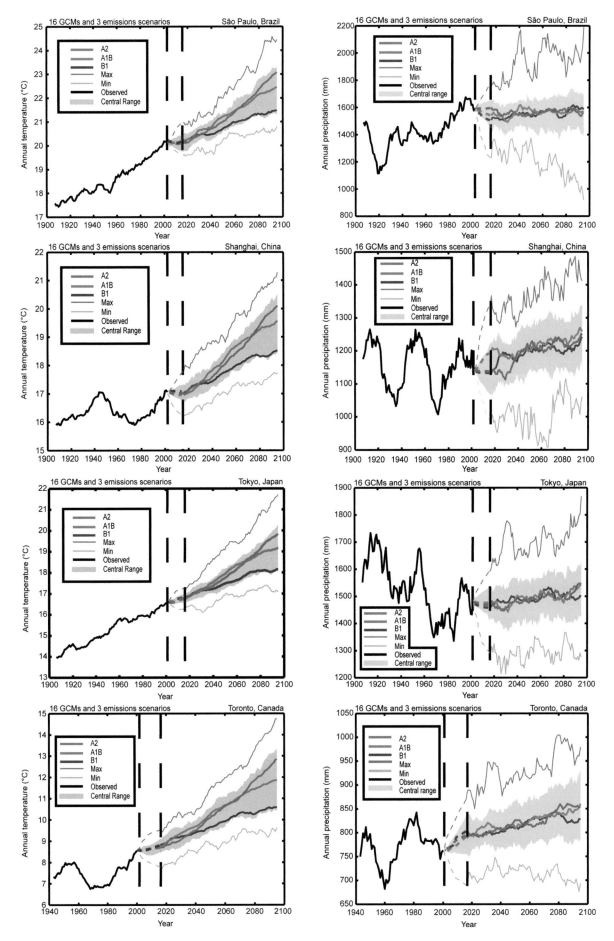

Figure 3.17: *(continued)*

2080s. Figure 3.17 shows the full range of GCM projections for annual mean temperature for Delhi. Only around the 2030s do differences between the emissions scenarios emerge.

3.5.2.2 Precipitation

There is greater variability in the direction of projections of percentage change in precipitation for these cities. Some cities are expected to see increases in precipitation while others are projected to experience sharp declines in precipitation by the 2080s. For cities in higher latitudes projected to see increased precipitation, such as Toronto and New York, most of the precipitation increase will be in the form of rainfall, as snowfall is likely to decline given the warmer temperatures.

Precipitation changes will differ dramatically by region. In general, mid-to-high latitude cities such as Toronto, New York, and Tokyo are expected to experience precipitation increases. However, some mid-latitude cities, such as London, are expected to experience significant summer-time drying, which could lead to net precipitation decreases. Other cities at the boundaries between the mid-latitudes and subtropics are expected to experience drying, including Harare and Melbourne. Some tropical cities are expected to experience more precipitation, while others are expected to experience less. In those cities where precipitation is strongly correlated with inter-annual variability, changes in the modes with climate change will be critical, and introduce an added element of uncertainty. These cities include Harare, Melbourne, Delhi, and São Paulo for ENSO, and Athens and London for the NAO.

It is important to note that in some cases the GCMs have trouble accurately simulating the baseline/observed precipitation in particular regions. This is especially true in cities that have strong seasonal precipitation cycles. With a poor handle on the baseline, the projections for the city are therefore skewed. The projections for Delhi, which has a monsoonal climate, provide an example of this, as extreme values for increased precipitation may overestimate how wet the climate may become. It is therefore important to focus on the direction of change, not so much the actual values. Because it was not possible to include all cities across the globe, the projections presented here can be used as a proxy for other cities with similar climate conditions.

Precipitation projections for Athens, Greece

Athens is one city where annual precipitation is projected to decrease over the next century. In the observed climate record for annual precipitation, there is no statistically significant trend. Slight decreasing trends in other precipitation variables, such as wet days, have been observed. What appears to be occurring in Greece, as well as many other parts of the globe, is an increase in extreme daily precipitation (Nastos and Zerefos, 2007). While total rainfall and the number of wet days show no change or a decline, when there is precipitation, it is more intense. This could potentially be most damaging, as these rain events can cause short-duration, flash flooding.

As shown in Table 3.4 Athens is projected to see drops in precipitation between 10 and 25 percent by the 2080s. Although seasonal projections are less certain than annual results, the climate models project much of the drying to occur during the summer months. Figure 3.17 shows the full range of GCM projections for annual total precipitation in Athens. Even using 10-year smoothing, there remains large historical variability in the observed trend. For the projections, only from approximately the 2040s onward does the B1 scenario produce smaller precipitation decreases compared to the other two scenarios.

3.5.2.3 Sea level rise

As the oceans warm and expand and land-based ice continues to melt, all the coastal cities analyzed here are expected to experience sea level rise this century. However, the rate will differ by city, for two primary reasons. First, the local height of the adjacent ocean can differ by city, due to the influence of ocean currents, water temperature, and salinity, and the influence of wind and air pressure. Second, local land height change can differ by city. Some cities such as Shanghai are sinking due to the effects of groundwater extraction and compaction of soils by the expanding built environment.

Projections for sea level rise are presented in a case study format for New York City. Described are the methods used to make these projections and the projections themselves. Most of the text in this section is from Horton and Rosenzweig (2010).

Sea level rise methods

The GCM-based methods used to project sea level rise for the New York City region include both global (global thermal expansion and meltwater from glaciers, ice caps, and ice sheets) and local (local land subsidence and local water surface elevation) components.

Within the scientific community, there has been extensive discussion of the possibility that the GCM approach to sea level rise may underestimate the range of possible increases, in large part because it does not fully consider the potential for land-based ice sheets to melt due to dynamical (motion-related) processes (Horton *et al.*, 2008). For this reason, the NPCC developed an alternative "rapid ice-melt" approach for regional sea level rise projection based on observed trends in melting of the West Antarctic (Velicogna and Wahr, 2006) and Greenland ice sheets (Rignot and Kanagaratnam, 2006) and paleoclimate studies of ice-melt rates during the most recent postglacial period (Fairbanks, 1989). Starting around 20,000 years ago, global sea level rose 120 meters and reached nearly present-day levels around 8,000–7,000 years ago. The average rate of sea level rise during this ~10,000–12,000 year period was 9.9 to 11.9 cm per decade. This information is incorporated into the rapid ice-melt scenario projections.

The GCM-based sea level rise projections indicate that sea level in New York City may rise by 5 to 13 centimeters in the

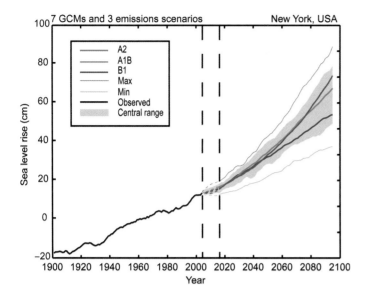

Figure 3.18: *New York City sea level rise. Combined observed (black line) and projected sea level rise is shown. Projected model changes through time are applied to the observed historical data. The three thick lines (green, red, and blue) show the average for each emissions scenario across the seven GCMs used for sea level rise. Shading shows the central range. The bottom and top lines, respectively, show each year's minimum and maximum projections across the suite of simulations. A 10-year filter has been applied to the observed data and model output. The dotted area between 2002 and 2015 represents the period that is not covered due to the smoothing procedure.*

Source: WCRP, PCMDI, and observed data from NOAA Tides and Currents.

Table 3.5: *Sea level rise projections including the rapid ice-melt scenario for New York City.*

New York City	2020s	2050s	2080s
Sea level rise[a]	+5 to 13 cm	+18 to 30 cm	+30 to 58 cm
Rapid ice-melt sea level rise[b]	~13 to 25 cm	~48 to 74 cm	~104 to 140 cm

[a] Based on seven GCMs and three emissions scenarios.

[b] "Rapid ice-melt scenario" is based on acceleration of recent rates of ice-melt in the Greenland and West Antarctic ice sheets and paleoclimate studies.

The rapid ice-melt sea level rise scenario addresses this possibility. It is based on extrapolation of recent accelerating rates of ice-melt from the Greenland and West Antarctic ice sheets and on paleoclimate studies that suggest sea level rise on the order of 9.9 to 11.9 cm per decade may be possible. Sea level rise projections for New York City in the rapid ice-melt scenario are shown in the bottom row of Table 3.5. The potential for rapid ice-melt is included in the regional projections for New York City because of the great socio-economic consequences should it occur. To assess the risk of accelerated sea level rise and climate change for the New York City region over the coming years, climate experts need to monitor rates of polar ice-melt, as well as other key indicators of global and regional climate change.

3.5.2.4 Extreme events

Some of the largest climate change effects on cities are associated with extreme events, such as heat waves, intense precipitation events, and coastal storms. The frequency, intensity, and duration of many extreme events are expected to increase with climate change. The following quantitative New York City example describes the types of extreme event threats faced by many cities, although each city will face slightly different extreme event and natural disaster risks.

Following the New York City case, we present an example of qualitative projections for tropical cyclones, an extreme event that impacts many cities included in this chapter. It should be noted that extreme climate events and natural disasters are intertwined, partly because climate hazards represent a major portion of all natural disasters. Furthermore, climate extremes (such as intense precipitation or drought-induced forest fires) may cause secondary natural disasters (such as landslides). The following section is a case of extreme event projections for New York City, from Horton and Rosenzweig, (2010) and Horton *et al.*, (2010).

Extreme events projections for New York City

Extremes of temperature and precipitation (with the exception of drought) tend to have their largest impacts at daily rather than monthly timescales. Because monthly output from climate models is considered more reliable than daily output, simulated changes in monthly temperature and precipitation were calculated; monthly changes through time from each of the 16 GCMs and three emissions scenarios described earlier in this section

2020s, 18 to 30 centimeters in the 2050s, and 30 to 58 centimeters in the 2080s. Sea level projections for the three emissions scenarios agree through the 2040s. Figure 3.18 shows that the B1 scenario produces smaller increases in sea level than the A1B and A2 scenarios beginning in the 2050s, and only around 2080 does the A2 scenario produce larger values than A1B. The divergence of A2 from A1B occurs approximately 10 years earlier for temperature than for sea level rise, in part reflecting the large inertia of the ocean and ice sheets relative to the atmosphere.

Sea level rise projections for the New York City region are higher than global sea level rise projections (generally by approximately 15 cm for twenty-first century projections) (IPCC, 2007). One reason is that the New York metropolitan region is subsiding by approximately 8 to 10 cm per century. The climate models also have a tendency to produce accelerated sea level rise along the northeast US coast, associated in large part with a projected weakening of the Gulf Stream (Yin *et al.*, 2009).

The model-based sea level rise projections shown in Table 3.5 are characterized by greater uncertainty than the temperature projections, due largely to the possibility that dynamic processes in polar ice sheets not captured by the GCMs may accelerate melting beyond currently projected levels. This uncertainty is weighted towards the upper bound: that is, the probability of sea level rise lower than that described in the GCM-based projections in Table 3.5 is very low, and the probability of sea level rise exceeding the GCM projections is relatively high.

[ADAPTATION]   Box 3.3   Adaptation to sea level rise in Wellington, New Zealand

Chris Cameron, Paul Kos and Nenad Petrovic

Wellington City Council

INTRODUCTION

Wellington is a coastal city, with an inner harbor and exposed southern coast. A pilot study focusing on the impacts of sea level rise on a low-lying city suburb has commenced to inform the adaptation approach across the city.

Part of the area included in this pilot study (Box Figure 3.3) has been identified as a key growth node for urban intensification. The area also contains a range of key infrastructure including a significant highway, an international airport, utilities, businesses, housing, and community facilities such as schools, pools, libraries, a marina, and a surfing beach.

METHODOLOGY

Sea level rise was viewed as one of the most critical climate change impacts on the study area, because it lies between only 1 m and 3 m elevation.

Box Figure 3.3: *The study area.*

The latest New Zealand guidance on coastal hazards[2] associated with climate change recommends that councils consider the impacts of a 0.8 m increase in sea level by 2090. However, given the considerable uncertainty in projections and the possibility of catastrophic events, an approach was taken based on testing infrastructure resilience and response via a range of scenarios. This will allow the development of strategies to manage the expected risk.

Three core scenarios were examined (0.5 m, 1 m, and 2 m) with each having an additional 0.5 m storm surge component within the harbour and a 1 m component on the exposed southern coast. These scenarios reflect the most recent scientific probabilities in the short term (50–100 years), while allowing for possible higher levels in the longer term.

Evaluation of the scenarios was carried out in an interdisciplinary cross-council workshop including water, drainage, roading, hazards, transport, coastal and recreational, and urban planning experts. For each asset the following information was gathered: description, ownership, criticality, condition, relocatability, economic value, proposed upgrades. Each asset was then tested against each sea level rise scenario to determine potential risks and impacts. Feasible response options were then proposed.

Mapping the scenarios

Sea level rise scenarios were mapped based on ground elevation data from LiDAR (Light Detection And Ranging). The LiDAR data were captured at 1 cm vertical intervals with +/–10 cm accuracy and were verified by field survey. The data were then used to create a digital terrain model. A local vertical datum was developed based on mean high water springs. This allowed for assessment of the highest likely sea level that includes mean tidal elements.

Infrastructure and key existing community facilities were mapped against the sea level rise scenarios so the impacts could be evaluated holistically.

RESULTS

Through qualitative assessment of likely impacts and appropriate responses to the sea level rise scenarios a number of issues were identified:
- Degradation of the level of service from the storm-water system
- Rising groundwater levels
- Need to evaluate response options for at-risk coastal areas across the city
- Need for early decision-making for response planning
- Interactions and interdependency between assets
- Need to prioritise adaptation responses across the city

The workshop highlighted that rising sea levels are likely to have some impacts on the storm-water system in the short

---

2   See www.mfe.govt.nz/publications/climate/preparing-for-coastal-change-guide-for-local-govt/index.html.

term (next 20–30 years). Some solutions were proffered that could be developed as part of normal asset management.

Low-lying parts of the study area may be susceptible to increased flooding due to a rise in the water-table. This was regarded as more urgent than "over-topping." Several likely options for responding were identified and will be consulted on with Councillors and the community. Further detailed work is required to examine the impact, behavior, and response of groundwater in the study area, together with the likely costs and benefits of each response option.

Different responses may be appropriate for the natural dune environment of the southern coast compared to the structurally modified northern coast. Maintaining a dune environment on the south coast would help the area retain its high aesthetic and amenity values. Moreover, this could be a more successful adaptation response, given the adverse effects that "hard engineered" structures can have on a beach.

The importance of taking into consideration linkages between infrastructure elements was recognised. For example, pumping stations require power and telecommunications, which must therefore be maintained through the area at all times.

Similar impacts may occur in other parts of the city, and an overall cost–benefit analysis cannot be completed in isolation within a limited area.

These findings will inform the proposed intensification plans and other ongoing development, maintenance, and asset management plans for the area. Findings will also be used for discussions within council and the community around prioritizing, costs, and residual risks.

CONCLUSIONS

This study gathered and evaluated key information needed to make an initial assessment of climate change impacts in a localized urban area. It has indicated where further detailed work could be undertaken to derive a more accurate assessment of costs and benefits. A modified approach based on this pilot study will be used across other coastal areas within Wellington City.

[ADAPTATION]   Box 3.4   Climate change adaptation in Kokkola, Finland

Philipp Schmidt-Thomé

Geological Survey of Finland

Juhani Hannila

Technical Service Centre, City of Kokkola

Kokkola, a medium-sized town on the west coast of Finland, was one case study of the Developing Policies & Adaptation Strategies to Climate Change in the Baltic Sea Region (ASTRA) project. The award-winning ASTRA project was co-financed by the European Regional Development Fund (ERDF) under the INTERREG IIIB Programme. The ASTRA (www.astra-project.org) and SEAREG (www.gtk.fi/dsf) projects are the predecessors of the currently ongoing BaltCICA project (www.baltcica.org). The aim of the projects is to support planners and decision-makers in understanding the potential impacts of climate change on regional development and to support the implementation of adaptation strategies. While SEAREG (2002–2005) mainly focused on awareness raising, ASTRA (2005–2007) went a step further to support the development of adaptation strategies. BaltCICA (2009–2012) builds on the results of the projects by implementing adaptation measures in cooperation with stakeholders. All three projects are managed by the Geological Survey of Finland (GTK) and count on partners from the Baltic Sea Region.

Kokkola was founded in 1620 adjacent to the Bothnian Bay (Baltic Sea). The postglacial rebound (land uplift) in this area amounts to 9 mm per year, and due to the retreating sea the original city center now lies about 2 km inland from the shoreline. Rising sea levels have made the glacial rebound less effective over recent years so that the net land rise has dropped to about 4–5 mm per year relative to the mean sea level. Normal sea level variation in Kokkola is between –1.0 m and +1.5 m relative to mean sea level. The city planning office of Kokkola was thus interested in sea level rise scenarios for the twenty-first century. The background is that Kokkola experiences a strong pressure for coastal housing development. If the coastal land uplift continued as at present, the coastal areas could be easily developed, even with the lower glacial rebound effect of 5 mm per year. But what if the sea level rise and storm flood events become stronger and flood patterns change?

Kokkola participated in the ASTRA project to better understand the climate change scenarios and to evaluate how these can be implemented into local planning. The uncertainties of the climate change scenarios especially played an important role in the assessments. Finally the city chose to use the MPIA2 "high case" sea level change scenario originally developed under the SEAREG project. According to these scenarios it is possible that the land uplift will be neutralized by sea level rise; in other words, the present coastline would not change over the twenty-first century. Consequently, the often predicted increasing wind speed peaks during storm events and the increase in heavy rainfalls would lead to flood-prone area changes. Two important locations for future housing development in Kokkola are the area of the 2011 housing fair and Bride Island.

The area for the housing fair, to be held in 2011, was planned several years ago. Such a housing fair is an important event in Finland, as the houses built for the fair are later to be used for housing, which is an important asset for the investors. The housing fair in Kokkola is planned in a new housing area on the sea shore. In the course of the ASTRA project the location of the housing fair 2011 was not changed, but the minimum elevation of the building ground above mean sea level was raised by 1.0–1.2 m compared to previous plans on the sea shore and is about 3.5 m (streets 3.0 m) above the mean sea level. In other words, the decision was taken that sea shore plots may be built up, but it has to be made sure that the lowest floor of the houses is well above potential flood levels. The cost of each plot and house was calculated, including the artificial elevation of the building ground, and the investors were willing to accept this extra cost because the demand for houses located on the sea shore is still rising.

The second example, Bride Island, is a very popular place for summer cottages in the close vicinity of the city. The current trend in many European countries is to improve these temporary summer homes into cottages that can be used all year round, and even to convert them into permanent homes. If the land were to continue to rise out of the sea as it has done so far, such a conversion from a temporary to a permanent home would pose no problem. The interest of the land owners is not only in converting the houses; the investments certainly would also be justified by rising land and house prices. The city of Kokkola, on the other hand, carries responsibilities if land use plans are changed and natural hazards start to threaten housing areas. Due to the scenarios used in the ASTRA project, land use plan changes for Bride Island were put on hold until improved climate change scenarios and observed trends have been analyzed. However, sea level change scenarios are taken into account in the minimum elevation of buildings above mean sea level, e.g., when old cottages are renovated. Lowest building elevation for the newest building permits has been raised to 2.5 m above the mean sea level – so, in building renovation the adaptation to sea level rise is put into practice in small steps or plot by plot.

Recommendations of the ASTRA report were also taken into account in building and city planning regulations. Several regulations are in place to protect from rising wind speeds, cold winds, and storms on the shore. For example, low houses with optimal roof pitch against wind turbulence on yards; inner courtyards towards the south; plantations and fencing against cold wind directions; and also directing main streets crosswise to the coldest winds to avoid wind tunnel effects.

The city of Kokkola is not taking part in the ASTRA follow-up project BaltCICA because, for now, all important decisions on climate change adaptation have been taken or are under discussion. Nevertheless, the city stays in close contact with the Geological Survey of Finland in order to be informed in a timely manner about the latest research results.

were then applied to the observed daily Central Park record from 1971 to 2000 to generate 48 time series of daily data.[3] This is a simplified approach to projections of extreme events, since it does not allow for possible changes in the patterns of climate variability through time. However, because changes in variability for most climate hazards are considered highly uncertain, the approach described provides an initial evaluation of how extreme events may change in the future. This level of information with appropriate caveats can assist long-term planners as they begin to prepare adaptation strategies to cope with future extreme events.

Despite their brief duration, extreme events can have large impacts on cities, so they are a critical component of climate change impact assessment. Table 3.6 indicates how the frequency of heat waves, cold events, intense precipitation, drought, and coastal flooding in the New York City region are projected to change in the coming decades. The average number of extreme events per year for the baseline period is shown, along with the central 67 percent of the range of the model-based projections.

The total number of hot days, defined as days with a maximum temperature over 32 or 38 °C, is expected to increase as the twenty-first century progresses. The frequency and duration of heat waves, defined as three or more consecutive days with maximum temperatures above 32 °C, are also expected to increase. In contrast, extreme cold events, defined as the number of days per year with minimum temperature below 0 °C, are expected to become rarer.

Although the percentage increase in annual precipitation is expected to be relatively small in New York, larger percentage increases are expected in the frequency, intensity, and duration of extreme precipitation (defined as more than 25 mm, 50 mm, and 100 mm per day). This projection is consistent both with theory – a warmer atmosphere is expected to hold more moisture, and while evaporation is a gradual process, precipitation tends to be concentrated in extreme events – and observed trends nationally over the twentieth century (Karl and Knight, 1998).

Due to higher projected temperatures, twenty-first century drought projections reflect the competing influences of more total precipitation and more evaporation. By the end of the twenty-first century the effect of higher temperatures, especially during the warm months, on evaporation is expected to outweigh the increase in precipitation, leading to more droughts, although the timing and levels of drought projections are marked by relatively large uncertainty. The rapid increase in drought risk through time is reflective of a non-linear response, because as temperature increases in summer become large, potential evaporation increases dramatically. Because the New York Metropolitan Region has experienced severe multi-year

---

3 Because they are rare, the drought and coastal storm projections were based on longer time periods.

Table 3.6: *Quantitative changes in extreme events.*

Extreme event	Baseline (1971–2000)	2020s	2050s	2080s
Heat waves and cold events				
Number of days/year with maximum temperature exceeding:				
~32 °C	14	23 to 29	29 to 45	37 to 64
~38 °C	0.4[a]	0.6 to 1	1 to 4	2 to 9
Number of heat waves/year[b]	2	3 to 4	4 to 6	5 to 8
Average duration (in days)	4	4 to 5	5	5 to 7
Number of days/year with minimum temperature at or below 0 °C	72	53 to 61	45 to 54	36 to 49
Intense precipitation and droughts				
Number of days per year with rainfall exceeding ~25 mm	13	13 to 14	13 to 15	14 to 16
Drought to occur, on average[c]	~once every 100 years	~once every 100 years	~once every 50 to 100 years	~once every 8 to 100 years
Coastal floods and storms[d]				
1-in-10 year flood to reoccur, on average	~once every 10 years	~once every 8 to 10 years	~once every 3 to 6 years	~once every 1 to 3 years
Flood heights (m) associated with 1-in-10 year flood	1.9	2.0 to 2.1	2.1 to 2.2	2.3 to 2.5
1-in-100 year flood to reoccur, on average	~once every 100 years	~once every 65 to 80 years	~once every 35 to 55 years	~once every 15 to 35 years
Flood heights (m) associated with 1-in-100 year flood	2.6	2.7 to 2.8	2.8 to 2.9	2.9 to 3.2

The central range (middle 67 percent of values from model-based probabilities) across the GCMs and greenhouse gas emissions scenarios is shown.

[a] Decimal places shown for values less than 1 (and for all flood heights).

[b] Defined as three or more consecutive days with maximum temperature exceeding ~32 °C.

[c] Based on minima of the Palmer Drought Severity Index (PDSI) over any 12 consecutive months.

[d] Does not include the rapid ice-melt scenario.

droughts during the twentieth century – most notably the 1960s "drought of record" – any increase in drought frequency, intensity, or duration could have serious implications for water resources in the region. Changes in the distribution of precipitation throughout the year, and timing of snow-melt, could potentially make drought more frequent as well. According to the IPCC, snow season length is very likely to decrease over North America (IPCC, 2007).

As sea level rises, coastal flooding associated with storms will very likely increase in intensity, frequency, and duration. The changes in coastal floods shown here are solely due to the IPCC model-based projections of gradual changes in sea level through time. Any increase in the frequency or intensity of storms themselves would result in even more frequent future flood occurrences. By the end of the twenty-first century, projections based on sea level rise alone suggest that coastal flood levels that currently occur on average once per decade may occur once every one-to-three years (see Table 3.6).

The projections for flooding associated with more severe storms (e.g., the 1-in-100 year storm) are less well characterized than those for less severe storms (e.g., the 1-in-10 year events), for multiple reasons. The historical record is not sufficiently long to allow precise estimates of the flood level associated with the once per century storm. Furthermore, the storm risk may vary on multi-decadal to centennial ocean circulation-driven timescales that are currently not well understood. Keeping these uncertainties in mind, we estimate that, due to sea level rise alone, the 1-in-100 year flood may occur approximately four times as often by the 2080s.

Tropical cyclones

One extreme climate event that impacts many cities around the globe is tropical cyclones. Of the cities used in the chapter, several are at risk of being impacted by these storms, which bring heavy rainfall, high winds, and coastal storm surge. Kingston is one city at risk of tropical cyclones and, in recent years, several storms have affected the island of Jamaica. These storms

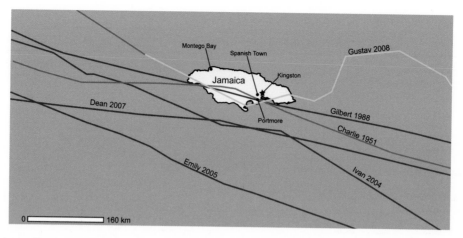

Figure 3.19: *Hurricane tracks near Kingston, Jamaica.*

Source: Adapted from NOAA.

include Ivan (2004), Emily (2005), Dean (2007), and Gustav (2008). Figure 3.19 shows the tracks of these storms along with two others, Charlie (1951) and Gilbert (1988).

Perhaps the most devastating storm to hit the island was Hurricane Gilbert in 1988, which had winds over 50 m/s as it passed over the island. Heavy rainfall and a storm surge close to 3 m

caused 45 deaths and over US$2 billion in damage (Lawrence and Gross, 1989).

There is much uncertainty as to how the frequency and strength of tropical cyclones will change with global climate change. Patterns of natural climate variability, including El Niño–Southern Oscillation (ENSO) and Atlantic Multidecadal

---

[VULNERABILITY]   Box 3.5   Lessons from a major climate event: Hurricane Katrina and New Orleans

Alexander S. Kolker

Louisiana Universities Marine Consortium, Chauvin, LA

Douglas J. Meffert

Center for Bioenvironmental Research, Tulane and Xavier Universities, New Orleans LA

Armando Carbonell

Department of Planning and Urban Form, Lincoln Institute of Land Policy, Cambridge, MA

Stephen A. Nelson

Department of Earth and Environmental Sciences, Tulane University, New Orleans, LA

Few cities in developed countries have felt the impacts of climatic processes as has New Orleans. On the morning of August 29, 2005, Hurricane Katrina made landfall in south Louisiana and again in Mississippi. It produced storm surges that ruptured levees on drainage and navigation canals that catastrophically flooded the City of New Orleans. Rapidly rising waters were met by an inadequate governmental response. In the process, well over 1400 people died, and many more lives were forever changed.

Katrina was a massive storm, fueled by very warm waters in the Gulf of Mexico; however, uncertainty exists as to whether

Katrina's strength was exacerbated by climate change. The northern Gulf of Mexico is historically hurricane prone, and has experienced large storms throughout the past 3000 years (Liu and Fearn, 2000). The power of Atlantic hurricanes appears to have increased over the past three decades (Emanuel, 2005), though this view is not accepted by all experts (Trenberth and Fasullo, 2008). However, climate models generally predict that global warming will increase the severity and/or intensity of tropical cyclones (IPCC, 2007), and Hurricane Katrina provides important lessons into how cities choose to adapt to and mitigate future climate change.

One lesson is that a city's physical landscape strongly affects its response to climate events. New Orleans developed along a series of natural levees at the banks of the Mississippi River and its former distributaries and the city's geography can be traced to alluvial processes. Rivers deposit their largest and heaviest particles closest to the main channel, and these natural levees were the high stable grounds where the city was originally settled (Coleman *et al.*, 1998; Coleman and Prior, 1980; Gould, 1970). As the city expanded in the late 19th and 20th centuries, lower-lying areas comprised of muddy, organic-rich sediments were developed. These areas were drained, and once dried out, subsided rapidly (Rogers et al., 2008; Nelson and LeClair, 2006; Kolb and Saucier, 1992). In the present city of New Orleans elevation, subsidence rates and depth of flooding following Katrina are roughly proportional to age, with younger regions typically being the lower, wetter, and more rapidly subsiding (Dixon *et al.*, 2006; Seed *et al.*, 2008; Russell, 2005).

New Orleans's geology also has important implications for the stability of the city's levees and its vulnerability to sea level rise. High subsidence rates in the delta plain contributed to one of the highest rates of relative sea level rise on Earth (Tornqvist *et al.*, 2008, Reed, 2002; Day *et al.*, 2007), which makes New Orleans a data-rich model for examining how cities respond to climatically driven sea-level rise. Subsidence lowered elevation of many of the city's levees while and peat and sand bodies below them allowed for subsurface water flow that undermined their stability (Rogers *et al.*, 2008, Seed *et al.*, 2008). Levee instability was also caused by engineering failures that are discussed below. Human activities have exacerbated wetland loss in the Mississippi Delta plain, at rates that now stand near 50 km^2 yr^{-1} (Day *et al.*, 2007; Barras *et al.*, 2003; Morton, Benier and Barras, 2006). A simple linear relationship between wetland area and storm surge does not exist, as surge magnitude is affected by factors including the storm's size, wind speeds, track and the shape of the continental shelf (Chen *et al.*, 2008). However, wetland loss has allowed storms surges to propagate further inland, increasing the city's vulnerable to storms over time.

The second lesson from Katrina is simply that the climate system is capable of delivering vast quantities of energy. The maximum wind speed of Hurricane Katrina exceeded 77 m s^{-1} (> 175 mph), the maximum eye wall radius reached 110 km and tropical storm force winds extended 370 km from the storm's center (McTaggart-Cowan *et al.*, 2007). Storm surges that reached 10.4 m in Biloxi, MS, and ranged between 5.6 – 6.9 m at Shell Beach east of New Orleans, and 3.95 – 4.75 m along the shore of Lake Pontchartrain at the northern edge of the city (Fritz *et al.*, 2008). Despite the vast power of Hurricane Katrina, it is important to recognize that the storm did not make landfall at New Orleans. This occurred at Buras, LA and again near Gulfport MS, which are 83 and 68 km from New Orleans, which was on the west, less intense side of the storm (Seed *et al.*, 2008).

These two lessons lead to a third, that climate disasters are often the product of an interaction between natural processes and human actions. A smaller storm would have produced a smaller storm surge and less pressure on the levees while properly constructed and maintained levees should have been able to withstand many of Katrina's surges in New Orleans. The catastrophe in New Orleans was exacerbated by inadequate local and federal governmental actions that include a poor assessment of the risks of flooding, an inadequate communication of these risks, poor levee design, poor levee maintenance, and an inadequate ability to evacuate people and provide for them in times of need (Seed *et al.*, 2008).

Looking to the future, New Orleans faces threats and opportunities. Global sea levels are predicted to rise, storms may increase in severity and frequency, lands surrounding New Orleans will continue to subside and wetland loss is likely remain problematic. While coastal restoration is needed on a massive scale, New Orleans finds itself hampered by jurisdictional difficulties: many coastal restoration decisions are made by a range of state, federal and local authorities that sometimes must balance restoration against other economic or environmental concerns (Carbonell and Meffert, 2009). While it may be impossible to restore the entire Mississippi Delta plain, sediment loads are high enough to substantially contribute to coastal restoration if managed wisely. Coastal progradation and wetland accretion will likely buffer storm surge and provide opportunities for carbon sequestration. Regional groups are also promoting a "multiple lines of defense" strategy (Lopez, 2006) that views flood protection as an integrated system of natural and man-made components, including barrier islands and beach ridges, wetlands, levees and evacuation plans. New Orleans was settled on high, stable grounds at the mouth of the continent's largest river and this strategic location made it desirable for nations looking to establish a claim to the continent's interior. Such strategic thinking relating a city to its broader environmental assets and liabilities is key to the future of this and perhaps other coastal cities.

Oscillation (AMO), have documented relationships with tropical cyclone activity. For example, when the AMO, a mode of natural variability of sea surface temperatures in the North Atlantic Ocean is in the cold phase, hurricane activity increases (Goldenberg *et al.*, 2001).

As far as changes in hurricane strength and frequency due to anthropogenic climate change go, there is no concrete evidence that global warming is having an influence. Although some scientific studies suggest that warming caused by increased greenhouse gases will increase hurricane intensity (Emanuel, 2005), the connection between the two is not conclusive. There are a number of issues that play into this debate. Sea surface temperatures and upper ocean heat content are very likely to increase in the North Atlantic's main hurricane development regions, and this increase alone will likely favor more intense hurricanes (Emanuel, 2005). However, changes in other key factors (not all of which are mutually exclusive) that influence hurricane number and intensity are more uncertain. These factors include: (1) vertical wind shear, which is dependent on uncertain changes in a range of factors (Vecchi and Soden, 2007) including the ENSO cycle (Gray, 1984; Mann *et al.*, 2009); (2) vertical temperature gradients in the atmosphere (Emanuel, 2007); (3) Saharan dust (Dunion and Velden, 2004; Evan *et al.*, 2006); (4) easterly waves and the West African monsoon (Gray, 1979; Donnelly and Woodruff, 2007); and (5) steering currents, which are influenced by a range of factors including highly uncertain changes in the NAO (Mann *et al.*, 2009).

3.6 Conclusions and key research questions

Climate change is expected to bring warmer temperatures to virtually the entire globe, including all 12 cities analyzed here. Heat events are projected to increase in frequency, severity and duration. Total annual precipitation is expected to increase in some cities, especially in the mid/high latitudes and tropics, and decrease in other cities, especially in the subtropics. Most cities are expected to experience an increase in the percentage of their precipitation in the form of intense rainfall events. In many cities, droughts are expected to become more frequent, more severe, and of longer duration. Additionally, rising sea levels are extremely likely in all the coastal cities, and are likely to lead to more frequent and damaging flooding related to coastal storm events in the future.

Climate change impacts on cities are enhanced by factors including high population density, extensive infrastructure, and degraded natural environments. Vulnerabilities will be great in many regions that currently experience frequent climate hazards, such as low-lying areas already exposed to frequent flooding.

Vulnerabilities will also be large among resource-poor populations, especially in developing countries, where infrastructure may be sub-standard or non-existent, governmental response to disasters may be limited, and adaptation options may be few for reasons including limited capital.

One implication from this chapter is the need for more and improved climate data, especially in cities of developing countries. In many cities, the historical record is either too short or the quality too uncertain to support trend analysis and climate change attribution. Without long historical records, the role of climate variability cannot be adequately described, and climate change projections will not have as strong a historical footing. However, even in cities that have a high quality, lengthy record of temperature and precipitation, there is a need for additional

---

[ADAPTATION] Box 3.6 Mexico City's Virtual Center on Climate Change

Cecilia Conde, Benjamin Martinez and Francisco Estrada

Centro de Ciencias de la Atmosfera, UNAM

Mexico City's Virtual Center on Climate Change (CVCCCM: www.cycccm-atmosfera.unam.mx/cvccccm/) was created in 2008, with the objectives of: (1) building an entity that concentrates and organizes the information regarding climate change effects on Mexico City, as well as coordinating research efforts on the subject; (2) supporting the continuous development of public policies that aim to increase adaptive capacity and reduce vulnerability of different social sectors; (3) creating an Adaptation, Vulnerability and Mitigation Policy Framework for Mexico City.

The Center aspires to support the development of "useful" science that must answer the questions concerning climate variability and change posed by policymakers of the city.

City authorities have had to deal with extreme events related to the urban heat island, such as heat waves, heavy rains and the resulting floods, and reduced water availability associated with severe droughts in the catchment basin from which the city satisfies part of its water demand.

Generally, it could be said that cities now experience what could occur under projected climate change scenarios. For example, systems outside of Mexico City might be resilient to an increase of 1 °C in mean temperatures, but in the city the warming process has already reached more than 3 °C (Jauregui, 1997).

As a consequence of the uncontrolled growth of the city's population and urbanization area, the increase in the occurrence of heavy rains since the 1960s has led to an increase in the frequency of severe floods. This provides an example of Mexico City's high sensitivity to climate change.

The Virtual Center must be seen as necessary to face the climatic problems that could increase in the future.

The research priorities selected by the City Government are:
1. Assessment of Mexico City air quality, the effect on the health of people exposed to allergenic bio-particles (pollen), and its relation to climate change
2. Effect of the interaction of temperature and ozone on Mexico City's hospital admissions
3. The impact of climate change on water availability in the Metropolitan Area of Mexico City
4. Vulnerability of potable water sources in Mexico City in the context of climate change
5. Energy consumption scenarios and emission of greenhouse gases produced by the transport sector in the Metropolitan Area of Mexico City
6. Assessment of the impacts in the Metropolitan Area of Mexico City related to solid waste under climate change conditions
7. Vulnerability of the ground of conservation of Mexico City to climate change and possible adaptation measures
8. Determination of the vulnerability of the conservation areas of Mexico City to climate change and possible adaptation measures

Of course, several problems have been detected during the development of this Virtual Center. In particular, "traditional" science has not yet been capable of achieving interdisciplinary research and being stakeholder driven. Currently, researchers are devoted to publishing papers instead of "translating" information for decision-making. Most of the policymakers' decisions are more focused on resolving the immediate problems than on designing long-term strategies.

The actors of this Virtual Center are aware of these barriers, and these problems are being confronted, for example, through periodic meetings with diverse authorities of Mexico City, which allow direct answers to their questions. This motivates a rich discussion from different points of view, resulting in joint strategies that facilitate the incorporation of the found solutions into public policies. It is expected that this critical stage of the research could be fulfilled during 2010, when partial results will be presented to Mexico City's authorities.

station data and climate variables at high temporal resolution to improve our understanding of the microclimates that help define the urban setting and climate risk. If these data can be integrated into monitoring systems and real time networks – which can be an expensive proposition for some cities – weather forecasting can be improved as well.

Furthermore, improved monitoring can help bridge the gaps between weather and climate, and climate regions and cities. This active area of research will enable analysts to use real-time ocean temperatures to better discern how the frequency of extreme events (including hurricanes) may vary by decade in the future. As predictions improve, adaptation strategies can be tailored to reflect these advances. Monitoring must include impact variables identified by stakeholders, such as water reservoir levels, frequency of power failures, and transportation delays. It is critical that this information not only be collected but also stored systematically in a unified database that facilitates the sharing of information and research results across agencies and cities.

Given the range of climate hazards and impacts described here, there is a critical need for climate change adaptation strategies that align with societal goals such as development, environmental protection (including greenhouse gas mitigation), and social justice and equity.

REFERENCES

Arrhenius, S. (1896). On the influence of carbonic acid in the air upon the temperature of the ground. *London, Edinburgh, and Dublin Philosophical Magazine and Journal of Science* (fifth series), **41**, 237–275.

Australian Bureau of Meteorology (2009). Climate information. www.bom.gov.au/climate/.

Balling, R. C. and S. W. Brazel (1987). Diurnal variations in Arizona monsoon precipitation. *Monthly Weather Review*, **115**, 342–346.

Barras, J. *et al.* (2003). *Historical and Projected Coastal Louisiana Land Changes*: 1978–2050. USGS.

Bornstein, R. and M. LeRoy (1990). Urban barrier effects on convective and frontal thunderstorms. *Fourth Conference on Mesoscale Processes*, American Meteorological Society, 2.

Burian, S. J. and J. M. Shepherd (2005). Effect of urbanization on the diurnal rainfall pattern in Houston. *Hydrological Processes*, **19**, 1089–1103.

Carbonell, A. and Meffert, D. J. (2009). *Climate Change and the Resilience of New Orleans: the Adaptation of Deltaic Urban Form*. Commissioned Research Report for the World Bank 2009 Urban Research Symposium, Marseilles, France.

Census of Canada (2006). Statistics Canada, accessed 2010, www.statcan.gc.ca/.

Changnon, S. A. (1968). The La Port weather anomaly – fact or fiction? *Bulletin of the American Meteorological Society*, **49**, 4–11.

Changnon, S. A. and N. E. Westcott (2002). Heavy rainstorms in Chicago: Increasing frequency altered impacts, and future implications. *Journal of American Water Resources*, **48**, 1467–1475.

Charlson, R. J., S. E. Schwartz, J. M. Hales, *et al.* (1992). Climate forcing by anthropogenic aerosols. *Science*, **255**, 423–430.

Chen, T. C., S. Y. Wang, and M. C. Yen (2007). Enhancement of afternoon thunderstorm activity by urbanization in a valley: Tapei. *Journal of Applied Meteorology and Climatology*, **46**, 1324–1340.

Chen, Q., Wang, L. and Tawes, R. (2008). Hydrodynamic response of northeastern Gulf of Mexico to Hurricanes. *Estuaries and Coasts* **31**, 1098–1116.

Christensen, J.H., *et al.* (2007) Regional Climate Projection. *Climate Change 2007: The Physical Science Basis. Contribution of Working Group I to the Fourth Assessment Report of the Intergovernmental Panel on Climate Change*, eds. S. Solomon, *et al.* Cambridge University Press, 849–940.

Coleman, J. M., Roberts, H. H. and Stone, G. W. (1998). The Mississippi Delta: An overview. *Journal of Coastal Research* **14**, 698–716.

Coleman, J. M. and Prior, D. B. (1980). Deltaic Sand Bodies. 171.

Collins, M. (2005). El Nino- or La Nina-like climate change? *Climate Dynamics*, **24**, 89–104.

Cunha, L. R. and M. Miller Santos (1993). The Rio reconstruction project: the first two years. In *Towards A Sustainable Urban Environment: The Rio de Janeiro Study*, World Bank Discussion Paper 195, Washington, DC, USA: World Bank.

Day, J. W. *et al.* (2007). Restoration of the Mississippi Delta: Lessons from Hurricanes Katrina and Rita. *Science,* **315**, 1679–1684.

Dixon, T. H. *et al.* (2006). Subsidence and flooding in New Orleans. *Nature* **441**, 587–588.

Dilley, M. (2000). Reducing vulnerability to climate variability in Southern Africa: the growing role of climate information. *Climatic Change*, **45**, 63–73.

Donnelly, J. P. and J. D. Woodruff (2007). Intense hurricane activity over the past 5,000 years controlled by El Nino and the West African monsoon. *Nature*, **447**, 465–468.

Dunion, J. P. and C. S. Velden (2004). The impact of the Saharan air layer on Atlantic tropical cyclone activity. *Bulletin of the American Meteorological Society*, **85**, 353–365.

Emanuel, K. (2005). Increasing destructiveness of tropical cyclones over the past 30 years. *Nature*, **436**, 686–688.

Emanuel, K. (2007). Environmental factors affecting tropical cyclone power dissipation. *Journal of Climate*, **20**, 5497–5509.

Emanuel, K. (2008). Hurricanes and global warming: results from downscaling IPCC AR4 simulations. *Bulletin of the American Meteorological Society*, **89**, 347–367.

Environment Canada (2009). Canada's National Climate and Weather Data Archive. www.climate.weatheroffice.gc.ca.

Eri, S., N. Aishali, and L. W. Horowitz (2009). Present and potential future contributions of sulfate, black and organic carbon aerosols from China to global air quality, premature mortality and radiative. *Atmospheric Environment*, **43**, 2814–2822.

Evan, A. T., J. P. Dunion, J. A. Foley, A. K. Heidinger, and C. S. Velden (2006). New evidence for a relationship between Atlantic tropical cyclone activity and African dust outbreaks. *Geophysical Research Letters*, **33**, L19813, doi:19810.11029/12006GL026408.

Fairbanks, R. G. (1989). 17,000-year glacio-eustatic sea level record: influence of glacial melting rates on the Younger Dryas event and deep-ocean circulation. *Nature*, **342**, 637–642.

Fernandes, E. (2000). The legalisation of favelas in Brazil: problems and prospects. *Third World Planning Review,* **22**(2), 167–188.

Fritz, H. M. *et al.* (2008). Hurricane Katrina storm surge reconnaissance. *Journal of Geotechnical and Geoenvironmental Engineering* **134**, 644–656.

Gaffin, S. R., C. Rosenzweig, R. Khanbilvardi, *et al.* (2008). Variations in New York City's urban heat island strength over time and space. *Theoretical and Applied Climatology*, **94**, 1–11.

Gao, X., L. Yong, and W. Lin (2003). Simulation of effects of land use change by a regional climate model. *Advances in Atmospheric Sciences*, **20**, 583–592.

Goldenberg, S. B., C. Landsea, A. M. Mestas-Nunez, and W. M. Gray (2001). The recent increase in Atlantic hurricane activity: causes and implications. *Science*, **293**, 474–479.

Gould, H. R. (1970), in *Deltaic Sedimentation* Vol. Special Publication 15, ed. J. P. Morgan. Society of Economic Paleontologists and Mineralogists.

Gray, W. M. (1979). Hurricanes: their formation, structure, and likely role in the tropical circulation. In D. B. Shaw (Ed.), *Meteorology Over the Tropical Oceans*, London, UK: Royal Meteorological Society, pp. 155–218.

Gray, W. M. (1984). Atlantic seasonal hurricane frequency. Part I: El Nino and 30 mb quasi-biennial oscillation influences. *Monthly Weather Review*, **112**, 1649–1668.

Grimm, A. M. (2000). Climate variability in southern South America associated with El Nino and La Nina events. *Journal of Climate*, **13**, 35–58.

Grimm, A. M. (2003). The El Nino impact on the summer monsoon in Brazil: regional processes versus remote influences. *Journal of Climate*, **16**, 263–280.

Grimm, A. (2004). How do La Niña events disturb the summer monsoon system in Brazil? *Climate Dynamics*, **22**, 123–138

Grimm, A. M. and R. G. Tedeschi (2009). ENSO and extreme rainfall events in South America. *Journal of Climate*, **22**, 1589–1609.

Gupta, P., S. A. Christopher, J. Wang, *et al.* (2006). Satellite remote sensing of particulate matter and air quality assessment over global cities. *Atmospheric Environment*, **40**, 5880–5892.

Gusmão, P. P., P. Serrano do Carmo, and S. B. Vianna (2008). *Rio Proximos 100 Anos*. Rio De Janiero, Brazil: Instituto Municipal de Urbanismo Pereira Passos.

Guttman, N. B. (1989). Statistical descriptors of climate. *Bulletin of the American Meteorological Society*, **70**, 602–607.

Hachadoorian, L., S. R. Gaffin, and R. Engelman (2011). Mapping the population future: projecting a gridded population of the World using ratio methods of trend extrapolation. In R. P. Cincotta and L. J. Gorenflo (Eds.), *Human Population: The Geography of* Homo Sapiens *and its Influence on Biological Diversity*, Heidelberg,Germany: Springer-Verlag.

Harshvardhan, S., E. Schwartz, C. M. Benkovitz, and G. Guo (2002). Aerosol influence on cloud microphysics examined by satellite measurements and chemical transport modeling. *Journal of the Atmospheric Sciences*, **59**, 714–725.

Hoerling, M. P., A. Kumar, and M. Zhong (1997). El Nino, La Nina and the nonlinearity of their teleconnections. *Journal of Climate*, **10**, 1769–1786.

Horton, R. E. (1921). Thunderstorm breeding spots. *Monthly Weather Review*, **49**, 193.

Horton, R. (2007). *An Observational and Modeling Study of the Regional Impacts of Climate Variability*, Columbia University.

Horton, R., C. Herweijer, C. Rosenzweig, *et al.* (2008). Sea level rise projects for the current generations CGCMs based on the semi-empirical method. *Geophysical Research Letters*, **35**.

Horton, R., V. Gornitz, M. Bowman, and R. Blake (2010). *Climate observations and projections*. Ann. New York Acad. Sci., 1196, 41–62, doi:10.1111/j.1749–6632.2009.05314.x.

Horton, R., and C. Rosenzweig (2010). *Climate Risk Information*. Ann. New York Acad. Sci., 1196, 147–228, doi:10.1111/j.1749–6632.2010.05323.x.

Howard, L (1820). *The Climate of London, Deduced from Meteorological Observations, Made at Different Places in the Neighbourhood of the Metropolis. 2*, London, 1818–1820.

Huff, F. A. and S. A. Changnon (1972a). Climatological assessment of urban effects on precipitation at St. Louis. *Journal of Applied Meteorology*, **11**, 823–842.

Huff, F. A. and S. A. Changnon (1972b) *Climatological Assessment of Urban Effects on Precipitation St. Louis: Part II. Final Report*, NSF Grant GA-18781, Illinois State Water Survey.

Hurrell, J. W. (1995). Decadal trends in the North Atlantic Oscillation and relationships to regional temperature and precipitation. *Science*, **269**, 676–679.

Hurrel, J. W., Y. Kushnir, G. Ottersen, and M. Visbeck (2003). An overview of the North Atlantic Oscillation. In J. W. Hurrell *et al.* (Eds.), *The North Atlantic Oscillation: Climatic Significance and Environmental Impact*, Geophysical Monograph 134, Washington, DC, USA: American Geophysical Union, pp. 1–35.

Ichinose, T. and Y. Bai (2000). Anthropogenic heat emission in Shanghai City. In *Proceedings of Annual Meeting of Environmental Systems Research*, Vol. 28, pp. 329–337.

IPCC (2007). *Climate Change 2007: The Physical Science Basis. Contribution of Working Group I to the Fourth Assessment Report of the Intergovernmental Panel on Climate Change*, Cambridge, UK: Cambridge University Press.

Jáuregui, E. (1997). Heat island development in Mexico City. *Atmospheric Environment*, **31**(22), 3821–3831.

Jauregui, E. and E. Romales (1996). Urban effects on convective precipitation in Mexico City. *Atmospheric Environment*, **30**, 3383–3389.

Jiong, S. (2004). Shanghai's land use pattern, temperature, relative humidity and precipitation. *Atlas of Shanghai Urban Physical Geography*.

Karl, T. R. and R. W. Knight (1998). Secular trends of precipitation amount, frequency and intensity in the United States. *Bulletin of the American Meteorological Society*, **79**, 231–241.

Kaufmann, R. K., K. C. Seto, A. Schneider, *et al.* (2007). Climate response to rapid urban growth: evidence of a human-induced precipitation deficit. *Journal of Climate*, **20**, 2299–2306.

Kolb, C. R. and Saucier, R. T. (1982). Engineering geology of New Orleans. *Review of Engineering Geology* **5**, 75–93.

Kothawale, D. R. and K. Rupa Kumar (2005). On the recent changes in surface temperature trends over India. *Geophysical Research Letters*, **32**.

Kratzer, P. (1937). *Das stadtklima*, Braunschweig: F.Viewg uE Sohne.

Kratzer, P. (1956). *Das stadtklima* (2nd edition), Braunschweig: F Vieweg uE Sohn (tranlated by the U.S. Air Force, Cambridge Research Laboratories).

Landsberg, H. (1956). The climate of towns. In W. L. Thomas (Ed.), *Man's Role in Changing the Face of the Earth*, Chicago, IL, USA: The University of Chicago Press.

Landsberg, H. E. (1970). Man-made climate changes. *Science*, **170**, 1265–1274

Lawrence, M. B. and J. M. Gross (1989). Annual summaries: Atlantic hurricane season of 1988. *Monthly Weather Review*, **117**, 2248–2459.

Lian, L. and J. Shu (2007). Numerical simulation of summer climate over center and east China using a regional climate model. *Journal of Tropical Meteorology*, **23**, 162–169.

Liu, K. B. and Fearn, M. L. (2000). Reconstruction of prehistoric landfall frequencies of catastrophic hurricanes in northwestern Florida from lake sediment records. *Quaternary Research* **54**, 238–245.

Lopez, J. A. (2007). *The Multiple Lines of Defense Strategy to Sustain Coastal Louisiana*. Lake Pontchartrain Basin Foundation, New Orleans.

Mann, M. E., J. D. Woodruff, J. P. Donnelly, and Z. Zhang (2009). Atlantic hurricanes and climate over the past 1,500 years. *Nature*, **460**, 880–883.

Marengo, J. A. and C. C. Camargo (2008). Surface air temperature trends in Southern Brazil for 1960–2002. *International Journal of Climatology*, **28**, 893–904.

McTaggart-Cowan, R., Bosart, L. F., Gyakum, J. R. and Altallah, E. H. (2007). Hurricane Katrina (2005) Part 1: Complex life cycle of an intense tropical cyclone. *Monthly Weather Review* **135**, 3905–3926.

Morton, R. A., Benier, J. C. and Barras, J. A. (2006). Evidence of regional subsidence and associated interior wetland loss induced by hydrocarbon production, Gulf Coast region, USA. *Environmental Geology* **50**, 261–274.

Nakicenovic, N. and Coauthors (2000). *Special Report on Emissions Scenarios: A Special Report of Working Group III of the Intergovernmental Panel on Climate Change*, Cambridge, UK: Cambridge University Press.

Nastos, P. T. and C. S. Zerefos (2007). On extreme daily precipitation totals at Athens, Greece. *Advances in Geosciences*, **10**, 59–66.

National Climatic Data Center, Global Historical Climatology Network: Version 2 (2008). www.ncdc.noaa.gov/oa/climate/ghcn-monthly/index. php.

Nelson, S. A. and LeClair, S. F. (2006). Katrina's unique splay deposits in a New Orleans neighborhood. *GSA Today* **16**, 4–10, doi: 10.1130/GSAT01609A.1.

Ohashi, Y., Y. Genchi, H. Kondo, *et al.* (2007). Influence of air-conditioning waste heat on air temperature in Tokyo during summer: numerical experiments using an urban canopy model coupled with a building energy model. *Journal of Applied Meteorology and Climatology*, **46**, 66–81.

Oke, T. R. (1987). *Boundary Layer Climates* (2nd edition), London, UK: Routledge.

Orr, J. C., Fabry, V. J., Aumont, O., *et al.* (2005). Anthropogenic ocean acidification over the twenty-first century and its impact on calcifying organisms. *Nature*, **437**, 681–686.

Parson, E., V. Burkett, K. Fisher-Vanden, *et al.* (2007). Global change scenarios: their development and use. Sub-report 2.1B of *Synthesis and Assessment Product 2.1 by the US Climate Change Science Program and the Subcommittee on Global Change Research*, Washington, DC, USA: Department of Energy, Office of Biological and Environmental Research, p. 106.

Pawan, G., C. A. Sundar, and W. Jun (2006). Satellite remote sensing of particulate matter and air quality assessment over global cities. *Atmospheric Environment*, **40**, 5880–5892.

Pielke, R. A. (2005). Land use and climate change. *Science*, **310**, 1625–1626.

Reed, D. J. (2002). Sea-level rise and coastal marsh sustainability: geological and ecological factors in the Mississippi delta plain. *Geomorphology* **48**, 233–243.

Ren, G. Y., Z. Y. Chu, Z. H. Chen, and Y. Y. Ren (2007). Implications of temporal change in urban heat island intensity observed at Beijing and Wuhan stations, *Geophysical Research Letters*, **34**, L05711, doi:10.1029/2006GL027927.

Rignot, E. and P. Kanagaratnam (2006). Changes in the velocity structure of the Greenland ice sheet. *Science*, **311**, 986–990.

Rogers, J. D. *et al.* (2008). Geological Conditions Underlying the 2005 17th Street Canal Levee Failure in New Orleans. *Journal of Geotechnical and Geoenvironmental Engineers* **134**, 583–601.

Ropelewski, C. (1999). The Great El Nino of 1997 and 1998: impacts on precipitation and temperature. *Consequences*, **5**, 17–25.

Rosenzweig, C., W. D. Solecki, L. Parshall, and S. Hodges (Eds.) (2006). *Mitigating New York City's Heat Island with Urban Forestry, Living Roofs, and Light Surfaces*, New York City Regional Heat Island Initiative, Final Report 06–06, New York State Energy Research and Development Authority.

Russel, G. (2005) in *The Times Picayune* Vol. 169 A1 (New Orleans, 2005).

Sanderson, E. and M. Boyer (2009). *Mannahatta: A Natural History of New York City*, New York, USA: Abrams Books.

Satterthwaite, D. (2008). Cities' contribution to global warming: notes on the allocation of greenhouse gas emissions. *Journal of Environment and Urbanization*, **20**, 539–549.

Seed, R. B. *et al.* (2008). New Orleans and Hurricane Katrina 1: Introduction, overview, and the east flank. *Journal of Geotechnical and Geoenvironmental Engineering* **135**, 701–739.

Selover, N. (1997). Precipitation patterns around an urban desert environment: topographic or urban influences? *Association of American Geographers Convention*, **2**.

Shepherd, J. M. (2005). A review of current investigations of urban-induced rainfall and recommendations for the future. *Earth Interactions*, **9**, 1–27.

Shepherd, J. M. (2006). Evidence of urban-induced precipitation variability in arid cliamte regimes. *Journal of Arid Environments*, **67**, 607–628.

Shepherd, J. M., H. Pierce, and A. J. Negri (2002). Rainfall modification by major urban areas: observations from spaceborne rain radar on the TRMM satellite. *Journal of Applied Meteorology*, **41**, 689–701.

Shu, J., J. A. Dearing, A. P. Morse, L. Yu, and C. Li (2000). Magnetic properties of daily sampled total suspended particulates in Shanghai. *Environment Science and Technology*, **34**, 2393–2400.

Shu, J., J. A. Dearing, A. P. Morse, L. Yu, and N. Yuan (2001). Determining the sources of atmospheric particles in Shanghai, China, from magnetic and geochemical properties. *Atmospheric Environment*, **35**, 2615–2625.

Simpson, M. D. (2006). *Role of Urban Land Use on Mesoscale Circulations and Precipitation*, North Carolina State University.

Stout, G. E. (1962). Some observations of cloud initiation in industrial areas. In *Air Over Cities*, Technical Report A62–5. Washington, DC: US Public Health Service. Tayanc, M. and H. Toros (1997). Urbanization effects on regional climate change in the case of four large cities of Turkey. *Climate Change*, **35**, 501–524.

Tornqvist, T. E. *et al.* (2008). Mississippi Delta subsidence primarily caused by compaction of Holocene strata. *Nature Geoscience* **1**, 173–176.

Trenberth, K. (1984). Signal versus noise in the Southern Oscillation. *Monthly Weather Review* **112**, 326–332

Trenberth, K. E. and J. M. Caron (2000). The Southern Oscillation revisited: sea level pressures, surface temperatures and precipitation. *Journal of Climate*, **13**, 4358–4365.

Trenberth, K. E. and Fasullo, J. (2008). Energy budgets of Atlantic hurricanes and changes from 1970. *Geochemistry, Geophysics, Geosystems* **9**, doi:2007GC001847.

UK Met Office and Hadley Centre (2009). Observations datasets. http://hadobs.org/.

UNFPA (2007). *The State of World Population 2007*. United Nations Population Fund, United Nations Publications.

UN-HABITAT (2006). *State of the World's Cities*, London, UK: Earthscan.

United Nations Statistical Division Demographic Yearbook (2010). http://unstats.un.org/unsd/demographic/products/dyb/dyb2007.htm.

Vecchi, G. A. and B. J. Soden (2007). Increased tropical Atlantic wind shear in model projections of global warming. *Geophysical Research Letters*, **34**, L08702, doi:08710.01029/02006GL028905.

Velicogna, I. and J. Wahr (2006). Acceleration of Greenland ice mass loss in spring 2004. *Nature*, **443**, 329–331.

Visbeck, M. H., J. W. Hurrell, L. Polvani, and H. M. Cullen (2001). The North Atlantic Oscillation: past, present, and future. *Proceedings of the National Academy of Sciences*, **98**, 12,876–12,877.

Wilby, R. L., G. O'Hare, and N. Barnsley (1997). The North Atlantic Oscillation and British Isles climate variability. *Weather*, **52**, 266–276.

Yin, J., M. E. Schlesinger, and R. J. Stouffer (2009). Model projections of rapid sea-level rise on the northeast coast of the United States. *Nature Geoscience*, **2**, 262–266.

Zhang, X., L. A. Vincent, W. D. Hogg, and A. Niitsoo (2000). Temperature and precipitation trends in Canada during the 20th century. *Atmosphere-Ocean*, **38**, 395–429.

Zheng Yiqun, Qian Yongfu, Miao Manqian, *et al.* (2002). The effects of vegetation change on regional climate I: Simulation results. *Acta Meteorologica Sinica*, **60**, 1–16 (in Chinese).

Part III

Urban sectors

4

Climate change and urban energy systems

Coordinating Lead Author:

Stephen A. Hammer (New York City)

Lead Authors:

James Keirstead (London), Shobhakar Dhakal (Tsukuba), Jeanene Mitchell (Seattle), Michelle Colley (Montreal), Richenda Connell (Oxford), Richard Gonzalez (New York City), Morgan Herve-Mignucci (Paris), Lily Parshall (New York City), Niels Schulz (Vienna), and Michael Hyams (New York City)

This chapter should be cited as:

Hammer, S. A., J. Keirstead, S. Dhakal, J. Mitchell, M. Colley, R. Connell, R. Gonzalez, M. Herve-Mignucci, L. Parshall, N. Schulz, M. Hyams, 2011: Climate change and urban energy systems. *Climate Change and Cities: First Assessment Report of the Urban Climate Change Research Network*, C. Rosenzweig, W. D. Solecki, S. A. Hammer, S. Mehrotra, Eds., Cambridge University Press, Cambridge, UK, 85–111.

4.1 Introduction

The energy systems that provide the "life blood" to cities are as complex and diverse as cities themselves. Reflecting local natural resource and economic conditions, supply chains that may extend globally, historic investments in technology, and cultural and political preferences, urban energy systems serve as either a key accelerator or brake on the vitality and prospects of a city or urban region. Because of this, the local energy system can be of great interest to policymakers in a city, and many have begun to develop plans that seek to change one or more aspects of this system over the coming decades.

Climate change concerns are increasingly a key driver behind these changes, with local authorities seeking to reduce their city's current level of contribution to global climate change. Climate is not the only reason local authorities engage on energy issues, however. In some cases, cities are seeking to ameliorate pollution attributable to local energy use, while in other cities, economic development is a key concern. The latter is particularly prominent in developing countries, where a lack of access to adequate, reliable energy services continues to impede the economic growth of many cities (UNDP/ WHO, 2009). In these situations, climate-related concerns are often secondary to efforts to improve access to modern energy services to reduce poverty, allow for new types of economic activity, and improve public health.

Looking to the future, climate change may stress local energy systems in many different and profound ways. The level of impact will vary significantly based on the age of the system, the ease with which the underlying technology or fuels can be changed or made more climate resilient, and the nature and severity of the climate change-related impacts likely to occur in that city. The capital-intensive nature of energy technology, and the decades-long lifespan of much of the energy supply and distribution infrastructure serving most cities, compounds the challenge of addressing climate change in a comprehensive manner. To date, relatively few cities have systematically explored how or whether their energy system must change to adapt to new climatic conditions.

This chapter explores these issues, weaving in examples and data from a range of types and sizes of cities around the world. The chapter begins with a generalized discussion of how cities obtain and use energy and govern energy matters. Although this discussion is somewhat lengthy, it is important to understand energy system and market fundamentals because they are so relevant to the subsequent discussion about climate change stresses and current policymaking efforts. The chapter concludes with commentary on areas for future research and potential policy changes that can help cities improve their management of local energy systems in the coming decades.

4.2 The urban energy system: technology choices, market structure, and system governance

Because there are so many issues that relate to how a city obtains or uses energy – including land use and mobility policies and practices, waste management collection and disposal practices, and the type and level of local economic activity – urban energy systems can be defined in either broad or narrow terms. Issues such as transportation and land use are taken up in other chapters of this assessment report, as they are significant enough to warrant detailed attention. Other topics by necessity fall outside the scope of this report given limited time and resources. For the purposes of this chapter, however, the analysis will focus on electricity and thermal energy supply and distribution systems, as these link to the bulk of the energy used in most cities.

4.2.1 System overview

"Centralized" electricity systems are commonplace in cities, involving large power plants generating power which is then distributed to users through a complex web of high- and low-voltage wires crossing a city. Centralized generation takes advantage of the economies of scale offered by large power plants, which can be fueled by a variety of different sources, including coal, natural gas, biomass, solid waste, or nuclear fuels. Even large renewable energy systems, including large wind farms, geothermal power plants, or concentrating solar "power towers" can be sized at scales equivalent to "traditional" power plants, allowing them to fit relatively easy into the central generation and distribution model.

Power plants linked to this system can be located either within a city's borders or at locations quite remote from the urban core. Locations within cities have proven less desirable in many locales because of concerns over the emissions from these plants, creating public health concerns and dampening real estate values in adjacent neighborhoods (Farber, 1998; Abt Associates *et al.*, 2000). The advent of comprehensive state, national, or trans-national grids has allowed many cities to increasingly rely on out-of-city power sources, lessening the severity of the problem and the political challenges associated with siting new in-city power plants; although the siting of transmission lines has become problematic in many locations as well.

"Distributed" forms of power generation and distribution (also known as DG) refer to systems with much smaller power production units that are located at or near the point of energy use. DG systems enjoy certain advantages, such as the fact that because they link directly to the electric wiring system within the host building, they tend to suffer from less "transmission loss." These losses occur due to Joule heating of power lines and when electricity voltage levels are "stepped" up or down at different points in the transmission and distribution network (Lovins *et al.*, 2002). DG systems may also allow a building or user to avoid certain design or service deficiencies involving the city-wide distribution grid, such as poor power "quality" or vulnerability to blackouts or other types of service disruptions (Lovins *et al.*, 2002). Finally, DG systems may allow buildings to utilize certain types of power more easily or cheaply, such as electricity generated from renewable sources such as wind or solar power or technologies such as combined heat and power (CHP) units that enjoy high rates of energy efficiency.

Thermal energy use in cities – that is, energy used for space, water, or process heating or cooling – can also be produced in a centralized or decentralized manner. Centralized (or "district") thermal energy systems, tend to be more common in cities with extreme temperatures in winter or summer months. For cost or pollution reasons, local authorities and/or utilities in many cities have found there are benefits to producing steam or hot (or cold) water centrally, and then distributing this thermal energy to users via a network of underground pipes (see Table 4.1). Because of the cost of installing and maintaining the pipeline network, some minimum population density or level of demand is necessary to make these systems cost effective (Gochenour, 2001). District heating and cooling systems can be fueled by a range of energy sources, such as coal, natural gas, biomass, nuclear power, and geothermal sources. In some cases, the plants producing the thermal energy may operate as co-generation facilities, simultaneously producing electricity for use around the city.

The alternatives to district thermal systems are building-based thermal technologies such as gas-powered stoves, boilers, furnaces, or ground source heat pumps. Some buildings employ combined heat and power technology, satisfying some or all of the building's thermal and electricity needs.

Although decentralized in function, building-sited thermal systems may nonetheless involve linkages to citywide fuel networks delivering natural gas or coal gas around a city. Building-sited thermal systems may also rely on fuel oil or liquid petroleum gas tanks located within the building that are refilled on an as-needed basis. Other buildings or homes rely on supplies of solid energy feed stocks, such as coal, kerosene, charcoal, biomass, or animal dung, which are burned in a boiler or cookstove to produce space or process heat. These latter systems may involve some type of formal supply chain, or less formal scavenging processes involving the building owner or dwelling occupant. In developing countries, these supply chains can create important opportunities for local economic development (Clancy *et al.*, 2008).

The thermal systems employed may have significant health impacts within or near the home, because of differing levels of smoke or other pollutants emitted while operational. Households in cities in developing countries are far more reliant on solid fuels for cooking than urban dwellers in developed countries (UNDP/WHO, 2009; see Figure 4.1) This is testament to both differing levels of energy infrastructure in these cities, the price of the different fuels, and difficulties obtaining interconnections to formal distribution networks (Dhingra *et al.*, 2008, Fall *et al.*, 2008).

A corollary to the thermal system discussion is the fact that, in many cities around the world, there may be heavy reliance on electric heating and cooling systems. Electric air conditioners are well-known features in many homes and businesses in warmer climates, but – especially in cities with historically cheap electricity sources such as nuclear or hydropower – there were many decades during the twentieth century when electric space

Table 4.1: *Selected urban district energy systems.*

City	Thermal application	Approx. energy production (GWh/yr)	Number of people served	Number of buildings served	Percentage of district served	Use co-generation?	Fuel Sources
Copenhagen, Denmark	Space heating	5,400	500,000	31,300	98%	Yes	Coal, natural gas, biomass
Seoul, South Korea	Space heating and cooling	10,600	>1,000,000	N/A	25%	Yes	Natural gas, oil, landfill gas
Austin, USA	Space heating and cooling	350	75,000	200	100%	Yes	Natural gas
Goteborg, Sweden	Space heating and cooling	4,000	300,000	N/A	64%	Yes	Natural gas, biomass, biogas
New York City, USA	Space heating and cooling	7,600	N/A	1,800	<10%	Yes	Natural gas, oil
Paris, France	Space heating	5,000	N/A	5,774	N/A	Yes	Natural gas, biomass, coal, oil

Sources: NYC SBD Task Force (2005), Toulgoat (2006), Elsman (2009), Goteborg Energi (2009), Ontiveros (2009), Won and Ahn (2009).

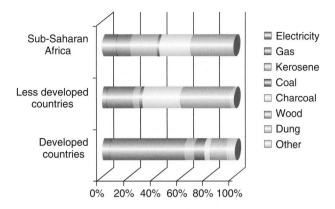

Legend:
- Electricity
- Gas
- Kerosene
- Coal
- Charcoal
- Wood
- Dung
- Other

Figure 4.1: *Share of urban population relying on different cooking fuels.*

Source: UNDP/WHO (2009).

heating, water heating, or cooking systems were aggressively promoted as preferred technologies (Hannah, 1979; Platt, 1991; Nye, 2001). Electric thermal systems remain popular in homes in many cities around the world, because of lack of access to gas supply lines or because builders did not want to incur the cost of installing the necessary feeder pipes.

Climatic conditions and the local economic base also figure decisively in local energy technology choices and usage levels. In China, for example, the colder and more heavily industrialized west has vastly different energy needs than the warmer and more service sector-focused coastal cities in the east and southeast (Dhakal, 2009). The "new" cities of coastal China also tend to have more modern buildings that are home to wealthier families, both of which contribute to different types of energy usage patterns (Chen *et al.*, 2010).

These technology and fuel choices made years ago in cities create a path dependency that shapes current climate change mitigation and adaptation policymaking efforts. The embedded system assets – the massive technology investments in power plants and pipes and wires – are costly to replace or upgrade, and they may provide energy to homes and businesses at a market price lower than newer, more "climate friendly" technologies (Unruh 2000, 2002). This raises important questions for local authorities about how aggressively to promote new technology adoption, including whether existing system assets should be replaced before the end of their useful life. In cities in less-developed countries, the question may focus on whether more advanced energy systems can be cost effectively deployed as part of the overall infrastructure system development efforts sought in these cities.

4.2.2 Energy market structure

"Modern" energy systems[1] involving gas and electricity supply chains were eventually recognized as operating most efficiently as natural monopolies, reducing the need for redun-

dant gas and electricity supply lines across a city (Hannah, 1979; Platt, 1991). In some cases, monopoly rights were expanded vertically with a single entity holding responsibility for both the supply and distribution of energy around some or all of a city. Ownership responsibilities for these system assets were largely dictated by state or national market regulatory preferences, with different assets owned and operated by either government, private firms, or some type of public–private partnership. Government ownership can take the form of nationalized utilities or municipally owned utilities, where local government has direct control over the management and operations of the local energy system.

In the 1980s and 1990s, when energy market liberalization efforts took off around the world, some cities began to see significant changes in who owned and operated these systems. In many cases, ownership responsibilities became more fragmented, with supply and distribution responsibilities broken apart in the name of competition and economic efficiency.

In cities with large informal settlements with less comprehensive or technically advanced energy system infrastructures, market structures may look very different. Many households are unable to afford clean-fuel cookstoves or appliances (UNDP/WHO, 2009); households may also be unaware of the ill effects caused by pollution from certain types of solid fuels (Viswanathan and Kavi Kumar, 2005). Supply chains may therefore focus on delivering fuels that can be used in very low-tech ways, satisfying heating and cooking needs. Developing country cities may also experience high levels of utility "theft," with homes and businesses illegally (and dangerously) tapping into local electricity distribution systems (USAID, 2004).

These different energy market circumstances all make addressing the issue of climate change a challenging one, as responsibilities for energy system planning – and payment for any system upgrades – may be divided among a very diffuse set of stakeholders.

4.2.3 Energy system governance

Technology choices, market structures, and ownership responsibilities are all important considerations as we look towards a future involving changing climatic conditions. The ability of cities to influence the design or operation of the local energy system varies widely, however, linked to a much broader set of governance questions discussed in Chapter 9 of this report.

In analyzing energy governance in cities, span of control (also known as agency or policy competency) is the critical factor. Span of control refers to the fact that energy policy is traditionally considered a supra-local issue, controlled at the state/provincial, national, or trans-national level (Bulkeley and Betsill,

---

1 Although cities have long relied on energy supply chains (primarily for biomass/charcoal imported to cities for use as a thermal energy source), we refer to the modern energy sector as the systems that began to develop widely in the mid to late 1800s, when gas and electricity supply and distribution markets began to become increasingly prominent in cities.

2002). This is a reversal of the situation in the mid to late 1800s, the era when gas and electricity use first became prominent in cities. At that time, electricity and gas utilities were frequently under local authority control, a function of the technology in use at the time. Over time, however, as local networks were linked into ever-larger systems serving entire states or countries and concerns arose over corrupt local oversight practices, regulatory oversight for these systems was transferred to state/provincial or central government agencies (Hughes, 1983).

As markets for certain forms of energy became global, and as energy-related pollution or other externalities crossed country boundaries, international agreements or treaties shifted certain policy control powers yet again, to trans-national organizations such as the European Union or United Nations (Bulkeley and Betsill, 2002).

Today, we are increasingly seeing a re-engagement on energy policy matters by local authorities in both developed and developing countries (Keirstead and Schulz, 2010). Capello et al.(1999) note that the focus is on land use planning and building regulations, energy conservation policies, market or behavioral stimulation programs (e.g., grants and information campaigns), and support for technical innovations. Cities also have significant control over energy use in local authority-owned buildings and in the type of energy or technology used in publicly managed services such as mass transit, waste disposal or treatment, and water supply systems.

Because key regulatory control powers still reside at the state or national level, however, most local authorities lack the ability to force fundamental changes in the technologies that utilities employ or their efforts promoting energy conservation or efficiency. Energy or carbon taxing powers also tend to be within the prerogative of national governments, and their availability locally varies significantly from city to city (for example, see European Commission, 2007b).

Other policy options may be unavailable to cities owing to the high costs of entry, such as funding for major research and development projects. Some areas where cities can act, such as planning and building codes, may be constrained by institutional capacity. An innovation such as the Merton Rule,[2] which promotes the increased use of renewables in buildings in London (House of Commons, 2007), requires adequately trained officials for plan approval and enforcement. Building codes exhibit this principle more generally; municipal governments in developing countries can often influence energy use through building codes, but the effectiveness of these measures is highly variable, depending on the resources available for application and enforcement. In China, enforcement of local buildings codes varies across cities and across different stages of design and construction (Shui et al., 2009).

4.3 Energy use in cities

Cities play a central role in driving global energy demand, but historically there has been relatively little information published on energy use in individual cities or urban areas. The OECD (1995) was one of the first organizations to estimate total urban energy demand around the world (74 percent), although the methodology supporting this estimate is unclear. More recently, the focus of aggregate urban-scale analysis has shifted to the level of greenhouse gas emissions attributable to cities, based largely on calculations of energy use in these cities.

In most cities around the world, data on local energy consumption or supply levels either have not been compiled or provide only a partial picture of the local situation. In the case of the latter, this is partly a function of the underlying goals of some of this research – it may be a residential sector-focused analysis, for instance – but it also reflects the many challenges inherent in obtaining these data. These include difficulties accessing proprietary market data held by the private companies serving a city and definitional questions related to what actually constitutes energy use resulting from activity in a city. Analyses may also focus narrowly on marketed fuels and technical energy, while fuels such as biomass and charcoal and non-technical sources (such as draft animals or other non-motorized transport modes in general) may be under-documented despite representing a large share of total local energy use.

There is also the issue of whether the urban system is defined as a spatial territory or functional unit and whether cities must account for all primary and/or embodied energy consumed within their borders (Parshall et al., 2010; Kennedy et al., 2009a). The notion that cities are ascribed responsibility for this use is considered problematic by some, arguing that such views diminish the energy efficiency benefits offered by urban lifestyles, given smaller dwelling sizes, reduced travel distances, increased access to public transportation, etc. (Satterthwaite, 2008; Dodman, 2009). The notion of holding a city accountable for local energy use can also be seen as problematic, as it is the behavior of individuals or institutions in cities that is at the root of this level of energy usage, rather than the city itself.

This argument speaks to the fact that analyses of urban energy usage are helpful primarily because of the spotlight they shine on the need for energy and climate policies that respect the unique attributes of urban areas. City-specific analyses focus even more directly on this point, using local energy supply and use data to inform local energy efficiency strategies or climate change mitigation or adaptation plans. In some cases, cross-city comparisons are employed because they provoke questions among local policymakers about how they can attain energy use or emission levels comparable to those in other cities. Per capita electricity

---

2   Named after the London borough where it was first established in 2003, the "Merton Rule" required that new development projects generate at least 10% of their energy needs from on-site renewable energy equipment. The Mayor of London subsequently adopted the rule as part of his climate change initiatives, along with many other local authorities around the UK. In 2008, the UK government published new planning guidance requiring all UK local planning authorities to adopt a "Merton rule" policy.

[MITIGATION] Box 4.1 Managing CO$_2$ emissions from buildings: Lessons from the UK, USA, and India

Rajat Gupta and Smita Chandiwala

Oxford Brookes University, UK

This synopsis is based on the research paper commissioned by the World Bank and presented in the 5th Urban Research Symposium on Cities and Climate Change held at Marseille in June 2009.

In 2002, buildings were responsible for 7.85 Gt, or 33 percent of all energy-related CO$_2$ emissions worldwide and these emissions are expected to grow to 11 Gt (B2 scenario) or 15.6 Gt (A1B scenario) by 2030 (IPCC, 2007). However, as the housing market in the UK, USA, and several other developed countries has gone into deep and prolonged recession, the opportunity for very substantial investment into improving the existing building stock has opened up. In fact according to the Fourth Assessment Report (AR4) of the IPCC (2007), approximately 29 percent of CO$_2$ emissions can be saved economically, or at a net benefit to society, even at zero carbon price. Mitigation measures in the residential and commercial sectors can save approximately 1.6 billion and 1.4 billion tons of CO$_2$ emissions, respectively, by 2020 (Urge-Vorsatz et al., 2007). While the magnitude of these large potentials that can be captured has been known for decades, many of these energy efficiency possibilities have not been realized. This is because of certain characteristics of markets, user behavior, and a lack of critical evaluation of the available tools and models that could be used by planners, building designers, and policymakers to measure, benchmark, target, plan, and monitor energy-related CO$_2$ emissions and forecast reductions from existing buildings. This research paper therefore comparatively evaluates the building-related CO$_2$ measurement, benchmarking, and reduction approaches available in the USA, UK, and India, to share the lessons learnt in implementing CO$_2$ reducing policies in each of these countries, by:

- Establishing what tools, approaches, and methodologies are available for measuring energy use and CO$_2$ emissions from existing buildings in the UK, USA, and India.
- Reviewing and comparing benchmarks of annual energy consumption (kWh/m^2 per year) and CO$_2$ emissions (kgCO$_2$/m^2 per year) from buildings-in-use in the case study countries.
- Developing more rigorous standards for existing buildings (to reduce their energy consumption), which could be adopted by developed and rapidly developing cities taking account of building type, local climate and occupancy.
- Evaluating various strategies and measures available for maximising CO$_2$ emission reductions in existing buildings (above 80 percent in developed countries) through improved energy efficiency, low and zero carbon technologies, as well as non-technical solutions (education and awareness, behavioral change), and to identify barriers for their implementation.

- Finally, recommending policy measures that would increase uptake of the selected CO$_2$ reduction strategies in existing buildings.

A comparative analysis is undertaken to evaluate the strengths and weakness of methods such as BREEAM/CSH in the UK, LEED in the USA, and TERI-GRIHA and LEED-India in India. Robust performance-based standards (in terms of kWh/m^2 per year or kgCO$_2$/m^2 per year) are recommended for reducing the energy consumption of existing buildings present in both developed and rapidly urbanising cities. A range of policy instruments and measures are suggested to remove or lower barriers and encourage uptake of various CO$_2$ reduction strategies in existing buildings. Among these are: appliance standards, building energy codes, appliance and building labelling, pricing measures and financial incentives, utility demand-side management programs, and public sector energy leadership programs including procurement policies. Because culture and occupant behavior are major determinants of energy use in buildings, these policy approaches need to go hand in hand with programs that increase consumer access to information, awareness, and knowledge. At present, however, there is particularly a lack of accurate information about exactly how much variation occupant behavior introduces to a building's energy consumption.

It is realized that the UK is world-leading in its CO$_2$ reduction policy for buildings but lacks good-quality bottom-up data sets of real energy consumption and CO$_2$ emissions in buildings. The USA on the other hand has excellent data sets by EIA and DoE, but needs to have national-level policies and targets for CO$_2$ reduction from buildings. India is working on both policy and data collection given that the energy data are quite polarized between the urban and rural. In fact the Bureau of Energy Efficiency is working with USAID's ECO-III program to benchmark a range of commercial and institutional buildings – although the focus is primarily on energy efficiency and not CO$_2$ reduction. Hopefully, robust targets for CO$_2$ reduction and policies to achieve those targets will be set soon.

The role of data and analysis is particularly emphasized, since the building sector is not considered as an independent sector and there is a lack of consistent data, which makes it difficult to understand the underlying changes that affect energy consumption in this sector. It is essential to make available comprehensive building energy information to allow suitable analysis and efficiently plan energy policies for the future. In fact, regularization of data collection and analysis for the building sector can help quantify technology performance, its cost-effectiveness, role of barriers, identification of beneficiaries, and targeting of government and industry policies, programs, and measures. In that respect, studies developed by the EIA on the energy consumption of residential and commercial buildings in the USA are a valuable reference (EIA, 2001, 2003, 2006, 2008a, b).

It is hoped that findings from this project will help to expedite the process of achieving significant reductions in energy use and CO_2 emissions from the existing building stock by formulating policies that address the conventional barriers to implementation and increase the uptake of low carbon systems (heat pumps, solar hot water, solar PV, micro-combined heat and power, micro-wind) in buildings and cities. There is a need for adaptive policies to be mainstreamed through all development and environmental policies such as retrofitting existing building stock to ensure that it remains resilient to climate change impacts. On the longer term, the data archive of this study will be of immense value to all those with a stake in a low-carbon future, be it policy, practice or academic understanding. No doubt the ultimate aim is to make the global building stock become low-energy, low-carbon and more resilient to climate change effects.

use is a metric commonly employed to highlight such comparisons between cities (see Figure 4.2), although such comparisons are most useful if they account for differences in climate or level of economic development.

Tracking urban energy consumption

Despite these challenges, there are several studies that have sought to examine urban energy consumption at different scales.

In 2008, the International Energy Agency (IEA) calculated global urban energy use, concluding that 67 percent of global primary energy demand – or 7,903 Mtoe – is associated with urban areas (IEA, 2008; see Table 4.2). By 2030, urban energy consumption is expected to increase to 12,374 Mtoe, representing 73 percent of global primary energy demand and reflecting dramatic anticipated growth in urban population levels around the world.

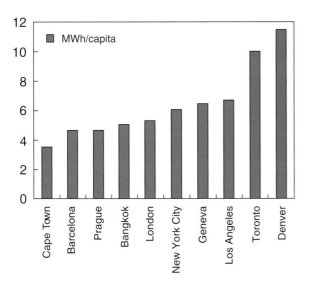

Figure 4.2: *Per capita electricity consumption in MWh/capita.*

Source: Kennedy et al., *(2009b).*

Regional or country specific analyses have also focused on aggregate urban scale energy consumption. Dhakal (2009) estimated that the urban share of total commercial energy use in China is 84 percent, while in the USA, urban areas are responsible for between 37 and 86 percent of national direct fuel consumption in residential, commercial, and industrial buildings, and between 37 and 77 percent of national on-road gasoline and diesel consumption (Parshall *et al.*, 2010).[3] Such studies frequently contrast energy use in urban and non-urban areas of a country. For example, a Brookings Institution study showed that large metropolitan areas in the USA have smaller per-capita energy consumption and carbon emissions compared with the national average (Brown *et al.*, 2008). By contrast, Dhakal (2009) found dramatically higher rates of energy use in Chinese cities compared to rural areas.

Other studies focus narrowly on detailing the fuel mix in specific cities. ICLEI (2009b) compiled energy use data in 54 South Asian cities, identifying the absolute quantities of each fuel type broken out by sector (see Table 4.3). Kennedy *et al.* (2009a, 2009b) compared energy use and emissions data in ten cities in Africa, Asia, Europe, and North America, while other individual local authority analyses have been published as part of each city's sustainability or climate initiatives or as part of ongoing public reporting efforts on different key performance indicators (for example, see Mairie de Paris 2007; Shanghai Municipal Statistics Bureau, 2008).

There is information that is less commonly available that is helpful when crafting city-specific mitigation and adaptation policies. New York City's sustainability plan breaks out building-related energy use by function and sector; that is, how much energy is expended on lighting versus heating, cooling, and other types of specific energy demand in different types of buildings (see Table 4.4). This information is useful because it can help a local authority prioritize its scarce time and financial resources when implementing a sustainability plan. Diurnal information, or a further breakdown of how energy is used by different applications (e.g., heating, lighting, etc.) over the course of the day, can also be helpful in highlighting opportunities to employ different types of energy efficient technology within a building or on a citywide scale (Parshall, 2010).

---

3   The wide range reflects different boundary assumptions employed in the analysis.

Table 4.2: *World energy demand in cities by fuel.*

	2006		2015		2030		
	Mtoe	Cities as percentage of global demand	Mtoe	Cities as percentage of global demand	Mtoe	Cities as percentage of global demand	2006–2030[a]
Coal	2,330	76%	3,145	78%	3,964	81%	2.2%
Oil	2,519	63%	2,873	63%	3,394	66%	1.2%
Gas	1,984	82%	2,418	83%	3,176	87%	2.0%
Nuclear	551	76%	630	77%	726	81%	1.2%
Hydro	195	75%	245	76%	330	79%	2.2%
Biomass and waste	280	24%	358	26%	520	31%	2.6%
Other renewable	48	72%	115	73%	264	75%	7.4%
Total	7,908	67%	9,785	69%	12,374	73%	1.9%
Electricity	1,019	76%	1,367	77%	1,912	79%	2.7%

Source: IEA (2008).
[a]Average annual growth rate.

4.4 Climate risks to urban energy systems

While the contribution of energy use to global climate change has been extensively studied, the literature on the impacts of climate change *on* urban energy systems is still in its infancy. What is clear, however, is that because cities are so reliant on energy sources and system assets based *outside* of the city, any discussion must examine climate change-related impacts at multiple levels of the energy supply and demand chain.

Figure 4.3 provides a schematic of how different climate change hazards projected for cities (Column 1) link to both physical (Column 2) and institutional risks (Column 3) for different segments of the energy system. All of these impacts could threaten the wider economic vitality of urban centers, as damaged or inoperable energy system assets jeopardize public health, commerce, and private property interests. Changing climate patterns may also have a big impact on energy system asset requirements and operating costs, driving up fuel prices, imposing changed maintenance regimes or operating practices, or requiring significant capital expenditures to adapt the system to these threats. This chapter does not address all of the impacts cited in Figure 4.3, but focuses on the areas of greatest significance.

4.4.1 Energy demand impacts of climate change

There is a rich literature detailing the link between climatic variables and changes in energy demand. Research has examined this issue at a variety of different scales, primarily for utility or government planning purposes, to assess the adequacy of overall energy system capacity to meet demand at different times of the year (Amato *et al.*, 2005).

A different set of literature examines the link between "urban heat islands" and energy demand (Taha *et al.*, 1999; Akbari and Konopacki, 2005; Rosenzweig *et al.*, 2009). Urban heat islands refer to the fact that cities are full of impervious surfaces that trap heat, leading to elevated air temperatures. In the summertime, heat island conditions can significantly increase local electricity demand for space cooling. For example, a study of Athens found that due to the heat island effect, a building in the urban core would have approximately twice the cooling load of an equivalent building located on the city's outskirts (Santamouris and Georgakis, 2003).

Global climate change is expected to increase temperature at the regional scale, which may exacerbate heat island conditions in cities. Whether a city's *net* energy demand increases or decreases will vary by location. Areas where energy demand peaks in the winter will experience a reduction in overall energy consumption if anticipated thermal load savings exceed anticipated summertime power load increases. Conversely, areas with summer peaks will see net demand increases if rising air conditioner loads exceed decreases in heating requirements (Scott and Huang, 2007).

No information was found that directly addressed the impacts climate change could have on other thermal loads in cities (e.g., cooking, water heating, or process energy requirements). However, to the extent these loads are supplied by the same fuels used for space conditioning, they will also be affected by changes in market supply conditions.

Looking strictly at electricity demand, it is important to distinguish between the impact on cumulative power requirements (i.e., the annual demand measured in MW or GW hours) and peak demand impacts (i.e., point-in-time demand measured in MW_p). Both can have significant financial implications around a city,

Table 4.3: *Energy use in selected cities in South Asia.*

		Agra, India	Chennai, India	Kolkata, India	Mysore, India	Chittagong, Bangladesh	Thimphu, Bhutan	Kathmandu, Nepal	Colombo, Sri Lanka
		Population 1.27 million	Population 4.34 million	Population 4.57 million	Population 750,000	Population 2.53 million	Population 90,000	Population 700,000	Population 640,000
Residential	Electricity (million kWh)	414	589	1,196	238	60	81	287	419
	LPG (metric tons)	10,014	197,748	75,997	2,398	N/A	N/A	25,386	6,876
	Kerosene (kL)	33,408		292,240	32,604				
	Fuel wood (metric tons)				12,400				
Commercial	Electricity (million kWh)	115	176	985	92	15	21	89	19
	LPG (metric tons)	N/A			8,349		N/A	1,148	N/A
Industrial	Electricity (million kWh)	53	78	503	380	102	4	54	455
	LPG (metric tons)				1,057		N/A	2,762	
	Coal/wood			2,929,348					
	Furnace oil (kL)								68,153
Transportation	Diesel (kL)	50,442	346,180	488,955	51,000	4,855	7,002	32,707	72,280
	Petrol (kL)	49,376	178,970	117,987	30,800	42,098	5,042	31,785	51,218
	CNG (kg)	930,271							
	Other (kL)					31,512 (octane)			8,224 (kerosene)
Other	Electricity (undefined user) (kWh)	318	1,850		273	35	25	62	219
	Fuel wood (metric tons)	7,200							

Source: ICLEI, 2009b.

Table 4.4: *Energy usage by building type in New York City as percentage of total energy in British Thermal Units (BTU).*

Building type	Heat	Hot water	Lighting	Appliances[a]	Cooling[b]	Other	Total
1–4 family residential	7.6%	2.6%	1.7%	2.2%	0.6%	0.0%	14.7%
Multi-family residential	7.4%	7.4%	3.0%	3.9%	1.2%	0.0%	22.9%
Commercial	8.5%	2.8%	10.2%	4.5%	4.5%	0.9%	31.4%
Industrial	2.6%	2.1%	4.0%	3.3%	1.1%	0.2%	13.3%
Institutional/government	6.3%	4.0%	3.6%	1.7%	1.4%	0.9%	17.9%
All types	**32.4%**	**18.9%**	**22.5%**	**15.6%**	**8.8%**	**2.0%**	**100.0%**

Source: City of New York (2007).
[a]Appliances including electronics and refrigerators as well as other appliances.
[b]Cooling includes ventilation as well as air conditioning.

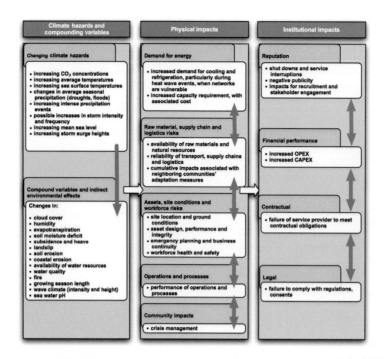

Figure 4.3: *Impacts of climate change on urban energy systems.*

Source: Adapted from Acclimatise (2008).

but they vary in terms of whether additional generation capacity must be deployed or if existing capacity is utilized more often. This occurs because warmer nights and longer cooling seasons can generally be served by a city's existing power generation capacity, which is sized to meet the highest summertime peak demand. By contrast, when peak demand growth outpaces total demand growth, spare capacity is in short supply, increasing the risk of blackouts and brownouts (Miller *et al.*, 2008).

To date, there have been several studies analyzing how climate change will affect energy demand, although most have examined this issue at a state or national scale (Smith and Tirpak, 1989; Baxter and Calandri, 1992; ICF, 1995; Franco and Sanstad, 2008). The analyses generally conclude that for the regions examined both total electricity demand and peak electricity demand will increase as a result of climate change, although peak demand will increase at a much faster rate.

There is a slimmer body of research examining how climate change will influence energy demand in cities. In Boston, climate change is expected to boost per capita energy demand in 2030 by at least 20 percent compared to the 1960–2000 average (Kirshen *et al.*, 2008). In New York, two different studies looked at climate change impacts on the wider metropolitan region. A 1987 analysis concluded climate change would increase peak electricity demand 8–17 percent by 2015, whereas overall demand would have a much slower 2 percent growth rate (Linder *et al.*, 1987). Hill and Goldberg (2001) looked at the peak demand impacts of climate change, projecting that by the 2020s climate change-induced demand growth would total 7–13 percent, reaching 12–17 percent by the 2080s. An analysis by Scott *et al.* (1994) examined the impacts of climate change on projected energy use

in commercial buildings in four US cities (Phoenix, Seattle, Minneapolis, and Shreveport), finding widely variable – but uniformly positive – demand increase in each city.

Scott *et al.*'s conclusion stems from the fact that (i) a sizable percentage of the energy demand in commercial buildings is tied to space cooling (EIA, 2009), and (ii) air conditioning deployment (aka "saturation") levels vary widely by city (see

Table 4.5: *Air conditioning (AC) saturation rates in US cities.*

City, state	Percentage of buildings with window AC units	Percentage of buildings with central AC systems	Total
Los Angeles, CA	27.3%	23.9%	51.2%
San Francisco, CA	6.0%	15.0%	21.0%
Sacramento, CA	22.1%	62.7%	84.8%
New York, NY	53.1%	10.1%	63.2%
Rochester, NY	25.6%	17.5%	43.1%
Buffalo, NY	16.3%	8.8%	25.1%
Columbus, OH	21.4%	56.1%	77.5%
Cincinnati, OH	32.3%	50.9%	83.2%
Cleveland, OH	25.7%	34.2%	59.9%
Houston, TX	14.7%	78.9%	93.6%
San Antonio, TX	27.8%	60.6%	88.4%
Dallas, TX	17.3%	78.5%	95.8%

Source: Sailor and Pavlova (2003).

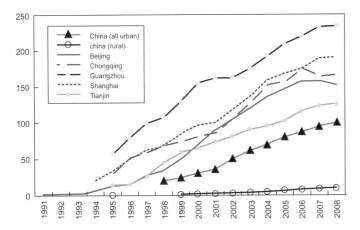

Figure 4.4: *Number of air conditioners per 100 households in selected Chinese cities.*
Source: CEIC (2010).

Table 4.5). Saturation levels are important because they hint at the level of demand growth that could occur as temperatures rise in a city. Cities with low saturation rates might experience higher rates of demand growth as buildings lacking air conditioning systems install them. Cities with high saturation rates could still experience demand growth, but at a slower pace as the percentage of buildings lacking air conditioning units approaches zero.

In China, air conditioner ownership rates have increased dramatically in the past 15 years, with rates of ownership in urban households exceeding an average of one unit per household in most urban areas. This contrasts markedly with rates of use in rural areas of China (see Figure 4.4). More research is necessary to tell us the actual level of use, however, or whether urban areas in China have markedly different saturation rates.[4]

4.4.2 Energy supply chain and operations risks and vulnerabilities

Climate change may affect the urban energy supply chain in three principal ways: through impacts on primary energy feedstock production or supply networks delivering these feedstocks to power plants; impacts on power generation operations; and via impacts on the energy transmission and distribution network. Our understanding of these risks varies widely, as does the severity each risk presents to cities around the globe.

4.4.2.1 *Energy resource production and delivery*

Primary energy fuel stocks tend to be found away from urban areas, but the impacts of climate change on the sourcing and processing of these materials would nonetheless be felt in urban areas, albeit in an indirect manner as cost impacts ripple across national or global economies.

For example, the US Climate Change Science Program (Bull *et al.*, 2007) has noted the vulnerabilities of oil and gas drilling platforms and refineries along the Gulf of Mexico coast to flooding and high winds associated with extreme weather events. Closure of these facilities and fuel terminals during and after Hurricane Katrina were linked to fuel price increases across the USA. Extreme weather events in non-coastal areas can also affect primary energy supply chains, as we saw in 2008 when heavy snows in central and southern China blocked rail networks and highways used for delivering coal to power plants in these regions. Seventeen of China's 31 provinces were forced to ration power, affecting hundreds of millions of people in cities across the country (French, 2008). Climate scientists have been wary of attributing these snowstorms to climate change (Perry, 2008), but others note they represent the type of extreme weather event related disruptions that may be more prevalent in the future (Pew Center on Global Climate Change, 2008).

Larsen *et al.* (2008) note that Arctic transport routes and energy infrastructure critical for moving oil and gas across Alaska are located across areas at high risk of permafrost thaw as temperatures rise. Potential vulnerabilities include structural failure and distribution problems as oil and gas pipelines fracture; reduced access and increased transport costs due to a shorter winter season for ice roads, and increases in repair and maintenance costs. System stresses such as these will produce *indirect* impacts on cities, generally in the form of higher energy prices.

In developing countries, areas heavily dependent on different types of biomass may be vulnerable to the extent certain climate change risks affect the availability of the material or the transport routes delivering this material to cities. For example, changing temperature levels may reduce biomass availability if plants reach the threshold of their biological heat tolerance or if storms or drought reduce plant or tree growth levels (Williamson *et al.*, 2009). The extent of these problems will be localized based on how biomass materials are sourced in different urban areas.

4.4.2.2 *Impacts on power generation*

Chapter 3 documents the risk to cities from rising sea levels and storm surges associated with climate change. There is little within this growing literature, however, that draws specific links between anticipated coastal threats and energy system assets located along or near threatened coastlines. Potential risks exist because many power stations were historically sited along waterways, a legacy of the need for cooling waters that were integral to the design of older thermoelectric power plants. Many facilities also relied on barge deliveries of their coal supply. Given the decades-long lifespan of most large power plants, they now face risks from anticipated sea level rise or more extreme weather events.

---

4   The Chinese government's approach of tracking air conditioning unit ownership rates tell us little about deployment patterns, as high-income households may deploy multiple units, skewing local saturation rates.

Power Plants along the East River, New York City

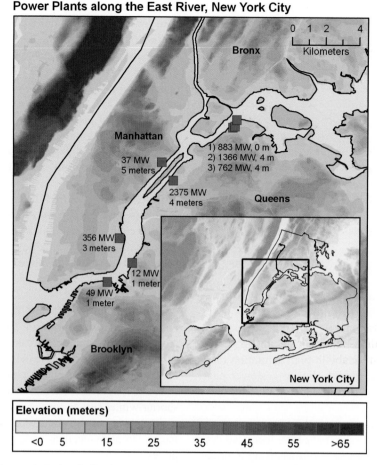

Figure 4.5: *Location and elevation of power plants along the East River in New York City. Power plant data for 2000 from eGRID (USEPA, 2002) to reflect with recently retired plants deleted. New York City digital elevation model is from the USGS (1999), which has a vertical error of approximately +/−4 feet.*

Whether a given power plant (or fuels temporarily stored on-site at or near these plants) is vulnerable to flooding problems is a function of the elevation of the facility, the facility design (e.g., surrounded by berms, etc.), and its proximity to any path that storm-linked tidal surges would follow during extreme weather events. Hurricanes Katrina, Rita, and Ike in the Gulf of Mexico were recent examples where coastal power plants were damaged by storm surges, with several facilities serving New Orleans, Houston, and Galveston forced to shut down due to anticipated flooding. Some remained closed for several days after the storms had passed, suffering extensive wind and water damage (Jovetski, 2006; McKinley, 2008).

Figure 4.5 displays the potential vulnerability of power plants along the East River in New York City to similar storm surges, highlighting the vulnerability of 5,840 MW of power generation capacity at an elevation of less than 5 meters, the height of the storm surge expected in some areas if a Category 3 hurricane directly hits the city.

A different type of risk arises from the fact that cooling waters needed to exhaust waste heat from older thermoelectric power plants may be less able to satisfy their cooling function in the future. Warmer ambient air temperatures and decreased stream flows attributable to climate change increase the risk that power stations will run afoul of rules restricting (ICF, 1995):

- The absolute temperature of water discharged from a power plant
- The absolute temperature of water downstream from power plants and/or
- The temperature rise of waters receiving cooling water effluent from power plants.

Water-cooled power plants are subject to one or more of these standards designed to protect aquatic life, with the exact rules varying by location. To prevent violations, power stations could be forced to scale back their operation or shut down entirely. Such was the case in Europe's deadly heat wave in 2006, when nuclear power plants in Spain and Germany were temporarily shut or forced to scale back operations due to high receiving water temperatures (Jowit and Espinoza, 2006). Little research has been done to date exploring the overall vulnerability of the energy system to this problem, and doing so – particularly from a city-level perspective – could be difficult, because climate models currently in use cannot be downscaled with pinpoint accuracy to a specific location on a river or bay.

Power plant operations may also be vulnerable to changes in air temperature and air density arising from climate change. The UK Met Office surveyed power plant operators around the UK in 2006, ultimately reporting operator concerns that combined-cycle gas turbines could experience decreased output as temperatures rise and air density decline (Hewer, 2006). Others discount the extent of this problem, however, noting that output reductions will be minor, totaling less than 1 percent under most climate scenarios (Linder et al., 1987; Stern, 1998; Bull et al., 2007), or concluding that it will require systems to be upgraded 1–2 years earlier than otherwise would have been required (Jollands et al., 2007). To the extent these problems do occur, they would apply to both central station power plants and district energy facilities. No data have been published thus far exploring such impacts on small-scale (<10 MW) cogeneration systems commonly deployed in many cities as a distributed power generation source.

Although this discussion has thus far focused on traditional thermoelectric power stations, power generation facilities reliant on renewable resources may also be affected by climate change. For instance, hydroelectric facilities fed by glacial and snow melt have historically benefited from the ability of glaciers to regulate and maintain water levels of rivers and streams throughout the summer – a time in many of these regions when precipitation-fed water sources often run low or dry. With increasing temperatures, however, snow levels are decreasing and glaciers are shrinking, jeopardizing the amount of hydroelectric production available to serve many urban areas (CCME, 2003; Markoff and Cullen, 2008; Madnani, 2009).

Changing climate patterns may also affect the timing and level of precipitation available to feed many hydropower systems. For example, although changing precipitation patterns are expected to increase hydropower production by roughly 15–30 percent in northern and eastern Europe by the 2070s, a 20–50 percent decrease in hydropower potential is projected for the Mediterranean region over the same period (Lehner et al., 2005). Problems could arise depending on whether the precipitation falls as rain or snow and at which elevation, because snow serves as a secondary water reservoir, gradually releasing water over the spring and early summer. The elevation at which precipitation occurs is key, because retention dams serve different functions (e.g., water supply, flood control, power generation) based on their elevation and thus have different water release rules. This could affect the availability of power at different times of the year (Linder et al., 1987; Aspen Environmental Group and M Cubed, 2005; Franco, 2005; Vine, 2008).

Whether cities are highly vulnerable to these problems is a function of the type and magnitude of impact of climate change on regional hydrologic conditions and their overall reliance on hydropower. The city of Seattle, Washington, obtains fully 50 percent of its electricity from a network of hydropower dams around the northwestern USA (Seattle City Light, 2005). Projections are that reductions in annual hydropower output in the region are likely by 2080 (Markoff and Cullen, 2008), putting

that city's power supply at risk. Seattle's municipally owned utility has already begun investing in wind farms on the state border with Oregon to hedge its power generation bets (Seattle City Light, 2005).

Even cities that do not directly rely on hydropower for the electric supply may feel the pinch of declining hydropower availability in their region. Hydropower is generally a low-cost power source, so decreasing availability means it will be replaced by higher cost forms of power, driving up prices as the regional supply market tightens during low-water months or years (Morris et al., 1996).

The impacts of climate change on two other important types of renewable power generation in cities – solar and wind power – are far less definitive. One study examining solar levels in the USA through 2040 projects increased cloud cover resulting from higher CO_2 level concentrations could cut solar radiation by 20 percent (Pan et al., 2004). A Nordic study estimates that a 2 percent decline in solar radiation levels could cut solar photovoltaic system output by 6 percent (Fidje and Martinsen, 2006). To the extent cities around the world are seeking to significantly expand deployment levels on local homes and businesses, this could be problematic as long-term power output levels could be less than anticipated. However, because current in-city solar deployment levels are so small compared to overall urban power demand, it will likely be some time before such a decline in solar production becomes significant enough to create major problems. Whether large new concentrating solar facilities recently installed in many parts of the world will suffer degraded output levels is unclear. Several utilities serving urban areas in Spain and the southwestern USA have invested in these projects in rural areas outside of the city (Philibert, 2004), dramatically boosting the level of renewable power feeding the local power system.

Wind patterns (wind speed, duration, and direction) may also change as a result of climate change, although the projected impacts will likely vary seasonally and differ widely from region to region. Research on the Baltic Sea region finds no clear signal on future wind resource levels (Fenger, 2007), while in the UK and Ireland, onshore wind speeds are expected to decrease in the summer and increase in winter (Harrison et al., 2008). Fenger (2007) notes the likelihood that system efficiency levels will increase in Scandinavia during the winter months because of reduced turbine blade icing attributable to warmer temperatures. Research on US wind patterns projects speeds will decline 1 to 15 percent over the next 100 years, depending on which climate models are used (Breslow and Sailor, 2002).

No studies have been identified to date that examine potential wind pattern changes in cities attributable to climate change. Cities are already recognized as being a challenging locale for deploying wind power systems due to the turbulence created by the built environment (Dutton et al., 2005); the extent to which this may change is unknown and a good area for future research.

[ADAPTATION]   Box 4.2   Cooling waters and nuclear power in the USA

Michelle Colley

Acclimatise

Morgan Herve-Mignucci

CDC Climat

An August 2007 heat wave forced the shut-down of a reactor unit at the Browns Ferry nuclear power plant in Alabama, USA, leading to international debate about the feasibility of nuclear power in warming temperatures. The Browns Ferry plant uses cooling water drawn from the Tennessee River to condense and cool the steam generated by the plant for its turbines. State environmental regulations impose a 90°F (32°C) cap on the river temperature downstream of the plant to minimize stress to aquatic ecosystems, and typically the plant increases the river's temperature by 5°F (3°C). During the heat wave, the upstream river temperature was often at, or above, 90°F and the plant then became constrained by regulatory limits preventing it from raising the river's temperature further.

As a result, one unit at Browns Ferry was shut down and power production from another two plants was decreased to reduce the quantity of process steam generated. This allowed the abstracted river water to condense the steam and pass back into the Tennessee River without violating the regulatory limits.

Plant operations were also affected by intake temperatures. In engineering terms, the plant can operate at 100 percent power output with river temperatures up to 95°F (35°C), leaving a 5°F margin between the environmental cap and the engineering threshold. As river temperatures rise, the river water's ability to condense the steam that drives the turbines drops rapidly, requiring the plant to operate at reduced power outputs. Heat waves increase demand for air conditioning, so Browns Ferry is prone to shut-down when electricity demand is highest.

Source: Fleischauer, E. (2007). 'Heat Wave Shutdown at Browns Ferry Stirs Nuclear Debate,' September 2, 2007. Available at the Climate Ark website: http://www.climateark.org/shared/reader/welcome.aspx?linkid=83238&keybold= climate%20blogs (accessed July, 2009).

4.4.2.3   Climate impacts on energy transmission and distribution

To the extent temperatures are expected to rise as a result of climate change, there may be impacts on the local power grid. In general, transmission and distribution lines and electrical transformers are "rated" to handle a maximum amount of voltage for a fixed period of time before they fail. Changing climatic conditions can lead to such failure by pushing power demand beyond equipment rating levels. In California, for example, a summer 2006 heat wave led to blackouts across the state, as sustained high nighttime temperatures prevented the transformers from cooling down before demand increased again the next morning. Insulation within the transformers burned and circuit breakers tripped, knocking out power for more than one million customers (Miller *et al.*, 2008; Vine, 2008).

Related to the previous discussion on the impacts of rising temperature levels on power plant output is the issue of how temperature changes will affect the power throughput of electricity transmission and distribution lines. When electric current flows through power lines, it encounters resistance from every system component it flows through, which produces heat and results in efficiency losses. These losses normally range from 6 to 15 percent of net electricity produced, depending on the age of the system and the degree of electric loading on the lines (EIA, 2009; Lovins *et al.*, 2002; IEC, 2007). The effect on above-ground lines is moderated by the cooler ambient air, while wires below the ground are cooled by moisture in the soil. As temperatures increase, the cooling capacity of the ambient air and soil declines, conductivity declines, and lines may begin to sag or fail altogether (Hewer, 2006; Mansanet-Bataller *et al.*, 2008). Because distributed generation systems tend to involve minimal wiring exposed to the elements, they may be less vulnerable to these problems than central station-based power networks commonly deployed around cities.

Transmission and distribution networks may conversely experience reduced vulnerability as a result of anticipated temperature increases during the winter months, when they are more subject to damage or failure caused by ice or snow storms. Whether cities will be affected by this change will hinge on their general vulnerability to snow and ice storms, the extent to which these patterns change, and their level of reliance on power delivered by these sources. The severe snowstorms plaguing China in 2008 also brought down power lines in many cities and rural areas, compounding the power outages brought about by diminishing fuel stocks (French, 2008).

Electricity transmission and distribution networks may also be vulnerable to storm surges, rising sea levels, and high winds associated with extreme weather events (McKinley, 2008). Whether cities employ above- or below-ground electric wiring systems is largely a legacy of investment or operating decisions made long ago; a blizzard downing wires across the city led local authorities in New York City to move to bury electric wiring in 1888 (New York Times, 1888). Although this tends to eliminate snow and icing problems, it does make the city's underground transformers and substations more vulnerable to flooding. The local utility in New York is moving to address this problem by installing salt-water submersible transformers in Category 1 flood zones around the city (New York State Department of Public Service, 2007).

Thermal power systems and fuel storage tanks located in buildings may also be vulnerable to sea level rise or storm surges, depending on where they are situated. In many areas, local authorities issue warnings during anticipated flooding events about the need to anchor fuel tanks so they do not shift or flip, spilling their contents and contaminating the building (for example, see State of Maryland (undated)).

4.5 Efforts by cities to reduce greenhouse gas emissions and adapt local energy systems

Local-level policy engagement on energy matters has historically ebbed and flowed, with most action resulting as a response to some tangible crisis or vulnerability. In many parts of the world, the 1970s were a period of heightened local energy policymaking activity, as cities sought to protect themselves from shortages and price increases brought on by the OPEC oil embargo. Many other cities took action in response to growing public engagement in the nascent environmental movement and its "think globally, act locally" mantra.

Other cities have sought to influence the local energy system because of concerns over fuel poverty; lack of public access to safe or reliable energy supply; the adverse impact local energy prices or energy system reliability are having on the city's economy; public health concerns related to local energy emissions; and concerns about the long-term energy security of the city (for example, see San Francisco Public Utility Commission, 2002; Greater London Authority, 2004; New York City Energy Policy Task Force, 2004; City of Cape Town, 2006; City of Amsterdam, 2007; City of Toronto, 2007, Mairie de Paris, 2007; Ciudad de Mexico, 2008).

Climate change began to influence local energy policy efforts in the early 1990s, an outgrowth of the Local Agenda 21 (LA21) movement emanating from the Rio de Janeiro "earth" summit. LA21 plans were conceived of as a means of rallying local support for policy and program initiatives designed to improve local "sustainability." Sustainability was broadly interpreted, but concerns about global climate change and the need to shift to alternative energy sources were clearly articulated as an important element of any local plan. The international non-governmental organization ICLEI was created at around this time, specifically with a goal of providing information and technical support to cities interested in developing these plans (ICLEI, 2000, 2009a).

Over the following decade, other similar organizations and initiatives were established, including Energie-Cités and the European Union-sponsored Covenant of Mayors in Europe; C40-Large Cities Climate Group; World Mayors Council on Climate Change; and climate-related programming by United Cities and Local Governments and Metropolis. The exact structure and purpose of these organizations and initiatives varies slightly, but most tend to provide policymaker education and training, information exchange, technical support, and recognition programs. Membership requirements, costs, and performance obligations vary widely across these initiatives.

Other information and technical support initiatives focused on local-level energy and climate change initiatives have also cropped up in recent years, sponsored by various non-governmental organizations, academic groups, international development banks, and private consultants (for example, see European Commission, 2007a; Natural Capitalism Solutions, 2007; ISET, 2009; Prasad et al., 2009; Suzuki et al., 2009; US Conference of Mayors, 2009). The visibility of these initiatives has grown considerably, particularly as local action has become identified as a counterweight to inaction on climate change by national governments (Grunwald, 2007; City of Copenhagen, 2009).

The number of cities enrolling in these initiatives is sizable,[5] although the results of this activity or participation are poorly documented.

4.5.1 Policy and program initiatives

The type of initiatives undertaken by cities around the world or endorsed by these technical assistance organizations varies widely, reflecting local climate, economic, and political conditions; available local authority resources; local authority political and policy preferences and span of control; the design of the energy system, including its age and breadth of geographic coverage; and sense of urgency by key stakeholders.

Structurally, these policy and program initiatives fall into three broad categories: energy and climate planning and governance, specific mitigation policies and programs focused on reducing the local energy system's contribution to climate change, and adaptation efforts dealing with the consequences of climate change. A review of climate action plans and other documentary evidence has found that the majority of cities engage in the first two categories of activities, emphasizing climate change planning and mitigation-focused initiatives. Adaptation planning is still a relatively new concept, more the exception to local authority efforts than the rule (Carmin et al., 2009).

In many ways, the programmatic emphasis seen to date mirrors the type of consultation and advice provided by many of the technical assistance initiatives cited above. Early handbooks and other guidance documents prepared to aid local authorities tended to be very mitigation focused. More recent initiatives appear to be more evenly balanced in their coverage or focus entirely on bolstering local resiliency to the impacts of climate change. Even in these documents, however, the sections on how local energy systems must adapt are still poorly developed, as will be discussed below.

4.5.1.1 Climate planning and governance efforts

Many cities treat energy and climate planning as a major initiative, involving a range of stakeholders in and outside of

---

5 For example, as of December 2010, 1,044 cities have signed on to the US Conference of Mayors Climate Protection Agreement (US Conference of Mayors, 2010), while 2,181 cities from around the world have signed on to the European Union's new Covenant of Mayors initiative (European Commission, 2010). Globally, ICLEI has more than 1,200 local authority members (ICLEI, 2010). The extent to which there is overlapping membership in these counts is unclear, as membership information is not made publicly available by ICLEI.

[MITIGATION]   Box 4.3   Seoul's efforts against climate change

Kwi-Gon Kim

Seoul National University

Young-Soo Choi

Climate Change Department, Seoul Metropolitan Government

In April 2007, the City of Seoul announced the Seoul Environment-friendly Energy Declaration, which promotes energy savings and efficiency, expansion of renewable energy and reduction of greenhouse gas emissions. The objective of the program is to actively cope with climate change, the energy crisis caused by excessive consumption of fossil fuels and the exhaustion of unstable energy supplies caused by oil price changes. Seoul is implementing a variety of measures to improve the city's self-reliance on energy, including establishing an objective of 15% energy reduction and 10% renewable energy use by 2020, compared to a year 2000 baseline.

To support this program, Seoul announced new Environment-friendly Building Criteria. Buildings use 57% of energy and produce 65% of greenhouse gas emissions in Seoul.

When the criteria are fully implemented, energy use and greenhouse gas emissions will decrease by at least 20% in new buildings and 10% in existing buildings. The city will require all public buildings to observe the criteria and actively encourage private buildings to follow the guidelines. Specific elements of the criteria include promotion of transit-oriented development (i.e., development near subway and bus transport); promotion of bike lane development; and the promotion of energy efficiency initiatives and the increased use of renewable energy.

In 2010, the Environment-friendly Building Criteria were enhanced to require more direct action. Buildings in Seoul must now achieve second class or higher performance in terms of energy efficiency. In particular, high-rise residential buildings (with 100 or more households) will only receive a construction permit if designed to satisfy 3% of the buildings' energy demand through the use of on-site renewable energy technology. Insulation for outer walls must also be enhanced, and automatic standby power cut off systems should also be installed on more than 80% of the facilities of the buildings.

Seoul is also undertaking a variety of other initiatives including:

- Developing a climate and energy map to show Seoul's climatic characteristics and energy usage patterns in each district around the city. The climate and energy map will aid local urban planning initiatives and provide an opportunity for the citizens to join activities to address climate change, such as energy saving.

- 79% of the city's transport-related greenhouse gas emissions are from automobiles. To address this problem, Seoul collects a 'congestion charge' from cars entering downtown Seoul through Namsan Tunnels No. 1 and No. 3. (There are several routes drivers can take to enter downtown Seoul, but Namsan Tunnels No. 1 and No. 3 offer the shortest routes into the city, saving time on the road.) The congestion charge allows smoother traffic flow, diminishes atmospheric pollution and makes efficient use of traffic facilities. The congestion fee is 2,000 Korean won (approximately US$1.60) per vehicle with less than 3 passengers. Ultimately, Seoul may seek to impose congestion charges on all routes entering downtown Seoul, but there is currently a lack of consensus among the citizens about this issue.

- Seoul also introduced a No-Driving Day scheme on weekdays. In this scheme, people voluntarily determine a day when he or she will abstain from driving his/her car. Compliance is monitored using a RFID system. There are currently 2.95 million vehicles registered in Seoul and 968,000 cars are participating in the No-Driving Day program. The program is estimated to reduce local CO_2 emissions by 246,541 tons annually. Participants in the program receive a 5% discount on vehicle tax, a 10% discount on the local congestion charge, and 30% discount on public parking fees. There are no penalties imposed on the drivers caught driving on their specified day, but drivers caught cheating more than three times per year will be banned from the benefits provided by the No-Driving Day program.

- In 2009 the Seoul Metropolitan Government launched the Eco-mileage program to encourage citizens to take an active role in reducing greenhouse gases. The program provides incentives to households that reduce their electricity, water, and gas consumption by 10% compared to their consumptions levels over the two prior years. As of the first half of 2010, 186,000 households have joined the program and 70,742 tons of CO_2 were reduced over the past 6 months. Participating households receive green consumer goods discounts worth US$50, tree planting vouchers, energy auditing services, etc. Groups (such as schools or apartment complexes) can receive subsidies for different greening projects worth approximately US$10,000.

- Seoul also established the *Low CO_2 Green Bank Account*, a green financial mechanism, to raise funds for use in responding to climate change. Citizens can be offered discount commission benefits when they open an account with Wooribank, a private Korean bank. This account is similar to other private bank accounts, except all profits yielded from the account will be directed to a new Seoul Climate Change Fund. This fund will then be used on a range of low carbon projects, including providing eco-friendly consumer goods to low income households.

local government. The effort is generally managed by local authority staff, although key support roles may be played by outside NGOs such as ICLEI-Local Governments for Sustainability, private consultants, academic researchers, or representatives from the general public or local private companies with special knowledge or interests in the outcome of the planning work. The extent of external involvement also varies depending on the desired scope of the planning exercise. Efforts focused on reducing energy use or greenhouse gas emissions solely from local authority operations (i.e., so-called "corporate" emissions) primarily involve officials from agencies responsible for these emissions. Broader efforts seeking to reduce emissions from other sources around the city (e.g., homes or local businesses) often include more external stakeholders. Another key scoping decision shaping local planning efforts is the decision over whether to target energy use or greenhouse gas emissions from new or existing buildings, as each requires a dramatically different policy orientation.

[MITIGATION] Box 4.4 Cities for Climate Protection (CCP) Campaign of ICLEI–Local Governments for Sustainability

Yunus Arikan

ICLEI-Local Governments for Sustainability

This case study summarizes information available on the ICLEI website, at www.ICLEI.org.

The Cities for Climate Protection (CCP) Campaign[6] assists cities and local governments to adopt policies and implement quantifiable measures to reduce local greenhouse gas emissions, improve air quality, and enhance urban livability and sustainability.

BACKGROUND

In 1993, at the invitation of ICLEI, municipal leaders met at the United Nations in New York, for the 1st Municipal Leaders Summit on Climate Change, and adopted a declaration that called for the establishment of a worldwide movement of local governments to reduce greenhouse gas emissions, improve air quality, and enhance urban sustainability. The result was the CCP Campaign, today recognized as the longest running climate change mitigation campaign globally.

FIVE MILESTONE PROCESS

The CCP Campaign follows a five-step course of action (milestones) providing a simple, standardized way to reduce greenhouse gas emissions and to monitor, measure, and report performance. The milestones allow local governments to understand how municipal decisions affect energy use and how these decisions can be used to mitigate global climate change while improving community quality of life. ICLEI has developed several software tools that help cities comply with the methodology.

THE FIVE MILESTONES ARE:

Milestone 1. Conduct a baseline emissions inventory and forecast. Based on energy consumption and waste generation, the city calculates greenhouse gas emissions for a base year and for a forecast year. The inventory and forecast provide a benchmark against which the city can measure progress.

Milestone 2. Adopt an emissions reduction target for the forecast year. The city establishes an emissions reduction target for the city. The target both fosters political will and creates a framework to guide the planning and implementation of measures.

Milestone 3. Develop a Local Action Plan. Through a multi-stakeholder process, the city develops a Local Action Plan that describes the policies and measures that the local government will take to reduce greenhouse gas emissions and achieve its emissions reduction target. Most plans include a timeline, a description of financing mechanisms, and an assignment of responsibility to departments and staff. In addition to reduction measures, most plans also incorporate public awareness and education efforts.

Milestone 4. Implement policies and measures. The city implements the policies and measures contained in their Local Action Plan. Typical policies and measures implemented by CCP participants include energy efficiency improvements to municipal buildings and water treatment facilities, streetlight retrofits, public transit improvements, installation of renewable power applications, and methane recovery from waste management.

Milestone 5. Monitor and verify results. Monitoring and verifying progress on the implementation of measures to reduce or avoid greenhouse gas emissions is an ongoing process. Monitoring begins once measures are implemented and continues for the life of the measures, providing important feedback that can be use to improve the measures over time.

ACHIEVEMENTS

Since its inception, the CCP Campaign has grown to involve more than 1,000 local governments worldwide that are integrating climate change mitigation into their decision-making processes, covering around 10 percent of the world's urban population and including approximately 20 percent of global urban anthropogenic greenhouse gas emissions.

Following the achievements in North America, Australia, and Europe in the early 2000s, CCP is recognized as the only local government climate mitigation campaign taking place in developing countries in Latin America, South Africa, South Asia, and Southeast Asia.

Specific regional/national achievements are:

SOUTH ASIA

In 2009 ICLEI South Asia published *Energy and Carbon Emissions Profiles of 54 South Asian Cities*, a comprehensive account of corporate and community emissions from 54 local authorities from India, Bangladesh, Bhutan, Nepal, and Sri Lanka, compiled by ICLEI South Asia. The report is recognized as one of the most comprehensive compilations of greenhouse gas emissions of cities in developing countries.

USA

In 2009 ICLEI USA welcomed Oklahoma City as the 600th member of its national CCP Campaign. Around 200 local governments have completed their greenhouse gas baseline inventory and at least 155 have committed to emissions reduction targets. The projected greenhouse gas emissions reduction from these targets is expected to add up to more than 1.36 billion tonnes CO_2-eq by 2020 – the equivalent of taking 25,000,000 passenger vehicles off the road for the next 10 years.

NEW ZEALAND

Communities for Climate Protection: New Zealand, Actions Profile 2009 summarizes greenhouse gas emissions reduction data from 34 councils covering 83 percent of the New Zealand population. The total of reported and quantifiable emissions reductions from CCP-NZ council activities, since councils' inventory base-years (starting from June 2004) to June 2009, has been conservatively calculated to be more than 400,000 tonnes CO_2-eq.

AUSTRALIA

CCP Australia was launched in 1997 and as of 30 June 2008 had 233 participating councils, representing about 84 percent of the Australian population. In 2007/2008 over 3,000 greenhouse gas abatement actions were reported by 184 councils across Australia. Collectively these actions prevented 4.7 million tonnes CO_2-eq from entering the atmosphere – the equivalent of taking over a million cars off the road for an entire year. Since the start of reporting in 1998/1999, 18 million tonnes CO_2-eq have been abated by Australian cities within the CCP Campaign.

IMPACTS

Following the success achieved through the implementation of CCP, ICLEI and local government networks are able to advocate for more ambitious greenhouse gas reduction policies at the national and international level. CCP enhances cities' access to carbon financing and improves standardization of urban greenhouse gas accounting. Based on the experience of the five-milestone process, ICLEI further developed innovative actions in adaptation to climate change at the local level.

The local climate planning process often begins with a data gathering and analysis exercise to track local greenhouse gas emissions. As noted in Chapter 8, this analysis can be difficult to conduct. Emission reduction targets set decades into the future are frequently established, in many cases at levels suggested by different climate policy networks or technical assistance initiatives. For example, the US Conference of Mayors Climate Protection Agreement launched in 2005 sought to convince mayors to commit to a 7 percent reduction in their city's emission levels by 2010, the same level called for by the USA as a whole under the Kyoto Protocol (US Conference of Mayors, 2010).

The lifespan of these planning initiatives varies. Some cities have designed them as ongoing initiatives, with regular reporting on results and updating of plans to reflect implementation progress, new knowledge, or changing local conditions.

4.5.1.2 *Greenhouse gas mitigation policies and programs*

Although many factors affect the exact policy and program prescriptions contained in local climate plans, it is common to see plans emphasizing specific policies and programs targeting the largest emission sources identified by the local greenhouse gas emission inventory. In London, for example, the local climate plan projects how specific proposed strategies will collectively shift the city from its current business-as-usual greenhouse gas emissions path (see Figure 4.6).

Existing technology choices directly influence the content of a city's plan. For example, cities receiving the bulk of their electricity supply from low- or non-greenhouse gas emitting sources such as hydropower or nuclear power (e.g., Paris) tend to focus their policy attention to thermal or transport-related energy use, essentially viewing electricity consumption as a less problematic issue.

Cities reliant on carbon-intensive power sources frequently emphasize fuel switching or technology switching as a means of driving down emission levels. Strategies include increasing the use of renewable power generated within or imported to the city, or the replacement of existing large power plants with more energy efficient turbine designs (San Francisco Public Utility Commission, 2002). Combined heat and power technology deployment or district energy system expansion may also be advocated, because these technologies can both heat homes and power steam chillers that replace electric powered air conditioning units. In New York City, it is estimated that such chilling units and other technologies connected to the district steam system displace nearly 375 MW of electric demand around Manhattan (City of New York, 2007).

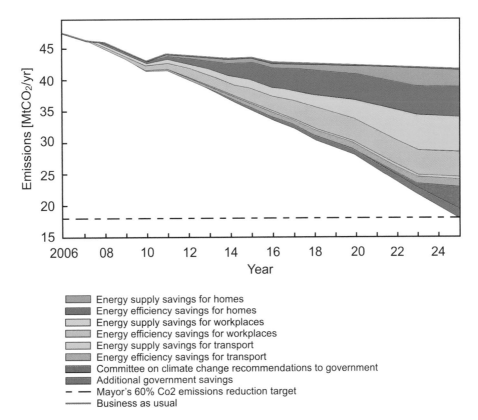

Figure 4.6: *Projected emission impacts of selected climate mitigation strategies contained in London Climate Change Action Plan.*

Source: Greater London Authority (2007).

Other strategies emphasize demand-side initiatives to reduce overall energy use around the city, including efforts promoting the use of more efficient lighting systems on roadways and in homes and businesses. Because peak power demand in many cities is linked to air conditioning use on hot days, efforts to reduce solar gain within buildings on hot days often find a home in local climate plans. Strategies promoted include green roofs, cool roofs, enhanced wall or ceiling cavity insulation, or landscaping programs planting trees on the sunny side of buildings. Electricity-intensive businesses or business units such as computer data centers may also be the target of policymaker interest.

There is a lengthy literature of technical assistance documents that have been developed over time advising cities on "best practice" initiatives (Natural Capitalism Solutions, 2007; US Conference of Mayors, 2009); some of these best practice claims are better documented than others. One of the biggest challenges local authorities face when considering these many ideas is assessing whether ideas deemed effective in one city can be effectively translated to a completely different local context. This problem is most pronounced in cities in less-developed countries, where the challenge of providing even the most basic type of energy infrastructure system has proven vexing (USAID, 2004).

A closely related point local authorities must consider is which type of policy instruments will prove most effective at delivering

on their policy goals. In many cases, cities lack relevant local evidence, and must rely on outcomes achieved in other cities where the underlying economic or policy conditions could be completely different. Policymakers may also have strong ideological preferences that shape their policy decisions, preferring mandates to incentives or vice versa.

Energie-Cités (2009) identifies ten different functions or roles for local government, each of which creates specific policy and program opportunities to influence local carbon emission levels.[7] Hammer (2009) argues that these functions can be collapsed into five unique types of policy levers that generally fall under mayoral control, each of which can contribute towards a comprehensive local mitigation policy:

- Rulemaking – regulatory or policymaking powers, including the ability to impose land use controls that reduce the need for private vehicle use; environmental standards; or clean energy technology requirements
- Regulatory oversight – responsibility for the enforcement of standards established by other governmental entities (such as building codes promoting energy efficiency)
- Direct expenditures/procurement – use of local purchasing powers to procure efficiency upgrades that reduce the local authority's own energy expenditures
- Financial incentives – tax breaks, permitting rule modifications, or cash subsidies or financial penalties designed to

---

7   These roles include consumer, service provider, model, planner, developer, regulator, advisor, motivator, producer, and supplier.

promote energy efficiency or clean energy investments and behavior

- Information/advocacy – use of local government's highly visible platform to speak out on local energy issues or convene stakeholder meetings to move projects forward.

Regardless of which instrument(s) local authorities choose to employ, officials must assess how long they will stick with a certain policy approach, and when they should shift to an alternative strategy if the original policy is proving less effective than first hoped. The fact that the greenhouse gas emission reduction targets established by most cities are aspirational, rather than obligatory, means there may be some laxness about shifting away from unsuccessful policy approaches. The long time horizon of most emission reduction targets also makes accountability difficult, as these targets generally extend far beyond the average mayor's term of office.

Their limited policy control powers also highlight the need for local authorities to structure their mitigation efforts in more holistic or cross-cutting ways that leverage support and involvement from other key stakeholders and levels of government. Some cities do this by articulating advocacy strategies designed to win changes giving local officials additional financial resources or more policy control powers (Greater London Authority, 2004). Others emphasize information and education campaigns that inform local energy users about financial or technical assistance resources available from state or central government.

4.5.1.3 Climate change adaptation policies and programs

Compared to efforts to mitigate the role of local energy systems in global climate change, efforts to adapt local energy systems to changing climatic conditions are much more difficult to identify. A scan of many local climate plans finds energy system adaptation rarely mentioned, or defined only in terms of a single type of climate risk, such as vulnerability to coastal flooding (Webster and McElwee, 2009).

Part of the problem is that local understanding of the climate impacts a specific city will face has historically been poor; efforts to downscale global climate models described earlier in this report are only now being employed in cities around the globe. Cities can use past extreme weather events as proxies, with the assumption that climate change will exacerbate the frequency or scale of these events. Even this approach, however, does not fully explain which of the many possible climate-related impacts identified earlier in Chapter 3 are most relevant to the local energy system over different timescales.

Another factor potentially impeding local authority engagement is their limited "ownership" of the solutions to this problem. The majority of the impacts cited in Chapter 3 affect long-embedded central energy system assets that fall outside of the direct jurisdiction of local officials. This fact forces local authorities to develop an advocacy, education, or partnership agenda, seeking to engage other key stakeholders such as utility owners and

---

[ADAPTATION/MITIGATION] Box 4.5 Adaptation and mitigation of climate change impacts in Kampala, Uganda

Shuaib Lwasa	Frank Mabiirizi
International Potato Center	*Independent Senior Consultant*
Cecilia Njenga	Paul Mukwaya
UN-HABITAT	*Makerere University*
Charles Koojo	Deogratious Sekimpi
URTC	*UNACOH*

Climate change, now a reality, is influencing realignment of global and country policies towards adaptation and mitigation (Prasad *et al.*, 2009). The effects of climate change are now being felt, with Africa as the most vulnerable region (UN-HABITAT, 2008). This is due to Africa's multidimensional unpreparedness, yet the continent is unequivocally urbanizing faster than any region globally, exposing inland and coastal cities to risks. Cities are both contributors to and vulnerable to climate change, but the effects of climate change are exacerbating the already grim environmental, social, and economic challenges heightening the risk to the urban poor (UN-HABITAT, 2008). Urban vulnerabilities are manifest in several areas including housing, energy, food security, water resources, health, transport infrastructure, environmental services, and

economic productivity. This box highlights findings of climate change effects, and strategies for mitigation and adaptation in Kampala. Under the Sustainable Urban Development network (SUD-Net), the Cities in Climate Change Initiative (CCCI) of the UN-HABITAT is aimed at raising awareness, developing tools, and building capacity for municipalities and intensification of adaptation and mitigation activities through demonstration projects. The CCCI is building on existing climate change mitigation and adaptation measures at national and city levels by providing frameworks for urban vulnerability assessment, identifying scalable adaptation and mitigation measures implemented at community to city levels through Local Climate Change Plans.

Applying a multi-faceted methodological approach that utilized geospatial analysis integrating demographic, social, economic, and environmental data complemented with meta-evaluation of climate change projects, findings show that the impacts are increasing. In Uganda, there has been recorded variation in average temperatures that correlates with an estimated increase of 1.5 °C in the next 20 years and by up to 4.3 °C by the 2080s, although recent scientific studies indicate that the globe could warm by 4 °C by 2050. Significant observed changes in rainfall patterns and temperature continue to pose vulnerabilities to urban areas in Uganda. The most significant impact to Kampala is flooding due to increased rainfall that is spread over relatively short or extended periods. Increase in runoff has made flooding the most serious threat to humans, livelihoods, the urban system, and the economy. On the other hand, changes in temperature regimes have affected urban livelihoods and food security.

Kampala city is the primary city, with 39.6 percent of the national urban population. Located along the shores of Lake Victoria, a region with evidence of increased precipitation, the challenge of surface runoff coupled with non-robust drainage systems has increased the vulnerability of Kampala city's infrastructure, housing, social services, and livelihoods. Between December 2006 and February 2007 there was serious damage to housing and schools and disruption of livelihoods from excessive rainfall. These vulnerabilities are felt variably in a city "region" of Kampala spanning an estimated surface area of 1,895 sq km with a spatial connectedness of economic, social, and environmental processes (Nyakaana et al., 2004). The various urban sectors of the city are affected in different ways, so that sector-specific vulnerability analysis provides better clues on mitigation and adaptation measures. Energy is an important sector with heavy reliance on biomass energy for domestic and institutional use. About 75 percent of Kampala's population use wood fuel and will use an estimated 535 metric tons annually by 2007 (Mukwaya et al., 2007). This is coupled with increases in motorized transportation and consumption of petroleum products leading to greenhouse gas emissions. Although the contribution of Uganda to CO_2 emissions is low, adapting urban transportation for energy efficiency is important. Another sector associated with energy is housing, with two roles: protection of inhabitants from climate change impacts; and contribution of buildings to emissions. Analysis shows that existing buildings are neither energy efficient nor protective to inhabitants. Low- or neutral-energy housing is needed and a housing code that is energy efficient is to be developed under the CCCI.

Urban water is an important sector vulnerable to climate change. Safe urban water supply reaches only 67 percent of the population in Kampala, with the large population left out being the urban poor. Climate change impacts around Lake Victoria have led to decrease in the water levels and affected supply for 2.5 years. Climate change is likely to worsen the situation for the urban poor. In respect to solid waste collection, transportation, and disposal, the principle of "generator pays" is the basis of solid waste management, but despite the initiative, solid waste management practices

are deplorable. The city has not benefited from the Clean Development Mechanism (CDM) of trading carbon credits from landfill gas capture. Local-level innovative ways of utilizing waste for energy with the potential to reduce landfill emissions is yet to be scaled up. The linkage between solid waste management, energy, and flooding has increased the vulnerability of the city's population to health hazards. Infectious diseases, especially water-related and air-borne, are prevalent in many of the neighborhoods of Kampala: disease outbreaks occurred in 1997, 1999, 2004, 2006, and 2008 due to the increased floods (KCC and BTC, 2008). With these impacts, urban health services become overstretched to meet the challenges of high service demand. The ecosystem of the city region is also under threat with wetland destruction, biodiversity loss and soil erosion augmented by clearance of vegetation, and ecosystem services decline. Ecosystem conservation and management remains an important component for climate change adaptation and mitigation. A gender perspective of climate change vulnerability has informed the initiative to be responsive by analyzing effects on different gender groups and strategies that address the needs of women and children.

SUD-Net CCCI has initiated awareness-raising campaigns, which will be followed by development of tools to enable different stakeholders to develop climate change plans. Drawing on the National Adaptation Program of Action (NAPA) and the Initial Climate Communication tool, the CCCI is enabling amplification of the role of urban areas in climate change adaptation and mitigation (Isabirye, 2009). A platform to enable engagement of stakeholders is envisaged to highlight vulnerability for policy action. Various demonstration projects, including city greening, alternative energy briquette utilization, clean wood fuel use, climate proofing of infrastructure, and designing energy efficient urban transport systems, are underway for long-term response to climate change. A key aspect of this program is building institutional resilience and adaptation to climate change by investing in action research that brings together different stakeholders. There is much needed knowledge to inform climate change policy, a wealth of which exists but is not widely disseminated. This necessitates innovations in enabling information flow for up- and outscaling of innovations. Thus, information sharing is important and provides an opportunity for communicating and networking on climate change. UN-HABITAT under the SUD-Net is supporting a Local Urban Knowledge Arenas (LUKAS) platform through which climate change information at city and national level will be exchanged.

regulators in a way that will advance the city's interest in a more robust and resilient energy system.

This is not to say there has been no progress on this issue. New York, London, and Chicago all have active energy system adaptation initiatives underway, working closely with key energy system stakeholders and regulators (Chicago Climate Task Force, 2007; Greater London Authority, 2008; City of New York, 2008). Other cities have also identified steps they would

like to take to adapt the local energy system to climate-related impacts, while researchers and non-governmental organizations are coming up with their own guidance documents (Neumann and Price, 2009; Williamson et al., 2009). Many of these emphasize the synergistic nature of mitigation and adaptation strategies, with system changes intended to reduce greenhouse gas emission levels whilst simultaneously enhancing the resilience of the system to climatic changes (Laukkonen et al., 2009).

It is possible to categorize adaptation initiatives in various ways, including those that reduce sensitivity, alter exposure, or increase resilience to changing conditions (Adger *et al.*, 2005). As discussed above, it also helps to categorize strategies as relating to either energy supply or demand. Examples of different energy system adaptation strategies are found in Table 4.6, broken out by these two classification systems. In some cases, individual policy and program strategies provide benefits in multiple impact categories.

Cities opting to pursue adaptation initiatives will likely find that grappling with uncertainty over the nature, scale, and timing of the impacts will be a significant challenge. Part of the problem arises from the fact that the energy system itself is constantly changing, reflecting technological and market innovation and growing demand levels. Each segment of the system also has a natural lifespan, creating opportunities to upgrade the system or enhance its climate resiliency as part of the natural life cycle

of the equipment (Neumann and Price, 2009). By themselves, these factors make system planning a highly complex endeavor; adding climate change to the mix only compounds the difficulty (Linder *et al.*, 1987; ICF, 1995; Scott and Huang, 2007). Equally challenging is the fact that many energy companies have a relatively short capital investment horizon, potentially limiting their interest or ability to take actions whose benefits may only be realized over a much longer timescale.

Local authorities seeking decision rules to rank their adaptation options or manage risk have several options. It may be helpful to apply scenario analysis to the problem, selecting the option(s) that result in the least sensitivity to future climate conditions (Lempert and Collins, 2007). Hallegatte (2008) suggests that "no-regret" or "reversible" policies be considered. No-regret strategies are those yielding benefits even if impact projections prove overblown; energy efficiency initiatives are "no-regret measures par excellence" (Mansanet-Bataller *et al.*,

Table 4.6: *Examples of energy system adaptation strategies.*

Impact category	Energy supply	Energy demand
Reduce sensitivity: alter the scale or type of local energy system assets or markets to minimize the effects of reduced system output or failure	• Reduce supply sensitivity to loss of hydropower availability by increasing reservoir system capacity (Adger *et al.*, 2005) • Install in-building supply systems (thermal or power) at elevations above anticipated flooding levels (Adger *et al.*, 2005) • Construct additional or redundant transmission or distribution line capacity to offset anticipated efficiency losses (Hill and Goldberg, 2001) • Establish new coastal power plant siting rules to minimize flood risk (Stern, 1998) • Install solar PV technology to reduce effects of peak demand (Franco and Sanstad, 2008)	• Install steam-powered chillers to reduce burden on local power system on hot days • Establish or expand demand-response programs which encourage consumers to voluntarily reduce power consumption during peak demand events (Stern, 1998)
Alter exposure: take steps that reduce opportunities for the local energy system to experience damage or problems resulting from climate change	• Upgrade local transmission and distribution network to handle increased load associated with higher temperatures (Hill and Goldberg, 2001) • Protect power plants from flooding with dykes/berms (Mansanet-Bataller *et al.*, 2008) • Expand hazard preparedness programs (Adger *et al.*, 2005) • Install solar PV technology to reduce effects of peak demand (Franco and Sanstad, 2008) • Require utilities to develop storm hardening plans on a regular basis (Neumann and Price, 2009) • Retrofit power plants so they use less cooling water (Neumann and Price, 2009)	• Install steam-powered chillers to reduce burden on local power system on hot days • Establish or expand demand-response programs which encourage consumers to voluntarily reduce power consumption during peak demand events (Stern, 1998) • Improve and rigidly enforce energy efficient building codes (Morris and Garrell, 1996)
Increase resilience: enhance ability of city to recover from losses by reducing overall need for energy services or enhancing speed with which system can recover	• Automate restoration procedures to bring energy system back on line faster after weather-related service interruption (Overbye *et al.*, 2007) • Expand refinery capacity in less vulnerable areas (Neumann and Price, 2009) • Provide additional support for distribution generation systems to spread climate risk over a larger area (Neumann and Price, 2009)	• Establish public education programs to promote lifestyles that are less energy-dependent • Employ passive building design strategies (e.g., larger windows, extra thick walls, flow-through ventilation, natural shading, etc.) to maintain minimum comfort or lighting levels even in situations where energy system losses occur (Commonwealth of Australia, 2007; Miller *et al.*, 2008) • Reduce or eliminate energy subsidies so prices reflect true cost (Stern, 1998)

2008) because they deliver cost-saving benefits regardless of what happens with climate change. Reversible policies allow a local authority to swiftly change course if anticipated problems do not arise or if the policy proves ineffective. Hallegatte (2008) also proposes that local authorities account for uncertainty over climate change by pursuing investments with a shorter projected lifespan. Such a strategy allows the local authority to exploit the replacement cycle for these investments, incorporating the latest scientific knowledge into the procurement process.

4.6 Conclusions, policy recommendations, areas for future research

This chapter makes clear the complexity of urban energy systems. Market structures vary across cities and countries, as do current-day economic and climatic conditions. Technology decisions made long ago that reflect past market and policy/regulatory realities continue to influence choices made today and plans looking toward the future.

The result is that urban energy consumption, the impacts of that consumption, and the vulnerability of urban energy systems to climate change will vary significantly across locales. This local context needs to be well understood, both to elucidate the vulnerabilities and challenges facing a particular city as well as to clarify the options available to combat these threats. Because of the unique circumstances facing each city, there is little evidence on how strategies promoted as "best practice" in one city can be effectively transferred from one city to another. It is also difficult to pinpoint one single energy system type as being more or less vulnerable to the impacts of climate change than another.

Localized climate change studies offer clear benefits, employing downscaled GCMs to establish a scientific justification for local action. Energy supply and demand or greenhouse gas emission inventories also set the stage for comprehensive policymaking efforts.

Local authorities drawing on these facts have many intervention strategies they can employ to influence local energy use or enhance the climate resilience of urban energy systems. Section 4.5 looked at initiatives including public information campaigns, building regulations, and market and policy changes. However, because cities often have a limited span of control, working with partners is vital, including the public, non-governmental organizations and other civil society groups, the private sector, and different scales of government. There is evidence that cities have already recognized this and such collaborations are becoming increasingly common.

4.6.1 Knowledge gaps

Much of the research cited in this report is quite recent. Work in this area continues to evolve and, while some of the basic issues are now well-understood, a number of knowledge gaps remain. These include:

- *Limits on structural or systemic change*: Section 4.2 highlighted the key drivers of urban energy consumption. Although many local action plans seek modest or incremental change to the current energy system, there is a larger question of whether cities can overcome their path dependency to implement large-scale overhauls, dramatically altering the way they make or use energy, and under what timescale this might be possible. In Denmark, cities in the Copenhagen region banded together to completely overhaul the way buildings in the city are heated, installing a comprehensive district energy system that reached into nearly every home and business in just a few years (Manczyk and Leach, 2002). Is such a model transferable to other cities, employing the same or other types of energy technologies? Research authoritatively evaluating all that was done to deliver this change in Copenhagen and its relevance to other cities might go far in helping local authorities move beyond their current energy or climate policymaking comfort zone.

- A corollary to that question is our lack of understanding of the point at which local features of climate, geography, and history are immutable facts that undercut our goals for system transformation. In other words, when do aspirational climate or energy goals become unattainable, and what can be done to identify the realistic limits on change so this can be directly woven into the local planning process?

- *Demand-side projections*: There is little evidence to date on what climate scenarios mean for local energy demand in different cities. Storm and flooding risks are more likely to be known, based on historical experience. Far less understood, however, is the issue of consumer behavior and local price elasticity of demand. If cities start to get hotter, at what point will consumers increase their adoption of air conditioning, and how much will they use it each day? This knowledge is critical because it leads to questions of market pricing, demand-side management, and (potentially) the need for new peak-load generation capacity.

- *Energy supply chains in developing countries*: Little is known to date about how climate change will affect the informal energy systems of developing countries. Because some of these cities are rapidly growing, it will be important to understand whether climate-related system vulnerabilities may be outpaced by a transition to cleaner fuels or by efforts to expand grid access in these cities.

- *Multi-level government policy coordination*: A gaping hole in the urban energy and climate literature is an understanding of the proper role of national and transnational governments in urban energy system governance. Local authorities have their own vision in terms of policy coordination and resource support (see City of Copenhagen, 2009), but more fundamental questions about power sharing or the devolution of power from central/state government have not been examined in meaningful ways. A related question involves market restructuring efforts. Increased competition in supply has led to significant changes in technology deployment and energy planning responsibilities, but transmission and distribution functions are still largely regulated monopolies. In both cases, are these regulatory systems structured in ways that they can meaningfully address that challenges presented by climate change? Moreover, are regulators informed about how climate change may manifest itself in different cities under their jurisdiction, and on what time frame? Have they begun to weave these facts or projections into their regulatory decisions, such as whether to allow utilities to receive rate recovery for climate resiliency investments?

- *Future-proofing*: Hallegatte's "no-regrets" strategy (Hallegatte, 2008) provides guidance on adaptation strategies cities can pursue with little concern for how climate change actually plays out, because the environmental or efficiency benefits of these strategies will always be valued. Uncertainty also exists, however, in the form of future national-level climate change mitigation policies and technological innovation. How can local authorities craft policies that will serve their long-term energy and climate interests without a full understanding of whether breakthrough technologies may fundamentally change how energy must be generated, imported, or used within a city? Local authorities would benefit from such guidance, particularly because it is likely that central and state governments will be increasingly active on climate change mitigation efforts in the coming years, leaving local authorities to play catchup.

4.6.2 Where do cities go from here?

While there are clearly a number of challenges facing urban governments, they need not respond passively to signals from higher level governance and broader trends. Indeed, there are a number of activities where cities can show leadership and promote better understanding in this area.

- *Climate planning and policy/program implementation*: Local authorities should establish or continue their efforts to both mitigate and adapt to the impacts of climate change. Working through local government networking organizations, such as ICLEI, Metropolis, the C40, and others, cities should share both their successes and failures as a means of advancing international knowledge and best practice on these issues.
- *Advocacy*: Local authorities need to continue their participation in both city networks and global discussions (e.g., the Mayor's Climate Summit at the Copenhagen Climate Summit in December 2009) as a reminder to national governments about their central global role and unique local risks and needs.
- *Data collection*: Understanding the problem is the first step. The IEA (2008) and other reports are helping to establish baseline estimates of total urban emissions, but data collection efforts should continue to facilitate comparative analyses and informed policymaking at the local, state/national, and international level. This might benefit from the adoption of clear urban energy and emissions accounting standards (see Kennedy *et al.*, 2009a).
- *Awareness*: Local authorities can begin or expand outreach to the public on these issues to heighten awareness of the unique challenges facing "their" city and the role that citi-

zens can play in ensuring the success of local energy and climate initiatives.

- *Academic research*: To support the broad research agenda outlined above, local authorities should work closely with research institutions to provide access to data and collaborate on long-term planning and assessment efforts.

REFERENCES

Abt Associates, ICF Consulting, and E. H. Pechan Associates (2000). *The Particulate-Related Health Benefits of Reducing Powerplant Emissions*. Prepared for Clean Air Task Force, Boston, MA, USA.

Acclimatise (2008). A practical approach to climate risk management in water and finance. Presentation by John Firth, CEO, Acclimatise, Conference: *Climate Change and the Caribbean: Strengthening the Science-Policy Interface*, Barbados, November 18–19, 2008.

Adger, W. N., N. W. Arnell, and E. L. Tompkins (2005). Successful adaptation to climate change across scales. *Global Environmental Change*, **15**, 77–86.

Akbari, H. and S. Konopacki (2005). Calculating energy-saving potentials of heat-island reduction strategies. *Energy Policy*, **33**, 721–756.

Amato, A. D., M. Ruth, P. Kirshen, and J. Horwitz (2005). Regional energy demand responses to climate change: methodology and application to the Commonwealth of Massachusetts. *Climatic Change*, **71**, 175–201.

Aspen Environmental Group and M Cubed (2005). *Potential Changes in Hydropower Production from Global Climate Change in California and the Western United States*. Prepared in support of the 2005 Integrated Energy Policy Report Proceeding (Docket #04-IEPR-01G). California Energy Commission.

Baxter, L. W. and K. Calandri (1992). Global warming and electricity demand: a study of California. *Energy Policy*, **20**, 233–244.

Breslow, P. B. and D. J. Sailor (2002). Vulnerability of wind power resources to climate change in the Continental United States. *Renewable Energy*, **27**, 585–598.

Brown, M., F. Southworth, and A. Sarzynski (2008). *Shrinking the Carbon Footprint of Metropolitan America*, New York: Brookings Institution.

Bulkeley, H. and M. Betsill (2002). *Cities and Climate Change: Urban Sustainability and Global Environmental Governance*, London, UK: Routledge.

Bull, S. R., D. E. Bilello, J. Ekmann, M. J. Sale, and D. K. Schmalzer (2007). Effects of climate change on energy production and distribution in the United States. In *Effects of Climate Change on Energy Production and Use in the United States. A Report by the U.S. Climate Change Science Program and the subcommittee on Global Change Research*, Washington, DC, USA.

Capello, R., P. Nijkamp, and G. Pepping (1999). *Sustainable Cities and Energy Policies*, Berlin: Springer-Verlag.

Carmin, J., D. Roberts, and I. Anguelovski (2009). *Planning Climate Resilient Cities: Early Lessons from Early Adapters*. Working paper presented at 5th Urban Research Symposium, Cities and Climate Change: Responding to an Urgent Agenda, Marseilles, France, June 2009.

CCME [Canadian Council of Ministers of the Environment] (2003). *Climate, Nature, People: Indicators of Canada's Changing Climate*, Climate Change Indicators Task Group of the Canadian Council of Ministers of the Environment, Canadian Council of Ministers of the Environment Inc., Winnipeg, Canada.

CEIC (2010). *Webceic Data Manager* (on-line database), China Premium Database, New York, USA: ISI Emerging Markets.

Chen, S., H. Yoshino, and N. Li (2010). Statistical analyses on summer energy consumption characteristics of residential buildings in some cities of China. *Energy and Buildings*, **42**, 136–146.

Chicago Climate Task Force (2007). Infrastructure. In *Climate Change and Chicago: Projections and Potential Impacts*, City of Chicago, November 2007.

City of Amsterdam (2007). *New Amsterdam Climate: Summary of Plans and Ongoing Projects*, Amsterdam Climate Office.

City of Cape Town (2006). *Cape Town Energy and Climate Strategy*, Cape Town Environment Planning Department.

City of Copenhagen (2009). *Mayor's Climate Summit Communique*. Accessed 23 January 2010, www.kk.dk:80/Nyheder/2009/December/~/media/B5A397DC695C409983462723E31C995E.ashx.

City of New York (2007). *PlaNYC 2030. A Greener, Greater, New York*, Office of Long Term Planning and Sustainability, Mayor's Office of Operations.

City of New York (2008). Press release: *Mayor Bloomberg launches Task Force to adapt critical infrastructure to environmental effects of climate change*, New York: Office of the Mayor, August 12, 2008.

City of Toronto (2007). *Climate Change, Clean Air and Sustainable Energy Action Plan: Moving from Framework to Action (Phase 1)*, Toronto Environment Office and Toronto Energy Efficiency Office.

Ciudad de Mexico (2008). *Programa de Accion Climatica de la Ciudad de Mexico 2008–2012*, Secretaria del Medio Ambiente and Gobierno del distrito Federal.

Clancy, J., O. Maduka, and F. Lumampa (2008). Sustainable energy systems and the urban poor: Nigeria, Brazil and the Philippines. In P. Droege (Ed.), *Urban Energy Transition: From Fossil Fuels to Renewable Power*, London, UK: Elsevier Science.

Commonwealth of Australia (2007). *Climate Change Adaptation Actions for Local Government*. Australian Greenhouse Office, Department of the Environment and Water Resources.

Dhakal, S. (2009). Urban energy use and carbon emissions from cities in China and policy implications. *Energy Policy*, **37**, 4208–4219.

Dhingra, C., S. Gandhi, A. Chaurey, and P. K. Agarwal (2008). Access to clean energy services for the urban and peri-urban poor: a case study of Delhi, India. *Energy for Sustainable Development*, **12**, 49–55.

Dodman, D. (2009). Blaming cities for climate change? An analysis of urban greenhouse gas emissions inventories. *Environment and Urbanization*, **21**, 185–201.

Dutton, A. G., J. A. Halliday, and M. J. Blanch (2005). *The Feasibility of Building-Mounted/Integrated Wind Turbines: Achieving Their Potential for Carbon Emission Reductions*, Energy Research Unit, Rutherford Appleton Laboratory, UK, Science and Technology Facilities Council.

EIA (2001). *Residential Energy Consumption Survey (RECS)*, Washington, DC: Energy Information Administration, U.S. Department of Energy.

EIA (2003). *Commercial Buildings Energy Consumption Survey (CBECS)*, Washington, DC: Energy Information Administration, U.S. Department of Energy.

EIA (2006). *International Energy Outlook 2006*, Washington, DC: Energy Information Administration, U.S. Department of Energy.

EIA (2008a). *2007 Buildings Energy Data Book*, Washington, DC: Energy Information Administration, U.S. Department of Energy.

EIA (2008b). *Annual Energy Review 2007*. Washington, DC: Energy Information Administration, U.S. Department of Energy.

EIA (2009). *Annual Energy Outlook 2009 Early Release*. Report DOE/EIA-0383(2009), Appendix A: Table A5 Commercial Sector Key Indicators and Consumption, Washington, DC: Energy Information Administration, U.S. Department of Energy.

Elsman, P. (2009). Copenhagen District Heating System, Application for the Global District Energy Climate Award, Municipality of Copenhagen, Denmark.

Energie-Cités (2009). *Save Energy, Save the Climate, Save Money: Guide for Local and Regional Governments*.

European Commission (2007a) *City Instruments: Monitoring, Evaluating and Transferring Instruments to Address Climate Change in Metropolitan Regions. Best Practice Catalog*.

European Commission (2007b). *State of European Cities Report: Adding Value to the European Urban Audit*. Accessed January 31, 2010, http://ec.europa.eu/regional_policy/sources/docgener/studies/pdf/urban/stateofcities_2007.pdf.

European Commission (2010). *Covenant Towns & Cities* (website). Accessed December 16, 2010, www.eumayors.edu/covenant_cities/towns_cities_en.htm.

Fall, A., S. Sarr, T. Dafrallah, and A. Ndour (2008). Modern energy access in peri-urban areas of West Africa: the case of Dakar, Senegal. *Energy for Sustainable Development*, **12**, 22–37.

Farber, S. (1998). Undesirable facilities and property values: a summary of empirical studies. *Ecological Economics*, **24**, 1–14.

Fenger, J. (Ed.) (2007). *Impacts of Climate Change on Renewable Energy Sources: Their Role in the Nordic Energy System. A Comprehensive Report Resulting from the Nordic Energy Research Project*. Norden.

Fidje, A. and T. Martinsen (2006). Effects of climate change on the utilization of solar cells in the Nordic region. In *European Conference on Impacts of Climate Change on Renewable Energy Resources*, Reykjavik, Iceland.

Franco, G. (2005). *Climate Change Impacts and Adaptation in California*, California Energy Commission.

Franco, G. and A. H. Sanstad (2008). Climate change and electricity demand in California. *Climatic Change*, **87** (supplement 1), S139–S151.

French, H. (2008). Severe Snow Storms Batter China. *New York Times, January 28*.

Gochenour, C. (2001). *District Energy Trends, Issues, and Opportunities: The Role of the World Bank*. World Bank, Technical Paper No 493, Washington, DC, USA: World Bank.

Goteborg Energi (2009). *Goteborg Energi's District Energy System*, Application for Global District Energy Climate Awards. Goteborg, Sweden.

Greater London Authority (2004). *Green Light to Clean Power: The Mayor's Energy Strategy*.

Greater London Authority (2007). *Action Today to Protect Tomorrow: The Mayor's Climate Change Action Plan*.

Greater London Authority (2008). *The London Climate Change Adaptation Strategy Draft Report*.

Grunwald, M. (2007). The New Action Heroes. *Time*, June 25, 2007.

Hallegatte, S. (2008). *Do Not Rely on Climate Scientists to Do Your Work*, Reg-Markets Center, Related publication 08–01.

Hammer, S. (2009). *Capacity to Act: The Critical Determinant of Local Energy Planning and Program Implementation*. Working paper presented at 5th Urban Research Symposium, Cities and Climate Change: Responding to an Urgent Agenda, Marseilles, France, June 2009.

Hannah, L. (1979). *Electricity Before Nationalisation: A Study of the Development of the Electricity Supply Industry in Britain to 1948*, London: MacMillan Press.

Harrison, G. P., L. C. Cradden, and J. P. Chick (2008). Preliminary assessment of climate change impacts on the UK onshore wind energy resource. *Energy Sources, Part A: Recovery, Utilization, and Environmental Effects*, **30**, 1286–1299.

Hewer, F. (2006). *Climate Change and Energy Management*. UK Met Office.

Hill, D. and R. Goldberg (2001). Energy demand. In C. Rosenzweig and W. D. Solecki (Eds.), *Climate Change and Global City: The Potential Consequences of Climate Variability and Change – Metro East Coast*, Report for the US Global Change Research Program, National Assessment of the Potential Consequences of Climate Variability and Change for the United States, Columbia Earth Institute, New York.

House of Commons (2007). *Local Energy: Turning Consumers into Producers*, First Report of Session 2006–7 (HC 257), Trade and Industry Committee, London: The Stationery Office.

Hughes, T. P. (1983). *Networks of Power: Electrification in Western Society 1880–1930*, Baltimore, MD, USA: Johns Hopkins University Press.

ICF (1995). *Potential Effects of Climate Change on Electric Utilities*, TR105005, Research Project 2141–11. Prepared for Central Research Institute of Electric Power Industry (CRIEPI) and Electric Power Research Institute (EPRI).

ICLEI (2000). *Best Practices for Climate Protection: A Local Government Guide*.

ICLEI (2009a). *Sustainability Planning Toolkit: A Comprehensive Guide for Local Governments on how to Create a Sustainability Plan*. ICLEI-USA in association with the City of New York's Mayor's Office of Long Term Planning and Sustainability.

ICLEI (2009b). *Energy and Carbon Emissions Profiles of 54 South Asian Cities*.

ICLEI (2010). *About ICLEI* (website). Accessed December 16, 2010, www. ICLEI.org/index.php?id=about.

IEA [International Energy Agency] (2008). Energy use in cities. In *World Energy Outlook*.

IEC [International Electrotechnical Commission] (2007). *Efficient Electrical Energy Transmission and Distribution*.

Isabirye, P. (2009). *Climate Change Impacts and Adaptation Strategies: Opportunities for Reduced Impacts*. Kampala, Uganda: UAIA.

ISET (2009). *Asian Cities Climate Change Resilience Network (ACCCRN): Responding to the Urban Climate Challenge*. Boulder, CO, USA: Institute for Social and Environmental Transition.

Jollands, N., M. Ruth, C. Bernier, and N. Golubiewski (2007). The Climate's Long-term Impact on New Zealand Infrastructure (CLINZI) project: A case study of Hamilton City, New Zealand. *Journal of Environmental Management*, 83, 460–477.

Jovetski, J. (2006). Preparation keyed Entergy's responses to Katrina, Rita. *Power*. May 15, 2006.

Jowit, J. and J. Espinoza (2006). Heatwave shuts down nuclear power plants. *The Observer*, July 30, 2006.

KCC and BTC (2008). *Baseline Survey for the Kampala Integrated Environmental Management Project, Bwaise III*. Kampala: Kampala City Council and Belgian Technical Cooperation: 81.

Keirstead, J. and N. Schulz (2010). London and beyond: taking a closer look at urban energy policy. *Energy Policy*, 38, 4870–4879.

Kennedy, C., J. Steinberger, B. Gasson, *et al.* (2009a). Methodology for inventorying greenhouse gas emissions from global cities. *Energy Policy*.

Kennedy, C., J. Steinberger, B. Gasson, *et al.* (2009b). Greenhouse gas emissions from global cities. *Environmental Science and Technology*, 43, 7297–7302.

Kirshen, P., M. Ruth, M., and W. Anderson (2008). Interdependencies of urban climate change impacts and adaptation strategies: a case study of Metropolitan Boston USA. *Climatic Change*, 86, 105–122.

Larsen, P. H., S. Goldsmith, O. Smith, *et al.* (2008). Estimating future costs for Alaska public infrastructure at risk from climate change. *Global Environmental Change*, 18, 442–457.

Laukkonen, J., P. K. Blanco, J. Lenhart, *et al.*(2009). Combining climate change adaptation and mitgation measures at the local level. *Habitat International*, 33, 287–292.

Lehner, B., G. Czisch, and S. Vassolo (2005). The impact of global change on the hydropower potential of Europe: a model-based analysis. *Energy Policy*, 33, 839–855.

Lempert, R. J. and M. T. Collins (2007). Managing the risk of uncertain threshold responses: comparison of robust, optimum, and precautionary approaches. *Risk Analysis*, 27(4), 1009–1026.

Linder, K. P., M. J. Gibbs, and M. R. Inglis (1987). *Potential Impacts of Climate Change on Electric Utilities*. New York State Energy Research and Development Authority, Edison Electric Institute, Electric Power Research Institute, US Environmental Protection Agency.

Lovins, A., E. K. Datta, T. Feiler, *et al.*(2002). *Small is Profitable: The Hidden Economic Benefits of Making Electrical Resources the Right Size*, Snowmass, CO, USA: Rocky Mountain Institute.

Madnani, K. (2009). *Climate Change Effects on High-Elevation Hydropower System in California*. Ph.D. Dissertation, Department of Civil and Environmental Engineering,University of California, Davis. Accessed, http://cee.engr.ucdavis.edu/faculty/lund/students/MadaniDissertation.pdf.

Mairie de Paris (2007). *Paris Climate Protection Plan* (English version).

Manczyk, H. and M. D. Leach (2002). *Combined Heat and Power Generation and District Heating in Denmark: History, Goals and Technology*. White paper. University of Rochester.

Mansanet-Bataller, M., M. Herve-Mignucci, and A. Leseur, A. (2008). *Energy Infrastructures in France: Climate Change Vulnerabilities and Adaptation Possibilities*, Mission Climat Working Paper, Paris, Caisse des Depots.

Markoff, M. S. and A. C. Cullen (2008). Impact of climate change on Pacific Northwest hydropower. *Climatic Change*, 87, 451–469.

McKinley, J. C. (2008). Crews from 31 states in Texas to restore power. *New York Times*, September 17, 2008, p. A18.

Miller, N., K. Hayhoe, J. Jin, and M. Auffhammer (2008). Climate, extreme heat, and electricity demand in California. *Journal of Applied Meteorology and Climatology*, 47, 1834–1844.

Morris, S. C. and M. H. Garrell (1996). Report of the Scenario Planning Group for Accelerated Climate Change: "Apple Crisp". In D. Hill (Ed.), *The Baked Apple? Metropolitan New York in the Greenhouse*, Annals of the New York Academy of Sciences, Vol. 790.

Morris, S. C., G. Goldstein, A. Singhi, and D. Hill (1996). Energy demand and supply in Metropolitan New York with global climate change. In D. Hill (Ed.), *The Baked Apple? Metropolitan New York in the Greenhouse*, Annals of the New York Academy of Sciences, Vol. 790.

Mukwaya, P. I., H. Sengendo, *et al.* (2007). *Energy Options in Uganda*, Kampala, Uganda: The Commonwealth People's Forum.

Natural Capitalism Solutions (2007). *Climate Protection Manual for Cities*.

Neumann, J. E. and J. C. Price (2009). *Adapting to Climate Change. The Public Policy Response: Public Infrastructure*. Resources for the Future, June 2009.

New York City Energy Policy Task Force (2004). *New York City Energy Policy: An Electricity Resource Roadmap*. New York State Department of Public Service (2007). In the Matter of Consolidated Edison Company of New York, Inc. Case 07-E-0523 September 2007. Prepared Testimony of Kin Eng, Utility Analyst 3, Office of Electric, Gas, and Water.

New York Times (1888). In a blizzard's grasp.

Nyakaana, J. B., H. Sengendo, *et al.* (2004). *Urban Development, Population and the Environment in Uganda: The Case of Kampala City and its Environs*. Kampala, Uganda.

NYC SBD Task Force [New York City Steam Business Development Task Force] (2005). *Steam Business Development Plan for the Consolidated Edison Steam System*. Steam Business Development Task force with the assistance of CCN Management Council and Thornton Energy Associates, August 2005.

Nye, D. (2001). *Electrifying America: Social meanings of a New Technology*. Cambridge, MA, USA: MIT Press.

OECD (1995). *Urban Energy Handbook: Good Local Practice*. Paris: Organization for Economic Cooperation and Development.

Ontiveros, J. (2009). Application for the First Global District Energy Climate Award. University of Texas, Austin, TX, USA.

Overbye, T., J. Cardell, I. Dobson, *et al.*(2007). *The Electric Power Industry and Climate Change: Power Systems Research Possibilities*. Power Systems Engineering Research Center.

Pan, Z., M. Segal, R. W. Arritt, and E. S. Takle (2004). On the potential change in solar radiation over the US due to increases of atmospheric greenhouse gases. *Renewable Energy*, 29, 1923–1928.

Parshall, L. (2010). Where is Distributed Generation Technically Feasible? Development of a Spatial Model and Case Study of New York. Doctoral thesis, Columbia University.

Parshall, L., K. Gurney, S. A. Hammer, *et al.*(2010). Modeling energy consumption and CO2 emissions at the urban scale: Methodological challenges and insights from the United States. *Energy Policy*, 38, 4765–4782

Perry, M. (2008). China's Snow Storms Not Climate Change: Scientists. *Reuters*.

Pew Center on Global Climate Change (2008). *Understanding Extreme Weather in China and Climate Change Impacts*. Washington, DC, USA.

Philibert, C. (2004). *International Energy Technology Collaboration and Climate Change Mitigation Case Study 1: Concentrating Solar Power Technologies*. International Energy Agency/OECD.

Platt, H. L. (1991). *The Electric City: Energy and Growth of the Chicago Area 1880–1930*. Chicago, IL, USA: University of Chicago Press.

Prasad, N., F. Ranghieri, F. Shah, *et al.*(2009). *Climate Resilient Cities: A Primer on Reducing Vulnerabilities to Disasters*. Washington, DC, USA: World Bank.

Rosenzweig, C., W. D. Solecki, L. Parshall., *et al.* (2009). Mitigating New York City's heat island: integrating stakeholder perspectives and scientific evaluation. *Bulletin of the American Meteorological Society*, **90**, 1297–1312.

Sailor, D. J., and A. A. Pavlova (2003). Air conditioning market saturation and long-term response of residential cooling energy demand to climate change. *Energy*, **28**, 941–951.

San Francisco Public Utility Commission (2002). *The Electricity Resource Plan: Choosing San Francisco's Energy Future*, San Francisco Public Utility Commission and San Francisco Department of the Environment.

Santamouris, M. and C. Georgakis (2003). Energy and indoor climate in urban environments: recent trends. *Building Services Engineering Research and Technology*, **24**, 69–81.

Satterthwaite, D. (2008). Cities contributions to global warming: notes on the allocation of greenhouse gas emissions. *Environment and Urbanization*, **20**, 539–549.

Scott, M. J. and Y. J. Huang (2007). Effects of climate change on energy use in the United States. In *Effects of Climate Change on Energy Production and Use in the United States*, Report by the US Climate Change Science Program and the Subcommittee on Global Change Research.

Scott, M. J., L. E. Wrench, and D. L. Hadley (1994). Effects of climate change on commercial building energy demand. *Energy Sources*, **16**, 317–332.

Seattle City Light (2005). *Annual Report*.

Shanghai Municipal Statistics Bureau (2007). *Shanghai Statistical Yearbook*, China Statistics Press.

Shui, B., M. Evans, H. Lin, *et al.*(2009). *Country Report on Building Energy Codes in China*, Prepared by Pacific Northwest National Laboratory for US Department of Energy.

Smith, J. B. and D. Tirpak (Eds.) (1989). *The Potential Effects of Global Climate Change on the United States, Report to Congress (EPA-230–05–89–050)*, Washington, DC, US EPA, Office of Policy, Planning and Evaluation, Office of Research and Development.

State of Maryland (undated). *Flooding and Fuel Tanks: A Guide to Anchoring Fuel Tanks in the Floodplain* (brochure), Maryland Department of the Environment.

Stern, F. (1998). Energy. In J. F. Feenstra, I. Burton, J. B. Smith, and R. S. J. Tol (Eds.), *Handbook on Methods for Climate Change Impact Assessment and Adaptation Strategies*, United Nations Environment Program/Institute for Environmental Studies.

Suzuki, H., A. Dastur, S. Moffatt, and N. Yabuki (2009). *Eco2 Cities: Ecological Cities as Economic Cities*, World Bank.

Taha, H. S., S. Konopacki, and S. Gabersek (1999). Impacts of large-scale surface modifications on meteorological conditions and energy use: a 10-region modeling study. *Theoretical and Applied Climatology*, **62**, 175–185.

Toulgoat, L. (2006). Postcard from Paris: sister companies heat and cool the 'City of Light'. *District Energy*, Third Quarter, 16–21.

UNDP/WHO (2009). *The Energy Access Situation in Developing Countries: A Review Focusing on the Least Developed Countries and Sub-Saharan Africa*, United Nations Development Programme and World Health Organization.

UN-HABITAT and ECA (2008). *The State of African Cities: A Framework for Addressing Urban Challenges in Africa*, Nairobi, Kenya: United Nations Human Settlement Programme, 206.

Unruh, G. (2000). Understanding carbon lock-in. *Energy Policy*, **28**, 817–830.

Unruh, G. (2002). Escaping carbon lock-in. *Energy Policy*, **30**, 317–325

Urge-Vorsatz, D., *et al.* (2007). Mitigating CO_2 emissions from energy use in the world's buildings. *Building Research and Information*, **35**(4), 379–398.

USAID (2004). *Innovative Approaches to Slum Electrification*, Bureau for Economic Growth, Agriculture and Trade, Washington, DC, USA.

US Conference of Mayors (2009). *Taking Local Action: Mayors and Climate Protection Best Practices*.

US Conference of Mayors (2010). *Mayors Leading Way on Climate Protection* (website). Accessed December 16, 2010, www.usmayors.org/climateprotection/revised/.

US EPA (2002). eGRID 2002 Archive. Available at www.epa.gov/cleanenergy/energy-resources/egrid/archive.html, accessed September 2008.

USGS (1999). *New York City Area Digital Elevation Model, 1/3 Arc Second*, US Geological Survey, EROS Data Center.

Vine, E. (2008). Adaptation of California's electricity sector to climate change. In *Preparing California for a Changing Climate*, Public Policy Institute of California.

Viswanathan, B. and K. S. Kavi Kumar (2005). Cooking fuel use patterns in India: 1983–2000. *Energy Policy*, **33**, 1021–1036.

Webster, D. and P. Mcelwee (2009). *Urban Adaptation to Climate Change: Bangkok and Ho Chi Minh City as Test Beds*. Working paper presented at 5th Urban Research Symposium, Cities and Climate Change: Responding to an Urgent Agenda, Marseilles, France, June 2009.

Williamson, L. E., H. Connor, and M. Moezzi (2009). *Climate-Proofing Energy Systems*, Paris, France: Helio-International.

Won, J. C. and I. R. Ahn (2009). Seoul Metropolitan District Heating Network. *Hot/Cold: International Magazine on District Heating and Cooling*, **4**, 5–7.

5

Climate change, water, and wastewater in cities

Coordinating Lead Authors:

David C. Major (New York City), Ademola Omojola (Lagos)

Lead Authors:

Michael Dettinger (San Diego), Randall T. Hanson (San Diego), Roberto Sanchez-Rodriguez (Tijuana)

This chapter should be cited as:

Major, D. C., A. Omojola, M. Dettinger, R. T. Hanson, R. Sanchez-Rodriguez, 2011: Climate change, water, and wastewater in cities. *Climate Change and Cities: First Assessment Report of the Urban Climate Change Research Network*, C. Rosenzweig, W. D. Solecki, S. A. Hammer, S. Mehrotra, Eds., Cambridge University Press, Cambridge, UK, 113–143.

5.1 Introduction

While many previous studies have looked at the worldwide changes and impacts of climate change and related variability on water resources, few have focused on an assessment of the specific effects and needed adaptation and mitigation for water systems in cities across the globe. The Intergovernmental Panel on Climate Change (IPCC) report on the water sector (IPCC, 2008) summarizes links between climate change and water through all of the physical elements of the terrestrial hydrologic cycle, ocean components, linkages to water supply, and global effects, but does not focus specifically on urban water systems. Similarly, the ADAPT Project (Aerts and Droogers, 2004) looked at adaptation for regional water management in seven typical watersheds across the world. However, most of this study was focused on surface water resources and their impacts on agriculture, food supply, energy production, and flood hazards, or on other impacts including groundwater resources, but did not focus on cities. There is thus an urgent need for a focused overview of the water supply and wastewater treatment sector in urban areas.

The range of challenges related to climate change and cities in regard to the water supply and wastewater treatment sector is very great, depending on geography, economics, administrative capacity, and demography. Many of the challenges are general, and some are more specific to particular cities. Accordingly, this chapter includes capsule descriptions of water supply and wastewater treatment in four cities that illustrate a variety of situations in which adaptation to climate change will be needed. These include a developed city with advanced climate adaptation planning (New York); a city with a formal water supply sector that will be under increasing pressure from climate change, with little wastewater treatment (Mexico City); a city with a very limited formal water supply system, and essentially no wastewater treatment (Lagos); and a city facing potentially serious water deficits from climate change as well as urban growth (Santiago de Chile). The situations of these four cities are representative of many other cities around the globe. In addition, there are two cross-cutting case studies of urban vulnerability in less developed countries that illustrates the centrality of water issues.

The chapter has five sections in addition to this Introduction: (5.2) Description of the water supply and wastewater treatment sector in cities, including sections on formal and informal water supply and wastewater treatment systems; (5.3) Vulnerabilities and impacts of climate change in urban areas; (5.4) Adaptation to climate change for urban water and wastewater systems; (5.5) Mitigation of climate change related to the water sector in cities; and (5.6) Policy considerations for urban water management, including gaps in knowledge. The policy considerations in Section 5.6 provide a range of focused approaches for urban policymakers with respect to the water and wastewater treatment sector, supplementing and extending the key messages of the chapter. Within the chapter, the kinds of issues and options facing cities are illustrated in the descriptions of conditions in New York, Mexico, Lagos, and Santiago de Chile described above.

5.2 Description of sector

Water and wastewater systems can be divided into two basic categories, those that are established formally in a city's governance and management structure and those that are established and that function informally. The informal systems typically have developed with little central planning or organization and often with only scattered and limited resources supporting them. The informal systems are operated and expanded in largely unregulated ways, often in slums, favelas, and other essentially impromptu settlements and extensions of the more organized city structures. Formal and informal systems have very different capacities to respond to the stresses that climate change is likely to impose on them. Both systems provide for delivery of water supplies to urban populations and the removal of wastewater. While formal systems may have more capital and knowledge to bring to bear on adaptation to climate change, they may not be able to respond to climate change because of inflexible governance structures. Given the limited monetary and planning resources supporting them, informal systems may be even less able to cope with the changes in both the supply and demand for water that climate change is projected to bring.

5.2.1 Formal water supply and wastewater sector

The functions of the formal urban water supply and wastewater sector include storage, supply, distribution, and wastewater treatment and disposal systems that provide organized water services to established urban areas. The infrastructure generally includes water and wastewater utility systems with large raw-water storage facilities, storm-water collection systems, trans-basin diversion structures, potable and wastewater treatment plant equipment, pipelines, local distribution systems, and finished-water storage facilities (Figure 5.1; California Department of Water Resources, 2008a). Urban supplies come from mixtures of surface water and groundwater as well as contributions from reuse of treated wastewater and desalination of seawater. Urban water infrastructures are large plumbing systems that tap and distribute water from these sources, and that commonly extend beyond the cities themselves to organize and draw from regional resources and services (Figure 5.2). Thus, for many cities, pipelines and other conveyances for the importation of distant waters provide access to major supplies. A city's internal distribution system can sometimes include subregions that are separately regulated.

Many of these facilities, structures, supply sources, and wastewater disposal mechanisms are vulnerable to adverse effects from climate variability and change (Case, 2008). Urban supplies can be affected by changes in water availability due to increases or decreases in precipitation, increases or decreases in temperature, sea level rise, and increases in climatic variability.

An important goal of urban water suppliers is to assure reliable supplies of high-quality potable drinking water in quantities

Typical water-use cycle for cities

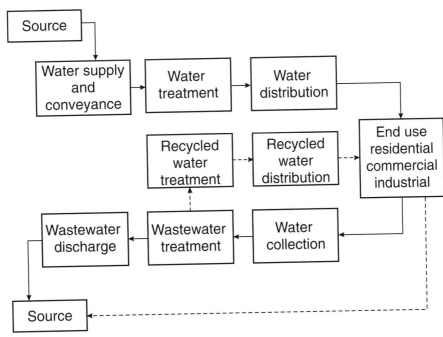

Figure 5.1: *Typical water-use cycle for cities and other developed supplies; dashed arrows indicate pathways that sometimes occur.*

Source: Modified from Klein et al. (2005a).

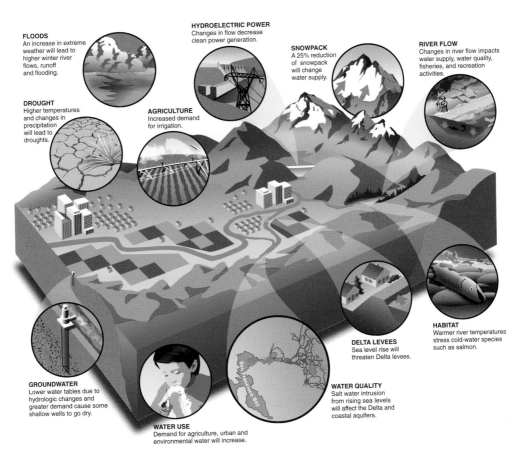

Figure 5.2: *Diagram showing the typical components of a coastal urban water sector as impacted by climate change.*

Source: Modified from CADWR (2008a).

that meet demands for municipal, commercial, and industrial uses. This most basic of tasks may not always be fully met, even in the absence of climate change. Demand has been estimated to exceed supply seasonally by as much as 25 percent (Queensland Department of Natural Resources, 2000) for some Australian cities to as much as 50 percent for a city like Calgary, Canada (Canadian Natural Resources, 2008), and, during droughts, by as much as 20 percent for urban coastal regions of California (Hanson et al., 2003). Thus, many urban water systems that are already challenged by unreliable supplies face additional challenges from growing demands driven by population growth and climate change.

The supply infrastructure of many cities has evolved, as demands driven by urbanization have grown, from systems that initially were supported mostly by streamflow and storage, to systems that now rely also on groundwater and imported water. Streamflow diversions are one of the primary sources of water for many urban centers throughout the world. The principal structures related to streamflow diversion are pipelines for direct distribution and storage in surface reservoirs. Such storage lessens the difference between supply and demand during seasonal and multi-year periods. As streamflow resources have been increasingly exploited and depleted, additional storage has been required to offset the growing demand. Groundwater resources have been exploited to fill demands and provide additional storage. In at least some settings where this exploitation has adversely impacted groundwater supplies and services, recharge has been augmented to partially replenish aquifers that are actively pumped or are subject to adverse conditions from overdraft such as seawater intrusion or land subsidence. As environmental constraints, increased competition among various land uses, and risks associated with surface reservoirs (e.g., dam failures and seismicity) limit options for additional surface storage in many urban systems, storage underground may emerge as one of the largest contributors to new storage.

Groundwater pumpage can be a primary source of water if surface water and its treatment are very expensive or unavailable, or if economic and regulatory conditions make this the preferred resource. Some urban centers, such as Shanghai, China (Zhang and Wei, 2005), Venice, Italy (Carbognin and Tosi, 2005; Gambolati et al., 1974), and Hanoi, Vietnam (Jusseret et al., 2009), are located on major river systems yet still rely predominantly on groundwater as a primary source of potable water and industrial supply. Many other intermediate and large cities such as Fresno, California, and Mexico City, Mexico, have relied on groundwater for many decades.

Where demand has exceeded supply from surface water and groundwater supplies within many urban areas, imported water has become a substantial source of supply in cities large and small. These sources are commonly constrained by transboundary agreements and competition that limit the amount or impacts of importation between multiple counties, provinces, states, or countries. The systems can also be limited by the physical capacity of conveyance systems. For instance, Kahrl

and Roland-Holst (2008) note that in 2003 the Metropolitan Water District of Southern California (Los Angeles, Orange, San Diego, Riverside, San Bernardino, and Ventura Counties) purchased water from agricultural growers in Sacramento but was unable to obtain the water because conveyance systems in the Sacramento–San Joaquin Delta were already operating at full capacity (CADWR, 2005).

In some cities, imported water has become the primary supply (e.g., San Diego, California; San Diego Foundation, 2008). The supply of water resources for many major metropolitan areas extends far beyond the city's watershed. For example, water has been imported to Los Angeles, California, from rural areas several hundred kilometers away in Owens Valley since 1913 and from the Colorado River since 1941. Large urban areas that rely on such regional-scale supply and distribution systems include several large metropolitan areas in southern and northern California, Tucson and Phoenix, Arizona, Denver, Colorado, and New York City. Generally the infrastructure and maintenance costs of establishing such long-distance supply systems, as well as political considerations, tend to restrict this strategy to use by large and prosperous urban centers.

Some urban distribution systems also include secondary distribution systems for reuse of treated wastewater, advanced treatment systems such as reverse osmosis or filtration, and multiple types of storage systems. Treated wastewaters may be distributed to meet irrigation and other non-potable needs, and with adequate treatment can be used to augment some drinking water supplies provided that communities are willing (Hurliman, 2007). Some cities such as Avalon on Santa Catalina Island, California, have implemented double plumbing systems with seawater used for the wastewater stream. In other cities, the notion of universal water-based sewage systems and sewage-treatment facilities that use water to transport waste streams is being questioned as impractical and potentially unhealthy (Brown, 2006). For example, Narain (2002) of the Centre for Science and Environment in India indicates that a water-based disposal system with sewage-treatment facilities is neither environmentally nor economically viable for India when an Indian family of five, producing 250 liters of excrement and using a water flush toilet, requires 150,000 liters of water a year to wash away its wastes.

The storage systems required for wastewater reuse can include local reservoirs, infiltration ponds for inducing groundwater recharge, as well as aquifer storage-and-recovery systems (Hanson et al., 2005, 2008). This type of replenishment is also used to prevent or reduce land subsidence and seawater intrusion owing to sustained groundwater pumpage as a primary source of water supply (Hanson et al., 2005). Treated wastewaters have long been injected into some coastal aquifers, e.g., in southern California, to deter seawater intrusion (Gleick, 2000). Wastewater management is thus integrated into many formal water systems and sometimes includes at least some reuse of treated wastewater. The treatment, distribution, and disposal of wastewater as well as reuse of wastewater are subject to the effects of climate change through increased energy costs and through increases in the volumes of wastewater and stormwater entering treatment facilities in areas where, and at times when, precipitation increases, and through increased needs for reuse where, and when, droughts become more prevalent.

Formal wastewater treatment systems in large cities are capital and operationally intensive. These systems receive wastewater from water supply systems and treat it to several levels, including primary, secondary, and tertiary treatment. In developed countries standards are set for these levels. Facilities related to the treatment of sewage include water pollution control plants, combined sewer overflow plants, wastewater pump stations, laboratories, sludge dewatering facilities, and transportation systems for sludge removal (New York City Municipal Finance Authority, 2004). A key issue for many cities, even in developed countries, is the existence of combined sewer and storm-water systems, which can result in combined sewer overflow events during heavy rains, and thus contribute to pollution in surrounding waterways. An example of a city with a developed formal water supply system but little wastewater treatment is given in the case study of Mexico City.

5.2.2 Informal urban water supply sector

For the most part, attention to the potential impacts of climate change on cities' water systems has been focused on formal water supply systems. However, there are increasing numbers of cities, especially in the developing world, where water supply systems for many or even most inhabitants and in most parts of the cities are informal. An example is given in the case study of Lagos. In these informal systems, water supply and treatment, and wastewater treatment and disposal, are not provided by large, centrally managed engineered systems under long-term plans, but rather are provided by a mixture of largely impromptu local supplies, informal water markets, and imports from outside an urban area through trucking and other means. The lack of large-scale central management leads to lack of planning and maintenance (United Nations Human Settlements Program, 2003). These limitations, in turn, suggest that the informal systems may be more vulnerable to climate change than the more carefully planned formal systems with their greater capital resources for infrastructure development and maintenance.

Figure 5.3: *Urban informal water supply: a water vendor's cart in Lagos, Nigeria.*

Photo by: Ademola Omojola

Even the formal water supplies in many of the developing world cities are inadequate or, at least, erratic due to high population growth rates not generally included in early planning, limited investments, and high operational costs of the regular systems. This situation has required finding alternative sources to supplement the original formal sector water supplies. In many cases, population growth in recent decades has been very rapid so that development of alternative supplies has lagged far behind demands.

Typical sources of water supply in the informal urban water system include water extraction from highly climate-sensitive shallow wells, deep groundwater extraction through boreholes, and the patronage of often-polluted urban fringe wetlands. Where even these sources are lacking or too limited, elaborate informal water markets develop to meet demands for drinking, cooking, and cleaning water with attendant high water prices, frequently poor quality and inadequate supplies (Lallana, 2003a; Davis, 2006; Gleick *et al.*, 2006). The distance to available water supplies is also an important factor in health and welfare (Howard and Bartram, 2003).

[ADAPTATION] Box 5.1 Lagos, Nigeria, megacity formal and informal water supply and wastewater treatment sectors

Ademola Omojola

University of Lagos

SYSTEMS DESCRIPTION

Lagos megacity in southwest Nigeria is one of the world's fastest growing urban centers. It covers an area of about 1,500 square kilometers of contiguously built-up area made up of twelve Local Government Areas (LGAs) in Lagos State and four LGAs in Ogun State. The total recorded population

of the LGAs of the trans-administration megacity in the 2006 census is over 10 million, which yields a population density of about 6,700 persons per square kilometer. There are indications and allegations that the figures are an underestimate. Lagos State for instance has been working with a population figure of about 17 million.

WATER SUPPLY

Urban water supply is the responsibility of the state government in Nigeria. The Lagos State side of the megacity

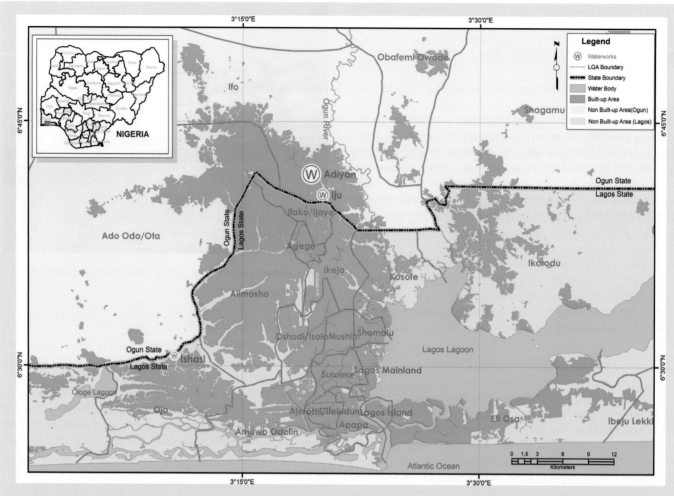

Box Figure 5.1: *Location map, Lagos megacity.*

is managed by the Lagos Water Corporation (LWC) and the Ogun State side by the Ogun State Water Corporation. Water supply to the megacity is both formal and informal. The formal sector is supplied mainly from three surface abstraction waterworks at Iju/Isheri (45 mgd) and Adiyan (70 mgd) on the Ogun River and Ishasi (4 mgd) on the Owo River. There are several mini-waterworks across the megacity mostly on the Lagos State side, relying on groundwater from boreholes. There are also some small non-water-corporation schemes of boreholes and overhead tanks built under different government programs in the megacity. These schemes usually supply consumers through localized public stand taps.

The LWC is basing the population to be served in the State on 17 million persons with an estimated demand of 600 mgd but currently with an installed capacity of 170 mgd out of which 119 mgd (66 percent) is from surface water and 51 mgd (34 percent) from groundwater. This translates into a 430 mgd (71 percent) supply gap. The water supply coverage in the Lagos State part of the megacity is about 40 percent through a pipeline network that runs north–south and mostly servicing the eastern part of the city and excluding a large sprawling population in the western part of the megacity. The water supply has significant capacity underutilization due to old

installations, erratic power supply, lack of maintenance, inadequacies in operational procedures, and poor funding among others. Also, water loss through leakage and poor service coverage due to limited pipeline network reticulation is very common. These problems are responsible for the inadequate supply in most of the areas serviced. The situation has forced many of the serviced residents to rely on redundant storage devices and on the informal sector to augment the current supply regime.

The informal water supply sector is very active in the megacity. This can be appreciated given that it serves about 60 percent of the residents in Lagos State and 67 percent in the Ogun State areas. This proportion may increase with the rapid sprawl and population growth of the megacity if proposed expansion programs are not promptly executed. The informal supply sources are mainly boreholes, shallow wells, and sometimes surface streams. The safe yield of the network of boreholes and their quality is yet to be adequately studied and monitored. This has grave implications on the sustainable use of the megacity's groundwater. For instance, there have been reported cases of groundwater contamination through seepages from the buried network of petroleum products in some parts of the megacity.

WASTEWATER MANAGEMENT

Wastewater management issues in the megacity are the responsibility of the Ministries of Environment and the Environmental Protection Agencies of Ogun and Lagos States. Pollution from wastewater is currently the greatest threat to the sustainable use of surface and groundwater in the megacity. Household, commercial, and industrial effluents and raw untreated sewage are often discharged into the open and fresh-water sources such as the Lagos and Ologe lagoons and the Ogun and Yewa Rivers.

There is no central sewage system in the megacity and less than 2 percent of the population is served with off-site sewage treatment plants that are currently in different states of neglect and disrepair. For example, the central sewage system in Festac Town – one of the few neighborhoods with such facilities – in the southwestern part of the megacity has collapsed for the past 15 years. Most of the sewer network has been connected directly to the storm-water lines. The direct spillage of sewage into the nearby rivers and roads from the sewage network and its consequent health hazard implications was published in the *Guardian* newspaper of December 22, 2008, page 73.

Besides the less than 2 percent with off-site sewage systems, only the toilet wastewater is connected to the septic tanks and soakaway systems, while the other household liquid wastes are discharged directly into the mostly open gutters in front of houses or on the streets in the high density areas. The wastewater eventually percolates or is washed into the water bodies by rainstorms. The stagnating pools of wastewater in the open gutters and on the roads often provide the breeding grounds for mosquitoes and habitat for several bacteria and viruses. In addition, wastewater pools contain hazardous contaminants such as oil and grease, pesticides, ammonia, and heavy metals (Saliu and Erutiya, 2006).

The water table in many parts of the megacity is very high. Consequently, the septic tanks and soakaway systems used in the collection of toilet wastewater readily contaminate and pollute the shallow groundwater that is a vital source of water supply to most low and middle income residents. Also, there is no septage treatment plant in the megacity and the evacuated untreated septage is mostly dumped in the Lagos Lagoon. The faecal contamination of the megacity's water system and the environment through the inadequate management of wastewater is a high health concern.

There is, however, a renewed awareness about the problem. For instance, the United Nations Environmental Programme (UNEP) has completed a training exercise with the environmental officers of Lagos State Environmental Protection Agency (LASEPA) on municipal waste management in coastal cities. Also, there is a renewed drive on storm drainage channel improvement and clearing to reduce incessant flooding in several parts of the megacity. Nonetheless, the 3rd Drainage Master Plan of the megacity, conceived in the 1990s, is yet to be implemented and flooding is still a common problem.

Box Figure 5.2: *Wastewater in open gutter with water supply pipeline*

Photo by: Ademola Omojola.

CLIMATE CHANGE INSTITUTIONS

Climate change awareness in the megacity is recent and can be associated with the Lagos State's participation in the C40 meetings. Lagos State has even gone ahead to establish a Climate Change Department in the Ministry of Environment. However, water and wastewater management in the megacity runs across administrations and agencies, which may be difficult to coordinate, most especially with respect to climate change mitigation and adaptation.

The assessment of climate change impact on water supply and wastewater treatment is currently not being given priority in the megacity. The issue of coping with the increasing water demands of the rapidly growing population in the megacity seems to be the topmost priority in water supply. Consequently, LWC has been reviewing its operational status within the framework of integrated urban water management such as reducing wastages due to leakages with the aim of increasing the resilience of the water supply. With respect to wastewater, the city has yet to fashion a coherent and comprehensive action of its management. This is, however, crucial for sustainable use of fresh water in the megacity.

METHODS AND IMPACTS

Climate change predictions for the coastal southwestern part of Nigeria suggest an increase in rainfall with increased intensity, sea level rise, flooding, coastal flooding from storm surge, and temperature increase and risk to coastal infrastructure (Ekanade *et al.*, 2008). It is therefore rational to accept that the strains on the supply and demand of Lagos megacity's water systems are likely to increase with such scenarios. However, unlike many cities that have been able to further provide background information and forecast data on the predicted impacts of climate change to local decision-makers, Lagos megacity does not currently have such background data for decision support.

Climate change scenario modeling for different emission levels and years for the megacity are yet to be carried out. To effectively plan for adaptation in a megacity such as Lagos, with low financial resources and adaptive capacity, there is the need to build scaled-down models of climate change to help in understanding the possible climate change scenarios and associated impacts and vulnerabilities in the city. This may eventually lead to an active input of climate change adaptation issues in managing this very vulnerable low-lying megacity.

CLIMATE CHANGE ADAPTATIONS

Climate change adaptations to water supply in Lagos have not been officially incorporated into its supply and demand. This may be due to the focus on the huge water supply infrastructural gap that the megacity is still trying to fill. There are, however, some visible adaptation options being carried out in the formal sector in the megacity that cannot be associated directly with climate change, but seen as a spin-off from actions taken to address more mainstream concerns of water supply management. For instance, at the management level, LWC has embarked on greatly reducing water loss through leakages, proper billing, and reducing water theft (unpaid

usage) to increase the resilience of water supply. There are also paid advertisements to sensitize residents on water wastage on the demand side. On infrastructure, the corporation is embarking on several water schemes for development and rehabilitation, while on policy it enacted a law in 2004 to encourage public–private partnerships (PPPs).

The biggest climate change adaptation challenge to water supply and wastewater treatment in the megacity is, however, in the informal sector. The uncoordinated and unregulated extraction of water from borehole and shallow wells, and the pollution of surface and groundwater from poor wastewater and solid waste management can all be worsened by the megacity's increasing population and climate change. Attempts at the LGA or community levels that are best positioned to monitor and implement policies and programs in the informal sector are obviously lacking. Thus, there is the need for policy formulation that will ensure monitoring and fashion out appropriate adaptation strategies in the demand and supply of water in the megacity. A new national policy on water has been sent to the National Assembly, but how well the policy recognizes the burgeoning number of people serviced by the informal water supply sector and the administrative capacity to implement the policy is another challenge.

Some of the dangers associated with urban water supply from such sources include:

- Most of the shallow wells are in areas with virtually no septic or wastewater treatment, which exposes the urban population to dangerous health risks (e.g., Kimani-Murage and Ngindu, 2007);
- Ad hoc solid waste dump sites with heavy leachates quite often share the same water-level dynamics as shallow wells;
- Massive uncontrolled extractions have been reported from some coastal cities' aquifers that in turn cause saline intrusion of the aquifers.

5.2.3 Future urban growth and the informal water markets

UN-HABITAT and other international organizations expect that future urban growth will take place predominantly in poor countries in the coming decades. Rapid urban growth in those countries has long been characterized by incomplete urbanization and severe shortages in the supply of key infrastructure (water, sewage, drainage, and electricity). In view of the difficulties already faced with rapid expansion of urban areas and populations, it is all the more difficult to see how urban areas in those countries will be able to meet future demands and stresses for water and wastewater management caused by higher temperatures and increased evaporative demand projected under the IPCC climate change projections (IPCC, 2007a). If the past is any indication, the role of informal water markets, with their challenges of adaptability and responsiveness to stresses of all kinds (Neuwirth, 2006), will likely become even more important to meet future increases in the demand for water.

5.3 Urban climate risks: vulnerabilities and impacts

5.3.1 Vulnerabilities

Water and wastewater treatment services in urban areas are vulnerable to direct impacts of climate changes such as changes in the amount and intensity of precipitation, increased temperatures and related evapotranspiration rates, changes in the intensity and timing of storm runoff, changes in both indoor and outdoor water demands, and, in coastal cities, sea level rise and storm surges. Vulnerabilities of both quantity and quality apply both to highly developed supply and treatment infrastructure and to less engineered and informal water supply and treatment systems. Warmer temperatures can also indirectly cause more severe weather (Cotton and Pielke, 2006) exacerbated by urban heat islands that could, in turn, result in additional convective thunderstorms, hail, cyclonic events (i.e., tornadoes, cyclones, and hurricanes), and higher winds that may exceed the design capacity of infrastructure. The centrality of water issues to urban planning more generally is indicated in the cross-cutting case study in this chapter: Urban vulnerability in the least developed countries (see Box 5.2).

Urban water supplies and wastewater systems are also vulnerable to climate change through less direct lines of connection. For example, warming trends may lead to increased demands for power production that, in turn, require power-plant cooling waters in competition with other water uses. Increased water demands associated with warming trends and, in some areas, reduced precipitation and runoff may lead to reliance on overdrafts from

[VULNERABILITY]   Box 5.2   Urban vulnerabilities in the least developed countries

David Dodman

International Institute for Environment and Development

The effects of climate change will be felt especially strongly in the towns and cities of the least developed countries (LDCs). These urban areas often have high concentrations of people and economic activities located in vulnerable physical settings, with their physical vulnerability exacerbated by poverty, the lack of appropriate infrastructure, and weak or inefficient systems of urban management. Yet this dense concentration can also provide the potential for effective adaptation, improved resilience, and the opportunity to meet broader development needs.

The Capacity Strengthening in the LDCs for Adaptation to Climate Change (CLACC) project is a group of fellows and international experts working on adaptation to climate change in twelve countries in Africa and three countries in Asia. The aim of the project is to strengthen the capacity of organizations in low-income countries and support their initiatives in sustainable development. In 2008, these fellows and experts prepared review documents highlighting climate-related problems (current stressors and potential climate change impacts) and their impacts for 15 cities (Box Table 5.1). These were deliberately selected to incorporate cities in a variety of physical settings – from low-lying coastal cities, to inland water-scarce cities, to high-altitude cities – and to include secondary urban centers as well as major cities and capitals as a means of demonstrating key vulnerabilities of cities in different positions in the urban hierarchy. This process was intended to identify key threats and major institutional stakeholders, and to provide the necessary information for communities and local governments to begin the process of climate change adaptation planning.

The CLACC project identified several key climate-related issues affecting urban areas in low-income countries, many of which will be exacerbated by anthropogenic climate change. LDCs are among the countries most affected by recent droughts. Drought is predicted to become more frequent and severe as a result of climate change, and many LDC cities are already badly affected. Zimbabwe has seen a decline in average rainfall of nearly 5 percent since 1900, with Harare and Bulawayo both affected by water stress. The Kariba hydropower plant that serves Harare has also been impacted by water scarcity, resulting in load-shedding by electricity providers. In Mali, Bamako is seeing widespread difficulties in accessing water throughout the city. Although 90 percent of families in the city have their own wells, the availability of water in these is declining as groundwater levels fall.

But even in towns and cities where overall rainfall totals are declining, precipitation is tending to occur in shorter, more intense bursts that can overwhelm urban drainage systems and lead to flooding. Frequent flooding has been affecting the congested slums of Kampala, particularly Kawempe, where almost half the houses are built on wetland. Heavy rainfall and flooding may also lead to landslides: in Kathmandu, Nepal, 207 mm of rainfall in a single day caused a landslide in nearby Matatirtha that killed 16 people.

Sea level rise will affect towns and cities in the LDCs particularly severely because a relatively large proportion of their populations live in the Low Elevation Coastal Zone – the continuous area along the coast lying less than 10 meters above sea level. Already, coastal erosion has damaged infrastructure (including houses and roads) in Cotonou (Benin) and necessitated heavy investment in coastal protection in Dar es Salaam (Tanzania).

Yet even within these cities, exposure to risk is distributed unevenly. In Mombasa (Kenya), low-lying areas vulnerable to coastal flooding are inhabited by low-income groups, for example in the coastal settlement of Tudor. In Khulna (Bangladesh), a mapping exercise showed substantial overlaps between slum settlements and areas that frequently suffer from waterlogging (Box Figure 5.3). And these spatial distributions are compounded by a variety of social phenomena: low-income groups are less able to move away from vulnerable sites; whilst the very young and very old are at greater risk from heat stress, and vector-borne and water-spread diseases.

Building urban resilience in the LDCs is important because of the vulnerability of large and growing urban populations to the hazards described above. But it is also important because of the potential economic costs without effective adaptation strategies. Successful national economies depend on well-functioning and resilient urban centers. Building urban resilience will require not only improving urban infrastructure, but also creating more effective and pro-poor structures of governance and increasing the capacity of individuals and communities to address these new challenges.

Box Table 5.1: *CLACC countries and cities.*

Asia	Eastern Africa	Southern Africa	West Africa
Bangladesh (Khulna)	Kenya (Mombasa)	Malawi (Blantyre)	Benin (Cotonou)
Bhutan (Thimphu)	Sudan (Khartoum)	Mozambique (Maputo)	Mali (Bamako)
Nepal (Kathmandu)	Tanzania (Dar es Salaam)	Zambia (Lusaka)	Mauritania (Nouakchott)
	Uganda (Kampala)	Zimbabwe (Harare)	Senegal (Diourbel)

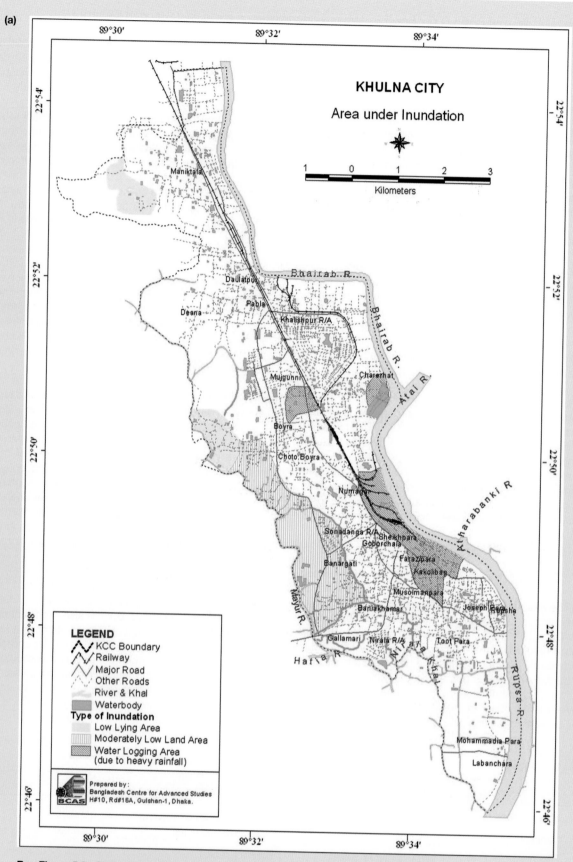

Box Figure 5.3: *Slums and waterlogging in Khulna, Bangladesh.*

Source: Bangladesh Centre for Advanced Studies.

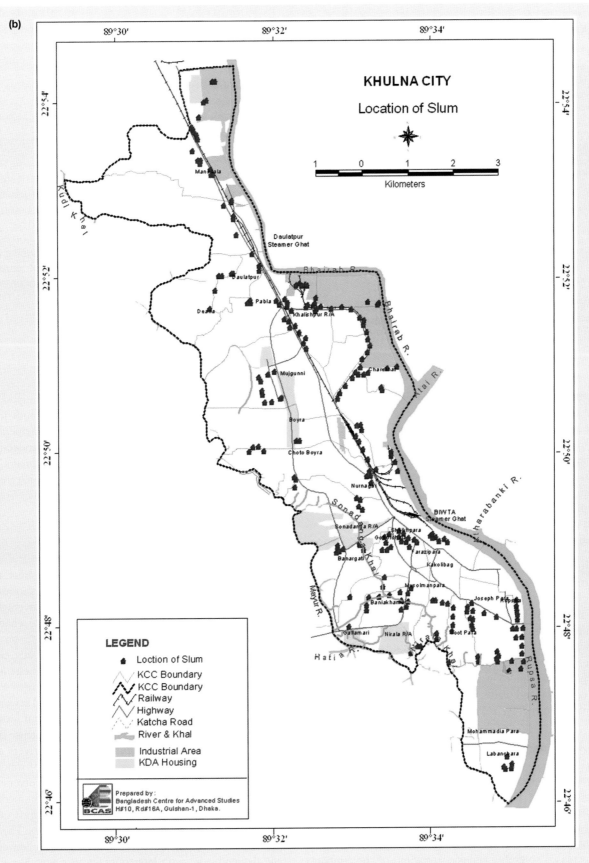

Box Figure 5.3: (*continued*)

Box Figure 5.4: *Low-lying informal settlement in Mombasa, Kenya*

Photo by David Dodman
Source: Reid, H. (2010). Building capacity to cope with climate change in the least developed countries. In J. Dodson (Ed.), Changing Climates, Earth Systems and Society, International Year of Planet Earth Series, Springer Science+Business Media.

Local and national governments have an important role to play, for they influence:

- the quality of provision of infrastructure for all areas within a city
- the quality of provision for disaster-preparedness and response
- the extent to which low-income groups can acquire safe housing in safe sites.

In addition, local governments strongly influence the environment within which local civil society action can take place. Community-based adaptation (CBA) is one such form of response. It is based on the premise that local communities have the skills, experience, local knowledge, and networks to undertake locally appropriate vulnerability reduction activities that increase resilience to a range of factors, including climate change. Facilitating and supporting this kind of response is likely to be key to building resilience in low-income urban centers.

For further information see Bicknell *et al.* (2009).

groundwater resources, land subsidence, and seawater intrusion. Increased overdraft and lower groundwater levels, in turn, can reduce or endanger the ability of wells to supply water without well retrofits such as lowering pumps and deepening wells. Finally, land subsidence can contribute to increased flooding as well as destruction of infrastructure resulting in additional leakage and reduced efficiency of distribution systems. These are examples of the complex connections and interdependencies that make essentially all water supplies and wastewater systems vulnerable to climate change at some level.

5.3.2 Impacts

These climate vulnerabilities lead to a wide range of potential climate impacts on urban water systems. Climate change and related increased climate variability threaten urban water infrastructures with disruption of service, reduced storage for potential emergencies, reduced water quality, and increased energy costs for operation and maintenance, at local and regional scales (see Chapter 4 on energy, and Chapter 7 on health).

5.3.2.1 Air temperature

Warmer temperatures, and especially more extreme temperature ranges, due to increasing greenhouse gas concentrations will accelerate the degradation of materials and structures in important urban water infrastructures. Warmer air temperatures can lead to biological and chemical degradation of water quality, e.g., by increased solubility and concentrations of contaminants in fresh water or enhanced growth of algae, microbes, parasites, and invasive species. Increased temperatures will result in higher evapotranspiration rates that will increase demands for

landscape irrigation and additional human consumption. Warmer temperatures will also result in additional demands for cooling water in arid and semi-arid regions. Warmer temperatures will result in greater summer peak demand and extended periods of increased demand during longer and drier summers, and may result in decreased reservoir or lake levels, which may require relocation of intake pipes that supply surface water from lakes or reservoirs (Thirlwell *et al.*, 2007).

5.3.2.2 Precipitation

More frequent intense rainfall leads to more street, basement, and sewer flooding and stormwater runoff to various disposal systems. In most parts of the world, whether average precipitation totals increase or decrease with climate change, more intense rainstorms are expected (see Chapter 3 on Climate). More intense rainstorms will increase nutrient loads, eutrophication, taste and odor problems, and loading of pathogenic bacteria and parasites (*Cryptosporidium* and *Giardia*) in reservoirs. More intense precipitation will lead to more combined sewer overflow events that, depending on the city, pollute coastal waterways or other nearby bodies of water. More frequent intense rainstorms will also increase the sediment load in some rivers and reservoirs, and this may decrease the water quality of water diverted for water supply or further restrict periods of diversion. More intense and frequent rainstorms also can result in more flooding and erosion, which will lead to destruction of infrastructure such as bridges and approach embankments to bridges. The timing of rainfall may change, causing further disparities between supply and demand, e.g., with later rainfalls in places like the Seattle region (Chinn, 2005). An example of a city facing future climate stress from direct precipitation and loss of glacier mass is presented in the case study of Santiago de Chile (Box 5.3).

[ADAPTATION]   Box 5.3   Santiago de Chile: Adaptation, water management, and the challenges for spatial planning

Jonathan Barton

Pontificia Universidad Católica de Chile

Dirk Heinrichs

German Aerospace Center

Santiago de Chile, with its population of six million concentrated in the Maipo river basin on the western flanks of the Andes, is regarded as a Latin American city that compares well with others in terms of poverty, security, economic activity, and other urban indicators. Although the country contributes little in terms of global greenhouse gases, and is highly active in CDM projects, it faces considerable adaptation challenges, e.g., due to the vulnerability of its agro-business sector and its coastal cities (CONAMA, 2005, 2008). Santiago's future is linked to these changes, but faces more specific local adaptation challenges. Perhaps the most important of these is water management. The catchment is fed year-round from Andean glaciers

since localized precipitation is highly concentrated in the June–July winter period. The projections to 2070 under an A2 scenario suggest a potential 40 percent reduction in precipitation, compounded by reductions in glacial flows and rising evapotranspiration tied to higher temperatures of 2–4 °C (CONAMA, 1999; CONAMA, 2006). Pressures will grow to change the current water management system and meet the adaptation challenge as a consequence of increasing conflicts over water access. The expected population by 2030 exceeds eight million people (MINVU, 2008). This is likely to correspond to urbanization processes that displace agricultural interests in the region as the metropolitan area expands into productive land, also areas of increased risk and areas that provide important environmental services for the watershed.

The adaptation challenge to tackle this scenario lies in three fields: water markets, equitable distribution and water conflicts, and climate change governance within spatial planning. National and local government has only partially addressed these concerns to date.

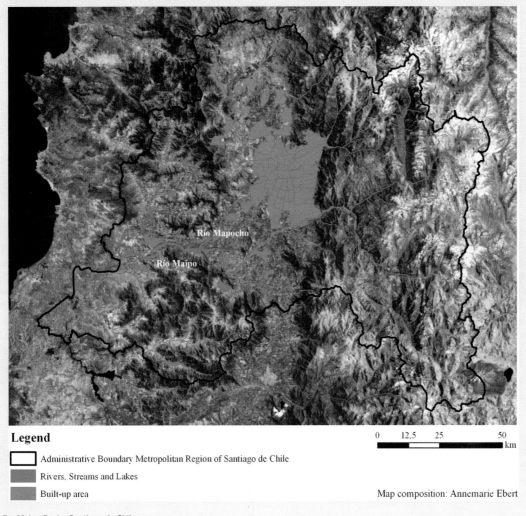

Legend

▢ Administrative Boundary Metropolitan Region of Santiago de Chile

▮ Rivers, Streams and Lakes

▮ Built-up area

0   12.5   25   50
km

Map composition: Annemarie Ebert

Box Figure 5.5: *Maipo Basin, Santiago de Chile.*

Box Figure 5.6: *San Carlos channel (early nineteenth century) entry into the Mapocho River, an artificial channel that draws water from the Maipo River and takes it north into the Mapocho.*

THE WATER MARKET

The water market is based on water rights that are purchased and transacted (Water Code, 1981, modified 2005), based on a minimum streamflow condition and a total availability calculated by the national water authority (DGA). There is currently insufficient supply for new consumptive rights to be made available in the Maipo basin (DGA, 2003). Meantime, there is pressure from more powerful interests to buy out smaller rights holders, such as small-scale irrigation associations. The market is also unable to respond to fluctuations in the hydrological cycle, including the El Niño phenomenon for example, since rights are fixed and are awarded in perpetuity. In consequence, water availability decreases, existing rights will not be able to be extracted, and no new uses will be catered for.

EQUITABLE DISTRIBUTION AND WATER CONFLICTS

The limitations of the existing market, its weakness to respond to the natural cycles in the water basin, and the anticipated scarcity due to climate change, present a major adaptation challenge. Conflicts will increase particularly between residential, agricultural, and mining demands and environmental services. Assuming that the high residential value of land will lead to wine investments and horticulture moving to other regions, the question remains as to how a potential 40 percent reduction in water availability will be met by a population in the metropolitan area that is 30 percent larger than at present. The city's location is in a Mediterranean biodiversity hot spot, where current levels of "green" space per habitant (3.2 m^2/cap; CONAMA, 2002) are well below the WHO recommendation of

9 m^2/cap. This raises the issue of water use for maintaining the region's ecosystems, and for increasing public spaces to enhance urban quality of life (particularly in the lower income municipalities of the city) and reducing, for instance, the heat island effect. This will require a significant shift in water management in many areas with:

- reduced agricultural irrigation capacity
- watering of, and species selection in public spaces and domestic gardens
- a stormwater drainage system that seeks to shift water downstream of the city as swiftly as possible during peak events (rather than capture and storage)
- broad-based demand reductions.

CLIMATE CHANGE GOVERNANCE WITHIN SPATIAL PLANNING

The 2005 national climate change strategy concentrates on productive sectors, particularly mitigation and CDM commercial opportunities, but fails to put much weight on adaptation issues (CONAMA, 2005). It also fails to explicitly consider urban centers in spite of over 80 percent of Chileans living in urban areas, with over 40 percent of the national population living in the Santiago metropolitan region (RMS). This has changed slightly with the publication of the 2008 national action plan (CONAMA, 2008). The plan focuses on seven fields for action, water being one of them.

Although urban change, except coastal city risk, is not an explicit focus of the plan, all seven issues relate to urban transformations. Their incorporation into planning instruments is going to be a primary challenge for climate change adaptation: development strategy, metropolitan and local regulatory plans, and local development plans. To date, these documents have not included climate change considerations explicitly, largely because of the sectoral approach to public sector management.

Climate change adaptation will demand a coordinated response from government agencies, within the context of a regional adaptation plan. Although the DGA manages the water market, the water planning dimension must be brought within the administration of the territorial authority, the Regional Government, as part of a strategy able to engage with the priorities of the national plan (less fisheries) and the multiple public and private actors who are direct stakeholders, from rights holders (agriculturalists, mining firms and others), to the environment commission, the housing and urbanization ministry, the public works ministry, and municipal authorities. It would appear that the limitations of this natural resource market for the climate change challenge to be faced this century are already evident.

More frequent and intense droughts may affect reservoir and groundwater storage, as well as rainwater capture systems. Reduced precipitation may also result in less groundwater recharge and lower summer streamflows (Pitre, 2005; Earman *et al.*, 2006). Reduced precipitation will also contribute to increased pumping costs due to deeper groundwater levels and will also contribute

to increased conflicts over water related to baseflow in streams, maintenance of water rights, and restriction of new water users (Pitre, 2005). Reduced snowfall results in less water stored in snowpack reservoirs that provide water to some cities, and thus will change the temporal patterns of flow to reservoirs and supply systems. Reduced snowfall can also challenge many water systems that implicitly or explicitly depend on seasonal storage in snow form, ultimately requiring the development of even more constructed storage to provide supply reliability. Reduced snowfall and warmer water temperatures may aggravate demands for various instream and non-consumptive water uses, such as maintenance of fisheries, ecosystems, river amenities, and recreation, as well as various industrial and cooling needs. Increased precipitation and related peak flows may affect coastal and inland shipping by seasonally reducing water depths in channels, reducing passage heights under bridges, and limiting passage through weirs and locks (Klein *et al.*, 2005b), which can impact the shipping of wastewater treatment system outputs.

5.3.2.3 Sea level rise and storm surges

Salt water will encroach on coastal surface water sources, groundwater, and ecosystems. For example, increased sea levels will result in higher pressures of the ocean on submarine and coastal outcrops of coastal aquifers that will result in additional seawater intrusion. An increase in sea level will lead to an increased probability of flooding of sewers and wastewater pollution control plants (WPCP) and a reduced ability to discharge combined sewer overflows (CSO) and WPCP effluent by gravity. High storm surge levels lead to more street, basement, and sewer flooding. Higher sea levels, when inundating polluted areas (brownfields), can cause harmful release of pollutants. Higher sea levels can inundate fresh and saline wetlands and threaten the stability of canals and levee systems, which can have impacts on water supplies, water quality, and flooding. For example, a projected sea level rise of about 1.5 m along coastal California by 2100 with related levee failures and obstruction of fresh water flow in the Sacramento Bay–Delta system would jeopardize the fresh water currently passed through the delta for irrigation and drinking water supply to the cities and farms south of the delta (CADWR, 2008a). Higher sea levels will also increase the probability of water and wastewater damage due to surge action. Rise in sea level will result in reduced sediment transport, and may require increased dredging of sluices, weirs, groynes, locks, and canals, filling of wetland areas and raising and reinforcing levees and embankments (Klein *et al.*, 2005b).

5.3.2.4 Surface-water impacts

These threats, in turn, may lead to further hydrological challenges, such as: loss of reservoir storage owing to competition for reservoir space for flood control, ecological flows, recreation, or agricultural supplies; reduced natural storage of water supplies on seasonal to decadal time scales due to declining ice and snowpack reserves; loss of inflow into the reservoirs owing to increased droughts; loss of storage owing to unscheduled releases from increased precipitation, runoff, or transition of runoff from snow to rainfall; reduction of diversion flows owing to competition with ecological flows during dry periods or droughts; and increased runoff (including urban runoff) preventing adequate water quality of streamflow diversions for water-supply needs through entrainment of increased total dissolved solids or agricultural and urban contaminants.

5.3.2.5 Degradation of groundwater aquifer systems used for urban supply

Climate change may eventually affect groundwater aquifers that supply water for cities through seawater intrusion, land subsidence, lateral and vertical migration, and capture of contaminants. Flow of degraded waters between different parts of aquifers, or different aquifers, tapped by boreholes and wells may compromise aquifers used for water supply, even where not all the aquifers are directly impacted by climate change. Climate change may also increase the need for injection systems and surface water delivery pipeline systems in lieu of coastal pumpage to prevent seawater intrusion (Hanson *et al.*, 2008).

5.3.2.6 Regional-scale changes

Because many urban systems are part of larger regional water systems, the effects of climate change will also yield regional-scale challenges, such as reduction in snow-melt in watersheds that provide crucial supplies or storage mechanisms for urban supplies; loss of groundwater storage in supply basins owing to urbanization or legal constraints as regional competition for water supplies and wastewater disposal options increase; and disruption of delivery or competition for imported water.

5.3.2.7 Impacts on informal urban water systems

Given the meager capital resources and lack of centralized planning associated with development and maintenance of most informal water systems, climate change will have additional impacts on urban informal water systems. Even small perturbations of water sources, informal conveyances, and wastewater disposal options by climate change are likely to challenge these informal systems, and larger disturbances by extreme climatic events such as storms and heat waves typically will not have been accommodated in informal system designs. Such climate change stresses will bring management challenges, since informal water supply systems are complex structurally and institutionally, and since decision-making tends to focus on the short-term. The biggest impacts of climate change on the informal water supply sector have to do with the maintenance of sources in terms of both quantity and quality. Here, adaptations go beyond the management of particular systems to larger issues faced by cities and regional institutions. For example, increasing temperatures may adversely affect the health of populations not served by organized sanitary systems.

5.3.3 Interactions of climate change, urban water, and other sectors

Water is a cross-cutting theme in urban life and function and, as such, it is at the nexus of many issues regarding how climate change will challenge cities. The effects of climate change on urban water will impact other urban sectors; equally, climate change effects on other sectors will, in many cases, impact the urban water systems.

[ADAPTATION]   Box 5.4   Mexico City's formal and informal water supply

Roberto Sanchez-Rodriguez

EL Colegio de la Frontera Norte, Tijuana

The importance and complexity of water problems in Mexico City make it particularly vulnerable to the negative impacts of climate change. Water management has been a critical factor in the evolution of Mexico City. This megacity of more than 18 million inhabitants grew in a hydrological basin composed of five shallow lakes. Historical urban growth was on the lower part of the basin on top of the lakes and it has extended to the slopes of the surrounding piedmonts. The city has overtaken most of the former lakebeds and it has suffered major floods throughout its history. Despite numerous efforts to control this problem, flooding continues to be a major hazard in Mexico City and its solution requires an integrated approach together with other water problems.

Water supply is a multidimensional problem in Mexico City. The city has gone from a high level of self-sufficiency to a high level of dependence on two external watersheds. Current water use in Mexico City is approximately 63 m³/s. Close to 66 percent (43.5 m³/s) is extracted from aquifers and the remainder is imported from the Lerma basin (6.0 m³/s) and the Cutzamala basin (13.5 m³/s) (Ezcurra *et al.*, 1999). Importing water to Mexico City from those two basins has had a significant impact on them. The mean annual input of rainwater into the basin in Mexico City is 23 m³/s. It is estimated that only 50 percent of that water recharges the aquifers. Deficiencies in the operation of the distribution system cause leaks estimated to be about 30 percent of the water managed by the city. Considering that some of that water makes it to the aquifers, the estimated total recharge of the aquifers is 28 m³/s, equivalent to approximately only 50 percent of the water extracted every year. Water extraction from aquifers has caused a subsidence problem since the early 1900s. The city has sunk at different rates in different parts, but it reaches its extreme in the old historical center, where some parts have sunk up to 9 m during the past century. Subsidence has caused severe maintenance problems with the urban infrastructure, building, and transport systems. It has also aggravated pollution problems of the aquifers, particularly in critical areas for their recharge. Monitoring of the water in the aquifers has shown deterioration in its quality due to overexploitation of groundwater, and high bacteria counts have been observed in some wells (Mazari *et al.*, 2000). The protection of critical recharge areas of the aquifers is also a critical problem in Mexico City. The rapid expansion of illegal settlements in those areas jeopardizes the recharge of the aquifers.

The last component of water problems in Mexico City is wastewater. The city has a complex sewage collection system where wastewater and rainstorm water are mixed. The capacity of the system is 57 m³/s, 42.8 m³/s for sewage and 14.2 m³/s for rainstorm water. The effluent is shipped to the Tula basin about 50 km north of the city. Twenty-seven treatment plants treat only 7 percent of the total sewage generated in Mexico City.

Box Figure 5.7: *Subsidence in Mexico City.*

Source: USGS.

The increasing volume of sewage generated by Mexico City during the last decades has compromised the capacity of the system to evacuate rainstorm water, increasing the risk of flooding. The subsidence of Mexico City has also created problems for the operation of the system. The slope of some of the major drains has been reversed, requiring the construction and operation of pumps to evacuate wastewater. The Mexican federal government and the local government in Mexico City initiated significant maintenance and repairs to the wastewater system to prevent floods during the rainy seasons in 2007 and 2008. Major additional works are still being considered for the near future.

Water problems in Mexico City represent a major challenge for present and future urban growth. Climate change will aggravate those problems. Some of the studies of potential climate change scenarios show an increase in precipitation

Box Figure 5.8: *Alfalfa, a fodder crop grown with wastewater from Mexico City in the State of Hidalgo. Health problems can arise from the use of untreated wastewater for crop production.*

Source: Flagstaffotos.

and temperatures in the city by 2025 and 2050 (Gay *et al.*, 2007). The challenge to secure water supply will increase in light of the expected increase in the demand, particularly during the dry season when temperatures are expected to increase. The risk of flooding, a chronic problem in the city, will also increase under a climate change scenario. Those impacts will not only create significant consequences for the water sector, but also for the energy sector and the health of the population. Mexico City needs a new strategy to address those problems. It will be critical to create integrated and multidimensional strategies recognizing the interactions among the different components of water in the city, as mentioned above. The federal and local governments have addressed each of those elements in isolation from the other, creating fragmented actions that have had limited success in solving a complex problem.

5.3.3.1　Energy

With warming, urban water demands and uses are likely to increase in many (perhaps most) cities. Future increases in the demand and use of water expected under climate change are likely to result in increased demands for energy (see Chapter 4 on Energy). The supply, treatment, and distribution of water supplies in urban areas require operation of pumps and other mechanical devices with attendant heavy energy use. Most sewage treatment plants operating in urban areas are mechanically operated and the collection, recycling, and outflow of sewage also frequently require the operation of pumps. If sewage flow is projected to increase under a given climate change scenario, an increase in the demand for energy must also be anticipated for both operation and capacity expansions.

5.3.3.2　Health

Urban water systems have close ties to many of the public health challenges associated with climate change (see Chapter 7 on Health). Water-borne diseases are a major health hazard in poor countries and emerging economies due to deficiencies in the supply of drinking water in their urban areas. Climate change can exacerbate those hazards by increasing gaps between drinking-water demands and supplies, and can stress sewage disposal systems and options beyond current conditions. Climate change may also aggravate public health challenges in urban settings by increasing the geographic ranges of some diseases and disease vectors (so that water facilities that did not sustain disease and vectors in the past may do so in the future); by increasing the opportunities for their propagation and development (e.g., by increasing reservoirs of standing water or promoting longer vector lives or more vector generations); and by generally reducing overall public hygiene and resistance to disease (as water supplies are challenged or limited).

5.3.3.3　Governance

The governance of a vital natural resource such as water is challenging in both poor and rich countries (see Chapter 9 on Governance). Climate change will stress further the political negotiations regulating access to water in formal and informal urban water markets even beyond often acrimonious historical levels. Whether privately or publicly owned, the governing structures of ownership, use, and sale of water resources may require redefinition if they are to be adaptable enough to accommodate growing and interacting pressures from rapid urbanization and climate change.

5.3.3.4　Land use

Urban demands for water often encourage or require land use changes in other areas that provide water supplies, storage, and conveyance corridors with the potential for severe negative social, economic, and environmental consequences. Increased urban water demands under climate change may create additional pressures to import water and to introduce land use changes in areas beyond the city. Water availability can dictate or limit land uses within urban areas, and consequently climate changes may redefine acceptable land uses within urban areas. Perhaps even more importantly in view of the rapid growth of cities expected in the twenty-first century, the forms, extensions, and types (particularly density) of future growth in and around cities will similarly depend on available options for provision of water and wastewater treatment and disposal, and how those options will be impacted by climate change.

5.3.3.5　Transportation

Water-borne transportation systems that are vital in many cities may be affected by changing climates, sea level rise, and changing streamflow timing and amounts (see Chapter 6 on Transportation).

5.4　Adaptation

Water supply and wastewater treatment infrastructures are typically long-lived, are ubiquitous elements in developed urban centers, and are subject to critical stresses from climate change, including sea level rise, higher temperatures, changes in precipitation patterns, and potentially more intense storms. In less-developed urban centers, water supply is more informal, and wastewater treatment may not be developed; yet these informal systems are also subject to climate change stresses. An orderly adaptation assessment process is needed to ensure the efficient use of scarce capital and operating funds over long time periods to meet these challenges in both formal and informal urban water systems. An example of a city with highly developed water supply and wastewater treatment systems, with an advanced adaptation planning process, is given in the case study of New York City (Box 5.6).

Climate change adaptations cover a wide range of actions in regard to urban water system operations and management, infrastructure, and policies. These will need to be developed for surface water, groundwater and rainfall-capture systems, and for wastewater treatment facilities for coastal and inland cities. Urban water adaptations will be needed for both highly developed supply and wastewater treatment infrastructure and for low-tech and informal water supply and treatment. They need to take into account rising populations, potentially rising and/or falling incomes, and changes in technology, and should be linked to co-benefits with investments for other purposes. Planning to address long-term needs for urban water system adaptation to climate change needs to be rigorous enough to justify major water-system investments but, given continuing uncertainties about the magnitudes and rates of the climate change challenges, will need to be flexible and ongoing (e.g., climate change scenarios for cities usefully could be updated on the order of every 5 years).

Some of the major urban areas that have aggressively begun to plan for, and adapt water systems and other infrastructure to, climate change are: Boston, USA (Kirshen *et al.*, 2004); Halifax, Canada (Halifax Regional Municipality, 2007); London, UK (Greater London Authority, London Climate Change Partnership, 2005); New York City, USA (New York City Department of Environmental Protection, 2008; see Box 5.6); Seattle, USA (University of Washington Climate Impacts Group and Washington Department of Ecology, 2008); and Toronto, Canada. Many adaptations for urban water systems have already been identified through the work of these "early adopters," but research is still required to develop and evaluate available options and capacities for adaptation. An example of needed work is the development of simulation tools for modeling climate change effects on reservoir system operations in order to evaluate changes in operating rules and storage capacity, as well as delivery and storage periods (New York City Panel on Climate Change, 2010). Much research on adaptation is needed for urban water systems in cities in the developing world. This need is illustrated by many local case studies such as that of Esmeraldas, Ecuador (see Box 5.5). There, urban areas are at risk from flooding and unstable hillsides, both problems that could become more serious with climate change.

In order to approach adaptation to urban water systems, a common assessment framework is needed to allow intercomparisons and coordination between and among cities and systems, in order to assist different jurisdictions to develop adaptations more efficiently. Such a framework is designed to encompass the full range of decision-making tools required to go from climate impacts and scenarios to adaptation project and program implementation, review, and monitoring (Table 5.1; Rosenzweig *et al.*, 2007). Potential climate change adaptations can be divided into operations/management, infrastructure, and policy categories, and assessed by their timeframes (immediate, medium, and long-term), the capital cycle, costs, and other impacts. The steps also need to adequately account for other changes that a city is likely to experience (such as population growth and changes in per capita water use), irrespective of climate change. Potential adaptations would ideally manage the combined risks of climate change and other predictable challenges to urban water supply and wastewater treatment facilities so as to provide an overall "coping strategy" (Ayers *et al.*, 2003).

[VULNERABILITY] Box 5.5   Urban expansion and vulnerability in the city of Esmeraldas, Ecuador

Christophe Lalande

UN-HABITAT

In 2001, Esmeraldas was the 12th largest municipality and its urban component the 15th largest city in Ecuador. It is a coastal city in the northwest of Ecuador. In most respects, the city is a typical Ecuadorian medium-sized city; its social and economic indicators are comparable to those of other cities in the same size group. However, unlike most cities in Ecuador, Esmeraldas experienced growth far below what was observed in most other cities in the group until the beginning if this decade.

The population estimate for the canton Esmeraldas for the year 2010 is 188,694, of which 66 percent is urban, up from 162,225 in 2001, an increase of 16 percent in nine years. As in most cities in Ecuador and Latin America, urban growth in Esmeraldas has largely been associated with illegal occupations of land in areas surrounding the consolidated city. Spatially the growth of the urban component of the canton in the past decade has concentrated in the south of the city, in new neighborhoods such as La Tolita, Tiwintza, San Rafael, and Los Pinos.

Esmeraldas, like most cities in Ecuador, has increasingly incorporated areas at higher risk to natural disasters as it grew. The first settlement of what today is Esmeraldas stood above the flood zone of the Esmeraldas River; some accounts place an earlier settlement several kilometres upstream. Until 1975, the city grew by occupying the hillsides surrounding the original settlement (see Box Figure 5.9) and later occupying the flood zones of the Teaone and Esmeraldas Rivers to the south, and the Piedad and de Prado Islands in front of the city. The hillsides surrounding Esmeraldas have proven to be unstable throughout the city's development, with the latest significant emergencies occurring during the strong rains associated with the ENSO (El Niño) event of 1998.

After this period, the risks the new settlements faced were primarily related to floods, in part due to significant infrastructure improvements in the hillside settlements, but also because new settlement areas avoided such locations.

Box Figure 5.9: *Overview of Esmeraldas, Ecuador.*

Source: UN-HABITAT

Box Figure 5.10: *Riverbank settlements in Esmeraldas, Ecuador.*

Source: UN-HABITAT

By 2007, almost 60 percent of the population lived in areas with medium to high risks of floods or landslides. Sixty-six percent of the city showed medium to high exposure to climate-related risks.

Climate change scenarios for the city of Esmeraldas

The variability and uncertainties associated with the climate change projections available for the Esmeraldas River Basin are consistent with those observed for Ecuador as a whole.

Five models and nine emission scenarios have been identified and analyzed at the local level. Models and scenarios project increases in temperatures of approximately 2–3 °C for the Esmeraldas River Basin. For the coastal region in and around Esmeraldas, precipitation projections vary from +30–50 percent to –30–50 percent (mm/day). There are several small, isolated watersheds in this area that would be affected severely by either extreme. Unlike large basins, and especially those with direct connections to the highlands, there is little room for compensating local increases or decreases in precipitation.

Exposure to climate change and tools for adaptation and mitigation in Esmeraldas

The impacts of climate change on the city of Esmeraldas identified by local stakeholders vary depending on the climate transition path guiding the analysis of potential adaptations and vulnerabilities. Current relatively high risk levels of landslides and floods (see Box Figure 5.10), both linked to current climate patterns, make Esmeraldas one of the riskier cities in which to live in Ecuador.

Under climate change scenarios predicting a path towards hotter and more humid climates, Esmeraldas would face even greater and more frequent disasters and more complex planning and management scenarios. Increased precipitation would certainly cause additional life and property losses.

In this respect, one of the key challenges the city faces is the ongoing expansion of informal settlements along flood zones of the Taeone and Esmeraldas Rivers and the low-lying Piedad and Prado Islands.

Under climate change scenarios predicting a path towards hotter and dryer climates, Esmeraldas could potentially experience lower risks of flooding, and lower stress on the water delivery systems. In contrast, stakeholders consider water shortages and price increases a major concern if the environment becomes dryer.

In both climate transition paths energy demands are expected to increase drastically, not only because of higher temperatures and larger populations, but also due to increased consumption levels throughout the city.

In this context, adaptation to climate change in and around the city of Esmeraldas requires a complex set of actions designed to compensate current vulnerabilities and to avoid expanding the range of risks associated with natural events.

Efforts to compensate for vulnerabilities or taking advantage of opportunities rely on actions and adaptations based on a combination of zoning, infrastructure modifications, energy shifts, capacity building, and improved governance.

According to stakeholder assessments, adaptations to climate change in Esmeraldas would consist of: structural adjustments, such as the construction of upstream water storage and flood control systems (e.g., dams, reservoirs), and levees to protect flood-prone neighborhoods; the consolidation of the existing drinking water and sewage systems, and their expansion into new settlement areas; and institutional tools, such as zoning plans and cadastral capacity, that improve governance. Economic diversification would also reduce vulnerability by facilitating the consolidation of marginal urban areas.

Source: Climate Change Assessment for Esmeraldas, Ecuador: a summary. Available at www.unhabitat.org/pmss/getElectronicVersion.aspx?nr=3005&alt=1.

Table 5.1: *Steps in adaptation assessment for water supply and wastewater treatment facilities.*

1. Conduct risk assessment inventory, including main potential climate change impacts

2. Apply future climate change scenarios

3. Characterize adaptation options:

   Operations/management

   Investments in infrastructure

   Policy

4. Conduct initial feasibility screening

5. Link to capital and rehabilitation cycles

6. Evaluate options

   Benefit/cost analysis

   Environmental Impact

   Legal mandates

7. Develop implementation plans, including timeframe for implementation

8. Monitor and reassess

Source: Rosenzweig et al. (2007).

The adaptation assessment steps in Table 5.1 are based on standard water-resource planning procedures (Goodman *et al.*, 1984; Orth and Yoe, 1997), with the significant addition of climate change (Step 2) and an explicit link to agency capital cycles to provide for efficient incorporation of adaptations during rehabilitation and replacement (Step 5). While these steps are broadly comprehensive, climate adaptations for particular circumstances may require additional steps (as, for example, securing external funding for adaptations in developing countries). These steps are further elaborated here, with examples of potential adaptation measures for formal and informal systems. This approach provides a way of framing the range of challenges and opportunities for adaptation of urban water supply and wastewater treatment systems to climate change. An application in slightly modified form is in New York City Panel on Climate Change (2010), Appendix B.

5.4.1 Step 1: Conduct risk assessment inventories

The inventory is designed to highlight the most significant potential climate change impacts on urban water systems. Suitable inventories of water supply and wastewater treatment systems are not always available for several reasons; and such inventories are fundamental to good adaptation planning. Since climate change is a new consideration for most urban jurisdictions, integrating it into water infrastructure

planning and operation has not typically been considered. In addition, water supply and wastewater treatment systems infrastructure is sometimes managed by several agencies, even in quite small urban jurisdictions, and each agency may have different record-keeping procedures. As a result, water supply and wastewater treatment system elements are not always identified by potentially limiting physical parameters that connect them to their changing environment (e.g., height above mean sea level and storm surge records, distance from shore, expected lifetime, rehabilitation cycle) relevant to climate change. Further, infrastructure is now often subject to planning for replacement over time periods much shorter than those relevant to long-term climate change. Examples of questionnaires are in New York City Panel on Climate Change (2010), Appendix B.

In conducting inventories, due attention should be paid to operational, financial, or physical relationships with neighboring urban areas (e.g., to address competition for water supplies, and to promote shared wastewater treatment plants and transportation facilities).

The risk inventory provides both a basis for focusing subsequent steps of the assessment framework and a basis for identifying major thresholds and tipping points beyond which the urban water systems are most likely to fail (Pielke and Bravo de Guenni, 2004). These thresholds then can provide clear and important criteria for planning and monitoring to avoid the most critical tipping points within the urban water systems, allowing planning and adaptation to proceed with greater clarity of purpose.

5.4.2 Step 2: Apply future climate change scenarios

Climate change scenarios are now typically developed from downscaled IPCC GCM model simulation runs (Rosenzweig *et al.*, 2007; Lettenmaier *et al.*, 2008). In some cases RCM models driven by GCM boundary conditions are used for this downscaling; statistical downscaling methods are at present more common than RCM modeling. Several approaches are possible, including using mean values of a variety of simulations with different emission scenarios, or providing a range of values, for example in the form of histograms, with suggestions for critical protection levels. (An even simpler approach is to assume certain levels of, e.g., sea level rise, and plan for these, e.g., Franco *et al.* (2008), although this is generally less instructive.) Ensembles of climate model projections can be combined with the ensemble of potential responses that can yield probabilities of occurrence of sea levels that exceed certain thresholds such as unacceptable levels of sea level rise and storm surges in coastal regions or seawater intrusion into coastal aquifers. This may result in redefinition of certain common historically based criteria such as design standards for siting and constructing urban structures, which are governed, in part, by flood frequencies that were previously based on historical streamflow records.

[ADAPTATION]   Box 5.6   Adapting New York City's water supply and wastewater treatment systems to climate change

David C. Major

Center for Climate Systems Research, Columbia University

SYSTEM DESCRIPTION

Water is supplied from upland reservoir systems north of New York City (NYC) with a total area of almost 2,000 square miles. Annual precipitation on the city's watersheds averages about 44 inches. The total storage capacity of the reservoir system is 547.5 billion gallons, with a safe yield of 1,290 million gallons daily (mgd). (There is a small additional 33 mgd of safe yield from well fields in the southeastern part of NYC.) Safe yield compares to daily system demand of about 1.1 bgd in NYC and upstate. Water from the system supplies 8 million people in NYC and an additional 1 million people in upstate counties. The NYCDEP sewer and wastewater treatment system includes over 6,600 miles of sanitary, storm, and combined sewer pipes. This system processes 1,500 mgd of wastewater at 14 water pollution control plants (WPCPs) located on the coast to allow for treated water discharge (Rosenzweig *et al.*, 2007; New York City Department of Environmental Protection, 2008; Box Figure 5.11).

CLIMATE CHANGE INSTITUTIONS

The New York City Department of Environmental Protection (NYCDEP) is the municipal agency responsible for water supply and wastewater treatment. It has taken a leading role in assessing the impacts of climate change on water supply and treatment facilities. This work began with the creation of the NYCDEP Climate Change Task Force in 2004, a joint NYCDEP, university, and engineering firm effort (Rosenzweig *et al.*, 2007). The most recent report is New York City Department of Environmental Protection (2008).

METHODS AND IMPACTS

Future climate scenarios for the 2020s, 2050s, and 2080s have been developed based on downscaled IPCC GCM simulations, using 16 models and 3 emissions scenarios in the most recent applications, and 7 models for sea level rise (SLR) (Box Figure 5.12) These methods are described in detail in New York City Panel on Climate Change (2010), Appendix A. Scenarios for the NYC region predict higher temperatures,

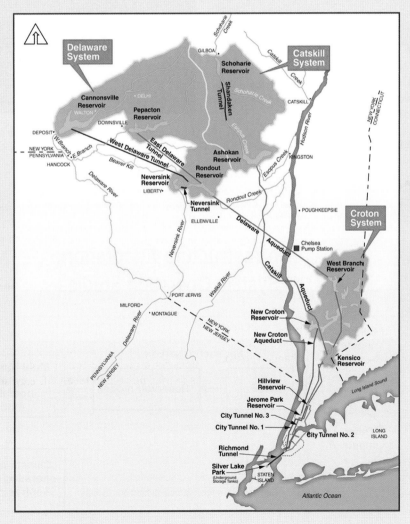

Box Figure 5.11: *New York City water supply system.*

Source: New York City Department of Environmental Protection.

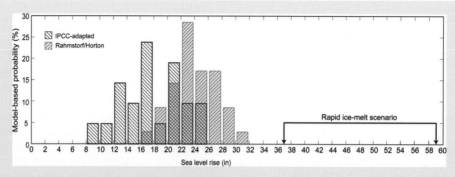

Box Figure 5.12: *Sea level rise scenarios, NYC 2050s.*

Source: New York City Panel on Climate Change (2010).

more precipitation, and sea level rise, with impacts including more precipitation, more frequent inland flooding, coastal flooding from SLR and storm surge, and water quality issues from higher temperatures and changing precipitation patterns (Box Figure 5.13). The most recent scenarios are in New York City Panel on Climate Change (2010), Appendix A.

CLIMATE CHANGE ADAPTATIONS

Using a multistep adaptation assessment process, wide-ranging adaptation studies are under way, including a study of the impacts of sea level rise on the system, and a study of reservoir operations using future climate scenarios in reservoir modeling. Potential adaptations include operating system changes (see the schematic in Box Figure 5.14), flood walls for WPCPs, relocation of facilities, improved drainage, and enhanced water quality treatment.

BROADER URBAN RELEVANCE

New York City initiated (2008) its Climate Change Adaptation Task Force, which is investigating climate change impacts on all of the City's critical infrastructure. The pioneering NYCDEP efforts are continuing within this broader

Box Figure 5.13: *Flooding at a coastal water pollution control plant (WPCP) in New York City.*

Source: New York City Department of Environmental Protection.

supportive framework (New York City Panel on Climate Change, 2010).

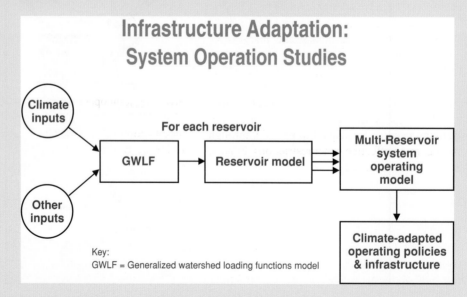

Box Figure 5.14: *NYCDEP reservoir modeling with climate change. In the figure, note that future climate inputs are among the drivers of the system.*

Source: New York City Panel on Climate Change (2010), Appendix B.

5.4.3 Step 3: Characterize adaptation options

Adaptation options may be usefully categorized as operations/management, investments in infrastructure, and policy. Many adaptations fall into more than one of these categories. A wide range of potential adaptations should be examined at the initial stage and evaluated and prioritized in later stages of the decision-making process.

5.4.3.1 Operations/management adaptations

These include a range of new criteria for system operations that reflect non-stationary hydrologic processes. As one example, scenario results can be incorporated into the design process for drainage structures by estimating new rainfall intensity–duration–frequency (IDF) curves based on projected precipitation patterns. This is a new challenge for planners and engineers (Frederick *et al.*, 1997). Studies are also needed of the operation of the sewer and wastewater treatment systems with rising sea levels and increases in storm intensity. As sea levels rise and storm surges increase, the operations of the systems will be increasingly compromised. Operational changes may be needed to deal with backsurge problems in the early stages of sea level rise, before infrastructure changes are required (Rosenzweig *et al.*, 2007).

Many operational adaptations may be suggested by the use of water system simulation models run with inputs from the climate scenarios, a process that has begun in NYC (see Box Figure 5.14 in Box 5.6 on New York City) (Moore *et al.*, 2004). This is especially important for both water supply and wastewater treatment systems, since in many urban regions of the world there are expected to be both more frequent floods and more frequent droughts, the result of increased precipitation, higher temperatures, and possibly more intense storms. Other operational changes may include further efforts to reduce consumption through changes in fixtures, pricing, and education. Demand reductions are themselves important adaptations to climate change because they can increase resiliency and margins-of-error in existing supplies. Examples of successful conservation programs can be found in Boston, USA (Massachusetts Water Resources Authority, 2001), and Melbourne, Australia. Another potential change is in drought rules: if droughts become more frequent, as is expected in many settings based on climate simulations, these rules may need to be adapted to provide, through regulation and pricing, more effective water restrictions. Finally, the incorporation of climate change considerations into a city's environmental review process is an important potential adaptation for planning growth or urban renewal.

Most of the operations/management adaptations described thus far are for developed (formal) systems. Evaluations of operation/management strategies to adapt informal systems to climate change are urgently needed, including strategies for entraining them into the formal water supply and wastewater treatment sectors so as to better adapt to dynamic climate conditions.

5.4.3.2 Investments in infrastructure

A wide range of potential adaptations is available for adapting water supply and wastewater treatment systems to climate change, and in some cases adaptations are under active consideration. These may entail substantial financial outlays, some of which may be included in future system expansions, while others may be required for actions undertaken specifically as climate change responses. There are also infrastructure changes undertaken for other reasons, such as the Thames (UK) (Aerts *et al.*, 2009) and Venice (Italy) tide surge barriers, which should prove useful in dealing with climate change stresses. Another such example is the New York City Department of Environmental Protection dependability study, designed to provide for continuing supplies of water for 9 million New York system customers if any element of the system goes off line (New York City Department of Environmental Protection, 2008). The redundancy measures contemplated in this study will also be helpful in future periods of increased droughts and floods.

Among infrastructure adaptations for sea level rise and storm surge, flood barriers for wastewater treatment plants and other coastal structures are possible, as is the construction or relocation of tide gates and the installation of pumping capacity. If more freshwater will need to be taken from estuaries or if less freshwater is expected to enter the estuaries, water intakes may have to be moved upriver to cope with encroaching salt fronts. New system interconnections will increase resiliency in both coastal and inland cities. Desalinization plants may become more attractive in changing climatic conditions, although capital and energy costs may be limiting. Artificial groundwater recharge by streamflow or imported water, aquifer storage and recovery (ASR), or seawater barrier injection systems to provide replenishment of subsurface stores of water may also be needed, and may expand options for more urban water banking and marketing.

5.4.3.3 Policy adaptations

Policy adaptations may be distinguished from operations and management decisions because they are made at higher levels of government. For example, policy decisions include those involving joint operation of systems run by different authorities, of which some examples exist already. These include a recent modification of operating rules at Lake Wallenpaupack, in the Delaware Basin (USA), to provide for changes in releases and more flexible joint operations involving a private utility, the Delaware River Basin, and New York City (DePalma, 2004). Further into the future, large potential adaptations involving joint operations and investments may include integrating New York City reservoir operations with Delaware River Basin facilities (Rosenzweig and Solecki, 2001). As policy focuses on water markets, and related distribution systems evolve near urban centers, a wider variety of water sources and applications of water reuse will become available. Conjunctive uses of groundwater and surface water supplies have economic benefits, or can be structured to create economic benefits (Reichard and Raucher, 2003) that can be used to promote alternative adaptations through incentives, reduced costs, and greater supply availability.

Policy adaptations may be very important in dealing with climate change in the informal water supply and wastewater treatment systems, including decisions to bring such systems under more adequate supervision. In much of the developing world, a policy shift in allowing public–private partnership (PPP) through water concessions and vendor-based supply, among others, has the capability to increase access to safe water supply; however, there are important pricing and equity considerations in such partnerships, which have not always been successfully implemented. Particularly in developing countries, it may be important to develop a new culture of water value, use, and consumption based on balanced perspectives of its economic, physical, ecological, social, political, and technical dimensions.

5.4.4 Step 4: Conduct initial feasibility screening

This is an important step to ensure, as soon as possible, that potential and proposed adaptations are at least feasible, during the time frame in which they might be implemented. Feasibility must be evaluated from the engineering, economic, environmental, legal, and other perspectives. In this step, it is important to err on the side of inclusion rather than exclusion, because many adaptations that may seem unfeasible according to current standards and expectations may become feasible as climate change conditions (and challenges) unfold, or advances in engineering, budget, and organizational possibilities occur.

5.4.5 Step 5: Link to capital and rehabilitation cycles

Adaptation strategies should be integrated with expected rehabilitation and replacement schedules. This will provide for potentially significant cost savings in implementing adaptations. In addition, there may be planned maintenance/operations and policy changes that will provide opportunities for efficient scheduling of adaptations. Revenue streams generated from use and sale of water and wastewater services may be linked to climate changes and can partially constrain when rehabilitation, expansion or adaptation can occur and can be funded.

5.4.6 Step 6: Evaluate options – benefit/cost analysis, environmental impact, legal mandates

Benefit/cost analysis has a long history in water resources and other infrastructure planning (classic treatments include Eckstein, 1958; Gittinger, 1972; current applications are discussed in, among many recent texts, Freeman, 2003 and Tietenberg, 2006). With reference to climate change, the focus of economic studies has been on general economic costs and impacts of climate change (e.g., Nordhaus and Boyer, 2000) and high levels of anthropogenic climate impacts (e.g., Mastrandrea and Schneider, 2004), although there have been some locally detailed studies (e.g., Yohe and Neumann, 1997). Despite the widespread

awareness of the importance of climate change for sectors such as water utilities (Miller and Yates, 2005), there have been few attempts to integrate the technique precisely into the decision frameworks of agencies. New applications should include three elements not typically integrated into benefit/cost applications:

1. Changing climate risks over time as represented by model-based probabilities
2. Capital cycle and capital programming for long-lived infrastructure
3. Planning and regulatory time-lags typical of, and often decisive for, urban infrastructure.

Environmental laws and other legal mandates will vary widely among jurisdictions; a first level of adaptation assessment should be the analysis (and revision where required) of current mandates to include climate change adaptation (Sussman and Major, 2010). Particularly in cities, environmental impact analyses should include the human environment as a priority.

5.4.7 Step 7: Develop Climate Change Adaptation Action Plans, including timeframe for implementation

Water and wastewater treatment agencies' Climate Change Adaptation Action Plans should include the program of identified adaptation strategies, outlining the resources committed to implement the plans, the resources that are still needed, and next steps to be taken (including areas that need to be researched further). All plans should include specific dates for implementation and metrics to measure success. When climate risk management has been fully incorporated into agency procedures, implementation plans will be incorporated into capital and operating budgets over the long run, and revisited regularly as uncertainties and eventualities associated with climate change unfold. Time horizons may be at different levels for mitigation and for adaptation and have parallel but separate metrics.

This step should also include coordination with other agencies, such as transportation agencies whose decisions may affect urban water flows (e.g., by changing surfaces or slopes). In addition, opportunities to link adaptation to mitigation should be explored.

5.4.8 Step 8: Monitor and reassess

Adaptation plans should be reviewed and updated on a regular basis to reflect changes in environmental conditions and climate change science. Agencies should continue to monitor their infrastructure and use updated climate risk information to determine further vulnerabilities and the adequacy of plans and efforts to date. Monitoring and reassessment is a normal long-term part of climate risk management, because the science, and the ability to prepare climate scenarios, progresses each year. In addition, every year there is additional information on actual climate changes, which should be used to refine and readjust adaptation programs on a regular basis. This step is crucial to the development of flexible adaptation pathways that are appropriate and realistic to the urban water systems of individual cities (New York City Panel on Climate Change, 2010). Using procedures similar to this Adaptation Assessment procedure (Rosenzweig *et al.*, 2007, Table 1), the state of California has since 1983 required every urban water supplier to develop and implement an Urban Water Management Plan (UWMP) (CADWR, 2008b) that includes urban water contingency analyses with six components to address drought, climate change, and other catastrophic shortfalls of supply:

1. A description of the stages of action an agency will take in response to water shortages
2. An estimate of supply availability under conditions of three consecutive dry years
3. A plan for dealing with catastrophic supply interruptions
4. A list of prohibitions, penalties, and demand reduction methods to be used
5. An analysis of expected revenue effects of reduced sales during shortages and proposed measures to mitigate those effects
6. A plan to monitor and document water cutbacks.

5.5 Mitigation

Under the IPCC definition (e.g., IPCC, 2007a), mitigation is action to reduce emissions of greenhouse gases in order to reduce overall climate change. Notably the IPCC Fourth Assessment Working Groups II and III conclude that "mitigation primarily involves the energy, transportation, forestry, and agricultural sectors, whereas actors involved in adaptation primarily represent a large variety of sectoral interests, including agriculture, tourism, and recreation, human health, water supply, coastal management, urban planning and nature conservation" (IPCC, 2007a). Most of the references in the Working Group III 2007 Assessment Report (IPCC, 2007c) to urban water and wastewater refer to improvements in wastewater management as an approach to mitigating emissions (primarily of methane). Nonetheless, in the urban-water sector, there are several important mitigation options that can be incorporated into planning and operations because, in urban areas, water and energy are inextricably linked.

5.5.1 Water conservation/demand reductions

Reductions in water use can have multiple benefits, including cost reductions, increased overall supply reliability, and mitigation of greenhouse gas emissions (e.g., Cohen *et al.*, 2004; Klein *et al.*, 2005a). Less water used can mean less water needing to be captured at, and drawn from, various reservoirs and aquifers, less water to be transported and lifted over obstacles, less water to be treated, less water to be heated, less wastewater to be treated, and less wastewater to be transported and disposed of (Figure 5.1). Each of these steps in the water system generally requires energy (e.g., Cohen *et al.*, 2004; Klein *et al.*, 2005a). For example, in California, the State Water Project,

which transports water from the wetter northern parts of the State to urban southern parts, is the largest single electrical energy user in the State (Cohen *et al.*, 2004). In many cities, the energy used for water supply, treatments, and disposal has come from burning fossil fuels that emit greenhouse gases into the atmosphere. In many – perhaps, most – urban systems and situations, water conservation and demand reductions can provide greenhouse gas emissions mitigation benefits.

Urban water-use demands for residential supplies are typically largest in cities where housing is most dispersed, because outdoor uses of water are frequently the largest demands (e.g., Mayer *et al.*, 1999). Thus, in many dense urban areas, achieving substantial demand reductions can be difficult. Nonetheless, one particularly important mechanism for controlling water waste in the cities of many developed and developing nations is reduction of large-scale leakage from the water-supply infrastructures. Lallana (2003b) compiled urban water-supply leakage estimates from 15 European nations, and found leakage rates ranging from about 4 percent of the total water supplies to 50 percent (Figure 5.4).

In informal urban settlements, planning and maintenance of water delivery systems are presumably less rigorous than in the formal areas, and leakage losses may be even larger. Leakage losses also represent opportunities for contamination of water supplies, so that efforts to reduce leakage will provide multiple benefits (Lallana, 2003b). Reduction of leakage is likely to depend on pressure in the water mains, soils, topography, and age of the water systems. Nonetheless, progress is possible and, indeed, being made in many urban water systems.

5.5.2 Water reclamation and recycling

Reclamation and recycling offer opportunities for reducing the energy used to provide water supplies (see e.g., Furumai, 2008). Recycled water generally still requires treatments that demand energy, but otherwise many of the initial extraction and transport energy demands can be reduced or eliminated because the reused water is already in the municipality (Figure 5.1).

5.5.3 Attention to energy efficiency of water supply expansion

More generally, most actions to expand or improve water supplies have ramifications in terms of overall energy use, which in turn need to be carefully assessed in terms of greenhouse gas emissions. Development of some urban water sources – such as groundwater pumpage or, more recently, desalination of brines or seawater – can require amounts of energy or conditions of energy development that may be problematic in terms of mitigating greenhouse gas emissions, especially if they are allowed to degrade (e.g., overdraft of aquifers with attendant increases in pumpage lift). Energy requirements for treatment of some water sources can also be decreased or increased depending on whether the water quality of the source is managed or mismanaged.

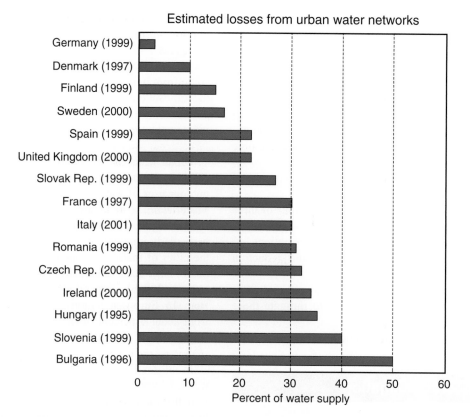

Figure 5.4: *Urban water supply leakage losses from 15 European nations. Note that leakage rates will vary by city within countries.*

Source: Lallana (2003b).

5.5.4 Hydropower and reservoirs

Hydropower and surface water reservoir-based water supplies can have implications for mitigation of greenhouse gas emissions, although those benefits remain difficult to specify. Most reservoirs emit varying amounts of greenhouse gases through processes involved in the natural carbon cycle (Battin *et al.*, 2008), although some reservoirs also absorb these gases. In particular, greenhouse gas emissions can be significant from shallow tropical reservoirs; e.g., Fearnside (1995) calculated emissions from two reservoirs in Brazil amounting to tens of millions of tons of CO_2 and tens of thousands of tons of CH_4 in a single year, three to six years after their initial inundation. It is believed that deeper, cooler reservoirs emit less gas (IPCC, 2008). However, wetlands and floodplains that were inundated in the establishment of reservoirs may also have been methane emitters, and those wetland emissions may have been substantially reduced by their inundation (Mata and Buhooram, 2007; IPCC, 2008). Thus, opportunities for greenhouse gas emissions mitigation may also be available in the planning, operations, and use of hydropower in general, hydropower as energy supplies for urban water and wastewater systems, and reservoir-based water supplies, but the extent of these opportunities generally needs more study.

5.5.5 Urban water heating

A significant amount of the energy associated with urban water supplies is dedicated to water heating for residential, commercial, and some industrial purposes. For example, in California, over half of urban water uses are residential, and over half of those uses involve water heating (Cohen *et al.*, 2004, Figures 3 and 4, and p. 26). Actions and opportunities that favor the expansion of solar water heating have been identified as a useful part of urban mitigation programs (e.g., Razanajatavo, 1995; IPCC, 1996b; Nadel *et al.*, 1998) in both developed and developing nations. Localized small-scale wind power could also help minimize the centralized energy needed to pump, distribute, and heat water for urban and domestic uses. Even if solar heating is not deployed, it is often possible to reduce greenhouse gas emissions by heating with fuels (or electricity) that are less carbon intensive, or by using tankless water heating.

5.5.6 Watersheds and river basins

Many traditional urban water systems tap into water supplies derived from broad hinterland watersheds and river basins (Kissinger and Haim, 2008; Broekhuis *et al.*, 2004). Many cities, recognizing the vulnerabilities of these hinterlands to contamination and disruption, are moving to institute better land use and watershed management practices in these resource areas. Land use and watershed management impact overall land- and water-surface emissions or sequestrations of greenhouse gases, and thus need to be assessed in terms of their mitigation impacts or benefits, along with other costs and benefits.

5.5.7 Wastewater

Cities are large and concentrated producers of wastewater (Satterthwaite, 2008). Methane emitted during wastewater transport, treatment, and disposal, including from wastewater sludge, amounts to 3 to 19 percent of global anthropogenic methane emissions (IPCC, 1996a). Globally, the major sources of the greenhouse gas nitrous oxide (N_2O) are human sewage and wastewater treatment (IPCC, 2007b). Methane emissions from wastewater are expected to increase by about 50 percent in the next several decades, and N_2O emissions by 25 percent. Thus, one of the most direct ways to mitigate greenhouse gas emissions is through improvements in collection and management of urban wastewaters, using technologies most appropriate to the economies and settings involved (IPCC, 2007b). Technologies already exist for reducing, and perhaps reversing, these emissions growth rates.

In cities in developed nations, wastewater treatment facilities are sometimes major greenhouse gas emitters but those emissions have been identified as important avenues for overall greenhouse gas emissions reduction (e.g., Rosenzweig et al., 2007). A large proportion of the greenhouse gas emission from urban wastewater, however, is expected to take place in developing countries (e.g., Al-Ghazawi and Abdulla, 2008) and from informal urban settlements. In many of those developing countries and informal urban settlements, rapid population growth and urbanization without concurrent development of sufficient wastewater collection, treatment, and disposal infrastructures results in very large and unmitigated greenhouse gas emissions. Non-existent sewer systems, open sewers, ponding, and unchecked releases of untreated wastewaters are a fact of life in the informal sectors of cities in both developed and developing countries (IPCC, 2007b; Foster, 2008). Improved sanitation facilities, infrastructures, treatment and disposal systems in these settings would not only mitigate emissions but would offer substantial public health benefits as well (e.g., Al-Ghazawi and Abdulla, 2008; IPCC, 2007b).

5.6 Policy considerations and knowledge gaps

5.6.1 Policy considerations

Some cities have implemented policies to encourage adaptation and mitigation in the face of the effects of increased urbanization combined with climate change and variability. These include, among others, Chicago, London, New York, Seattle, and Toronto (Parzen, 2008). A broad range of policies is available for implementation, including operational/management changes, infrastructure investments, and new policy including institutional changes. Diversity and redundancy promote the reliability of water supply and wastewater systems and allow for a broader platform of policies that promote the implementation of adaptation and mitigation.

Approaches to managing urban and regional water resources for long-term sustainability have become more complicated in recent years as our understanding of the interconnections within hydrologic systems has increased. The fact that surface water and groundwater resources are not separate resources, but instead are generally linked, is increasingly recognized as both constraint and opportunity in the development and management of urban water resources (Winter et al., 1998). The concept of sustainable yields (Alley and Leake, 2004; Alley, 2006) has grown from capture of recharge to include capture of discharge, groundwater storage depletion, and capture of streamflow. With these and other linkages, achieving sustainability is becoming more complex as urban centers attempt to reconcile supplies and demands in the face of changing climates, runoff, and recharge.

Following are some important options for urban adaptation of water and wastewater systems to the impacts of climate change that emerge from this assessment. These options provide a range of focused approaches for urban policymakers with respect to the water and wastewater treatment sector, supplementing and extending the key messages of the chapter.

1. *Fix leaks.* In many urban areas, both in developed and emerging economies, water leakage from collection and distribution systems amounts to large fractions of the overall flows of water through the urban systems and, indeed, amounts to large absolute volumes of water wasted. Programs to control leakage, and thus to reduce demands and increase resilience to climate change, are among the first and most cost-effective adaptations that should be undertaken. Moreover, the control of leakage reduces pumping and thus energy costs, providing opportunities for both adaptation and mitigation.

2. *Integrate systems.* Innovative policies and strategies to ensure that new investments produce benefits across the integrated water systems (supply, access, quality, treatment, and recycle) are needed to help urban areas respond more effectively and efficiently to the new challenges created by climate change and to improve equity in the access and use of water. Integration of entire ranges of hydraulic components, such as artificial recharge/water banking with incentives for replenishment, reuse, and conservation, will produce efficiency and cross-cutting benefits.

3. *Improve institutions.* The roles of institutions managing formal and informal water resources in urban areas need to be analyzed and reassessed to insure that institutions are appropriate to changing challenges, including climate change impacts. Potential issues may include the investigation of the benefits of increased regionalization of systems to promote redundancy.

4. *Capture rainwater.* The capture of rainwater may be an important conservation adaptation to reduce pumped groundwater and related energy use as well as to reduce potential urban flooding and provide a supplemental source of water for irrigation of urban landscaping.

5. *Reduce demand.* Programs for the implementation of indoor demand reductions, such as low-flow toilets, shower heads,

and other efficiency measures, and outdoor demands such as landscaping, would increase the reach and resiliency of water supplies in both formal (and to the extent relevant) informal water supply systems.

6. *Reuse water.* Policies to increase water reuse would increase options for adaptation by reducing overall demand for original system water, thus making the system more robust, and they also can contribute to mitigation depending on the balance of energy required for reuse and original source.

7. *Establish water marketing.* Water marketing can provide mechanisms and opportunities to increase efficiency and improve system robustness. Water marketing also may facilitate integration of multisector use in situations where urban, agricultural, and environmental uses could be enhanced by combined management.

8. *Increase water banking.* Water banking may also provide important options for adaptation as a way of hedging against uncertainties and improving system robustness.

9. *Incorporate climate change into planning.* Optimal scheduling of adaptations for long-lived infrastructure, as a regular part of planning, can help to ensure that investments are made at the most efficient times in terms of climate change and other system drivers.

10. *Rationalize water rights.* Revisions of the structure and distribution of water rights to more fully encompass conjunctive use of groundwater and surface water may be needed. In the same context, "use-or-lose" policies that do not promote efficiency, conservation, and reuse are likely to become more and more problematic in the face of climate change induced growth of demands and stresses.

11. *Develop public–private partnerships.* In much of the developing world, a policy shift to allow public–private partnerships (PPP) through water concessions, vendor-based supply, and other measures may help to increase access to safe water supply as the impacts of climate change are felt. However, equity considerations and the urgent needs of the poor are also important considerations (e.g., Argo and Laquian, 2007), and there have been cases in which such arrangements have been problematic.

5.6.2 Gaps in needed understanding of urban water systems and climate change

Effective responses to the many challenges that climate change poses to urban water and wastewater systems require a strong understanding of the challenges themselves, of the workings and limits of the urban systems under stresses for which they may not have been designed, of the range of options that are feasible in engineering, economic, and political terms, and of the interactions between the water sector and other sectors also striving to accommodate climate change. Many gaps in our understanding currently limit our ability to develop those responses.

The largest gap in understanding the implications of climate change for urban water systems is the limited and short history

of monitoring, evaluation, and prediction of vulnerabilities of water supplies and wastewater management in informal settlements. Much more information is needed in order to manage these systems well and to encourage the incorporation of these systems into the formal sector. All "urban watersheds" need to have a data monitoring structure that will allow the computation and monitoring (of supply and demand as well as potential climate change) of the hydrologic budget (i.e., inflows and outflows).

In both formal and informal urban areas, a key knowledge gap is lack of intercomparability among measures of water and wastewater systems describing the systems from city to city. Regular, intercomparable data describing water uses, stormwater rates and the economics of water and wastewater are needed to prioritize urban water/climate issues at regional to international scales, and to identify early successful management responses in both formal and informal sectors.

The hydrologic cycle in urban settings is less well understood and quantified than those in many less intensely populated regions (e.g., Grimmond *et al.*, 1986; van de Ven, 1990; Gumbo, 2000; Marsalek *et al.*, 2007; Furumai, 2008). Better, more site-specific quantification and knowledge of the urban water cycle will provide baselines for interpreting the influences of climate change on urban waters and environments.

The impacts of untreated wastewaters (both residential and industrial) on ecosystems and downstream water supplies (Gleick *et al.*, 2006) are not well understood, and the interaction of climate change with such impacts is even less well characterized at present (Nelson *et al.*, 2009). Significant research is required on these subjects.

One set of specific data that needs to be compiled is information about which of the world's coastal cities' water and wastewater systems are threatened by different levels of sea level rise, accounting for local subsidence and ocean conditions. This is essential both at a city-by-city level, and at international and global levels to understand the magnitude and prioritization of required adaptation investments.

The role of water management in greenhouse gas emissions mitigation has not been well addressed to date. These mitigation issues are complex and interlinked (Satterthwaite, 2008); more research in the area of how water management interfaces with mitigation is needed.

REFERENCES

Aerts, J. C. and J. H. P. Droogers (Eds.) (2004). *Climate Change in Contrasting River Basins: Adaptation Strategies for Water, Food and Environment*, Oxfordshire, UK: Commonwealth Agricultural Bureaux (CAB) International Press.

Aerts, J., D. C. Major, M. Bowman, P. Dircke, and M. A. Marfai (2009). *Connecting Delta Cities: Coastal Cities, Flood Risk Management, and*

Adaptation to Climate Change, Amsterdam, the Netherlands: Free University of Amsterdam Press.

Al-Ghazawi, Z. D. and F. Abdulla (2008). Mitigation of methane emissions from sanitary landfills and sewage treatment plants in Jordan. *Clean Technology and Environmental Policy*, **10**, 351–350.

Alley, W. M. (2006). Another water budget myth: The significance of recoverable ground water in storage. *Ground Water*, **45**, 251.

Alley, W. and S. A. Leake (2004). The journey from safe yield to sustainability. *Ground Water*, **42**, 12–16.

Argo, T. and A. A. Laquian (2007). Privatization of water utilities and its effects on the urban poor in Jakarta Raya and Metro Manilla. In A. A. Laquian, V. Tewari, and L. M. Hanley (Eds.), *The Inclusive City: Infrastructure and Public Services for the Urban Poor in Asia*, Woodrow Wilson Press, Johns Hopkins University Press.

Ayers, G, P. Whetton, R. Jones, and K. Hennessy (2003). *The Australian Climate: Present and Future*. National Symposium & Short Course Climate Change & Health, Research Methods & Policy Issues, Australian National University, Canberra, ACT: ANU.

Battin, T. J., L. A. Kaplan, S. Findlay, *et al.* (2008). Biophysical controls on organic carbon fluxes in fluvial networks. *Nature Geoscience*, **1**, 95–100.

Bicknell J., D. Dodman, and D. Satterthwaite (2009). *Adapting Cities to Climate Change: Understanding and Addressing the Development Challenges*, London,UK: Earthscan.

Broekhuis, A., M. de Bruijn, and A. de Jong (2004). Urban–rural linkages and climatic variability. In A. J. Dietz, R. Ruben, and A. Verhagen (Eds.),*The Impact of Climate Change on Drylands: With a Focus on West Africa*, Dordrecht, the Netherlands: Kluwer Academic Publishers, pp. 301–321.

Brown, L. R. (2006). Plan B 2.0: Rescuing a planet under stress and a civilization in trouble. In *Designing Sustainable Cities*, New York, USA: W.W. Norton & Co. Accessed June 28, 2009, http://earth-policy.org/Books/PB2/PB2ch11_ss5.htm.

CADWR, *see* California Department of Water Resources

California Department of Water Resources (CADWR) (2005). California water plan update. *CADWR Bulletin* 160–05, 1, Sacramento, California. Accessed June 28, 2009, www.waterplan.water.ca.gov/docs/cwpu2005/vol1/v1complete.pdf.

California Department of Water Resources (CADWR) (2008a). *Managing an Uncertain Future: Climate Change Adaptation Strategies for California's Water*. CADWR Fact Sheet, Sacramento, California.

California Department of Water Resources (CADWR) (2008b). *Urban Drought Guidebook 2008*, updated edition. CADWR Report.

Canadian Natural Resources (NRC) (2008). Geoscape Canada: *Bow River Basin Waterscape Urban Water*. Accessed June 28, 2009, http://geoscape.nrcan.gc.ca/h2o/bow/urban_e.php.

Carbognin, L. and L. Tosi (2005). Interaction between climate changes, eustacy and land subsidence in the northern Adriatic region, Italy. *Marine Ecology*, **23**, 38–50. Accessed June 28, 2009, www3.interscience.wiley.com/journal/119192005/abstract?CRETRY=1&SRETRY=0.

Case, T. (2008). Climate change and infrastructure issues. AWWA Research Foundation, Drinking Water Research, *Climate Change Special Issue*, **18**, 15–17.

Chinn, A. (2005). The future ain't what it used to be: problem definition in the municipal water supply sector. *Global Climate Change and Its Potential Effects on Seattle's Water Supply*, 2005 Climate Change Conference, King County, Seattle, WA, USA. Accessed June 28, 2009, http://dnr.metrokc.gov/dnrp/climate-change/conference-2005-results/municipal-water-supply/pdf/presentation-chinn.pdf.

Cohen, R., B. Nelson, and G. Wolff (2004). *Energy Down the Drain: The Hidden Costs of California's Water Supply*. Natural Resources Defense Council, Oakland, CA, USA. Accessed June 28, 2009, http://www.nrdc.org/water/conservation/edrain/edrain.pdf.

CONAMA (Comisión Nacional del Medio Ambiente) (1999). *Primera Comunicación Nacional bajo la Convención Marco de las Naciones Unidas sobre el Cambio Climático* (Santiago).

CONAMA (Comisión Nacional del Medio Ambiente) (2002). *Áreas Verdes en el Gran Santiago* (Santiago).

CONAMA (Comisión Nacional del Medio Ambiente) (2005). *Estrategia de Cambio Climático* (Santiago).

CONAMA (Comisión Nacional del Medio Ambiente) (2006). *Estudio de la variabilidad climática en Chile para el siglo XXI* (Santiago).

CONAMA (Comisión Nacional del Medio Ambiente) (2008). *Plan de Acción Nacional de Cambio Climático, 2008–2012* (Santiago).

Cotton, W. R. and R. A. Pielke, Sr. (2006). *Human Impacts on Weather and Climate* (2nd edition). Cambridge, UK: Cambridge University Press,.

Davis, M. (2006) *Planet of Slums*. London, UK: Verso Publishing.

DePalma, A. (2004). Do fish have water rights? *New York Times,* June 25.

DGA (Dirección General de Aguas) (2003). *Evaluación de los Recursos Hídricos Superficiales en la Cuenca del Rio Maipo* (Santiago).

Earman, S., A. R. Campbell, B. D. Newman, and F. M. Phillips (2006). Isotopic exchange between snow and atmospheric water vapor: Estimation of the snowmelt component of groundwater recharge in the southwestern United States. *Journal of Geophysical Research*, **111**, D09302, doi:10.1029/2005JD006470.

Eckstein, O. (1958). Water-resource development: *The Economics of Project Evaluation*, Cambridge, MA, USA: Harvard University Press.

Ekanade, O., A. Ayanlade and I. O. O. Orimogunje (2008). Geospatial analysis of potential impacts of climate change on coastal urban settlements in Nigeria for the 21st century. *Journal of Geography and Regional Planning*, 1(3), 049–057, www.academicjournals.org/JGRP.

Ezcurra, E., M. Mazari-Hiriart, M. Pisanti, and A. Aguilar (1999). *The Basin of Mexico: Critical Environmental Issues and Sustainability*, Tokyo, Japan: United Nations University Press.

Fearnside, P. (1995). Hydroelectric dams in the Brazilian Amazon as sources of 'greenhouse gases'. *Environmental Conservation*, 7–19.

Foster, V. (2008). *Overhauling the Engine of Growth: Infrastructure in Africa*. Africa Infrastructure Country Diagnostic, World Bank Executive Summary. Accessed June 28, 2009, http://siteresources.worldbank.org/INTAFRICA/Resources/AICD_exec_summ_9–30–08a.pdf.

Franco, G., S. C. Moser, and D. Cayan (2008). *The Future Is Now: An Update on Climate Change Science, Impacts, and Response Options for California*. California Energy Commission Public Interest Energy Research Program report 500–2008–077. Accessed June 28, 2009, www.energy.ca.gov/2008publications/CEC-500–2008–077/CEC-500–2008–077.PDF.

Frederick, K. D., D. C. Major, and E. Z. Stakhiv (Eds.) (1997). *Climate Change and Water Resources Planning Criteria*. Dordrecht, the Netherlands: Kluwer Academic Publishers, pp. 43–270. Published concurrently as a special issue of *Climatic Change*, **37**(1).

Freeman, A. M., III. (2003). *The Measurement of Environmental and Resource Values: Theory and Method* (2nd edition). Resources for the Future.

Furumai, H. (2008). Urban water use and multifunctional sewerage systems as urban infrastructure. *Urban Environmental Management and Technology*, **1**, 29–46.

Gambolati, G., P. Gatto, and R. A. Freeze (1974). Predictive simulation of land subsidence of Venice. *Science*, **183**, 849–851.

Gay, C., Estrada, F., and Conde, C. (2007). Some implications of time series analysis for describing climatologic conditions and for forecasting. An illustrative case: Veracruz, México. *Atmósfera*, **20**(2), 147–170.

Gittinger, J. P. (1972). *Economic Analysis of Agricultural Projects*, Baltimore, MD, USA: Johns Hopkins University Press.

Gleick, P. (2000). *The World's Water, 2000–2001: The Biennial Report on Freshwater Resources*, Washington, DC, USA: Island Press.

Gleick, P., H. Cooley, D. Katz, E. Lee, and J. Morrison (2006) *The World's Water, 2006–2007: The Biennial Report on Freshwater Resources*, Washington, DC, USA: Island Press.

Goodman, A. S., with D. C. Major *et al.* (1984). *Principles of Water Resources Planning*. New York, USA: Prentice-Hall.

Greater London Authority, London Climate Change Partnership (2005). *Adapting to Climate Change: A Checklist for Development*. Greater London Authority report. Accessed June 28, 2009, www.london.gov.uk/lccp/publications/docs/adapting_to_climate_change.pdf.

Grimmond, C. S. B., T. R. Oke, and D. G. Steyn (1986). Urban water budgets, 1, A model for daily totals. *Water Resources Research*, **22**, 1397–1403.

Gumbo, B. (2000). Mass balancing as a tool for assessing integrated urban water management. In *Proceedings, 1st WARFSA/WaterNet Symposium – Sustainable Use of Water Resources*, Maputo, Uganda.

Halifax Regional Municipality (2007). *Climate Change: Developer's Risk Management Guide*, Halifax Regional Municipality report. Accessed June 28, 2009, www.ccap.org/docs/resources/394/DevelopersGuidetoRiskManagment.pdf.

Hanson, R. T., P. Martin, P., K. M. Koczot (2003). *Simulation of Groundwater/Surface Water Flow in the Santa Clara–Calleguas Basin, California*, US Geological Survey Water-Resources Investigation Report 02–4136. Accessed June 28, 2009, http://water.usgs.gov/pubs/wri/wri024136/text.html.

Hanson, R. T., Z. Li, and C. Faunt (2005). *Documentation of the Santa Clara Valley Regional Groundwater/Surface Water Flow Model, Santa Clara County, California*, US Geological Survey Scientific Investigations Report SIR2004–5231. Accessed June 28, 2009, http://pubs.usgs.gov/sir/2004/5231/.

Hanson, R. T., W. Schmid, J. Lear, and C. C. Faunt (2008). Simulation of an aquifer-storage-and-recovery (ASR) system using the Farm Process in MODFLOW for the Pajaro Valley, Monterey Bay, California. In *Proceedings, MODFLOW and More: Ground Water and Public Policy*, Golden, CO, USA, pp. 501–505.

Howard, G. and Bartram, J. (2003). *Domestic Water Quantity, Service Level And Health*, WHO Report WHO/SDE/WSH/03.02.

Hurliman, A. (2007). Is recycled water use risky? An urban Australian community's perspective. *Environmentalist*, **27**, 83–94.

Intergovernmental Panel on Climate Change (IPCC) (1996a). *IPCC Guidelines for National Greenhouse Gas Inventories: Reference Manual*. National Physical Laboratory, New Delhi, India, pp. 6–15.

Intergovernmental Panel on Climate Change (IPCC) (1996b). *Technologies, Policies and Measures for Mitigating Climate Change*. IPCC Technical Paper 1, 94 p. Accessed June 28, 2009, www.ipcc.ch/pdf/technical-papers/paper-I-en.pdf.

Intergovernmental Panel on Climate Change (IPCC) (2007a). *Climate Change 2007: The Physical Science Basis. Contribution of Working Group I to the Fourth Assessment Report of the IPCC*, Cambridge, UK: Cambridge University Press.

Intergovernmental Panel on Climate Change (IPCC) (2007b). *Climate Change 2007: Impacts, Adaptation and Vulnerability. Contribution of Working Group II to the Fourth Assessment Report of the IPCC*, Cambridge, UK: Cambridge University Press.

Intergovernmental Panel on Climate Change (IPCC) (2007c) *Climate Change 2007: Mitigation of Climate Change. Contribution of Working Group III to the Fourth Assessment Report of the Intergovernmental Panel on Climate Change*, Cambridge, UK: Cambridge University Press.

Intergovernmental Panel on Climate Change (IPCC) (2008). *Climate Change and Water*, IPCC Technical Paper VI. Accessed June 28, 2009, www.ipcc.ch/pdf/technical-papers/climate-change-water-en.pdf.

IPCC, *see* Intergovernmental Panel on Climate Change

Jusseret, S., T. Vu Thanh, C. Baeteman, and A. Dassargues (2009). Groundwater flow modeling in the central zone of Hanoi, Vietnam. *Hydrogeology Journal*, **17**, 915–934.

Kahrl, F. and D. Roland-Holst (2008). *California Climate Risk and Response*. University of California, Berkeley, Department of Agriculture and Resource Economics, Research Paper 08102801. Accessed June 28, 2009, http://are.berkeley.edu/~dwrh/CERES_Web/Docs/California%20Climate%20Risk%20and%20Response.pdf.

Kimani-Murage, E. W. and A. M. Ngindu (2007). Quality of water the slum dwellers use: the case of a Kenyan slum. *Journal of Urban Health*, **82**, 829–838.

Kirshen, P., M. Ruth, W. Anderson, *et al.* (2004). *Climate's Long-term Impacts on Metro Boston (CLIMB) Final Report*. Civil and Environmental Engineering Department, Tufts University, www.clf.org/uploadedFiles/CLIMB_Final_Report.pdf.

Kissinger, M. and A. Haim, A. (2008). Urban hinterlands: the case of an Israeli town ecological footprint. *Environmental Development and Sustainability*, **10**, 391–405.

Klein, G., M. Krebs, V. Hall, T. O'Brien, and B. B. Blevins (2005a). *California's Water–Energy Relationship*. California Energy Commission Final Staff Report CEC-700–2005–011-SF.

Klein, H., K. J. Douben, W. van Deursen, and E. R.van Stevenick (2005b). Increasing climate variability in the Rhine Basin: business as usual? In J. C. Aerts and P. Droogers (Eds.), *ADAPT: Climate Change in Contrasting River Basins*, Manchester, UK: CABI Publishing, pp. 133–155.

Lallana, C. (2003a). *Water Prices*, European Environment Agency Indicator Fact Sheet WQ05. Accessed June 28, 2009, http://themes.eea.europa.eu/Specific_media/water/indicators/WQ05%2C2003.12.

Lallana, C. (2003b). *Water Use Efficiency (in Cities): Leakage*. European Environment Agency Indicator Fact Sheet WQ06. Accessed June 28, 2009, http://themes.eea.europa.eu/Specific_media/water/indicators/WQ06,2003.1001/WatUseEfficiency_RevOct03.pdf.

Lettenmaier, P., D. Major, L. Poff, and S. Running (2008). *Water Resources. The Effects of Climate Change on Agriculture, Land Resources, Water Resources, and Biodiversity in the United States*, US Climate Change Science Program Synthesis and Assessment Product 4.3, pp. 121–150.

Marsalek, J., B. Jimenez-Cisneros, M. Karamouz, *et al.* (2007). *Urban Water Cycle Processes and Interactions*, UNESCO-IHP Urban Water Series 2, London, UK: Taylor and Francis.

Massachusetts Water Resources Authority (2001). *Summary Report of MWRA Demand Management Program, Fiscal Year 2001*, Boston, MA, USA.

Mastrandrea, M. D. and S. H. Schneider (2004). Probabilistic integrated assessment of "dangerous" climate change. *Science*, **304**(5670), 571–575.

Mata, L. J. and J. Buhooram (2007). Complementarity between mitigation and adapation: the water sector. *Mitigation and Adaptation Strategies for Global Change*, **12**, 799–807.

Mayer, P. W., W. B. DeOreo, E. Opitz, *et al.* (1999). *Residential End Uses of Water*, American Water Works Association Research Foundation, Denver, CO, USA. Accessed June 28, 2009, http://books.google.com.

Mazari, M., E. Cifuentes, E. Velazquez and J. Calva (2000). Microbiological groundwater quality and health indicators in Mexico City. *Urban Ecosystems*, **4**, 91–103.

Miller, K. A. and D. Yates (Eds.) (2005). *Climate Change and Water Resources: A Primer for Water Utilities*. American Water Works Association Research Foundation.

MINVU (Ministerio de Vivienda y Urbanismo) (2008). *Propuesta de Modificación del Plan Regulador Metropolitano de Santiago* (Santiago).

Moore, K., C. Rosenzweig, R. Goldberg, *et al.* (2004). Impacts of projected climate change on phosphorous and sediment loadings for a New York City water supply reservoir. In *Proceedings of the 29th International Congress of Limnology*, Lahti, Finland.

Nadel, S., M. Rainer, M. Shepard, M. Suozzo, and J. Thorne (1998). *Emerging Energy-Saving Technologies and Practices for the Buildings Sector*. American Council for an Energy Efficient Economy, Washington, DC, USA.

Narain, S. (2002). The flush toilet is ecologically mindless. *Down to Earth*, 28–32.

Nelson, K. C., M. A. Palmer, P. L. Angermeier, *et al.* (2009). Forecasting the combined effects of urbanization and climate change on stream ecosystems: from impacts to management options. *Journal of Applied Ecology*, doi:10.1111/j.1365–2664.2008.01599.x.

Neuwirth, R. (2006). *Shadow Cities: A Billion Squatters, a New Urban World*. London, UK: Routledge.

New York City Department of Environmental Protection, Climate Change Program (2008). *Assessment and Action Plan*, Report #1.

New York City Municipal Finance Authority (2004). *Water and Sewer System Revenue Bonds, Fiscal 2005 Series B*.

New York City Panel on Climate Change (2010). *Climate Change Adaptation in New York City: Building a Risk Management Response*, W. Solecki and C. Rosenzweig (Eds.), New York Academy of Sciences.

Nordhaus, W. D. and J. Boyer (2000). *Warming the World: Economic Models of Global Warming*. Boston, MA, USA: MIT Press.

Orth K. D. and C. E. Yoe (1997). *Planning Primer*. US Army Corps of Engineers, Institute for Water Resources Report 97-R-15. Accessed June 28, 2009, www.au.af.mil/au/awc/awcgate/army/97r15.pdf.

Parzen, J. (2008). *Lessons Learned: Creating the Chicago Climate Action Plan*. Report to the Mayor of Chicago by the Global Philanthropy Partnership. Accessed June 28, 2009, www.chicagoclimateaction.org.

Pielke, R. A., Sr. and L. Bravo de Guenni (Eds.) (2004). How to evaluate vulnerability in changing environmental conditions. In *Vegetation, Water, Humans and the Climate: A New Perspective on an Interactive System*, Global Change: The IGBP Series, P. Kabat *et al.* (Eds.), Berlin: Springer, pp. 483–544.

Pitre, C. (2005). Will the well go dry: Potential impacts of climate change on groundwater systems used by municipal water suppliers. In *2005 Climate Change Conference*, King County, Seattle, WA, USA. Accessed June 28, 2009, http://dnr.metrokc.gov/dnrp/climate-change/conference-2005-results/municipal-water-supply/pdf/presentation-pitre.pdf.

Queensland Department of Natural Resources (QDNR) (2000). *Improving Water Use Efficiency in Queensland's Urban Communities: Executive Summary*. Consultants report, Montgomery-Watson.

Razanajatavo, M. (1995). Solar water heaters in the Seychelles. *Renewable Energy Development*, **8**, 5–6.

Reichard, E. G., and R. S. Raucher (2003). Economics of conjunctive use of groundwater and surface water. In *Water: Science, Policy, and Management*, Water Resources Monograph, **16**, 161–176.

Reid, H. (2010). Building capacity to cope with climate change in the least developed countries. In J. Dodson (Ed.), *Changing Climates, Earth Systems and Society*, International Year of Planet Earth Series, Springer Science+Business Media.

Rosenzweig, C. and W. D. Solecki (2001). Climate change and a global city: learning from New York. *Environment*, **43**, 8–18.

Rosenzweig, C., D. C. Major, K. Demong, *et al.* (2007). Managing climate change risks in New York City's water system: assessment and adaptation planning. *Mitigation and Adaptation Strategies for Global Change*, doi:1007/s11027–006–9070–5.

Saliu, J. K. and Eruteya, O. J. (2006). Biodiversity of gutters in Lagos Metropolis, Nigeria. *Journal of Biological Sciences*, **6**(5), 936–940.

San Diego Foundation (2008). San Diego's changing climate: a regional wake-up call. *Summary of the Focus 2050 Study*. Accessed June 28, 2009, www.sdfoundation.org/news/pdf/Focus2050glossySDF-ClimateReport.pdf.

Satterthwaite, D. (2008). Cities' contribution to global warming: Notes on the allocation of greenhouse gas emissions. *Environment and Urbanization*, **20**, 539–549.

Sussman, E., D. C. Major, *et al.* (2010). Law and regulation. In W. Solecki and C Rosenzweig (Eds.), *New York City Panel on Climate Change, 2010: Adapting to Climate Change: The New York City Experience*, New York Academy of Sciences.

Thirlwell, G. M., C. A. Madramootoo, I. W. Heathcote, and E. R. Osann (2007). Coping with climate change: short-term efficiency technologies. *Canada-US Water Conference*, Washington, DC, USA.

Tietenberg, T. (2006). *Environmental Economics and Policy*. New York, USA: Addison-Wesley.

United Nations Human Settlements Program (2003). *Water and Sanitation in the World's Cities: Local Action for Global Goals*, London, UK, and Sterling VA, USA: Earthscan.

University of Washington Climate Impacts Group and State of Washington Department of Ecology (2008). *Sea Level Rise in the Coastal Waters of Washington State*, January.

van deVen, F. H. M. (1990). Water balances of urban areas. In *Hydrological Processes and Water Management in Urban Areas, Proceedings of the Duisberg Symposium*, IAHS Publication, 198, pp. 21–32.

Winter, T. C., J. W. Harvey, O. L. Franke, and W. M. Alley (1998). *Ground Water and Surface Water: A Single Resource*. US Geological Survey Circular 1139.

Yohe, G. and J. Neumann (1997). Planning for sea-level rise and shore protection under climate uncertainty. In K. D. Frederick, D. C. Major, and E. Z. Stakhiv (Eds.), *Climate Change and Water Resources Planning Criteria*, Dordrecht, the Netherlands: Kluwer Academic Publishers, pp. 243–270. Published concurrently as a special issue of *Climatic Change* **37**(1).

Zhang, A. and Z. Wei (2005). Prevention and cure with Shanghai land Subsidence and City Sustaining Development. In Zhang Agen, Gong Shilang, L. Carbognin, and A. Ivan Johnson (Eds.), *Proceedings of the Seventh International Symposium on Land Subsidence (SISOLS)*, Shanghai, China: Shanghai Scientific & Technical Publishers, pp. 10–17.

6

Climate change and urban transportation systems

Coordinating Lead Authors:

Shagun Mehrotra (New York City, Delhi), Benoit Lefevre (Paris), Rae Zimmerman (New York City)

Lead Authors:

Haluk Gerçek (Istanbul), Klaus Jacob (New York City), Sumeeta Srinivasan (Cambridge)

The authors would like to thank Irune Echevarría, Masahiko Haraguchi, Young-Jin Kang, and Somayya Ali, for their excellent research assistance. Clark Murray provided exceptional and timely inputs. The initial contributions of Deborah Salon are also gratefully acknowledged.

This chapter should be cited as:

Mehrotra, S., B. Lefevre, R. Zimmerman, H. Gerçek, K. Jacob, S. Srinivasan, 2011b: Climate change and urban transportation systems. *Climate Change and Cities: First Assessment Report of the Urban Climate Change Research Network*, C. Rosenzweig, W.D. Solecki, S.A. Hammer, S. Mehrotra, Eds., Cambridge University Press, Cambridge, UK, 145–177.

6.1 Introduction

Cities are key hubs of the transportation sector. According to the C40 Cities Climate Leadership group (www.C40Cities. org), cities contribute 75 percent of greenhouse gas emissions. Based on the IPCC assessments, "petroleum... supplies 95% of the total energy used by world transport. In 2004, transport was responsible for 23% of world energy-related GHG emissions with about three quarters coming from road vehicles" (Kahn-Ribeiro et al., 2007, p. 325). Conditions in developing countries pose additional challenges on transportation systems – demand far exceeds supply, particularly for the growing number of urban poor. For instance, the United Nations predicts: "by 2030, the towns and cities of the developing world will make up 80 percent of urban humanity" (UNFPA, 2007) p. 1. And in these developing country cities, transportation systems are already severely undersupplied. In addition, geographical location poses additional challenges. Nicholls et al. (2008) p. 8 estimate that, by 2070, the "top 10 cities in terms of population exposure to climate change (including environmental and socio-economic factors)" will be located in developing countries of south and east Asia.[1] These cities have transportation systems that are currently navigating the challenges posed by mixed land use and a large proportion of the population living in poverty. In response to such diverse challenges posed by a changing climate to transportation systems, this chapter focuses on the construction and maintenance of the physical assets that account for the bulk of urban transportation investment and climate associated risks. Since urban transportation systems are built and managed by both the public and private sectors, the chapter also considers institutions and organizational structure, regulation, governance, and economic issues that play a role in the development of urban transportation and their response to climate change.

The quality of transportation planning and management is critical for the functioning of a city, and thus issues of urban climate change adaptation and mitigation require attention. Yet, in practice, climate change impacts and associated responses significantly vary by city characteristics and conditions. In addition to providing an overview of urban transport and climate interactions, this chapter has a three-fold purpose: (1) to define climate risk as it pertains to this sector and what general considerations create variations by city conditions; (2) to develop an understanding of pragmatic adaptation and mitigation strategies that cities can adopt and indeed are adopting; and (3) to derive policy lessons for the urban transport sector in the context of climate change in cities. The distinguishing attribute of this chapter is that it primarily focuses on urban transport. Further, a diverse range of country conditions – both developing and developed – are addressed. Finally, it presents a combination of adaptation as well as mitigation strategies allowing the articulation of co-benefits, of addressing mitigation and adaptation together as well.

6.1.1 Description of urban transportation sector

Transportation can be categorized based on what is being transported, the mode of transportation, and by its regulation and other institutional dimensions. Regarding what is being transported, the three subsectors in transportation – moving passengers, freight, or information – make different demands on transportation systems. Impacts on greenhouse gas emissions as well as mitigation and adaptation measures to reduce those emissions can vary widely depending on how passengers, freight, and information are transported. Further, transportation is expressed in terms of modes of travel, which are categorized broadly as occurring by land, air, and water. Within that broad categorization, transportation modes may also be classified in terms of the physical infrastructure that is used and include those that use rail, road, ships, and airplanes, each of which can be subdivided further. Land-based transportation systems are generally those with the highest usage in urban regions, and can be divided into rail- and road-based systems. According to Kahn-Ribeiro et al. (2007, p. 328), "road vehicles account for more than three-quarters" of total transport energy use and thus associated greenhouse gas emissions. This combination of what (or who) is being transported and the mode by which it is transported is significant because it provides a measure of the amount of greenhouse gas emissions by modes and types of uses. This measure of emissions in turn helps in devising adaptation and mitigation strategies. Finally, the nature of regulation and other forms of oversight and management of the transport sector have a critical effect on emissions.

This chapter primarily focuses on the movement of people or passengers, and where relevant, issues of freight or information are referenced. Many transportation systems, especially urban mass transit, particularly in developing countries, are predominantly publicly owned and operated, but other systems (such as air and water-based transport) are more often privately owned; as are cars, vans, and trucks owned and operated by individuals. However, the ownership and management patterns for transportation systems vary by city and country and are an important factor in the design of institutional arrangements to formulate and implement mitigation and adaptation strategies. The ability to mitigate and adapt to climate change related scenarios depends on ownership. For instance, a publicly owned facility may have access to direct and indirect subsidies and large-scale public investments that are unavailable for private sector operators. This issue of public and private sectors is further discussed in Section 6.5.

6.1.2 Role of transportation in climate change

In general, the influence of transportation systems on global climate change has been well documented by the Intergovernmental Panel on Climate Change Working Group III (Kahn Ribeiro et al., 2007), and the impacts of climate change on the

---

1 The only exception is Miami, which is located in the United States.

transportation sector as a whole have likewise been summarized by the IPCC Working Group II (IPCC, 2007a). Bradley *et al.* (2007) p. 27 report that "transport accounts for about 14 percent of global greenhouse gas emissions, of which road transport accounts for the largest share, at 72 percent of sector and 10 percent of global greenhouse gas emissions". Some of the relevant findings are summarized below.

6.1.2.1 Contribution of transportation sector to greenhouse gas emissions

Greenhouse gas emissions vary according to types of transportation systems, geographic location, scale, and time period. Based on IEA estimates for 2004, Kahn Ribeiro *et al.* (2007, p. 328) find, "the transport sector was responsible for about 23% of world energy-related greenhouse gas emissions" (as distinct from total greenhouse gas emissions) and in addition, "The 1990–2002 growth rate of energy consumption in the transport sector was highest among all the end use sectors." Bradley *et al.* (2007, p. 27) note that in this period, although "transport-related emissions grew 20–25 percent in most industrialized countries, growth rates were higher in many developing countries". Their calculations suggest that the "fastest growth was in South Korea, Indonesia, and China, where transport emissions doubled over the 12-year period" (Bradley *et al.*, 2007, p. 27).

Different motorized transportation modes – automobile, transit, or two wheelers – have different carbon footprints, which are measured in tons of emitted carbon per passenger mile, or per ton-miles, respectively, depending on whether people or goods are transported. In cities with concentrated and distinct urban employment centers, mass transit is generally the most efficient urban transport system, with rail-based systems including subways or elevated rail systems usually outperforming bus systems, in terms of minimizing greenhouse gas emissions per passenger mile (also see Bertaud *et al.*, 2009). The choice of dominant transport systems, and in particular of urban mass transit with its intended capture of a large fraction of the total passenger miles traveled in an urban setting, can be an important contributor to reducing greenhouse gas emissions. According to the IPCC (2007a, p. 329), "the world automobile fleet has grown with exceptional rapidity – between 1950 and 1997, the fleet increased from about 50 million vehicles to 580 million vehicles, five times faster than the growth in population." It is noteworthy that "between 1999 and 2004, China's motor vehicle production increased more than 175 percent, approaching half of Japanese levels by 2004" (Bradley *et al.*, 2007, p. 27). The 2007 IPCC report also notes that "two-wheeled scooters and motorcycles have played an important role in the developing world", with a current world fleet of a few hundred million vehicles. The report further notes that buses, "though declining in importance against private cars in the industrialized world", are increasing their role especially in developing country urban areas where they account for a substantial proportion of the modal share of trips. However, individual case studies suggest that transit mode share may be declining in some cities as incomes grow. For example, a study by Gakenheimer and Zegras (WBCSD, 2004, p. 162) notes, for

Chennai, which they identify as having a relatively low GDP per capita, "Chennai's public transport mode share declined by 20 percent in the 25 years preceding 1995, largely due to a rapid rise of the number of motorized two-wheelers."

6.1.2.2 Impacts of climate change on transportation sector

Transportation not only affects climate (Kahn Ribeiro *et al.*, 2007), but it is also affected by climate change. The Transportation Research Board (2008, p. 2) citing IPCC identified "increases in hot days and heat waves, arctic temperatures, sea level, intense precipitation events and hurricane intensity" as the climate change characteristics having the most significant adverse impacts for transportation. The list of impacts is extensive, covering structural and material damages of many different types and associated disruption of transportation services for users depending on the type of transportation facility, its location relative to waterways, and the types of materials and design used (Hunt and Watkiss, 2007; Transportation Research Board, 2008). A number of local transportation inventories exist and have been identified in climate change studies for particular types of impacts. For example, in New York City an extensive set of facilities has been identified in connection with sea level elevations (USACE, 1995; Zimmerman and Cusker, 2001; Zimmerman and Faris, 2010). Further, Rossetti (2002) investigates potential impacts of climate change on railroads for various timescales.

6.1.3 Urban transportation and land use

The relationship between transportation and the spread of urban settlements is interactive. The building of railroads and highways has influenced urban development, and conversely the growth of urban areas has influenced the development of road, air, and rail networks that facilitate travel within and across urban areas (Chomitz and Gray, 1996). As urban areas become vulnerable to climate change, addressing transportation issues in adaption and mitigation involves addressing the interactions between the sector and land use in cities.

Urban land use planning is based on functional designation of land for different human purposes including economic and leisure activities. Categories of land use typically include residential, commercial, industrial, recreational, natural protection, institutions, parking, vacant land, and transportation or utilities (for details see Chapter 8 on land use). Wegener (2009, p. 4) suggests that "the distribution of infrastructure in the transport system creates opportunities for spatial interaction." This "can be measured as accessibility". Further, "the distribution of accessibility in space co-determines location decisions" and results in changes in land use. And a combination of energy intensity as well as demand for transportation systems determines the degree of carbon emissions. Lower levels of fuel efficiency and increase in demand for transportation result in greater greenhouse gas emissions.

Individual land uses within the urban fabric, including transport infrastructure, in great measure determine the urban form.

Urban form can be measured through various density measures – for dimensions such as population or economic activity – and the spatial location of activities and households. Newman and Kenworthy (1989) find that urban population density is closely related to vehicle miles traveled (VMT). VMT is a key indicator for greenhouse gas emissions, and conversely, the greenhouse gas emissions from transport vehicles are related to VMT. However, the relationship is far more complex and other measures including the price of fuel, employment levels, trip origins and destinations and the size of cities have much influence on vehicle miles traveled. A recent study in Indonesia (Permana *et al.*, 2008) also suggests that residents of locations with mixed land uses consume less energy than those who live in suburban locations even after controlling for income.

Cities show considerable variation in terms of urban form depending on the planning and economic goals that they choose to implement in the long term. Geurs and van Wee (2006, p. 139) in "a broad evaluation of relevant land use, transport, accessibility and related societal and ecological impacts" … "between 1970 and 2000 note that without policies to encourage compact city development, urban sprawl in the Netherlands was likely to have been much greater". Stone *et al.* (2007, p. 404) "found the densification of urban zones to be more than twice as effective in reducing vehicle miles of travel emissions as the densification of suburban zones, suggesting compact growth to be better for air quality" in the United States. The study finds densification of urban zones to be more than twice as effective in reducing vehicle miles of travel and emissions as the densification of suburban zones, suggesting compact growth to be better for air quality than historical patterns of growth in most cities of the United States.

The intensity of urban land use also results in urban heat islands that modify the local climate (see Chapter 3 for details on climate change process and projections). Thus, the relationships between urban land use, transportation infrastructure, and climate involve feedback loops that are significant in multiple physical scales – local, regional, and global – and temporal terms – long and short term (Figure 6.1). Just as accessibility is a goal for addressing the land use and transport feedback loop, the goal of sustainability addresses the multiple feedbacks between urban form, transportation infrastructure, air quality, health, and climate change. Research associating land use change with transportation and air quality in developed and developing countries suggests that sustainable urban transportation and land use policies may be vital to achieving greenhouse gas mitigation (see, for example, Iacono and Levinson, 2008; Kinney, 2008; Rogers and Srinivasan, 2008).

In Figure 6.1, the flow (or movement of goods, services, and people) pattern in the urban transportation system is determined by both the transportation system and the land use patterns. The current flow pattern causes changes over time in land use through the type of transportation services provided and through the resources

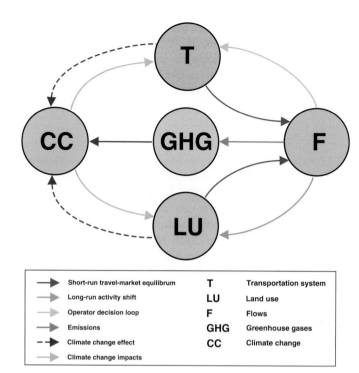

Figure 6.1: *Urban transportation, land use, and climate change interactions.*

Source: H. Gercek, adapted from Manheim (1979).

consumed in providing that service. The current flow pattern also causes changes over time in the transportation system in response to actual and anticipated flows, as entrepreneurs or governments develop new transportation services or modify existing ones. These interactive patterns in the transportation system and land use determine the degree of greenhouse gas emissions. Furthermore, transportation system facilities and land use also contribute to local climate change through urban heat islands. In sum, climate change will have impacts on both the urban transportation system and urban land use and vice versa.

6.1.4 Responding to climate change: adaptation and mitigation[2]

Adaptation to climate change means minimizing the potential impacts on the transportation system from climatic changes such as rising average temperatures, increased intensity of storms, rising sea levels, and increases in overall climatic variability. Adaptation of transportation and land use to climate change may involve changes in individual travel behavior as well as macro-scale urban development policy. Mitigation, in the context of urban transport, implies reduction in greenhouse gas emissions resulting from movement of goods, services, and people in cities. For instance, mitigation efforts may involve reducing VMT through a range of incentives and regulation.

Effective response to climate change requires that transportation plans consider both adaptation and mitigation to climate

---

2 Adaptation is "An adjustment in natural or human systems in response to actual or expected climatic stimuli or their effects, which moderates harm or exploits beneficial opportunities" (IPCC, 2007b, p. 869). Mitigation is "An anthropogenic intervention to reduce the anthropogenic forcing of the climate system; it includes strategies to reduce greenhouse gas sources and emissions and enhancing greenhouse gas sinks" (IPCC, 2007b, p. 878).

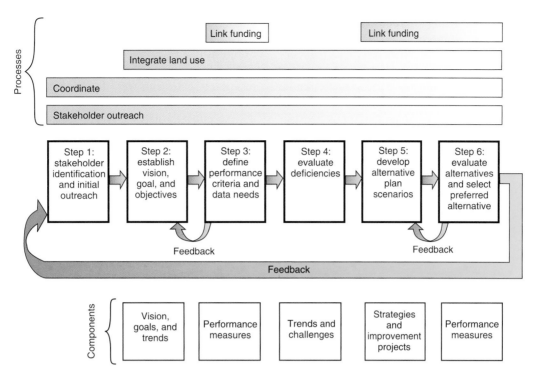

Figure 6.2: *Opportunities to integrate long-term transportation planning with climate change.*

Source: Federal Highway Administration and ICF International (FHWA, 2008).

change as well as establishing a process by which climate change can be integrated into long-term transportation planning processes that include land use planning as well as stakeholder outreach. A summary of the approach adopted by the United States Federal Highway Administration is summarized in Figure 6.2, where various processes and components of the transportation planning and management are outlined, each offering the opportunity to integrate aspects of climate change adaptation and mitigation. Details of such adaptation and mitigation strategies are presented below.

6.2 Risk management as a framework for adaptation and mitigation

Adaptation and mitigation of urban transportation systems to climate change can be defined as a form of risk management. Mitigation is a global mandate, while adaptation is a local necessity. The goal of risk management is to minimize future damages at affordable costs. The social context is: What risks are managed at whose cost and for the benefit of whom? Besides considering potential material damages, urban transportation managers have a broad array of cultural, social, quality of life, and ecological issues to consider in regard to climate change risks. The risk profiles of the transportation sectors in cities of developed countries are radically different from those in underdeveloped or rapidly developing countries. For instance, in several developing countries existing private and public urban transportation assets and systems are substantially undersupplied and less resilient to

extreme events, and the vast majority of the passengers are poor people, many of whom live in informal settlements.

There are two fundamental options for risk management in the transportation sector. One is by mitigation measures in cities around the world that reduce *globally* the climate hazard factor by reducing greenhouse gas emissions. Mitigation measures implemented by a single city are insufficient to reduce the hazard factor for that city and thereby reduce climate risk. However, the cumulative effect of mitigation measures in cities around the world is likely to reduce total greenhouse gas emissions, and by extension climate hazards as a whole, including changes in extreme temperature, precipitation, sea level rise, and the like. The global hazard reduction can be achieved through the cumulative effects of local measures such as land use planning, zoning, and placing of assets. Mitigation affects the hazard factor only. The lower the hazard, regardless of the value of assets, the lower the vulnerability, and hence the lower is the risk for transportation systems. Hence, aggregate mitigation measure by cities around the world can reduce climate risk in cities. See Section 6.5 for mitigation strategies for urban transportation systems.

The second option in reducing risk is by adaptation. Zoning and land use planning can affect the spatial arrangement of assets and protective structures in an urban setting and hence modify their exposure to the spatially varying hazards. Risk is critically dependent on where assets are placed with respect to the spatial distribution of hazards. Settlements and transportation systems develop together, but often one precedes the other. Whichever comes first tends to occupy the less hazardous areas, and the latecomer ends up with the more hazardous areas. Commuter rails, bus routes, or tunnel entrances should not be located in high-hazard flood zones. If space is constrained, then the more valuable, essential, and critical assets should be placed in the safest areas, while the less valuable, ordinary, or non-essential assets may be placed in the more hazardous locations. In practice this rarely is the case in developed-country cities. Usually residences tend to occupy relatively safer areas in developed country cities. Some exceptions are along coasts where the ocean view and proximity are considered valuable assets, as is the proximity of forests that are prone to wildfires – and transportation routes and facilities tend to occupy the higher-risk locations in which residential settlements were deemed too risky to build. The converse tends to be true in developing countries (although there are exceptions), where the poor live in slums and tend to occupy the most vulnerable locations (see Chapter 8 on land use and

---

[ADAPTATION]   Box 6.1   London, UK, storm surge barriers

Klaus Jacob

Columbia University

Levees and barriers can radically change the character and functioning of a city and its transport system. Multiple adaptation paths may be pursued, where one path may be effective up to a certain climate threshold level, and then another adaptation path may have to be chosen, since the former may become gradually ineffective (Box Figure 6.1). London's Thames storm surge barriers are expected to become ineffective in a few decades due to rising sea level, and another new and larger barrier system is planned farther downstream, for later this century (Box Figure 6.2).

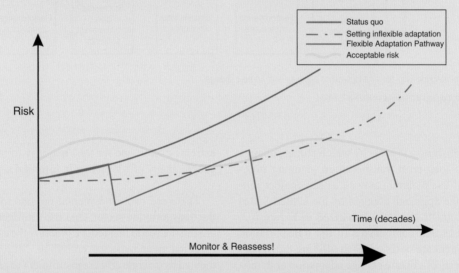

Box Figure 6.1: *Flexible adaptation pathways.*

Box Figure 6.2: *The London Barriers are the centerpiece of today's Thames Tidal Defenses that protect London and the Thames Estuary corridor where a significant proportion of England's wealth is produced (1.25 million people live and work in the flood risk area). The system provides at least a 1:1000 per year protection (to 2070). The system allows for a combination of risks (high tide/surge/freshwater floods). It comprises 9 tidal barriers (including the Thames Barrier), as well as, 35 major gates, 400 minor gates and over 300 km of tidal walls and embankments.*

Source: Adapted from NPCC (2010)

Chapter 9 on governance). Cumulatively, adaptation options related to residential and other forms of land use fall under the category of land use planning and zoning.

With respect to adaptation of transport systems in cities, Leonard, *et al.* (2008, pp. 5–21) have identified and discussed in detail three types of objectives for adaptation strategies: *protect, accommodate, or retreat*. Accordingly, one option is for climate hazards to be addressed by *protective* measures, some of which can be on a regional rather than local scales. For instance, as in the case of the London storm surge barriers, the flood hazards can be modified by adjustable protective structures such as levees, dams, flood barriers, and pumping stations, aimed at protecting large areas of cities, including the transportation systems (see Box 6.1). However, the experience of Hurricane Katrina hitting New Orleans has shown that, when the hazard exceeds the design specifications of the protective structure, or if the structure cannot be upgraded to keep up with local subsidence and sea level rise, or if the protective structure is not well maintained, such conditions can cause severe damage, loss of life, and assets. Due to the mismatch between perceived and actual protection, the individual firms and households, and also transportation agencies, may tend to neglect additional localized protective or adaptive measures behind these regional protection structures. Such additional local protection provides some localized resilience when regional protection structures fail.

Another way to minimize risk via adaptation is by reducing the vulnerabilities of individual assets. This adaptation approach comprises setting and enforcing adequate *engineering* and *performance standards*, including preparing, disseminating, and implementing technical guidelines, providing updated design and construction codes, and setting common climate protection levels, as is presently under consideration in New York City. Enforcing these, developing protective operational procedures, and putting them into practice are important to achieve the risk management targets. When this is successfully done, transportation infrastructure vulnerabilities to a given hazard at a given place can be reduced by good engineering, construction quality control, code enforcement, retrofitting (that could include elevating assets in their current place in case of flood hazards), reinforcing protective walls, increasing pumping capacity, and increasing drainage capacity of culverts, to name a few.

As capital improvements are made over time, transportation infrastructure may be raised in place to higher elevations, although in plan they may continue to be located in flood zones. The Taipei subway in Taiwan has raised subway entrances to avoid flash-flood and tidal inundations, and its high-speed intercity trains, similar to the Shinkansen in Japan, run largely on elevated tracks to avoid river flooding during frequent typhoons. Vulnerability reduction may include disaster preparedness as well as increase resilience to future effects of climate change, comprising installation of warning systems and having procedures in place that temporarily move the rolling stock out of harm's way.

6.3 Assessing climate risks to urban transportation

This section describes methods for assessing climate risks to urban transportation systems and then highlights specific hazards to the systems posed by climate change.

6.3.1 Defining climate risks and methods of risk assessment

Risk in general is defined as "the product of the likelihood of an event occurring and the magnitude of consequence should that event occur. For the purposes of this document, likelihood is defined as the probability of occurrence of a climate hazard" (NPCC, 2009, p. 62). In the context of urban transportation systems, the consequences are largely the direct damage to the transportation systems, the associated losses to the transportation systems' operators or owners, but they also include the impact on, and losses to, society at large from hazardous climate conditions or events. For instance, even when there is no damage to the transportation system, as during a heat wave, heat stroke fatalities occur, say, in a subway or underground system because of inadequate ventilation or cooling, it constitutes a climate-related transportation risk. Risk refers to the probability of future impacts, damages, or losses.

In sum, there are different approaches to risk assessment. Some are narrowly defined in engineering terms, and others are much broader and address the complex social context. The former tend to be quantitative, the latter qualitative because of the inherent complexity of social systems. Several quantitative definitions of risk have been put forward, the most general being: Risk (R) is the probability (P) of hazardous events (H) times their cnsequence (C) and can be expressed in an equation ($R = P(H) \times C$). For an urban transportation system, climate risk can be defined more specifically as the spatially integrated sum of the local products of the following two, spatially varying factors: First, probability (P) that the local climate hazards (H) of various magnitudes will occur; and second, the magnitude of the consequences (C).

The consequences C are in turn the product of the *vulnerability* of the transportation assets times the *replacement value* of the assets that are exposed to the climate hazards. In engineering terms, vulnerability is often called *fragility* and is the *probability of failure* resulting in various *states of damage* such as light, modest, severe, or total damage; a measure of the outcome of these damage states can be, for instance, the *fraction of the replacement value* of the transport system that was damaged by the hazard. Component or system fragility is a function of the hazard magnitude. Modest flooding, for instance, will do only light damage to transportation systems with minor consequences, while severe flooding could totally destroy a given transport system, resulting in total loss of the assets and of their functionality, causing in turn major economic damages to society.

[ADAPTATION]   Box 6.2   Hyderabad, India, infrastructure adaptation planning

Diana Reckien, Matthias Lüdeke, Fritz Reusswig,
Oleksandr Kit, Lutz Meyer-Ohlendorf and Martin Budde

Potsdam Institute For Climate Impact Research

In Hyderabad, Andhra Pradesh in South India, we investigate the magnitude and the impacts of climatic changes, and we develop planned adaptation options. The work is part of the research network Climate and Energy in a Complex Transition Process towards Sustainable Hyderabad: Mitigation and Adaptation Strategies by Changing Institutions, Governance Structures, Lifestyles and Consumption Patterns.[3] The project addresses the challenges of climate change and resource depletion in the context of so-called "megacities" by taking into account their complex social and economic characteristics. The aim of the project is to develop a Perspective Action Plan in close cooperation with the state government that will establish Hyderabad as a Low Emission City in Asia within 30 years.

Approaching adaptation in cities is a vital necessity to meet climate change and development goals, but it remains largely unnoticed by governmental as well as local climate policy, particularly in developing countries such as India. Hyderabad is one of the fastest growing cities in the country with a population increase of 27 percent from 1991 to 2001. With 5,530,000 million people (2001) it is the sixth largest city in India (Census of India). Population scenarios for the Hyderabad urban agglomeration (HUA) indicate a population of 7.7 million people in 2011 and 10.8 million in 2021 (GHMC, 2005). After Bangalore, Hyderabad is the second most important economic center for the Indian IT industry. Another strand of the local economy lies in the biotechnology and pharmaceutical industries (GHMC, 2005).

To assess the magnitude of potential climatic changes for Hyderabad (item (1)), we compare two AOGCM outputs (ECHAM5 and GFDL, more to follow) and their statistical downscaling to the region of Hyderabad. We focus on four climate variables, which are derived from expert interviews (with local authority officials, NGOs, community self-help groups) and which are candidates for inflicting damages in the city under present conditions: the frequency of extreme daily precipitation and the probability and duration of heat waves (indicators of extreme weather events), the total annual precipitation and the mean annual temperature (indicators of gradual climate changes).

Frequency of precipitation greater than 80 mm/day (medium to very high impact intensity and potential damages) is likely to double until 2100 (reference period 1980–2000) (B1 in comparison to A2 will only buy some time with stronger rain events occurring later in the century), and very strong rain events will increase significantly (160 mm/day events to potentially increase four-fold). With respect to heat waves, we looked at the number of days/year with night temperatures above 27°C. They will approximately triple until 2050, relatively independent of the SRES scenario. In 2100, A2 will lead to +560(±50) percent and B1 to +240(±50) percent increases with respect to the reference period. The frequency of heat waves longer than one week will double to triple until 2050 and increase further until 2100. Total annual precipitation is likely to change by −4 to +17 percent, whereby the difference between the AOGCMs is larger than between the SRES scenarios. The mean annual temperature is projected to increase monotonically up to 5 °C above present (A2 in 2100).

Accounting for an improvement in the city's current performance to climate occurrences will help to prepare for future impacts related to these climate projections, contribute to reaching development goals, and constitutes one way of adaptation. With respect to the current climate impacts (items (2) and (3)), we adopt a systems dynamics approach to urban development (Forrester, 1969) and draw so-called impact nets. We look at five currently affected subsectors of the urban system: food security and health, water supply and energy security, as well as transport.

After a literature review, interviews with local stakeholders, NGO representatives, and community self-help groups, as well as more than 200 newspaper articles published since April 2009, we drew impact nets (see Box Figure 6.3).

Box Figure 6.3 illustrates how climate and certain climate events affect current subsystems of the city.

The key impacts in Hyderabad are generally situated in the area of supply and demand of a particular resource, be it water, energy, or goods and services for industry (van Rooijen, 2005). Extreme flood and drought events severely reduce the availability of quality water by either contaminating existing water resources or generating severe surface and groundwater scarcity. Droughts also lead to a reduction in hydropower generation, although the majority of electricity in Andhra Pradesh is produced by thermal power plants. With a further rise in average temperatures, the demand for energy rises too, as do shortages during heat waves (Sivak, 2009).

Climate extremes adversely impact transport infrastructure in Hyderabad in a way that makes it either inaccessible or uncomfortable. Floods directly result in infrastructure damage, the breakdown of transport/communication networks, and the slowdown of services (Shukla *et al.*, 2003). Heat waves cause direct damage to electronic/electric devices and make public transport a very uncomfortable service to use, whereas a gradual temperature increase works in a more concealed way, slowly damaging railway and road infrastructure.

---

3   To this end, a consortium of Indian and German Research Institutes, partners from the public and private sectors, as well as NGOs has been formed. Lead partner and project coordinator is the Division of Resource Economics at the Humboldt University of Berlin. The project is funded by the German Ministry of Education and Research (BMBF) in the program "Research for Sustainable Development of the Megacities of Tomorrow" and will run over five years until May 2013. See www.sustainable-hyderabad.in.

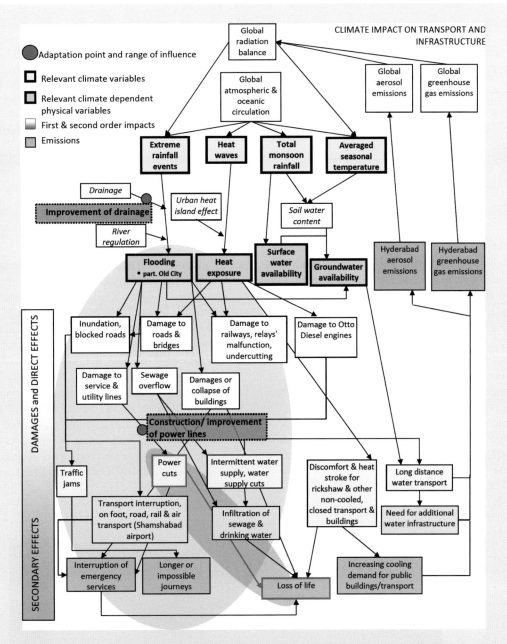

Box Figure 6.3: *Impact nets on different urban subsectors in Hyderabad. Key: Yellow – climate and climate related events; Gray – multi-layered (primary/secondary...) impacts; Blue – feedback on the climate system; Green – possible adaptation points and their ranges of influence.*

Impacts of climate change on health relate to the direct exposure to heat and flooding as well as to second-order impacts. A direct effect of flooding and droughts is the interruption of food supply in some areas. Secondary impacts include the long-term influence on water availability for irrigation and of heat-related crop diseases on food production, malnutrition, and hunger. Hyderabad's ability to provide sufficient food to its inhabitants partly depends on the food availability in the surrounding areas (Smith *et al.*, 2007). As a second-order impact, the contamination of fresh water with bacteria, chemicals, or other hazardous substances (Young *et al.*, 2004) has to be identified. Their consumption can result in diarrheal diseases, cholera, and toxic effects.

The situation is particularly severe in areas of Hyderabad where sewage flows in open ditches close to water distribution pipes (Vairavamoorthy, 2008) and where people live in industrial areas close to factories (Kovats and Akthar, 2008). Climate-sensitive diseases such as malaria, dengue, and chikungunya might increase due to favorable climatic and breeding conditions for insects (Bhattacharya *et al.*, 2006). Intermittent rainfalls, such as in June and July in Hyderabad, provide perfect breeding grounds in stagnant urban water such as wells, car tires, bottles, and cans. In particular, dengue cases have reached alarming levels in Hyderabad; it is increasing in the city while its trend is going down in the state as a whole.

The diagrams serve various purposes: (1) communicating aspects of climate change and initiating discussion, (2) understanding the impacts of climate change and picturing them accordingly, (3) generating adaptation options and assisting in their ranking and evaluation, and in a later stage of the project (4) laying the basis for a quantitative, computer-aided, assessment tool. Referring to (1) and (2), the diagrams will be further discussed with local and regional stakeholders, experts, and climate-related scientists, with other experts in the field (e.g., our project partners in Germany, currently underway), and at a later stage of the project with officials from policy and plan making in Hyderabad. With respect to point (3), the green-colored dots and shapes in Box figure 6.3 depict potential adaptation points and their range of influence. These adaptation points are placed at different locations within the nets and therefore have a different "size" of impact (shown as light green shapes). We assume that such an illustration can actively support decision-making for adaptation, particularly in developing countries where development goals and climate change adaptations have to be harmonized.

Source: Division of Resource Economics, Humboldt University, Berlin, and www.sustainable-hyderabad.in

These terms (risk, hazard, and fragility) have been used extensively for quantitative risk assessments in combination with Geographical Information Systems (GIS) as an organizing and computational tool. In the USA, for instance, a GIS-based risk assessment tool is HAZUS-MH, where MH stands for multi-hazards, since it allows the quantification of three different risk-producing natural hazards: floods, winds, and earthquakes (FEMA, 2009). The methodology used for these risk assessment tools is discussed below in more detail. It should be noted, however, that these quantitative engineering approaches tend to neglect or inadequately capture the *social risks* to vulnerable societies, which are more complex and much harder to quantify than the physical climate threats to the built environment (Wisner *et al.*, 2004; Birkmann, 2006).

In quantifying risks to transportation systems, maps of climate hazards, for instance flood-zone maps or wind speed maps, related to a specific annual exceedance probability, are often used as input. These hazard maps (e.g., Figure 6.3) typically express the hazard of interest at a pre-selected annual probability, for instance, the flood height for the annual flood probability of 1 percent per year. This implies an *average* recurrence period of that flood height on the order of once every 100 years. Wind speeds are often mapped in km/h and associated with an annual probability of 2 percent per year, which in turn implies an average recurrence period for a particular wind speed on the order of once every 50 years. Alternatively, scenario events, such as a specific flood or storm scenario, with spatially distributed hazard amplitudes such as

Figure 6.3: *Mapping climate risks to urban transport. Shaded areas depict worst-track storm surge flood zones for Saffir-Simpson Category-1 in red, SS2 in brown, SS3 in yellow, and SS4 in green. Shaded lines are subways, black lines are rail sytems.*

Source: Lamont-Doherty Earth Observatory, Google Earth, and New York State Emergency Management Office (NYSEMO), New York City Transit Authority subway lines.

flood depths or wind speeds, can be defined as input for the risk calculations.

Similarly on the asset side, the vulnerability and monetary replacement values of the transportation systems can be mapped in their geographical locations relative to, or superimposed upon, the mapped hazards. In the case of transportation systems, their vulnerability is, for instance in case of flood hazards, a function of their location (and elevation) inside or outside of flood zones. This "mapping" and vulnerability assessment method can be applied to rail systems, stations, bridges, tunnels, roads and highways, bus terminals, or maintenance shops, etc. The mapped vulnerability is then used for the quantitative risk (loss) assessment, as an attribute of the indicated assets exposed to the mapped hazard conditions. The latter can be defined either for a given scenario event, or for a given exceedance probability of the hazard, for which the risk or expected losses are being determined. If such risk computations are made for many different probability levels (or average recurrence periods), then these various contributions from rare large events to frequent small events can be annualized (divided by their respective recurrence period) and added up to provide a total *annualized loss* to a city's transportation (or any other) system. Without such quantitative assessment methods, decision-makers are not aware of the impending climate risks and therefore may not be willing to invest in preventive measures. For a more comprehensive redefinition of a framework for city climate risk assessment where risk is defined and quantified as a function of climate hazards, physical and social vulnerabilities, and institutional adaptive capacity see Chapter 2 on cities and climate risk framework and Mehrotra *et al.* (2009).

6.3.2 Climate hazards to urban transportation systems

The climate hazards that may pose the greatest challenges to urban transportation systems are different for different kinds of physical hazards (see Table 6.1), cities in different geographical environments, and for different modes of transportation.

Physical hazards from global climate change that affect urban transportation systems can be expressed in terms of changes in average values and often higher variability of temperature, precipitation, storm frequency and severity, coastal storm inundations especially in conjunction with sea level rise, and other climate processes. The vulnerability of transportation assets to these physical hazards, stemming from the existing spatial and economic organization of the city and its transport system, depends on a range of factors, including: (1) design and spatial layout of transport infrastructure – flood hazards tend to be worse for underground infrastructure; storms affect land, marine, and air travel differently; (2) basic urban form – high-density settlements with mass transit systems, or mostly motorized transportation with low densities; (3) availability of resources to keep transportation systems functioning in both disaster and non-disaster conditions.

Table 6.1: *Impacts of climate hazards on urban transportation systems.*

Physical hazard	Vulnerability of transport systems
Water	
Floods	Inundation of surface and subsurface infrastructure
	Undermining of support structures such as bridge footings
	Corrosion from salt water, where applicable
	Increased scour around bridge footings
Storms	Physical damage to roads and rail networks and vehicles from high winds and wind-driven rain and debris
Sea level	Similar to floods
	Clearances of some bridges might be diminished
Other	
Heat	Destruction or deterioration of materials
Wind	Physical damage to roads and rail networks and vehicles from high winds and wind-driven debris
Secondary hazards: fire from drought, landslides from rainstorms	Facility destruction and disruption of services

Source: From Special Report 290: Potential Impacts of Climate Change on U.S. Transportation, Annex 3–1 (table), pp. 117–123. Copyright, National Academy of Sciences, Washington, DC, USA, 2008. Reproduced with permission of the Transportation Research Board.

6.3.3 Examples of climate change risk assessment of urban transport systems

A number of nations and cities have undertaken risk-related analyses of the impact of climate change on their transportation infrastructure. Revi (2008) provides a comprehensive climate change risk survey combined with an adaptation and mitigation agenda for cities in India. As part of a sequence of general risk assessments from climate change for the New York metropolitan region and discussion of adaptation and response strategies (Rosenzweig and Solecki, 2001; NPCC, 2010) several assessments were directly focused on its transportation systems (Jacob *et al.*, 2000, 2008; MTA, 2009). However, preparedness and response remain a challenge. For instance, on August 8, 2007, a windstorm combined with an intense downpour caused urban flash floods, bringing large portions of the New York City mass transit systems to a near standstill (MTA, 2007).

Likewise, there are multiple analyses available on the impact of hurricanes Katrina and Rita on the transportation systems of New Orleans and other communities (Transportation Research Board, 2008). For the Greater London area, climate adaptations reports contain individual chapters with a focus on transportation (Greater London Authority, 2005, 2008). National perspectives on challenges posed by climate change, and guidelines on how to adapt transportation systems to these changes in the context of the United States are offered by the Transportation Research Board

(Transportation Research Board, 2008) addressing a wide range of geographic conditions, from Alaska to Florida, and modes of transport. While equivalent studies in the context of developing countries are sparse (an example for Jakarta, Indonesia, is given by Aerts *et al.*, 2009; also see Box 6.3 for preliminary efforts), the work of the Transportation Research Board offers a benchmark, with generic implications transferable to other cities threatened by climate change.

6.4 Adaptation of urban transportation systems

Adaptation of urban transportation systems to the challenges of climate change implies making these systems optimally suited to operate safely; to experience minimal interruptions and losses from the immediate effects of extreme climate events; and to be designed and modified in order to make these adaptations possible over the long term. Urban transport managers need to meet both types of climate challenges – changes in long-term mean trends and short-term extremes – while meeting all the normal challenges of providing transportation that is reliable, accessible, affordable, cost-effective, equitable, and environmentally sound. In addition, mainstreaming adaptation requires attention to existing transportation assets and planned investment (Trilling, 2002).

6.4.1 Adaptation planning

Managing climate risks of urban transportation systems requires more than quantitative risk assessment and risk reduction. They require an assessment of the basic mitigation and adaptation options available to the community, as well as their respective costs and benefits, both fiscal and social. When such adaptation measures are planned, the timing as mandated by the changes in hazard levels needs to be considered. So do the hazards' changing spatial distributions as a function of time. These demands may best be met by developing a *climate change adaptation plan* that balances the technical adaptation options and their costs against the benefits they provide to the community in terms of risk reduction. Several examples of such plans that address transportation were given in Section 6.3.3 and they are also covered in the NYC Panel on Climate Change study (NPCC, 2010), and include plans created for London, King County in Washington State, and the State of Maryland. The plan should also serve other objectives such as improving quality of life, infusing economic vitality, and ensuring long-term sustainability. Planning bodies typically set strategies on 20–30 year time frames, and operationalize these with shorter-term, 2–3 year implementation plans that are consistent with long-term goals (Lindquist, 2007). A major challenge is devising clear plans to maintain the basic functionality of urban transport systems in the face of potentially increasing

---

[ADAPTATION] Box 6.3 Cities in Climate Change Initiative: Maputo, Mozambique

Paulo Junior and Bridget Oballa

UN-HABITAT

The city of Maputo (Mozambique) is one of the four cities around the world where UN-HABITAT is providing capacity building and technical support to the implementation of the pilot phase of the Cities in Climate Change Initiative (CCCI). (The other three cities are Kampala (Uganda), Sorsogon (Philippines), and Esmeralda (Ecuador).) The main focus areas of the CCCI are: (i) awareness, advocacy, and policy dialogue; (ii) tool development and tool application; (iii) piloting climate change mitigation and adaptation measures; and (iv) knowledge management and dissemination.

Maputo is the capital city of Mozambique, it is located at the extreme south of the country, along the coast (Box Figure 6.4). The city is highly vulnerable to the impacts related to climate change since it is facing the Indian Ocean and is the most densely populated urban area in Mozambique. According to the 2007 census (INE, 2007) the city has about 1.1 million inhabitants; however, the metropolitan area Maputo–Matola–Marracuene shows a fluctuating population between 2.5 and 3 million. Maputo, like other African cities, is experiencing rapid population growth causing an increasing demand for housing and infrastructure (UNFCCC, 2006), especially in the peri-urban slum areas. Consequently, the risk of severe impact on the urban poor will increase along with their inability to adapt or relocate to safer areas.

A preliminary assessment on climate change impacts in the urban areas of Maputo city was carried out, and key vulnerable sectors and areas have been identified:

- Coastal zones and ecosystems
- Human settlements and infrastructure
- Health, food security, and waste management
- Transportation system
- Wetlands and urban agriculture

The main climate-related hazards with destructive consequences for these sectors are floods, droughts, rising sea levels, and storms (cyclones).

The predicted sea level rise related to global warming may result in flooding of the lowest topographical areas of Maputo, which are the most populated and where slum dwellers are concentrated. This prediction is also supported by the Mozambique National Adaptation Plan of Action to Climate Change prepared in 2007. The National Institute for Disaster Management's (INGC, 2009) study on the impacts of climate change in Mozambique shows that in the next three decades most of the coastal area of Maputo, including its harbor and other important infrastructure will be affected by sea level rise if no adaptation and mitigation measures are adopted, resulting in high economic and social costs (see Box Figure 6.5).

The assessment identifies the following for establishing a climate change adaptation strategy for Maputo city:

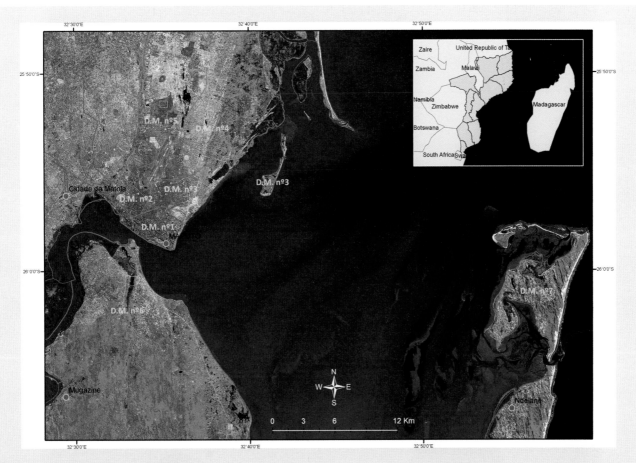

Box Figure 6.4: *Maputo City seen from satellite.*

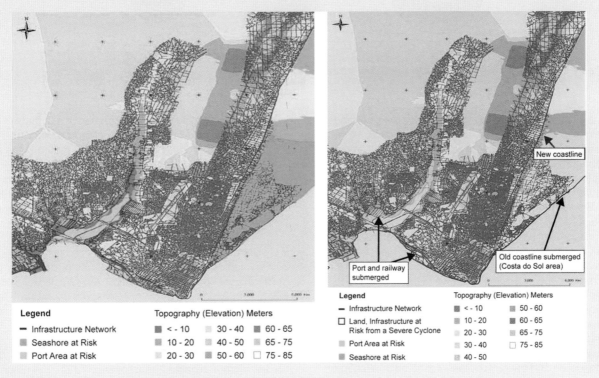

Box Figure 6.5: *Map showing areas that can be affected by predicted sea level rise at 5 meters.*

Source: INGC (2009).

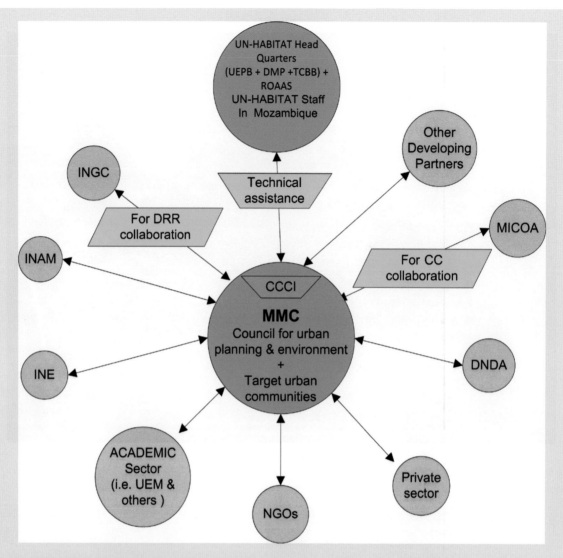

Box Figure 6.6: *Diagram showing the communication mechanism of the CCCI.*

1. Actively involving key stakeholders from the public sector, private sector, academia, civil society, and development partners in the process of raising awareness about the impacts of climate change at all levels.
2. Establishing institutional arrangements between city and central governments through the Ministry of Coordination of Environmental Affairs (MICOA) and the National Institute for Disasters Management (INGC), to ensure effective management and implementation of the climate change risk reduction plans.
3. Establishing communication mechanisms to ensure participatory and inclusive processes in the identification and implementation of sustainable solutions, including the creation of a Natural Disasters Risk Reduction and Climate Change Unit at local level.
4. Preparing an in-depth assessment of the impacts of climate change in Maputo city, in order to determine the required adaptation and mitigation measures to be implemented.
5. Developing methods and tools for the analysis of climate change effects in order to facilitate the financial planning

and decision-making and preparation of a Climate Change Adaptation and Mitigation Plan of Maputo city, which identifies priority interventions to be implemented in the short, medium, and long term.
6. Creating synergy and coordination mechanisms with new initiatives, and ongoing projects, to jointly identify potential sources of funding, and ensure continuity of operations.

A CCCI inception workshop was organized in Maputo in May 2009 which was attended by stakeholders dealing with climate change impacts in Maputo: MICOA, INGC, National Institute of Meteorology (INAM), Maputo Municipal Council (MMC), representatives from academia, the private sector, civil society, NGOs, and development partners.

An outcome of the workshop was development of a stakeholder communication mechanism, with the municipal authority taking on the lead role in coordinating projects (see Box Figure 6.6).

Box Figure 6.7: *Threatened mangrove areas identified for immediate demonstrative adaptation/mitigation actions under CCCI.*

The threatened mangroves surrounding the Costa do Sol neighborhood (Box Figure 6.7) were identified as a pilot project site for initiating adaptation and mitigation responses. The aim of these immediate interventions is to map the mangrove area, and incorporate responses into the existing

Master Plan, which includes special provisions for the protection of endangered species.

Source: Cities in Climate Change Initiative (CCCI), UN-HABITAT, www.unhabitat.org.

frequency and strength of storms. Managers can plan in advance to have contingency plans to maintain movement of people and goods in the urban regions when major highways or rail lines become impassable due to climate-related disasters.

6.4.2 Specific adaptation measures

The adaptation of transportation systems to climate change is a critical challenge since many of these systems are already in place and are rigid. The exceptions are some of the very large transportation projects underway in the planning stages. Key factors affecting the choice of adaptation measures are whether or not the hazards and threats are intermittent or continuous and whether they are acute or long term. Rosenzweig and Solecki (2001) consider short-term protective measures with local engineering, regional mega-engineering, and long-term land use change. The 2001 study concludes that the most effective and sustainable measures are via land use changes. According to the New York City Panel on Climate Change (2009), adaptation

measures may be grouped into three categories: operations and management; capital investment in infrastructure; and policy. Adaptation actions for operations and management of transportation include changing travel routes; altering (reducing) travel behavior to avoid the congestion brought about by altering travel routes; altering repair cycles to anticipate ongoing repairs to damaged infrastructure. Capital investment adaptation consists of measures such as retrofitting existing infrastructure that is susceptible to climate changes – installing pumps to reduce flooding of vulnerable facilities; construction of permanent barriers in the case of water or wind to prevent exposure of transport systems to these forces. Policy oriented adaptation measures may include incorporating climate projections into the siting of transportation projects, e.g., avoiding the construction of roads and rails in areas vulnerable to climate change induced floods or storm surges; or adapting land use to store rain water and reduce wind speed in proximity to transport routes. While impacts vary by location and types of transportation systems, policy measures need to take into account national assessment of the consequences of climate

Table 6.2: *Climate change impacts on urban transport and adaptation mechanisms.*

Climate hazard	Transport system component	Impacts targeted	Adaptation measure
Temperature	Roads, rail, transit vehicles, private vehicles	Adverse effect on speed	Milling out ruts; laying of more heat resistant materials such as asphalt for roads and more heat tolerant metals for rail and rail connections
Floods (sea, river, precipitation)	Drainage system; roads, rails, subways; transit vehicles, personal/private vehicles	Congestion, accidents, delays	Use of remote sensing technology to detect damaging water levels and trends Bring in dikes and barriers Better management of drainage system, detours
Storms	Roads, bridges, rails, airports, and subways; transit vehicles, personal/private vehicles	Accidents, delays, cancelled trips, limitations of routes	Ability to shelter vehicles by moving them into garages Improving emergency evacuation planning Building protective barriers Higher clearance bridges
Sea level	Roads, rails, subways, and airports; transit vehicles, personal/private vehicles	Limits speeds and routes	Bring in dikes and barriers Move vehicles to higher ground Heavier use of pumps Elevation of land and structures to minimize the impacts of flooding
Wind	Surface structures and facilities, such as overhead transit electric lines and signaling systems and road support systems such as lighting and traffic lights	Visibility, signaling system	Weighting down of vulnerable structures Readiness in the form of repair equipment and maintaining a replacement inventory

variability and change. For example, the United States assessments incorporate three types of analysis: regional, sectoral, and national overviews (MacCracken, 2002, p. 60). Table 6.2 provides a brief summary of the impacts of climate change on urban transportation systems and their components and specific adaptation measures. A comprehensive summary is presented in Annex 6.1. The table contains adaptation measures that many different transport systems, modes, and components share in common. For heat resistance, being able to rapidly install structures and facilities with materials to extend the temperature range of tolerance is critical. For water-related impacts from flooding, having barriers, mobile pumps, and drainage facilities that can be deployed quickly to flooded areas will reduce the impacts.

Adaptation measures for urban transportation systems for specific climate hazards are discussed here.

Numerous technologies exist to protect transportation from flooding and have been used routinely. At the extreme, where some roads and rail lines are subject to continuous inundation, some options for operations and management adaptation are: providing alternative travel choices; modifying operational procedures such as reversing the direction of roads and being able to backup trains; avoiding the added congestion on a more limited number of routes due to a shift in travel patterns by

encouraging and in some cases mandating other travel choices to reduce the number of trips altogether (which will have mitigation as well as adaptation benefits); and using continuous pumps and drainage technologies (including clearing existing drains) to reroute water. Capital investment adaptation may entail construction of permanent barriers to keep water away from these routes. Policy measures include encouraging land use patterns that complement reduced travel (such as transit oriented development) and modifying land uses to include those that entrain or trap water, relying on porous surfaces and water absorbing uses such as parks.

Potential adaptation actions for transport via rivers might include: additional dredging in shallow areas, limiting the number and weight of barges, releasing more water from upstream sources (recognizing that this can interfere with other water uses such as hydropower generation, ecological resources, agriculture, municipal, industrial, and recreational) and finding alternate navigation routes or modes of transportation (DuVair, *et al.*, 2002, p. 131). For detailed considerations of how transportation systems along the Great Lakes are adapting see Quinn (2002). Additionally, Quinn identifies benefits from climate change to the transportation systems, such as the potential gains from year-round vessel utilization due to decrease in ice cover may exceed the costs imposed by lower water levels in the lakes.

Continuous episodes of high temperatures combined with stress from roadway usage by heavy vehicles can cause deterioration of transportation materials such as steel, concrete, and asphalt. Although the physical and chemical properties of these materials may indicate considerable resilience, the environmental conditions in which they operate, particularly the heat-related effects of climate change, can reduce their resilience. Asphalt can lose its stability in persistent heat. Steel can buckle, and a number of heat-related instances of rail tracks buckling have been documented in the United States and worldwide. Concrete can also buckle above certain temperatures. The adverse impacts of rising temperatures on transportation materials pose operational risks: "Longer periods of extreme heat in summer can damage roads in several ways, including softening of asphalt that leads to rutting from heavy traffic. Sustained air temperature over 90 °F is a significant threshold for such problems. Extreme heat can cause deformities in rail tracks resulting in speed restrictions and, at worst, causing derailments" (Karl *et al.*, 2009, p. 65). Adaptation measures require retrofitting of existing assets and incorporation of higher tolerance to heat stress in ongoing and planned transportation investments.

If high wind speeds are intermittent, existing transportation systems can be retrofitted to resist wind damage, and usage during high wind events, when foreseen, can be reduced or eliminated. If high winds are continuous, permanent weighting down of structures may be in order. For example, structural work was added to several bridges in the United States after the Tacoma Narrows bridge collapsed due to high winds that resonated with and amplified the movement (Petroski, 1982 (1992 edition), pp. 164–165). Furthermore, land use can be adapted in a manner to shield transportation systems from high winds, by creating both natural and artificial wind barriers.

Where damage is discrete in space and time, namely it does not extend over large areas, section repairs are possible as an adaptive measure and one that can be programmed institutionally into repair cycles. This approach has been adopted in previous events, such as earthquakes and accidental infrastructure network collapses involving the destruction of bridge, road, and rail segments, providing analogies to what can happen with climate change. However, large-scale damage and complete collapse of structures requires the exploration of revision of design codes as well as, in some cases, relocation of assets. This has been the case in California and Japan for earthquakes, where local authorities in both of those areas have incorporated earthquake resistant structural requirements into construction specifications. Similarly, in areas subject to climate change impacts, equivalent changes in codes and other transportation design standards may be relevant.

6.4.3 Policy and economic considerations for adaptation

Adaptation includes modifications of both the transportation system and land use planning changes needed to reduce the vulnerability of urban mobility to climate-induced hazards. Transport infrastructure is capital intensive and has long gestation periods. Therefore, climate adaptation plans need to be coordinated with fiscal planning. Climate adaptation is different for different actors within the urban transportation system. For example, policymakers help adapt a city's infrastructure and its management. While individuals may change their behavior, the role of local authorities is to regulate and coordinate land use and transport systems such that passengers are less vulnerable to the hazards of climate change and therefore do not need to rely solely on expensive personal adaptation. Further, city-specific adaptation assessment of transport systems is essential as the adaptation response will vary. In the case of the Boston Metro Area, Suarez *et al.* (2005) conducted a system-wide analysis of the impacts of climate-induced stress such as riverine and coastal flooding on transportation network performance. Through modeling projected land use changes and demographic shifts and their impacts on transport demand, the study identifies future climate impacts that are likely to cause a "doubling in delays and lost trips." However, the study concludes that system-wide adaptation is not cost effective; instead, adaptation efforts should be limited only to some critical transport segments.

According to the Federal Highway Administration, at present, similar adaptation needs assessments and strategies with regards to transport infrastructure are lacking in most countries. Furthermore, the overall institutional response to the expected impacts of climate change on urban transportation systems is absent, causing most Metropolitan Planning Organizations and Departments of Transportation to withhold adaptation planning until better assessments are conducted.

In the fiscal year 2006–2007, four of the fifty State Departments of Transportation in the United States – California, Oregon, Washington, and Connecticut – and some of the seventy Metropolitan Planning Organizations surveyed mentioned climate change as a consideration. The few departments that were considering climate change focused on mitigation. Counter-intuitively, despite the long-term planning by Metropolitan Planning Organizations and Departments of Transportation, climate impacts on transportation and associated mechanisms for adaptation were neglected (Lindquist, 2007). However, increasingly some sub-national Departments of Transportation in the United States have stressed the need to fill this gap in adaptation assessments and have begun conducting needs assessments of climate impacts and adaptation planning for the shipping industry, such as relocation or protection of ports due to sea level rise, or alterations to transportation planning and management to incorporate climate risk assessments (Lindquist, 2007).

According to the Federal Highway Administration (FHWA, 2008), a growing number of Metropolitan Planning Organizations and Departments of Transportation are participating in or leading inter-agency initiatives on adaptation needs assessments and response planning. By 2008, the number of Departments of Transportation that were working to define transportation policy with regards to adaptation at the state and regional level had more than doubled, from four to ten states of the total fifty states surveyed. New York, Chicago, and San Francisco have initiated research into the effects of sea level changes and storm surges on infrastructure, and thus are establishing partnerships with regional agencies to facilitate coordination among local jurisdictions and establishing protective measures. Despite the lack of analysis, the Federal Highway Administration is encouraging city transportation agencies to take the lead by incorporating adaptation mechanisms into long-range transportation planning (FHWA, 2008).

When it comes to adaptation of urban transport systems to climate-related hazards, mass-transit systems – such as railway tracks, highways, and bridges – face political and fiscal constraints that differ from privately owned individual vehicles. This adaptation handicap is because urban mass-transit systems require lumpy investments – for instance, a subway system or a bus rapid transport system and highways may require hundreds of millions of dollars to initiate adaptation through retrofitting for interventions that have benefits long after the electoral cycle, making such decisions politically challenging, while individuals are free to make personal investments in their own vehicles as each deems appropriate (within the boundaries of regulatory requirements for inspection and maintenance, for example). Furthermore, because climate change impacts have implications in the long term, transport systems such as tunnels, bridges, rights of way, tracks, and roads, which all have long lifespans, are more susceptible to climate-induced risks and uncertainty. The same climate risks are considered to be less of a concern for personal vehicles such as cars and trucks due to the shorter lifespans of these vehicles. Large transit systems, especially low-carbon-emitting rail-based systems, have a greater challenge to adapt to the climate demands over the long term, given the political context in which investments are made. Thus, incremental adaptations as opposed to one-off interventions deserve attention. For instance, when new transportation systems are planned with expected lifespans of up to a hundred years (as is the case for subway systems), adaptation measures may be difficult to introduce quickly. In contrast, cars have a short lifespan of ten years. Therefore, car manufacturers adapt by improving technology with each new generation of cars, yet the design modifications have generally not been in a direction that prevents widespread damage to individual vehicles in disasters that are similar to those expected from climate change. Hence, adaptation in automobiles is occurring in manageable increments at small costs, commensurate with increments in improved climate change projections. In sum, a differentiated and complementary approach to adaptation of individually owned and more flexible transportation systems versus those that are large and lumpy mass transit systems is required.

6.5 Mitigation of greenhouse gas emissions due to urban transportation

This section identifies and categorizes mitigation strategies, discusses potential constraints on their effective implementation, and describes relevant policies and financial mechanisms.

6.5.1 Mitigation strategies

The transport sector represents around 30 percent of global CO_2 emissions, with urban transportation comprising more than half of this amount. This is the type of greenhouse gas emissions "that is expected to grow the fastest in business-as-usual scenarios, increasing at an annual rate of 2–3 percent" (Zegras, 2007, p. 112). The largest part of this growth is expected to happen in developing countries. Price et al. (2006, p. 122) forecast a worldwide average annual growth rate between "2.2 and 3.4 percent over the next 30 years.[4] During this 30 year period, the share of developing countries, in world transportation co_2 emissions, is projected to grow substantially (Price et al., 2006). However, developing countries account for about five times the population of the developed countries. Thus, per capita urban transportation emissions in developing countries remain many folds lower than developed country cities. The aim of mitigation strategies is to reduce the transport carbon footprint of the city. Because of the complex interrelationships between transport, land use and climate change, reducing greenhouse gas emissions requires a two-pronged approach to tackle urban energy consumption, combining both transportation policy and land use policy. Linkages with other sectoral policies also are critical. For instance, see Chapter 4 on energy. For transportation, motorized vehicles can be made more efficient, carbon content of fuels reduced, use of private vehicles discouraged, and efficient non-motorized and public transport promoted. For land use, urban planning and land use regulation, property taxes need to be adapted to facilitate the concentration of private investment in areas of high accessibility, generated by the implementation of mass transport systems. This will reduce the need for mobility due to higher density and diversity of urban functions. Non-motorized commuting can also be encouraged through appropriate urban design and the articulation of different types of transportation. Table 6.3

Table 6.3: *Mitigation strategies for urban transportation systems and related land use.*

	Demand	Supply
Transport	Transport demand management; speed limits; congestion pricing; fuel tax; public transport subsidy; promotion of non-motorized transportation; road tolls; parking fees; provision of eco-driving schemes.	Investment in mass transit system. Regulation and incentives for improvement of vehicle energy yields or low emission fuels. Facilitate inter-modal linkages application of information technology.
Land use	Land use planning; provision of basic services; property tax regimes to discourage sprawl.	Zoning regulation; town planning schemes; incentives for high density urbanization, regulation to discourage sprawl.

---

4 In comparison, Price *et al.* (2006) forecast an annual growth of 1.2–1.4 percent for OECD countries.

shows strategies for reducing greenhouse gas emissions in urban transportation systems, through a combination of demand and supply management policies in transportation systems and land use policies.

Regulatory instruments are applied in various forms; for instance, limiting the number of days a vehicle can be on the road, as is the case in Beijing, Bogota, and Mexico City's "hoy no circula" (one day a week, without a car) or quantitative restrictions on ownership, as in Singapore. However, unintended market distortions from such interventions require attention (for details see Bertaud et al., 2009). Efficient fuels and technology choices are alternate mechanisms to reduce CO_2 emissions. For instance, compressed natural gas (CNG) operated automobiles emit between 20 and 30 percent less CO_2 than automobiles operating on a regular gasoline engine (Ministry of Environment, Government of Japan, 2008). In this regard, over the five-year period from 1998 to 2002, in Delhi all public transport buses were converted to CNG operated systems largely due to a verdict by the Supreme Court of India (Mehrotra et al., 2009, p. 21, also see Box 6.2). Moreover, the Delhi Metro Rail Corporation introduced measures to reduce greenhouse gas emissions through the use of regenerative braking to capture the energy

during deceleration and feed it back into the electrical system. Delhi Metro Rail is considered the first railroad company to obtain carbon credits through such an effort (Mehrotra et al., 2009, p. 24).

Pricing instruments (Table 6.4) modify consumer incentives; for instance, relative prices between private vehicles such as cars and mass-transit modes such as commuter rails. Cities around the world deploy different types of pricing instruments – fixed tolls and congestion pricing as in the case of Singapore, London, and Stockholm; fuel tax as in Bogota, Singapore, Chicago; parking charges as in New York, Sheffield, Edinburgh (Bertaud et al., 2009). Some of these pricing efforts aim to reduce market distortions. Pricing congestion and parking, for instance, aims at adjusting the price of using a highway or of a parking space to reflect its economic value, including externalities due to congestion.

Pricing instruments also include subsidies. Subsidies are often aimed at redistribution. For instance, many transit fares are subsidized, as in Los Angeles, San Francisco, Mumbai, Delhi. Transit fare subsidies are aimed at increasing the mobility of low-income households, allowing them to fully participate in

Table 6.4: *Environmental benefits of congestion pricing.*

City	Environmental benefits, including decline in carbon dioxide emission (per year)	Source
London (2002–2003)	Within the congestion pricing zone: 19.5% carbon dioxide emission reduced; 12% decline in oxides of nitrogen; 12% reduction of suspended particulate matter (PM10, particles <10 micrometers in diameter; 15% drop in vehicle kilometers travelled.	Beevers and Carslaw (2005)
	12% decrease in traffic	Transport for London (2004)
	2.3–2.5 million pounds in savings from carbon dioxide emission reduction; decline of 211–237 million vehicular miles travelled	Evans (2007)
	Reduction of 35% in pollution; total environmental benefits: €4.9 million	Prud'homme and Bocarejo (2005)
Stockholm (January–July 2006)	13% carbon dioxide emission reduced (or 36,000 tons in saved emissions); 8.5% decline in oxides of nitrogen; 14% drop in carbon monoxide levels;13% reduction of PM10; avoidance of 27 premature deaths; 22% reduction of vehicle passages in congestion pricing zone.	Johansson et al. (2008), Lundqvist (2008)
Singapore (1998, 1992, 1975)	75% reduction of car traffic during morning peak hours; in 1992 car volume was 54% of the pre-1975 level; in modal split, share of cars dropped from 48% to 29% immediately congestion pricing was introduced.	Olszewski (2007)
	1998: Elasticity of passenger cars –0.106 within congestion pricing zone (–0.21 in the short run, –0.30 in the long run); 15% drop in daily traffic volumes.	Olszewski and Xie (2005), Olszewski (2007), Menon (2000)
	1975: Traffic volumes in morning peaks reduced by 45%; car entries decreased by 70%	Willoughby (2000)
Milan (2008)	9% carbon dioxide emission reduction (or 150,000 tons per year reduced); 19% reduction of PM10-emissions; savings of €3.3 million; 37% decline in ammonia (NH_3) emissions; 11% drop in oxides of nitrogen emissions; traffic reduced by 14.4%	Milan municipality (2009)
Durham	Number of vehicles declined by 50–80%	Santos and Fraser (2005)

Source: Adapted from Lefèvre and Renard (2009).

a unified metropolitan labor market. Transit fare subsidies are also an incentive for car commuters to opt for a modal switch to transit. Although, this is not a very effective manner to increase transit mode share in the long run (Bertaud *et al.*, 2009).

Pay-As-You-Drive (PAYD) programs offer another mechanism to reduce vehicular miles traveled. With encouragement from public authorities, insurance companies are charging insurance premiums based on driving records and other traditional risk factors but are broken down into per-mile charges. Motorists have the opportunity to lower their insurance costs by driving less. When PAYD insurance is offered to a large percentage of California drivers, it may reduce vehicle miles traveled and associated greenhouse gas emissions (Lefèvre and Renard, 2009). Such strategies have limited application in a developing country city context, where car ownership is limited to a small fraction of urban households and most trips are by walking, bicycling, and other two wheelers, complemented by severely constrained mass transit systems. Instead, Perera and Permana (2009) present alternate strategies appropriate for developing countries using the case of Bandung City, Indonesia.

Bicycling is also being encouraged in developed country cities with the aim of reducing automobile dependency and associated greenhouse gas emissions. Strategies take the form of bicycle rental stations, being used in a number of European cities, and provision of bike lanes. On the other hand, while non-motorized transport accounts for a large proportion of commuter trips within the developing country city, the challenge with rising incomes is to facilitate the retention of such low emissions modes as well as complement with demand-responsive mass transit, as opposed to the present trend of low-cost motorized personal transport. Curitiba in Brazil has been a noteworthy example of the use of bus rapid transit in South America, though examples are now widespread throughout the world from Mexico City to recent efforts in Delhi. Shaping land use is also another way to improve accessibility while mitigating emissions; however, the degree of public control on land use varies significantly by jurisdictions and its efficacy remains debatable.

On categorizing energy consumption in urban transport in developing countries by the degree of land use planning: (1) controlled residential and commercial areas, (2) unplanned peri-urban areas, (3) planned satellite towns, Permana *et al.* (2008) find that, in Bandung City, Indonesia, households in "controlled residential and commercial areas" use transport systems – including walking and bicycling – that consume less energy than households in unplanned or planned areas. However, there are several confounding covariates, such as income and type of employment, which correlate with the type of land use (a proxy for house prices) and modal choices, that need to be analyzed further for relevance in alternate geographies.

New York City's mitigation efforts in transportation largely center on some key initiatives within PlaNYC (2007) and the Metropolitan Transport Authority's (MTA, 2009) Blue Ribbon Task Force. PlaNYC and subsequent regulatory and management initiatives to support the plan's goals, including environmental ones, emphasize anti-idling laws and parking restrictions. The city's congestion pricing initiative did not get the support of the New York State legislature. In the area of transit, the MTA, which is the primary provider of transit in the city, has incorporated a strategy of greening its stations and supporting facilities such as maintenance yards.

6.5.2 Assessments of mitigation potential and cost

There are some efforts to assess urban transportation sector mitigation potential and cost, but this is still an incipient trend, based on a city-by-city analysis. There is no worldwide urban transportation sector mitigation potential and cost assessment in the literature.

Until now, the issue of cost-effectiveness has been successfully applied to international negotiations, such as the European Emissions Trading Scheme (EU-ETS), and to national policies. Energy-economy or sectoral energy models have made it possible to simulate the economic impact of different policies and especially to build sets of marginal abatement cost curves. These mechanisms are efficient tools for analyzing different aspects of climate policies, particularly seeking to reduce the global cost through a certain leveling of the marginal costs of sectoral initiatives (Lefèvre and Wemaere, 2009). The development of marginal abatement cost curves for urban transportation aims to inform methodological efforts to measure and prioritize the actions to inform policymakers' choices.

For instance, the Siemens study of London's transport system (Siemens, 2008) estimates a reduction in transport emissions "by about one-quarter, from 12.1 million tons of CO_2 in 2005 to 9 million tons in 2025". The study identifies better fuel efficiency in cars as a cost-effective means of reducing carbon emissions from transport. In addition, hybrid cars and some biofuels hold abatement potential, albeit at higher costs, given present technology. London could save 0.3 million tons of CO_2 by 2025 by switching to hybrid buses and optimizing road traffic management. For similar calculations on how much *becaks* (human powered tricycle transport) and *ojeks* (motorcycle taxis) can contribute to reducing CO_2 in Bandung City see Permana *et al.* (2008). Increased use of biofuels could cut emissions by 0.5 million tons – assuming biofuels with low greenhouse gas emissions are used.

As cities are complex, a project-based approach is insufficient to reduce urban transportation carbon emissions. Instead an incremental programmatic approach is more likely to be highly cost-effective such that local climate action plans apply a systemic approach to innovations in spatial organization and transportation planning in the broader context of city development and management. For a case study of such an incremental approach see the case of Bandung City (Perera and Permana, 2009).

6.5.3 Constraints to mitigation in urban transportation systems: prospects for green technology diffusion

In the near future, "emergence and large deployment of viable green individual transport technologies is limited" (Pridmore, 2002; Cabal and Gatignol, 2004; Assmann and Sieber, 2005). A study by Heywood *et al.* in 2003, as cited by Zegras (2007), assesses the potential for advancements in passenger vehicle technology "in the United States over a 30-year horizon and concludes that a combination of technological improvements and demand management will be required to reduce transportation energy consumption". Furthermore, according to Assmann and Sieber (2005) the additional time needed for a well-established technology in developed countries to penetrate the market in developing countries is around 10 years. Consequently, according to some estimates a new "green" car, launched today in the developed world, will take 40 to 45 years to reach a significant share of the market in poor countries (Cabal and Gatignol, 2004). However, the pace and scope of global technological diffusion remains a subject of great debate and these estimates need to be revisited as empirical data on technology transfer from developed to developing countries and vice versa becomes available for the urban transport sector.

Mitigation of urban transportation systems requires multilevel-governance arrangements – city level, regional, national, and in some cases global. For instance, while a city can ensure that all vehicles used for city operations are fuel efficient (including taxis), it requires federal legislature to set fuel economy standards and enforce compliance by automakers, as was the case in Santiago (Chile) and Bogota (Colombia). At the national scale, the U.S. has instituted Corporate Average Fuel Economy (CAFE) standards for automobile manufacturers for each model year. (NHTSA 2010). Additionally, measuring carbon emissions at the city scale is challenging for several reasons – the city's jurisdiction does not necessarily overlap with the urban agglomeration, embedded carbon in goods consumed within the city but produced at long distances, or emissions due to transit passengers all pose accounting challenges. In the case of Bandung city, as in many other cities, land use, energy, and transportation policies lack horizontal interagency coordination and vertical intra-sector collaboration (Perera and Permana, 2009).

Public policies that accomplish greenhouse gas reduction in the urban transportation sector are challenging in part because climate change competes with other pressing priorities. Policymakers, especially those in developing countries, face the challenge of ensuring sustainable development of their transportation sector in order to meet the demands of rapid urbanization, economic growth, and global competition. The green agenda is limited to local environmental challenges, especially local air quality, a classic example being Delhi. However, co-benefits from these efforts offer positive externalities in emission reduction as well as capacity-building for institutional response to combat climate change (Mehrotra *et al.*, 2009).

Non-point sources imply diffused emission, making it difficult to collect baseline data for monitoring and evaluation of emissions and their reduction. Additionally, a life-cycle analysis is required if greenhouse gas emissions in the urban transport sector are to be fully accounted. Due to diffuse emission sources it is difficult to involve the key actors necessary to influence the level of greenhouse gas emissions for a given urban transportation system. Finally, due to lack of capacity and willingness of local institutions, enforcement is ineffective.

6.5.4 Mitigation policies

Policy objectives for greenhouse gas emission reductions in the urban transportation sector may include both demand-and-supply-side initiatives (see Table 6.3 for example). Demand-side interventions include reducing the need for transportation through land use planning and incentives for decreasing vehicle miles traveled. Some instruments include congestion pricing, charging user fees for parking, and fuel tax to internalize the social cost of the transport sector into pricing to correct for market distortions that presently do not price environmental degradation due to carbon emissions; incentivizing use of clean fuel vehicle technology, and the like. Supply-side interventions include enhancing the provision of energy-efficient mass-transit systems; regulating land use to preserve and enhance carbon sinks (forests, wetlands) when considering locations of new infrastructure facilities as well reducing the need for transit through better land use management; and coordinating land use and transport policies to exploit synergies (Schipper *et al.*, 2000).

The key stakeholders for mitigation response are various levels of government, firms, and households. While the state takes the lead on supply-side initiatives, including regulation and provision of transport systems and providing incentives for behavioral change, private and public vehicle producers respond by supplying various degrees of fuel efficiency in transport modes – cars, rails, and buses. Consumers – both households and firms that consume transport services – lead demand-side initiatives, including consumer choice of transport modes and the like. However, the various levels of government, types of vehicle producers, and consumers of various transport services have a complex set of overlapping and conflicting interests. For instance, while transit-oriented local governments may create incentives for mass-transit systems, state governments with car manufacturing bases may oppose such initiatives. Likewise transit may be welcomed by non-car-dependent urban communities and resisted by car-dependent suburban communities and vice versa. The effectiveness of such mitigation instruments also varies between developed and developing countries. In many developing countries, as car ownership is limited and demand for transportation services is rapidly growing, creating incentives for fuel-efficient private and public transportation systems can yield substantial gains in reduction of the growth of greenhouse gas emissions.

6.5.5 Financial tools and incentives

Two key incentive mechanisms for mitigation of greenhouse gas emissions from urban transport systems are intergovernmental transfers and carbon markets. Federal transfers for management of ecological goods and services, which are public goods with positive externalities beyond local jurisdictions, are in practice. For example, since 1996, the "German advisory council on the environment has called for the integration of ecological indicators into intergovernmental fiscal transfers" and performance indicators. For instance, financing is determined partially on the basis of improvements in rural land management and protection of nature reserves (Perner and Thöne, 2005). In India, the thirteenth finance commission advised that 7.5 percent of fiscal transfers to states and union territories be based on percentage of forest cover (Kumar and Managi, 2009). While these environmental grants are performance-based the scope of the projects is limited, but such federal grant conditionals can potentially include broader concerns of climate change, including urban transport mitigation, as in the case of Klimp, a Swedish investment grants scheme targeted at sub-national level governments to address climate change.

Likewise, carbon markets are underutilized for energy-efficient urban transportation. Only two out of twelve hundred clean development mechanism (CDM) projects that have been registered by the UNFCC's Executive Board address urban transportation projects. These two projects are TransMilenio, Bogotá's bus rapid transit, and Delhi subway's regenerative braking system (Mehrotra et al., 2009). Together these two urban transport projects represent less than 0.13 percent of the total CDM project portfolio.

The underutilization of CDMs for urban transport projects is due to three key factors. First, there is a mismatch between the local expertise and global requirements. Local government priorities and associated skills are aimed at developing transport projects to ease severe mobility constraints in developing countries. In contrast, the global institutional requirements of CDMs and the associated standards and fees for screening applications and fulfilling requirements are often beyond the scope and abilities of local governments. Second, due to the diffused emissions in the transportation sector, the cost of aggregating data is high. Thus CDM's "act and gain money" incentive has limited impact. Third, basic CDM project requirements are difficult to fulfill for urban transportation: project boundaries are difficult to define due to up- and down-stream leakages, establishing credible baselines is difficult due to constraints in data collection, and data constraints render monitoring methodologies unreliable.

Enhancing the proportion of urban transport projects within the existing CDM framework may require focusing on short gestation and high-return energy-efficient technologies – technology switch to energy-efficient engines and fuels switch; mass-transit systems; information technology for transport systems optimization, such as smart traffic light systems bundled together for programmatic CDMs. Mehrotra et al. (2009) elaborate on the efforts of the city of Delhi, where additional opportunities for utilizing CDM related to urban forestry, street lighting, and landfill are illustrated. These have co-benefits for urban transportation sectors that are yet to be explored. Like the CDMs, the urban transport sector has yet to adequately utilize the Global Environment Facility (GEF), established in 1991 to support developing countries in tackling climate change mitigation and adaptation (Colombier et al., 2007). Thus, project preparation and development support remain a critical gap in linking urban transport to broader climate change and environmental initiatives. In addition to carbon markets, the Partnership on Sustainable Low Carbon Transport, among others, is exploring alternate strategies for financing low-emission urban transport systems through pooling of public and private resources as well as sector-wide approaches to low carbon transport.

[MITIGATION/ADAPTATION] Box 6.4 A sustainability framework tailored for transportation and applied to Sydney, Australia

Ken Doust

Asset Management & Sustainability Assessment, Atkins Global

John Black

School of Civil and Environmental Engineering, University of New South Wales

CUSCCRR (Coalition of Urban Sustainability & Climate Change Response Research) Urban transport system characteristics and system vulnerability provide the basis for a climate change scenario in the context of a sustainability framework as shown in Box Figure 6.8. Climate change impacts are drawn in from physical infrastructure characteristics, network relationships, and behavioral changes, which can vary for different cities. The sustainability metrics incorporate the capability to mitigate and resilience for adapting to climate change impact.

Sydney, Australia, is used to illustrate and visualize the metrics and the environmental sustainability measure (Pillar 1), formulated from known fuel consumption of vehicles (see Cosgrove, 2003, p. 342) with speed and used to calculate carbon dioxide equivalent (CO_2-eq) footprints for motor vehicles between each trip origin/destination pair.

Sydney's transport system primarily consists of tolled motorways, arterial roads, and an extensive suburban heavy rail system with a heavy reliance on cars in the suburbs. Estimated annual transport greenhouse gas emissions for Sydney Metropolitan area rose from 9.8 million tons in 1990 to 12.2 million tons in 2001, and emissions are forecast to rise to 16.8 million tons by 2020 on current trends. Car

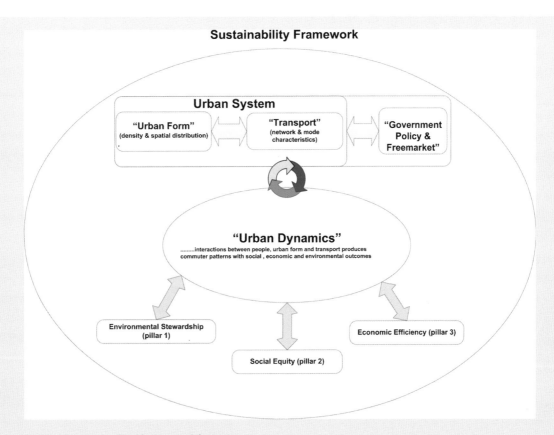

Sustainability Framework

Urban System

"Urban Form" (density & spatial distribution)

"Transport" (network & mode characteristics)

"Government Policy & Freemarket"

"Urban Dynamics"
..........interactions between people, urban form and transport produces commuter patterns with social , economic and environmental outcomes

Environmental Stewardship (pillar 1)

Social Equity (pillar 2)

Economic Efficiency (pillar 3)

Box Figure 6.8: *A sustainability framework – tailored for transport infrastructure.*

usage contributes 67 percent of these emissions (www.bitre. gov.au/publications/93/Files/r107.pdf, Tables 3.13 and 3.14, accessed August 18, 2009).

A quantifiable measure of greenhouse gas mitigation effectiveness was developed from detailed operational methods, using transport planning building block techniques (Doust, 2008, Chapter 4). The scenario is based on the combination of two concepts, accessibility and environmental sustainability.

Accessibility has been identified as a useful measure in social and economic aspects of sustainability (see Expert Group on the Urban Environment, 1996; Warren Centre for Advanced Engineering, 2003; Kachi *et al.*, 2005, 2007). In the Sydney Case Study, accessibility measures were derived (Doust, 2008, Chapter 4) for each travel zone pair. Separate operational methods were developed to generate worker and employer focused accessibility measures. These are measures that are relatable to social equity (Pillar 2) and economic efficiency (Pillar 3) respectively.

Environmental sustainability measure is defined as the inverse of CO_2 emissions from the total Journey to work trips between zone pairs, including an allocation of emissions from manufacture of vehicle and road infrastructure. This is calculated as a sum of the CO_2-eq per unit trip km at the average speed with the shortest path trip length and number of trips. The CO_2-eq is calculated as the sum of the quantity of greenhouse gas and the Global Warming Potential Index (AGO, 2005, Appendix 3).

These metrics can also be applied in a way that expresses sustainability performance in terms of sustainability risk. High risk, where sustainability performance is poor, is indicated by low metric values. Low risk, where sustainability performance is satisfactory, is indicated by a higher metric value, above a community accepted minimum target. The grid concept can be likened to a risk matrix allowing each zone pair to be assigned a sustainability risk rating (Box Figure 6.9). The sustainability risk boundaries are specific to each city, and influenced by the population's estimated resilience.

This sustainability risk rating can then be plotted onto geographic space using geographic information system (GIS) thematic mapping. Box Figure 6.10 illustrates such a visualization in geographic space.

Each of these visualizations provides insight into the position, spread, and internal distribution trends for a city's urban sustainability pillars of environmental stewardship, social equity, and economic efficiency. For community and decision-makers these visual differences give a simple snapshot of overall sustainability performance for each scenario being considered. It is straightforward to change the scenario, use the building block techniques, and produce a new metric plot to see the sustainability effect of the policies embedded in the scenario. Stakeholders can see measurable change for their communities in relation to sustainability goals. The process provides another dimension to visioning and sustainability strategy development by adding the means by which a community can measure and judge one infrastructure and urban form scenario with another.

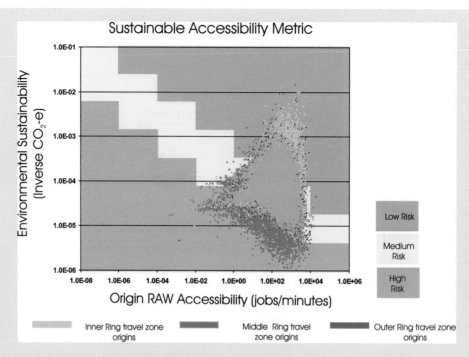

Box Figure 6.9: *Sustainability performance risk overlay to Sydney Inner, Middle, and Outer Ring results in 2001.*

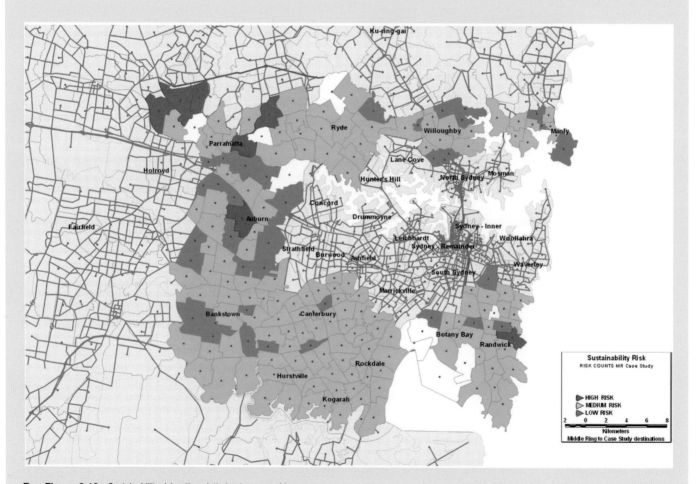

Box Figure 6.10: *Sustainability risk rating plotted onto geographic space.*

A particular strength of using the sustainability framework, and the metrics demonstrated here, is that they are derived from data sets that have been commonly used by urban and regional planners for many years. Visualizations of this type can be used to inform decision-makers (community and government agencies) in the process of choosing climate change policies and programs for a city.

Under climate change, scenario systems are also at risk of failure. The current process of determining the metrics under an operational system state is to be extended to also estimate the metrics under failed system states due to transport infrastructure system vulnerabilities to climate change impact. The methodology discussed in this case study provides a useful tool for each city beginning to understand how they should respond to climate change. Every city will have its own unique set of urban response scenarios

to choose from in mitigating greenhouse gases, and each of these will need to occur in a future that does involve adaptation to climate change impacts of some degree. The more there is understanding of the effectiveness of each of these urban response scenarios, inclusive of the adaptive capability, in balance with other sustainability pillars, the greater the likelihood of real outcomes being realized for each of our cities.

This Case Study is based on the original research titled "Metrics of Environmental Sustainability, Social Equity, and Economic Efficiency in Cities" Doust, K. (2008). Papers from this research have been published in the following peer reviewed journals and conference proceedings: Black and Doust (2008); Black et al. (2010); Doust (2010); Doust and Black (2008); Doust and Black (2009); Doust and Parolin (2008); Doust and Parolin (2009); Nakanishi et al. (2009).

6.6 Key uncertainties, research needs, and information gaps

Many levels of uncertainties drive research needs and define information gaps, these range from the identification and degree of certainty of climatic factors and consequences specific to transportation to the performance of transportation technologies for both mitigation and adaptation and their relationship to other infrastructure.

The IPCC (2007a) has identified three macro-scale uncertainties in assessing transportation adaptation and mitigation potentials for any city. The first uncertainty relates to the international price of oil and associated demand for alternate fossil fuels. The second uncertainty relates to the pace of innovation in alternate energy sources such as biofuels, externalities associated with their large-scale consumption: food prices, water shortages, and the like, and batteries. The third uncertainty is the timeline on which policies on greenhouse gas emissions reduction will be adopted and implemented by developed and developing countries.

Other critical issues pertain to the availability of, and access to, information at the international level. First is the lack of data on transport energy consumption in developing countries now and projected for the future, and thus uncertainties in the magnitude of emissions that will occur as a result of energy consumption overall and the use of alternatives to oil. With a few exceptions, the geographical focus in the present research is on developed country transport systems, especially in the United States and Europe. There is a gap in the literature on the implications of climate change on urban transport policy and planning in developing counties.

Second is the absence of a broad framework to assess global opportunities and costs of reducing greenhouse gas emissions in the transportation sector. Consistency and coordination is

needed among international agencies in their support of climate-friendly transportation projects and longer-term goal setting. The availability and distribution of information about adaptation and mitigation vary from local to global levels in both developed and developing countries, and this variability contributes to problems of coordination and consistency.

Potter and Savonis (2003) and Hyman et al. (2008) outline some of the important research needs and challenges to prepare transportation system for the impacts of climate change. Regional and local-scale climate projection models that incorporate unique attributes of the urban transportation systems are lacking. But downscaled assessments are a prerequisite for transportation planners to identify facilities and locations that are vulnerable to the impacts of climate-related events. Further, impacts need to be disaggregated into implications for operations, maintenance, and safety of transportation systems for the short- and long-term effects.

City managers require new tools to evaluate the benefits and costs of a range of response options that incorporate future uncertainties into modal choices, siting infrastructure, design and engineering standards, and the like. For example, "[it] is unlikely that infrastructure improvements such as realignment of roadways, many of which run through river valleys, can be justified on a cost-benefit basis" in the Boston area (Tufts University, 2004, p.155). Additionally, there is a need for improved techniques for assessing risk and integrating climate information into transportation planning and management within the broader context of city planning and management.

On top of these uncertainties and gaps in knowledge are those within climate science itself, discussed extensively in other chapters. One example is the extent of ice melting and its impact on sea level rise at specific geographic locations. Another is the intensity and frequency of storms in light of the difficulty of predicting cloud formation and, more recently, the influence of changes in monitoring protocols of storm frequency and intensity. These climate uncertainties combine with uncertainties associated with the consequences of climate conditions specifically for transportation at any given place and time. Finally, the performance of technologies in reducing the adverse consequences adds another layer of uncertainties which reflects gaps in knowledge. For example, construction standards and the designs of transportation infrastructure will adapt to climate change impacts depending on environmental factors that affect the strength of the new systems.

Research gaps for adaptation include the need for knowledge on how climate hazards will affect transport infrastructure, including the engineering design and performance standards for urban transport systems and associated infrastructure assets. Furthermore, research on relationships between climate hazards and social and economic impacts related to urban transport systems is much needed. For instance, research on quantifying the expected impacts of climate change on types of transport systems and their users is lacking. Further, there is need for research analyzing climate impacts on planned investments in the transport sector, as most research focuses on climate impacts on existing transportation systems. Finally, intra-and-inter sectoral co-benefits of adaptation and mitigation need attention (Lindquist, 2007). For example, transportation is dependent on electric power and telecommunications as well as water and environmental services and will be affected by adaptation and mitigation efforts in these sectors. Likewise, changes in the the delivery of goods and services for transportation industry supply chains will affect the transportation systems as well.

Research needs and information gaps for mitigation include the evaluation of specific mitigation policies for effectiveness in reducing greenhouse gas emissions from transport in a city; and economic costs of mitigation in the urban transport sector with careful consideration of areas in which mitigation measures for transport might conflict with others and actually contribute to greenhouse gas rather than reducing it. For instance, there is a need for comparative studies on the cost of producing and switching to clean fuels and new vehicle technology so as to allow urban transport planners and stakeholders to make informed choices on new technology options for low-emission transportation systems. Finally, as with adaptation, the feasibility for mitigation of various options in light of the extent and role of the interdependencies within various transportation sectors (Lindquist, 2007) and between the transportation sectors and other sectors and activities with which transportation interacts or affects must be studied. For instance, there is a need for further research on methodologies for standardizing inventories from transportation emissions and criteria for including emissions from national, regional, and local transport systems that extend beyond the jurisdiction of cities and are interconnected with other sectors such as telecommunications, energy, and water. There is also a need to explore the interaction between the energy intensity resulting from modal choices and their interaction with types of land uses and other urban infrastructure systems.

Thus, large uncertainties will continue to exist and new ones will continue to emerge as summarized above in the science and technology that justify and support moving forward on both adaptation and mitigation. In light of this, some have argued that the key strategy is to move forward on reasonably robust measures and to evaluate the performance of those measures over time, rather than to await the resolution of uncertainties before acting (Dessai *et al.*, 2009). For transportation, adopting such a policy with its accompanying strategy suggests moving ahead with a multi-pronged approach that emphasizes the availability and use of multiple modes of travel that avoid greenhouse gas emissions and at the same time are flexible and resilient to the impacts of climate change, that is, made of more resistant materials and able to withstand potentially prolonged flooding and the intensity of storms.

6.7 Conclusions

Generally, public actions aim to anticipate and frame market-based urban development toward a more energy-efficient city. That can involve analyzing market dynamics – transport markets, real-estate markets, and housing markets – and integrating them in local urban planning. A challenge is to adopt a planning model that works to create dynamic middle- and long-term urban development trajectories, rather than static or "one-off" systems.

The combination of adaptation and mitigation policy instruments to be implemented for the urban transportation system is a city-specific issue combined with overarching global policies as a guide. Transport policies are easier to implement, but their potential to reduce greenhouse gases may be lower than for other activities. Land use policies can be stronger levers of action to reduce greenhouse gas emissions due to urban transportation, but they may be harder to implement.

For cities in developing countries, the challenge is more to keep their mixed land use and high-density settlements, and transport systems in which low-emission modes are still dominant, without compromising efforts for future economic development and urban poverty reduction, both of which rely on expanding effective and efficient transportation systems and associated modal choices.

Public education and effective communication on policies that aim to reduce greenhouse gas emissions are important aspects for successful promulgation of adaptation and mitigation policies in urban transportation systems. These efforts can emphasize local co-benefits in and seek to gain – and secure over the long-term – public support for these measures.

Finally, several elements have emerged for climate change adaptation and mitigation policies to be successful for urban transport planning and management. These include strong leadership and even championing climate change mitigation; an effective outreach and public education campaign; risk assessment, benefit-cost analysis, and management – particularly for adaptation; mainstreaming of climate concerns in transport planning and policymaking; higher-level government or international community provision of incentives to mainstream climate – not only through funding, but through international recognition as well; technology transfer; regional–local governance coordination; assessment of the capacity to act among all actors – all levels of government, private sector and individuals; and overall, attention to reduction in vehicle miles traveled through land use and other strategies that may accomplish this. Science-based policy is crucial.

Annex

Annex Table 6.1: *Impacts of climate change on transportation.*

Climate change	Infrastructure impact	Operations impact	Adaptation measures	Sources
Temperature-related				
Temperature increase	Pavement damage; asphalt rutting	Traffic speed	Frequent maintenance; milling out ruts; laying of more heat-resistant asphalt	Andrey and Mills., 2003 (Canada); Wooler, 2004 (United Kingdom); Soo Hoo, 2005 (Seattle)
	Deformation and "deterioration of road and rail infrastructure from buckling and expansion" (CRI, p. 59)	Potential for derailment of trains; decreased travel speed	Improved monitoring of rail temperatures and more frequent maintenance track; speed restrictions	OFCM, 2002 (Mid-Atlantic, U.S. Amtrak derailment incidence 2002, pp.1–7); Caldwell et al., 2002 (p. 11); Wooler, 2004 (United Kingdom)
		Heating of underground cars and lack of ventilation.		Wooler, 2004 (London)
Extended period of growing season for trees and vegetation	Obscuring signs; slipperiness (by fallen leaves) on roads Potential for derailment of trains; decreased travel speed	Better management of foliage; better management of trees that grow alongside the transportation corridors; planting slower growing plants to reduce leaf fall		Wooler, 2004
Temperature increase in winter	New passages for marine transportation (Northwest Passage)	Less spending on winter maintenance for snow and ice control; less pavement damage from frost		Andrey and Mills, 2003 (Canada); Kinsella and McGuire, 2005 (New Zealand); Infrastructure Canada, 2006
Freeze–thaw cycle frequency increase	Premature damage of pavement, roads, runways, railroads and pipelines.			Caldwell et al., 2002 (p.11); Mills and Andrey, 2002 (p. 79)
Thawing of permafrost	Damage to roads, rail lines, pipelines, and bridges; affect northern latitude (Alaska) more severely because it depends more heavily on frozen roads for freight movements	Road capacity to sustain transportation is reduced	Need for different construction methods, such as installment of cooling machineries	Caldwell et al., 2002 (Alaska region, p. 10); Infrastructure Canada, 2006 (Manitoba region, p. 13)
Other	Reduction in ice loads on structures (bridges and piers)	Extended transport-related construction season due to warmer temperature		Andrey and Mills, 2003 (p. 246); Lockwood, 2006.

Annex Table 6.1: (*continued*)

Climate change	Infrastructure impact	Operations impact	Adaptation measures	Sources
Water-related				
Increase in precipitation amount and frequency	Erosion and decay of the physical structure in the track subgrade.	Instability of the tracks for the transit of heavy engines.	Improvement of remote sensing technology that allows detection of water bodies and air pockets	Wooler, 2004 (London, Liverpool); Kinsella and McGuire, 2005 (Western half of New Zealand, p. 6); Kafalenos, 2008 (pp. 4–20) (Gulf coast region); CRI, 2009 (New York City)
		Flooding of roads, basement and sewer will overload drainage systems more frequently, resulting in more wear and tear on equipment and infrastructure	Increased congestion, accidents, and delays	Better management of drainage system, detours
		Snow accumulation on roadways	Disrupts traction control, visibility, and increases crash risks	Improve advisory system and perform pretreatments on roadways
Decrease in precipitation amount and frequency	Likelihood of drought which affects growth of roadside vegetation.	Less disruption to construction and maintenance activities; mobility benefit		Mills and Andrey, 2002; Kinsella and McGuire, 2005 (Eastern half of New Zealand, p. 6)
Sea level rise	Progressive damage caused by flooding to the infrastructure that lacks a fouling-resistant design against salt water.	Limits speeds and routes	Frequent maintenance, relocation, construction of flood-defense mechanisms, such as dikes; elevation of land and structures to minimize the impacts of flooding; heavier use of pumps;	Titus, 2002; Wooler, 2004 (London, Liverpool); Kinsella and McGuire, 2005 (New Zealand coastal highway); Infrastructure Canada, 2006; Kafalenos, 2008 (Gulf coast, New Orleans buses and streetcar, pp. 4–17); CRI, 2009 (New York City)
Increase in the frequency or intensity of extreme weather events	Storm surge (and/or wave crests) – damaging roadways, bridges, rails, airports, and subways; structural damage to street, basement and sewer infrastructure due to wave action; degradation of road platform	The reduction in routes caused by the power outage increases the need for the implementation of emergency plans, to reduce traffic delays and travel rescheduling.	Preventive design of emergency systems (as well as evacuation plans) include protective barriers, relocation of physical supplies, ability to generate alternative road routes and higher bridges over water surface.	Mills and Andrey, 2002; Tufts, 2004 (Boston); Kinsella and McGuire, 2005 (New Zealand); Jacob *et al.*, 2007 (New York City); Kafalenos, 2008 (pp. 4–15, Gulf coast); Meyer, 2008 (p.6, Gulf coast, bridges); CRI, 2009; Zimmerman and Faris, 2009
		High winds and lightning could damage overhead cables, vehicles, trees, signs, etc.	Debris blown onto the roadway impacts visibility; produces power outage, fires, and disrupts signaling system	Adapt transportation systems to defy wind damage (as was done by applying a second layer to several bridges after the Tacoma Narrows bridge collapse)
Decrease in storm frequency or intensity during winter		Facilities of transportation for operators and users		Mills and Andrey, 2002 (p.77)
Other				
Humidity increase (fog)		Reduces visibility and increases crash risks	Signs, speed control, monitor	Lockwood, 2006 (Annex A)

Source: Mehrotra and Kang (2009) with inputs from Hyman et al. (2008) and Lockwood (2006).

Annex Table 6.2: *Indirect impacts of climate change as a result of direct impacts outlined in Annex Table 6.1.*

Economic impacts	
Increased temperature	"Increase energy demand, resulting in more frequent power outages and requiring energy restrictions on use of HVAC and other systems." CRI, 2009 (New York City); "Increase the number of passengers overheating while waiting for trains" CRI, 2009 (New York City); "This is a public health concern, but this could also lead to decreased demand for trains, sales of all the railway-related goods and services. Possible adaptation measures include installing air conditioners" (Wooler, 2004)
Increasing temperatures in northern regions	Northern ice roads may thaw earlier than usual and trucks may have to reduce their loads
Decreasing inland waterway levels	"Milder winters could lengthen the ice-free shipping season by several weeks, increasing vessel utilization and reducing the costs of icebreaking" (Caldwell, 2002, p.10); "Falling water levels on the lakes will decrease water depths, necessitating shallower draft vessels, and therefore less tonnage capacity per trip. … Past instances of low water levels on the Great Lakes hint at the seriousness of the problem. Most recently, in 2000, low water levels forced carriers into 'light loading,' reducing their cargo tonnage by five to eight percent" (Caldwell, 2002, p.10); Similar study by Quinn (2002, p. 120) was cited in Hyman *et al.* (2008); St. Lawrence Seaway and the Great Lakes are good examples
All adverse weather impacts on roadways that lead to traffic delays	As of 2002, congestion costs Americans $78 billion a year in wasted fuel and lost time – up 39 percent since 1990. In Houston, traffic jams cost commuters on the Southwest Freeway and West Loop 610 an average of $954 a year in wasted fuel and time. In New Jersey's Somerset County, congestion costs the average licensed driver $2,110 a year (US News and World Report, 2001); The Federal Highway Administration projects that, over the next 10 years, the number of vehicle-miles traveled is estimated to increase by 24 percent. In 20 years, it is expected to increase by 53 percent (FHWA, 2002a)*;
Environmental impacts	
Reduced winter maintenance	Reduced use of road salt (and other de-icing chemicals) will lead to less salt corrosion of vehicles and salt loadings in waterways, which in turn will positively impact the environment (Warren *et al.*, 2004, p. 139)
Air condition	"Transportation-related activities are major sources of NOx, VOCs, CO, and particulate matter. The surface and upper air conditions (warm temperatures; stagnant anticyclonic air masses) that promote the occurrence of high concentrations of these pollutants may become more frequent and of longer duration under certain climate change scenarios" (Mills and Andrey, 2002, p. 82)
Increased marine transportation in the Arctic region	Increase the probability of hazardous spills (Mills and Andrey, 2002, p. 82)
Dredging	"Dredging of waterways – in response to falling water levels – could have unintended, harmful environmental impacts." (Hyman *et al.*, 2008, p. 16) – Great lakes study Sousounis (2000) was cited
Demographic impacts	
UK Climate Impacts Programme Report on the West Midlands noted, "higher temperatures and reduced summer cloud cover could increase the number of leisure journeys by road." (Entec UK Ltd, 2004)	
Security impacts	
Increased shipping activities will raise security, ownership, maintenance, and safety concerns	

Source: Mehrotra and Kang (2009) with inputs from Hyman et al. (2008) and Lockwood (2006).

REFERENCES

Aerts, J., D. C. Major, M. J. Bowman, P. Dircke, and M. F. Marfai (2009). *Connecting Delta Cities*, Amsterdam, the Netherlands:VU University Press.

AGO (2005). *AGO Factors and Methods Workbook*, Australian Greenhouse Office, Australian Government, Canberra.

Andrey, J. and B. N. Mills (2003). Climate change and the Canadian transportation system: vulnerabilities and adaptations. In J. Andrey and C. Knapper (Eds.), *Weather and Transportation in Canada*, Department of Geography, University of Waterloo, Ontario, Canada.

Assmann, D. and N. Sieber (2005). Transport in developing countries: Renewable energy versus energy reduction? *Transport Reviews*, **25**(6), 719–738.

Beevers, S. and D. Carslaw (2005). The impact of congestion charging on vehicle emissions in London. *Atmospheric Environment*, **39**, 1–5

Bertaud, A., B. Lefèvre, and B. Yuen (2009). *Greenhouse Gas Emissions, Urban Mobility and Efficiency of Urban Morphology: A Hypothesis*, World Bank Commissioned paper, Marseille Symposium, July 2009.

Bhattacharya, S., C. Sharma, R. C. Dhiman, and A. P. Mitra (2006). Climate change and malaria in India. *Current Science* **90**(3), 369–375.

Birkmann, J. (Ed.) (2006). *Measuring Vulnerability to Natural Hazards: Towards Disaster Resilient Societies*, Tokyo, Japan, and New York, USA: United Nations University Press.

Black, J. and Doust, K. (2008). Metrics of environmental sustainability, social equity and economic efficiency for employment location and commuting. In T. Gilmour, E. J. Blakely and R. E. Pizarro (eds), *Dialogues in Urban Planning: Towards Sustainable Regions*. Sydney: University of Sydney Press, Chapter 2.

Black, J., Cheung, C., Doust, K. and Masuya, Y. (2010) *Land-Use/Transport Scenario Assessment Model: Optimization Algorithms and Preference Functions, Accessibility and Green House Gas Trade-offs*. 12th WCTR, Lisbon.

Bradley, R., K. A. Baumert, B. C. Staley, T. Herzog, and J. Pershing (2007). *Slicing the Pie: Sector Based Approaches to International Climate Agreements*, Washington, DC, USA: World Resources Institute.

Brandon, C. and K. Hommann (1995). *The Cost of Inaction: Valuing the Economy-wide Cost of Environmental Degradation in India*. Washington, DC, USA: World Bank.

Cabal, C. and C. Gatignol (2004). *Rapport sur la définition et le simplications du concept de voiture propre*, Office parlementaire d'évaluation des choix scientifiques et technologiques.

Caldwell, K., H. Quinn, J. Meunier, J. Suhrbier, and L. Grenzeback (2002). Potential impacts of climate change on freight transport. In *The Potential Impacts of Climate Change on Transportation Workshop, October 1–2, 2002*, Center for Climate Change and Environmental Forecasting, US Department of Transportation, Washington, DC, USA.

Census of India. http://censusindia.gov.in/.

Chomitz, K. M. and D. A. Gray (1996). Roads, land-use, and deforestation: a spatial model applied to Belize. *World Bank Economic Review*, **10**(3), 487–512.

Colombier, M., H. Kieken, and M. Kleiche (2007). From Rio to Marrakech: development in climate negotiations. In *Energy and Climate Change: the Main Analyses of Regard sur la Terre*. IDDRI-AFD, pp. 53–62.

Cosgrove, D. (2003). *Urban Pollutant Emissions from Motor Vehicles: Australian Trends to 2020*, Final Draft Report for Environment Australia, Bureau of Transport and Regional Economics (BTRE), Australian Government, Canberra.

Cropper, M. L., N. B. Simon, A. Alberini, and P. K. Sharma (1997). *The Health Effects of Air Pollution in Delhi, India*, World Bank Policy Research Working Paper No. 1860. Retrieved from http://ssrn.com/abstract=604994.

Dessai, S., M. Hulme, R. Lempert, and R. Pielke, Jr. (2009). Do we need better predictions to adapt to a changing climate? *Eos*, **90**(13), 111–112. Accessed August 25, 2009, http://sciencepolicy.colorado.edu/admin/publication_files/resource-2720–2009.08.pdf.

Doust, K. (2008). Metrics of environmental sustainability, social equity, and economic efficiency in cities. Unpublished Ph.D. thesis, University of New South Wales.

Doust, K. (2010) *Responding to Climate Change in Cities: Techniques for Minimising Risk through Transport System and Urban Form Planning and Asset Management*. HKIE Civil Conference 2010 Infrastructure Solutions for Tomorrow, Hong Kong

Doust, K. and Black, J. (2008). *Visual Sustainability Metrics : A Community Discussion and Decision Making Tool*. 2nd Sustainable Development Conference: Making Tommorrow Today, Milan.

Doust, K. and Black, J. (2009). A Holistic Assessment Framework for Urban Development and Transportation with Innovative Triple Bottom Line Sustainability Metrics. In Eva Kassens (ed.) *Sustainable Transportation – An International Perspective*. MIT Journal of Planning, vol. 9, pp. 10 – 27.

Doust, K. and Parolin, B. (2008). *Performance Metrics for Sustainable Urban Transport: Accessibility and Greenhouse Gas Emissions*. 31st Australasian Transport Research Forum, Gold Coast.

Doust, K. and Parolin, B. (2009). *Enabling city sustainability through transport systems: moving from vision to reality*. State of Australian Cities National Conference, Perth

DuVair, P., D. Wickizer, and M. J. Burer (2002). Climate change and the potential implications for California's transportation system. In *The Potential Impacts of Climate Change on Transportation Workshop, October 1–2, 2002*, Center for Climate Change and Environmental Forecasting, US Department of Transportation, Washington, DC, USA, pp. 125–134.

Entec UK Ltd (2004). *The Potential Impacts of Climate Change in the West Midlands*, Sustainability West Midlands, UK, January.

Evans, R. (2007). *Central London Congestion Charging Scheme: Ex-post Evaluation of the Quantified Impacts of the Original Scheme*, Prepared for London Congestion Charging Modelling and Evaluation Team, Transport for London. Accessed August 31, 2009, www.tfl.gov.uk/roadusers/congestioncharging/6722.aspx.

Expert Group on the Urban Environment (1996). *European Sustainable Cities Report*. Brussels, Belgium: European Commission.

FEMA (2009). *HAZUS Overview*. Accessed August 25, 2009, www.fema.gov/plan/prevent/hazus/hz_overview.shtm.

FHWA (2002). *Federal Highway Administration FY2002 Performance Plan and FY2000 Performance Report*, Budget Submission Draft, Washington, DC, USA: US Department of Transportation. Accessed August 28, 2009, www.fhwa.dot.gov/reports/2002plan/index.htm.

FHWA (2008). *Federal Highway Administration: Integrating Climate Change into the Transportation Planning Process*, Final Report, July 2008. Accessed August 25, 2009, www.fhwa.dot.gov/HEP/climatechange/climatechange.pdf.

Forrester, J. W. (1969). *Urban Dynamics*. Portland, OR, USA: Productivity Press.

Geurs, K. T. and B. van Wee (2006). Ex-post evaluation of thirty years of compact urban development in the Netherlands. *Urban Studies*, **43**(1), 139–160.

GHMC (Greater Hyderabad Municipal Corporation) (2007). *City Development Plan*, Hyderabad: GHMC.

Greater London Authority (2005). *Climate Change and London's Transport Systems*, Summary report prepared by Mayor of London with the London Climate Partnership, September 2005. Accessed December 11, 2009,www.london.gov.uk/lccp/publications/docs/climatetransport-sept05.pdf.

Hyman, R. C., J. R. Potter, M. J. Savonis, V. R. Burkett, and J. E. Tump (2008). *Why Study Climate Change Impacts on Transportation*? In M. J. Savonis, V. R. Burkett, and J. R. Potter (Eds.), *Impacts of Climate Change and Variability on Transportation Systems and Infrastructure: Gulf Coast Study, Phase I*, A Report by the US Climate Change Science Program and the Subcommittee on Global Change Research, Department of Transportation, Washington, DC, USA.

Hunt, A. and P. Watkiss (2007). *Literature Review on Climate Change Impacts on Urban City Centres: Initial Findings*, OECD, Paris.

Iacono, M. and D. Levinson (2008). *Predicting Land-use Change: How Much Does Transportation Matter?* Working Papers 000045, University of Minnesota, Nexus Research Group, http://ideas.repec.org/p/nex/wpaper/predictinglandusechange.html.

Infrastructure Canada (2006). *Adapting Infrastructure to Climate Change in Canada's Cities and Communities*. Accessed August 23, 2009, www.infc.gc.ca/altformats/pdf/rs-rr-2006–12_02-eng.pdf.

INE (National Statistics Institute) (2007). www.ine.gov.mz/censo2007.

INGC (2009). *Study on the Impact of Climate Change on Disaster Risk in Mozambique*, INGC Climate Change Report, K. Asante, R. Brito, G. Brundrit, *et al.* (Eds.), INGC, Mozambique.

IPCC (2007a). *Climate Change 2007, Synthesis Report, An Assessment of The Intergovernmental Panel on Climate Change*. Accessed August 25, 2009, www.ipcc.ch/.

IPCC (2007b). *WG II Glossary*, Accessed August 25, 2009, www.ipcc.ch/pdf/assessment-report/ar4/wg2/ar4-wg2-app.pdf, also see www.ipcc.ch/.

Jacob, K. H., N. Edelblum, and J. Arnold (2000). Risk increase to infrastructure due to sea level rise: sector report. In *Infrastructure for Climate Change and a Global City: An Assessment of the Metropolitan East*

Coast (MEC) Region. http://metroeast_climate.ciesin.columbia.edu/reports/infrastructure.pdf. A shortened version of this report is included in Rosenzweig and Solecki (2001).

Jacob, K. H., V. Gornitz and C. Rosenzweig (2007). Vulnerability of the New York City metropolitan area to coastal hazards, including sea level rise: inferences for urban coastal risk management and adaptation policies. Chap. 9, In L. McFadden, et al.(Eds.), Managing Coastal Vulnerability: Global, Regional, Local, Amsterdam, London and New York: Elsevier, pp. 141–158. Accessed August 25 2009, http://pubs.giss.nasa.gov/abstracts/2007/Jacob_etal.html and http://pubs.giss.nasa.gov/docs/2007/2007_Jacob_etal.pdf.

Jacob, K. H., C. Rosenzweig, R. Horton, D. Major, and V. Gornitz (2008). MTA Adaptations to Climate Change: A Categorical Imperative, White paper for the MTA BRCS. Accessed August 25, 2009, www.mta.info/sustainability/pdf/Jacob_et%20al_MTA_Adaptation_Final_0309.pdf.

Johansson, C., L. Burman, and B. Forsberg (2008). The effects of congestions tax on air quality and health. Atmospheric Environment, 42, 1–12.

Kachi, N., Kato, H., Hayashi, Y., and Black, J. (2005). Making cities more compact by improving transport and amenity and reducing hazard risk. Journal of the Eastern Asia Society for Transportation Studies, 6, 3819–3834.

Kachi, N., Hayashi, Y., and Kato, H. (2007). A computable model for optimizing residential relocation based on quality of life and social cost in built-up areas. Journal of the Eastern Asia Society for Transportation Studies, 7, 1460–1474.

Kafalenos, R. S., K. J. Leonard, D. M. Beagan, et al. (2008). What are the implications of climate change and variability for Gulf Coast transportation? In M. J. Savonis, V. R. Burkett, and J. R. Potter (Eds.), Impacts of Climate Change and Variability on Transportation Systems and Infrastructure: Gulf Coast Study, Phase I, Report by US Climate Change Science Program and the Subcommittee on Global Change Research, Department of Transportation, Washington, DC, USA.

Kahn-Ribeiro, S., S. Kobayashi, M. Beuthe, et al. (2007). Mitigation. In B. Metz, O. R. Davidson, P. R. Bosch, R. Dave, and L. A. Meyer (Eds.), Climate Change 2007. Contribution of Working Group III to the Fourth Assessment Report of the Intergovernmental Panel on Climate Change, Cambridge, UK and New York, USA: Cambridge University Press.

Karl, T. R., J. M. Milillo, and T. C. Peterson (Eds.) (2009). Global Climate Change Impacts in the United States. Cambridge, UK: Cambridge University Press.

Kinney, P. L. (2008). Climate change, air quality, and human health. American Journal of Preventive Medicine, 35(5), 459–467.

Kinsella, Y. and F. McGuire (2005). Climate Change Uncertainty and the State Highway Network: A Moving Target, Transit New Zealand, Auckland, New Zealand.

Kovats, S. and Akhtar, R. (2008). Climate, climate change and human health in Asian cities. Environment and Urbanization, 29(1), 165–175.

Kumar, S. and S. Managi (2009). Compensation for environmental services and intergovernmental fiscal transfers: the case of India. Ecological Economics, 68(12), 3052–3059.

Lefèvre, B. and V. Renard (2009). Financing Sustainable Urban Policies, OCDE Background Paper.

Lefèvre, B. and M. Wemaere (2009). Fitting Commitments by Cities into a Post-2012 Climate Change Agreement: Political, Economic, Technical and Legal Aspects, Ideas Iddri.

Leonard, K. J., J. H. Suhrbier, E. Lindquist, et al. (2008). How can transportation professionals incorporate climate change in transportation decisions? In M. J. Savonis, V. R. Burkett, and J. R. Potter (Eds.), Impacts of Climate Change and Variability on Transportation Systems and Infrastructure: Gulf Coast Study, Phase I, Report by US Climate Change Science Program and Subcommittee on Global Change Research, Department of Transportation, Washington, DC, USA, pp. 5–1–5F-5.

Lindquist, E. (2007). Climate Change and Adaptive Strategies in Subnational Transportation Planning Agencies in the United States, Institute for Science, Technology and Public Policy. Accessed on August 25, 2009, www.2007amsterdamconference.org/Downloads/AC2007_Lindquist.pdf.

Lockwood, S. A. (2006). Operational Responses to Climate Change Impacts. PB Consult, December 29. Accessed August 19, 2009, http://onlinepubs.trb.org/onlinepubs/sr/sr290Lockwood.pdf.

Lundqvist, L. (2008). Can congestion charging support sustainable development of metropolitan areas? The Case of the Stockholm trial on congestion charging, Paper presented at 47th Annual Meeting of the Western Regional Science Association, Hawaii.

MacCracken, M. C. (2002). National assessment of the consequences of climate variability and change for the United States. In The Potential Impacts of Climate Change on Transportation Workshop, October 1–2, 2002 Center for Climate Change and Environmental Forecasting, US Department of Transportation, Washington, DC, USA, pp. 65–76.

Manheim, M. L. (1979). Fundamentals of Transportation Systems Analysis, Volume 1, Basic Concepts, Cambridge, MA, USA: MIT Press.

Mehrotra, S. and Y. J. Kang (2009). Impacts of Climate Change on Transportation. New York, USA: Columbia University.

Mehrotra, S., C. E. Natenzon, A. Omojola, et al. (2009). Framework for City Climate Risk Assessment, Commissioned research, World Bank Fifth Urban Research Symposium, Marseille.

Menon, A. (2000). ERP in Singapore: A perspective one year on. Traffic Engineering and Control, 41, 40–45.

Meyer, M. D. (2008). Design standards for U.S. transportation infrastructure: the implications of climate change. In The Potential Impacts of Climate Change on U.S. Transportation, National Research Council, National Academy of Science, Transportation Research Board and Department of Earth and Life Sciences, Washington, DC, USA.

Milan Municipality (2009), in Milan Today Newsletter, www.apcom.net/milano_today/latestnewsmilan.

Mills, B. and J. Andrey (2002). Climate change and transportation: potential interactions and impacts. In Potential Impacts of Climate Change on Transportation, Proceedings of the U.S. Department of Transportation Research Workshop, Washington, DC, USA, October 1–2.

Ministry of Environment Government of Japan (2008). Teikogaisha Guidebook 2008 [Less-emission Vehicle Guidebook 2008].

MTA [Metropolitan Transportation Authority] (2007). Report to the NYS Governor and NYC Mayor, Appendix 2: Climate Component, 8.8.07 MTA Task Force Report, prepared by C. Rosenzweig, R. Horton, D. C. Major, V. Gornitz, and K. Jacob, Columbia Center for Climate Systems Research. Accessed August 25, 2009, http://www.mta.info/mta/pdf/storm_report_2007.pdf.

MTA [Metropolitan Transportation Authority] (2009). Greening Mass Transit & Metro Regions: The Final Report of the Blue Ribbon Commission on Sustainability and the MTA, New York, NY, USA. Accessed August 25, 2009, http://mta.info/environment/pdf/SustRpt Final.pdf.

Nakanishi, H., Black, J. and Doust, K. (2009). Metrics of Environmental Sustainability, Social Equity and Economic Efficiency. IHDP 7th Open Meeting, 7th International Science Conference on the Human Dimensions of Global Environmental Change, 26-30 April 2009, Bonn, Germany.

Newman, P. and J. Kenworthy (1989). Cities and Automobile Dependence: An International Sourcebook, London, UK: Gower.

Nicholls, R. J., S. Hanson, C. Herweijer, et al. (2008). Ranking Port Cities with High Exposure and Vulnerability to Climate Extremes: Exposure Estimates, OECD Environment Working Papers, No. 1, OECD. Accessed August 25, 2009, http://lysander.sourceoecd.org/vl=8846437/cl=26/nw=1/rpsv/workingpapers/19970900/wp_5kzssgshj742_long_en.htm.

NPCC (2009). Climate Risk Information, New York, NY: City of New York, February 17. Accessed August 25, 2009, http://www.nyc.gov/html/planyc2030/downloads/pdf/nyc_climate_change_report.pdf.

NPCC (2010). Climate change adaptation in New York City: building a risk management response. Annals of the New York Academy 1183, in press. Accessed December 11, 2009, www.nyas.org/Publications/Annals/Detail.aspx?cid=ab9d0f9f-1cb1-4f21-b0c8-7607daa5dfcc.

OFCM (2002). Weather Information for Surface Transportation: National Needs Assessment Report. Office of the Federal Coordinator for Meteorological Services and Supporting Research. Accessed August 28, 2009, www.ofcm.gov/wist_report/wist-report.htm.

Olszewski, P. (2007). Singapore motorisation restraint and its implications on travel behaviour and urban sustainability. *Transportation*, **34**, 319–335.

Olszewski, P. and Xie, L. (2005). Modelling the effects of road pricing on traffic in Singapore. *Transportation Research Part A: Policy and Practice*, **39**, 755–772.

Perera, R. and Permana, A. (2009). The nexus of land use, transport, energy and environment as a workable policy arena to synergize co-benefits: evidence from Bandung City, Indonesia. In E. Zusman, A. Srinivasan, and S. Dhakal (Eds.), *Low Carbon Transport in Asia: Strategies for Optimizing Co-benefits*, London, UK: Earthscan.

Permana, A. S., R. Perera, and S. Kumar (2008). Understanding energy consumption pattern of households in different urban development forms: a comparative study in Bandung City, Indonesia, *Energy Policy*, **36**, 4287–4297.

Perner, A. and M. Thöne (2005). *Naturschutz im Finanzausgleich; Erweiterung des naturschutzpolitischen Instrumentariums um finanzielle Anreize für Gebietskörperschaften*, FiFo-Reports, No.3, University of Köln.

Petroski, H. (1982) *To Engineer is Human*. New York, NY, USA: Vintage Books, a division of Random House, Inc. [1992 edition].

PlaNYC (2007). *A Greener, Greater, NY.* Transportation. Accessed August 25, 2009, www.nyc.gov/html/planyc2030/downloads/pdf/report_transportation.pdf.

Potter, J. R. and M. J. Savonis (2003). Transportation in an age of climate change. *TR News*, **227**, July–August 2003, 26–31.

Price, L., S. de la rue du Canm, J. Sinton, *et al.* (2006). *Sectoral Trends in Global Energy Use and Greenhouse Gas Emissions*, LBNL-561444, Ernets Orlando Berkeley National Laboratory, Environmental Energy Technologies Division, Berkeley, CA, USA.

Pridmore, B. (2002). *The Role of Hydrogen in Powering Road Transport*, Tyndall Centre for Climate Research, Working Paper.

Prud'homme, R. and J. Bocarejo (2005). The London Congestion Charge: A Tentative Economic Appraisal. *Transport Policy*, **12**, 279–288.

Quinn, F. H. (2002). The potential impacts of climate change on Great Lakes transportation. In *The Potential Impacts of Climate Change on Transportation Workshop*, October 1–2, 2002. Center for Climate Change and Environmental Forecasting, US Department of Transportation, Washington, DC, USA, pp.115–124.

Revi, A. (2008). Climate change risk: an adaptation and mitigation agenda for Indian cities. *Environment and Urbanization*, **20**(1), 207–229.

Rogers, P. and S. Srinivasan (2008). Comparing sustainable cities: examples from China, India and the USA. In M. Keiner (Ed)., *Sustainable Urban Development in China: Wishful Thinking or Reality?*, Munster, Germany: Verlagshaus Monsenstein und Vannerdat OHG.

Rosenzweig, C. and W. D. Solecki (Eds.) (2001). *Climate Change and a Global City: the Potential Consequences of Climate Variability and Change – Metro East Coast*. Report for the US Global Change Research Program, National Assessment of the Potential Consequences of Climate Variability and Change for the United States, Columbia Earth Institute, New York. Accessed August 25, 2009, http://ccsr.columbia.edu/cig/mec/index.html.

Rossetti, M. A. (2002). Potential impacts of climate change on railroads. In *The Potential Impacts of Climate Change on Transportation Workshop*, October 1–2, 2002. Center for Climate Change and Environmental Forecasting, US Department of Transportation, Washington, DC, USA, pp209–221.

Santos, G. and G. Fraser (2005). *Road Pricing: Lessons from London*, Paper for the October 2005 Panel Meeting of Economic Policy, London.

Schipper, L., C. Marie-Lilliu, and R. Gorham (2000). *Flexing the Link Between Transport and Greenhouse Gas Emissions: A Path for the World Bank*. International Energy Agency.

Shukla, P. R., S. K. Sharma, N. H. Ravindranath, A. Garg, and S. Bhattacharya (Eds.) (2003).*Climate Change and India. Vulnerability Assessment and Adaptation*. Hyderabad: Universities Press.

Siemens (2008). *Sustainable Urban Infrastructure, London Edition: A View to 2025*. w1.siemens.com/entry/cc/features/urbanization…/study_london_en.pdf.

Sivak, M. (2009). Potential energy demand for cooling in the 50 largest metropolitan areas of the world: implications for developing countries. *Energy Policy*, **37**(4), 1382–1384.

Smith, N., J. Garrett, and V. Vardhan (2007). *Food and Nutrition in Hyderabad, Current Knowledge and Priorities for Action in an Urban Setting. Research Report 1, Sustainable Hyderabad Project.* www.nexusberlin.com/megacity/uploads/documents/Microsoft+Word+Megacity+Report-1.pdf.

Soo Hoo, W., M. Sumitani, and S. Cohen (2005). *Climate Change Will Impact the Seattle Department of Transportation*, Office of City Auditor, Seattle.

Stone Jr., B., A. C. Mednick, T. Holloway, and S. N. Spak (2007). Is compact growth good for air quality? *Journal of the American Planning Association*, **73**(4), 404–418.

Suarez, P., W. Anderson, M. Vijay, and Lakshmanan T. R. (2005). Impacts of flooding and climate change on urban transportation: a systemwide performance assessment of the Boston Metro Area. *Transportation Research Part D*, **10**(3), 231–244.

Titus, J. (2002). Does sea-level rise matter to transportation along the Atlantic coast? In *The Potential Impacts of Climate Change on Transportation: Workshop Summary*, US Department of Transportation, Workshop, 1–2 October. Accessed August 28, 2009, http://climate.volpe.dot.gov/workshop1002/.

Transport for London (2004). *Congestion Charging Central London: Impacts Monitoring: Second Annual Report*, London, UK.

Transportation Research Board (2008). *Potential Impacts of Climate Change on U.S. Transportation*, TRB Special Report 290, Washington, DC, USA: National Academy of Sciences. Accessed August 25, 2009, http://onlinepubs.trb.org/onlinepubs/sr/sr290.pdf.

Trilling, D. R. (2002). Notes on transportation into the year 2025. In *The Potential Impacts of Climate Change on Transportation Workshop*, October 1–2, 2002, Center for Climate Change and Environmental Forecasting, US Department of Transportation, Washington, DC, USA, pp. 65–76.

Tufts University (2004). *Climate's Long-Term Impacts on Metro Boston: Final Report*. Environmental Protection Agency, EPA Grant Number: R.827450–01. Accessed August 28, 2009, www.tufts.edu/tie/climb/.

UNFCCC [United Nations Framework Convention on Climate Change] (2006). Impacts, vulnerability and adaptation to climate change in Africa. Background paper for *African Workshop on Adaptation Implementation of Decision 1/CP.10 of the UNFCCC Convention*, Accra, Ghana, 21–23 September.

UNFPA (2007). *State of World Population 2007: Unleashing the Potential of Urban Growth*. Accessed August 25, 2009, http://www.unfpa.org/webdav/site/global/shared/documents/publications/2007/695_filename_sowp2007_eng.pdf.

UN-HABITAT (2008). *State of the World's Cities 2008/2009: Harmonious Cities*, United Nations Human Settlements Programme, Nairobi, Kenya.

USACE [United States Army Corps of Engineers] (1995). *Metro NY Hurricane Transportation Study*, Interim Technical Data Report, New York: Federal Emergency Management Agency, National Weather Service, NY/NJ/CT State Emergency Management.

USDOT (United States Department of Transportation), NHTSA (National Highway Traffic Safety Administration) (2010) "CAFE Standards - Frequently Asked Questions," Accessed Nov. 28, 2010, http://www.nhtsa.gov/cars/rules/cafe/overview.htm

Vairavamoorthy, K., S. D. Gorantiwar, and A. Pathirana (2008). Managing urban water supplies in developing countries: climate change and water scarcity scenarios. *Physics and Chemistry of the Earth*, **33**, 330–339.

van Rooijen, D. J., H. Turral, T. W. Biggs (2005). Sponge city: water balance, mega-city water use and wastewater use in Hyderabad, India. *Irrigation and Drainage*, **54**, 581–591.

Warren Centre for Advanced Engineering (2003). *The Sustainable Transport in Sustainable Cities Project*, University of Sydney, CD-ROM.

WBCSD (World Business Council for Sustainable Development) (2004). *Mobility 2030: Meeting the Challenges to Sustainability*, The

Sustainable Mobility Project, World Business Council for Sustainable Development. Accessed Nov. 28, 2010, http://www.wbcsd.org/web/publications/mobility/mobility-full.pdf.

Wegener, M. (2009). Possible future transport and land-use strategies for sustainable urban development in European cities. *Proceedings of the CIB-W 101 & GCOE Workshop Urban Infrastructure and Land-use Control*, University of Tokyo, 13 June 2009. Accessed September 2009, www.spiekermann-wegener.de/pub/pdf/MW_Tokyo_130609.pdf.

Willoughby, C. (2000). *Singapore's Experience in Managing Motorization and Its Relevance to Other Countries*, World Bank Discussion Paper TWU-43, Washington, DC, USA.

Wisner, B., P. Blaikie, T. Cannon, and I. Davis (2004). *At Risk: Natural Hazards, People's Vulnerability, and Disasters.* London, UK: Routledge.

Wooler, S. (2004). *The Changing Climate: Impact on the Department for Transport*, London, UK: Department for Transport.

Young, S., L. Balluz, J. Malilay (2004). Natural and technologic hazardous material releases during and after natural disasters: a review. *Science of the Total Environment*, **322**, 3–20. Zegras, C. (2007). As if Kyoto mattered: The clean development mechanism and transportation. *Energy Policy*, **35**.

Zimmerman, R. and M. Cusker (2001) Institutional decision-making. In C. Rosenzweig and W. D. Solecki (Eds.), *Climate Change and a Global City: The Potential Consequences of Climate Variability and Change – Metro East Coast*, New York, USA: Columbia Earth Institute and Goddard Institute of Space Studies, pp. 9–1 to 9–25 and A11–A17.

Zimmerman, R. and C. Faris (2010). Infrastructure impacts and adaptation challenges. In *Climate Change Adaptation in New York City: Building a Risk Management Response, New York City Panel on Climate Change 2010 Report*, edited by C. Rosenzweig and W. Solecki. Annals of the New York Academy of Sciences, **1196**, New York, NY, NY Academy of Sciences, 63–85.

7

Climate change and human health in cities

Coordinating Lead Authors:

Martha Barata (Rio de Janeiro), Eva Ligeti (Toronto)

Lead Authors:

Gregorio De Simone (Rio de Janeiro), Thea Dickinson (Toronto), Darby Jack (New York City), Jennifer Penney (Toronto), Mizanur Rahman (Dhaka), Rae Zimmerman (New York City)

This chapter should be cited as:

Barata, M., E. Ligeti, G. De Simone, T. Dickinson, D. Jack, J. Penney, M. Rahman, R. Zimmerman, 2011: Climate change and human health in cities. *Climate Change and Cities: First Assessment Report of the Urban Climate Change Research Network*, C. Rosenzweig, W. D. Solecki, S. A. Hammer, S. Mehrotra, Eds., Cambridge University Press, Cambridge, UK, 179–213.

7.1 Introduction

Current climate extremes and projections for future changes to climate have resulted in growing attention to the health effects of climate on all human populations, urban and rural (McMichael *et al.*, 2003; Confalonieri *et al.*, 2007; Costello *et al.*, 2009). Indeed, almost all the impacts of climate change have direct or indirect consequences for human health. However, for a number of reasons, city dwellers – especially those in low- and middle-income nations – are especially vulnerable to the health impacts of climate change.

In this chapter we present issues and case studies relevant to human health in cities under climate change conditions. In Section 7.2, we discuss the most relevant conclusions from the health chapter of the IPCC Fourth Assessment Report (Confalonieri *et al.*, 2007), looking at its application to cities. In Section 7.3, we present an overview of urban health outcomes and their climate-related drivers. In Section 7.4 the factors that can modify the impacts of climate change on human health in urban areas are discussed. Section 7.5 presents examples and case studies of adaptations that protect city residents from some of the health impacts and risks posed by climate change. A brief discussion of needed changes in energy, transportation, and other sectors to reduce emissions of harmful pollutants and provide co-benefits for human health is found in Section 7.6. The urgent need for better, more urban focused and targeted research is discussed in Section 7.7. Conclusions are presented in Section 7.8.

7.1.1 Why health and climate change for cities?

Urban populations are increasing in absolute numbers and relative to rural populations in every part of the world. According to the United Nations Population Fund (2007) the world is undergoing the largest wave of urban growth in history. In 1800 there were only 2 cities larger than a million inhabitants. By 1950 there were 75 cities of this size and by 2000 there were 380 "million-cities," half of these in Asia (Satterthwaite *et al.*, 2008). Cities have also grown larger. In 2000, the largest 100 cities had an average of 6.3 million inhabitants. In 2008, more than half of the world's population was living in cities and their surrounding areas. By 2030 this number is projected to reach almost 5 billion, with urban growth concentrated in Africa and Asia. While megacities are important a great deal of the new growth will occur in smaller towns and cities, which tend to have fewer resources (Bicknell and Dodman, 2009).

Because cities concentrate populations, extreme weather events such as intense precipitation, cyclonic storms, or storm surges affect a much larger number of people than when they strike less populated regions. Such damages may be exacerbated in the future, since extreme weather events are expected to increase in number and intensity under climate change (IPCC, 2007). Cities also concentrate poor populations who are especially vulnerable

to the effects of climate change because of the conditions in which they live. In low- and middle-income countries, poor slum dwellers can make up 50 to 60 percent of the urban population, living in precarious structures, often with little access to water, sanitation, electricity, health care, or emergency services (Huq *et al.*, 2007).

The concentration of populations in urban areas also tends to lengthen the supply lines for essentials such as water, food, and energy sources, and makes them more dependent on waste collection (and more susceptible when waste is not collected). Storms, floods, or droughts that disrupt these urban lifelines can have serious consequences for the health of city dwellers (McBean and Henstra, 2003).

Many cities are located in areas that are vulnerable to both existing and projected climate hazards. Most of the world's mega-cities were originally established on seacoasts or beside major rivers that enabled trade and commerce or territorial control (Huq *et al.*, 2007). Cities such as Venice (Italy) and Mumbai (India), located on low-elevation seacoasts, are particularly vulnerable to sea level rise and storm surges. Of 180 countries with populations in low-elevation coastal zones, about 70 percent have large urban areas extending into that zone (McGranahan *et al.*, 2007). As cities have grown, many have expanded from their original, secure locations onto river deltas and floodplains, marshlands, or up steep hillsides and into other areas that are poorly suited for human habitation and are vulnerable to weather extremes. In many cases the expansion of the city itself has created hazards by filling in water courses or cutting down adjacent forests, increasing the risk of floods and landslides.

The ways in which cities are constructed – reducing vegetation, covering large areas with impermeable surfaces, and obstructing natural drainage channels – make many city dwellers more vulnerable to heat waves, heavy precipitation, and other extreme weather events, which are already increasing, likely as a result of climate change.

Many cities, even in high-income countries, are exposed to multiple stresses not related to climate change. Such stresses include lack of financial resources to meet the needs and demands of a growing population; aging, poorly maintained, inadequate, or non-existent infrastructure; poor land use planning and enforcement; inadequate resources for disaster preparedness; self-serving political institutions or outright corruption that divert resources from pressing problems; and increasing income disparity and growing numbers of impoverished families living in unplanned, unserviced settlements and slums (Satterthwaite *et al.*, 2008). Severe weather events can combine with some or all of these stresses to create conditions of extreme hazard as demonstrated recently by:

- flash floods and landslides in Caracas, which killed 30,000 people in 1999 (Satterthwaite *et al.*, 2008)
- floods in Shanghai in 1998, which killed 3,000 and displaced 16 million in the Yangtze basin (de Sherbinin *et al.*, 2007)

- extreme heat in Paris in the summer of 2003, which contributed to an estimated 2,085 deaths out of a total of 15,000 for the whole of France (le Tertre *et al.*, 2006)
- monsoon floods in 2005 in Mumbai, which killed more than 1,000 (de Sherbinin *et al.*, 2007)
- intense rainfall combined with high-tide conditions that submerged more than 100 square kilometers of Dhaka in 1988, 1998, and 2004 (Alam and Rabbani, 2007)
- hurricane Katrina, which devastated New Orleans in 2005 and killed more than 1,800 people (Glantz, 2008).

In each of these instances, more than 1,000 people died and many more suffered serious health effects. Unless more action is taken to address the risk factors from current extreme weather events and from future climate change, events of this kind are very likely to become more common, and their health impacts will be even more severe.

7.1.2 Ways in which climate change will affect the health of city dwellers

Climate change is expected to exacerbate a number of existing threats to human health and well-being rather than to introduce new health effects (Costello *et al.*, 2009). Both direct and indirect impacts on human health are beginning to be observed under current climate conditions and are predicted to be amplified in the coming decades. The health of city dwellers is expected to be affected in the following ways:

- direct physical injuries and deaths from: extreme weather events such as tropical cyclones and other major storms with high winds; storm surges; intense rainfall that leads to flooding; or ice storms that damage trees and overhead structures and produce dangerous transport conditions
- illnesses resulting from the aftermath of extreme weather events that destroy housing, disrupt access to clean water and food and increase exposure to biological and chemical contaminants
- water-borne diseases following extended or intense periods of rainfall, ground saturation and floods and saline intrusion due to sea level rise
- food-borne diseases resulting from bacterial growth in foods exposed to higher temperatures
- illnesses and deaths from the expanded range of vector-borne infectious diseases
- respiratory illnesses due to worsening air quality related to changes in temperature and precipitation resulting in the formation of smog
- morbidity and mortality, especially among the elderly, small children, and people whose health is already compromised, as a result of stress from hotter and longer heat waves – which are aggravated by the urban heat island effect
- malnutrition and starvation among the urban (and rural) poor who have reduced access to food as a result of drought-induced shortages and price rises
- uprooting and migration of populations negatively affected by climate events to areas that are unable to provide the services they need.

7.1.3 The challenge for cities

As a primary climate change prevention measure, cities need to contribute more effectively to reducing their greenhouse gas emissions (Frumkin *et al.*, 2008) (See Chapter 4 on energy). While emission reductions will not prevent many of the damaging climate changes that are already underway and likely to continue in the next 50 to 100 years, it remains a critical task for high-income cities and countries, and for some middle-income cities and countries whose emissions are rapidly growing. Emission reductions will moderate the expected impacts of climate change over the long term. However, as Huq *et al.* (2007) and Satterthwaite *et al.* (2008) argue, low-income cities and their most vulnerable residents contribute relatively little to the worldwide complement of greenhouse gas emissions and thus the focus of their efforts will likely be more on identifying vulnerabilities and preparing adaptation strategies.

Because the world has failed so far to prevent the build-up of greenhouse gases in the atmosphere and thus to prevent health impacts that climate change is likely to inflict on city dwellers, a key task is to alter existing urban conditions that may combine with climate change and result in deaths, injuries, and illnesses in the population. In low- and middle-income cities and countries this requires a focus on meeting basic development needs – adequate housing; provision of infrastructure that supplies clean water, sanitation, and energy; education; and primary health-care services. Provision of these services will reduce vulnerability to many of the health impacts of climate change and increase the capacity of the most vulnerable to withstand some of the impacts that cannot be avoided. High-income cities already provide these services to most of their population, but often do this in a way that interacts with weather extremes to worsen impacts – by paving much of the urban landscape, for example, which exacerbates heat and flooding. Whether building new structures and services or modifying existing ones, cities need to alter the urban characteristics that worsen the impacts of extreme weather events and turn at least some of them into health and economic disasters. Anticipated climate change over the lifetime of the structures should also be taken into account. Otherwise such investments may be jeopardized by climate change.

In addition to long-term vulnerability reduction, cities need to undertake disaster preparedness planning, and develop early warning and emergency disaster relief systems (Few *et al.*, 2006). This will necessitate identifying, assessing, and monitoring disaster risks. The World Bank (2008) *Climate Resilient Cities* primer recommends preparing a disaster history and a city hazard profile map to identify areas vulnerable to natural hazards, an essential task for disaster planning. In addition to helping with longer-term adaptation planning, these actions will aid in the development of shorter-term plans to avoid disasters in areas that have proved to be prone to recurrent calamities. Improved urban planning and linking it to disaster risk reduction in climate change adaptation strategies will be necessary (Revi, 2008; GFDRR, 2009). Recovery and reconstruction will benefit from preparatory measures.

All these tasks are a challenge, though they are easier for higher-income cities, which are reasonably well-organized and resourced. A great many cities have committed to reducing emissions – and some have very ambitious targets. A smaller number of cities and urban regional governments have developed and are in the early stages of implementing plans to adapt to climate change (Penney, 2007; Bicknell and Dodman, 2009). The Cities of Stockholm and Toronto and the Greater London Authority have explicitly incorporated health concerns in their plans and activities (Ekelund, 2007; Toronto Environment Office, 2008; Mayor of London, 2008).

Cities in low-income countries face a much larger hurdle to prepare for climate change, for they need to develop basic structures and services to support adequate housing, water, sanitation, energy distribution, transportation, education, and health-care services and at the same time consider how to do this in a way that will increase adaptation and resilience for their residents and citizens. Some researchers express skepticism about the likelihood that municipal governments will act "to protect the populations within their jurisdiction from risks arising from climate change when they have shown so little inclination or ability to protect them from other environmental hazards" (Satterthwaite *et al.*, 2008, p. 2). They emphasize the need for support from national governments to provide the legal and institutional basis for reducing the risks from climate change and for support from the international community to provide the financial resources. They also discuss successful examples of engaging communities at risk in the process of determining how to adapt and, in some cases, where to move risky settlements (Huq *et al.*, 2007).

7.2 Health-related findings of the IPCC Fourth Assessment Report

Climate change and human health issues are addressed primarily in Chapter 8 of the IPCC's Working Group II report *Impacts, Adaptation and Vulnerability* (Confalonieri *et al.*, 2007). The authors of the chapter reviewed more than 500 scientific publications related to climate change and health in the preparation of their report. They expressed high confidence that climate change has already altered the seasonal distribution of some allergenic pollen species, and medium confidence that climate change has already modified the distribution of select infectious disease vectors, and that it has increased heat wave-related deaths. The group also expressed high confidence that in future climate change-related exposures would lead to: increased malnutrition and related development effects; increased numbers of deaths, diseases, and injuries from heat waves, floods, storms, fires, and droughts; and increased cardio-respiratory morbidity and mortality from greater ground-level ozone. They also suggested that diarrheal diseases would increase (medium confidence) and more people would be at risk of dengue (low confidence). There may be health benefits from reduced cold weather. The report concluded that adverse health impacts would be greatest in low-income countries but that adaptive capacity needs to be improved everywhere.

The Chapter 8 of the IPCC Fourth Assessment Report outlined three main mechanisms by which climate change may affect human health:
1. **Direct exposures to extreme climatic events.** These affect health through influences on human physiology (e.g., heat waves) or by provoking physical traumas caused by natural disasters such as storms and floods.
2. **Indirect effects from changes to the determinant factors of human health.** Relevant examples are the effects of the climate on the production of food, on the quality of the water and the air, and the ecology of vectors of infectious diseases (e.g., mosquitoes).
3. **Effects of climatic events on social welfare by disruption of social and economic systems.** For example, migration of populations dependent on subsistence farming to urban areas due to prolonged droughts, which can create a burden for resources and social safety mechanisms of the receiving communities.

There are a variety of modifying influences that can change exposures to extreme weather and climate change impacts or amplify or dampen their health effects. The IPCC chapter distinguished three kinds of modifying factors – environmental conditions, social conditions, and health system conditions (Confalonieri *et al.*, 2007). Environmental conditions include such attributes as location, available water resources, and status of ecological systems. Social conditions include income, knowledge, and the capacity to plan, responsiveness of authorities, and organization of vulnerable populations. Health system conditions refer to the underlying health status of the population (although this could also be characterized as a social condition), and the availability of primary health care and public health and emergency services.

In Figure 7.1 below, adapted from Confalonieri *et al.* (2007), we have added two additional modifying factors that are important for cities. These are density and demographics (see Section 7.4.4), and local infrastructure (see Section 7.4.2). The density of cities can increase vulnerability to the local manifestations of extreme weather and climate change in several ways: reducing green space and mature trees that have a cooling effect on ambient air, for example; creating barriers for natural air and water flow; and producing large quantities of biological waste that can contaminate floodwaters. The location, quality, and upkeep of local infrastructure – water supply, sanitation and drainage, roads and related structures such as culverts and bridges – can all play an important role in modifying health and other impacts of climate change. Ultimately, the level of vulnerability of these modulating systems will be important in the mitigation of adverse impacts, and should be considered when developing adaptation strategies for urban centers.

The IPCC chapter (Confalonieri *et al.*, 2007) described general health risks expected from climate change related to heat and cold; wind, storms, and floods; drought, nutrition, and food

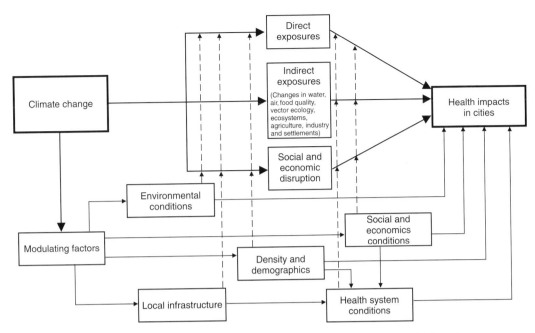

Figure 7.1: *Three main mechanisms of how climatic processes may affect health in cities and its modulating factors.*

Source: Adapted from Confalonieri et al., (2007).

security; food safety; water; air quality; aeroallergens; and vector-borne, rodent-borne, and other infectious diseases. For the most part the IPCC's health chapter did not address specific urban health issues, though it did mention some urban examples. Since the Fourth Assessment was written, a growing number of authors have explored the impacts of climate change on cities, and more specifically the effects of climate change on the health of urban populations (e.g., Campbell-Lendrum and Corvalán, 2007; de Sherbinin *et al.*, 2007; Huq *et al.*, 2007; Satterthwaite *et al.*, 2008; Gosling *et al.*, 2009; Kovats and Akhtar, 2009; Romero Lankao and Tribbia, 2009). This chapter uses the broad typology of the IPCC chapter, but focuses on the health challenges that climate change appears likely to pose for urban centers.

7.3 Urban health outcomes and their drivers

Cities and city dwellers face many of the same climate-related health issues as the rest of the world's population, but because of the location of many cities, concentration of their populations, the density and impermeability of hard surfaces, the urban heat island effect, increased pressure on local ecosystems for water, concentration of wastes, and the existence of large areas of un- or underserviced urban slums, cities have particular vulnerabilities that require assessment and redress.

As the following sections make clear, we expect that climate change will, on balance, increase the overall burden of illness and these deleterious effects are likely to be substantially magnified in urban areas, primarily due to high population densities.

There will be considerable local variation in the health impacts of climate change among cities in different regions, and between cities in the same region. The changes in temperature, precipitation, storminess, and other characteristics of climate change will vary significantly among regions with diverse geographical characteristics, weather patterns, and demographics. But the effects of climate change will also depend on the social, economic, and institutional conditions of each urban center that will determine their initial level of resilience (WHO, 2009; see Section 7.4).

Table 7.1 provides a summary of the drivers and health outcomes linked to climate change. In the discussion of each driver, this chapter looks specifically at how these health concerns might play out specifically in cities.

7.3.1 Heat and cold (temperature extremes)

Mean temperatures are increasing around the world, and the frequency and intensity of heat waves also appear to be on the increase (Della-Marta *et al.*, 2007). Climate models predict increased frequency and intensity of heat waves for the future as well (Meehl and Tebaldi, 2004; IPCC, 2007). This poses direct threats to health through heat stress, especially for the elderly, young children, and those with pre-existing health problems. Heat waves are exacerbated in urban environments by the urban heat island effect, caused by the concentration of concrete and asphalt surfaces, reduction of vegetation, and anthropogenic heat sources. Urban temperatures can be as much as 8 °C higher than the surrounding countryside, though typically the increase is more in the order of 3–4 degrees (Oke, 1997). Figure 7.2 provides an illustration of an urban heat profile compared to surrounding suburban and rural areas.

Table 7.1: *Climate change-related drivers and outcomes for urban health.*

Drivers	Health endpoint	Impact on cities	Degree of uncertainty	Key references
Temperature extremes	Mortality via heat exhaustion. Elderly most vulnerable	Pronounced (via urban heat islands)	Low (extensive evidence exists)	Kovats and Hajat, 2008; Bell *et al.*, 2008
Wind, storms, and floods	Mortality via physical trauma, drowning	Pronounced (population density in vulnerable areas)	Low (extensive evidence exists)	Guha-Sapir *et al.*, 2004; Ahern *et al.*, 2005
Drought	Malnutrition	Ambiguous	Medium (adaptive capacity is poorly understood)	Confalonieri *et al.*, 2007
Water quality	Diarrheal disease	Pronounced (flush contaminants, may overwhelm city water systems)	High (limited research)	Hunter, 2003; Kistemann *et al.*, 2002
Air quality	Respiratory illness	Pronounced (population density, pollution sources)	High (limited research, likely to vary greatly across cities)	Kinney, 2008; Holloway *et al.*, 2008; Steiner *et al.*, 2006
Aeroallergens	Allergies, asthma	Ambiguous	High (limited research, likely to vary greatly across cities)	USEPA, 2008
Vector-borne diseases	Malaria, dengue, others	Diminished – few vectors thrive in urban environments	High (scholarship on urban effects)	Lindsay and Birley, 1996

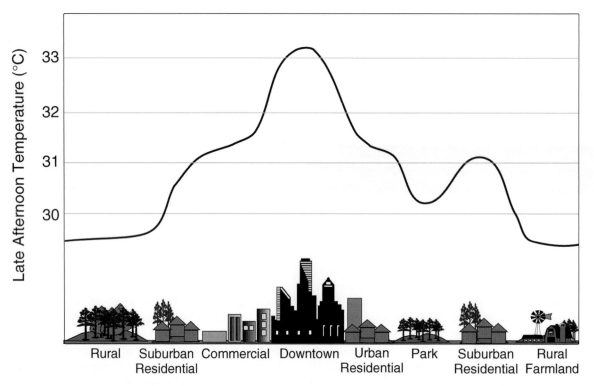

Figure 7.2: *Urban heat island profile.*

Source: Natural Resources Canada, Climate Change Impacts and Adaptation Division. Accessed at: http://adaptation.nrcan.gc.ca/perspective/health_3_e.php.

Increasing heat also contributes to smog formation and worsening air quality, both of which tend to be more problematic in urban than rural environments. In urban environments, the effects of heat on air pollution may have a much larger effect on mortality and morbidity than the direct effects of heat stress (Pengelly et al., 2007).

The empirical literature shows that extreme heat events are associated with transient, but potentially substantial, increases in daily mortality (Kovats and Hajat, 2008). The 2003 European heat wave, which is estimated to have killed 70,000 people (Robine et al., 2007), illustrates the deadly potential of temperature extremes. In Paris, where maximum temperatures exceeded 38 °C for 6 days, the mortality rate tripled for at least 3 days (Kalkstein et al., 2008). New York, Chicago, Shanghai, and other cities have also experienced higher death rates during heat waves when temperature values exceed about 30 °C and nighttime values exceed 25 °C (Davis et al., 2003; Laaidi et al., 2006; Tan et al., 2007). The effects of heat waves are sometimes magnified by other forms of environmental stress. For example, heat in combination with increased aerosols can produce pulmonary and cardiovascular diseases (CCSP, 2008).

Almost all epidemiological evidence linking heat waves to mortality draws on urban data. This is borne out in research from high-income countries (e.g., Baccini et al., 2008) and for the limited data available from the developing world (e.g., Bell et al., 2008). While the primary reason for this urban emphasis appears to be data availability, evidence suggests that the urban heat island effect renders city dwellers without access to cooling particularly vulnerable to heat waves (Kunkel et al., 1996).

The ability to combat heat stress varies over different segments of the population. The elderly, for example, have more difficulty acclimatizing to heat, especially when they are first exposed (Zimmerman et al., 2007). Demographic changes may lead to urban populations that are even more vulnerable in the future, as the elderly population grows in absolute and relative numbers. In developed countries, 20 percent of the population in 2005 was aged 60 years or over, and by 2050 that proportion was projected to grow to 32 percent (UN, 2005). The elderly population in developed countries was greater than the number of children (under 14 years of age). The UN projects that by 2050 there will be two elderly persons for every child, and that in the developing world, the proportion of the population aged 60 or over will rise from 8 percent in 2005 to close to 20 percent by 2050 (UN, 2005).

Extreme cold temperatures also affect mortality and morbidity in those latitudes that experience winter. In most cities in Canada, the United States, and parts of Eurasia, there are higher death rates in the cold season (Davis et al., 2003). The increase in global mean temperatures worldwide is expected to reduce cold-related mortality and morbidity, though expected decreases in cold-related morbidity have not yet been observed in Quebec (Gosselin et al., 2008). There is no clear distinction between the impacts of cold on urban versus rural residents.

7.3.2 Wind, storms, and floods

Climate change is likely to bring increases in the frequency and intensity of heavy precipitation events and in the intensity of tropical cyclonic storms with larger peak wind speeds and heavier precipitation due to warmer sea surface temperatures. The frequency of tropical cyclonic storms is projected to stay constant or decline, although the IPCC is not confident in this conclusion, noting, "the apparent increase in the proportion of very intense storms since 1970 in some regions is much larger than simulated by current models for that period" (IPCC, 2007 Summary for Policymakers). Extratropical storm tracks are expected to expand poleward (IPCC, 2007).

A recent series of articles in *Environment and Urbanization* investigated existing and future threats of storms to Cotonou (Benin), Dhaka, Mumbai, Shanghai, and other cities, concluding, "it is within urban centers and urban governments that so much of the battle to prevent climate change from becoming a global catastrophe will be won or lost" (Huq et al., 2007, p. 14). Storms surges and resultant floods pose particular threats to coastal areas, where most cities are located.

These extreme events create direct and indirect effects on health. Direct effects originate from the physical force of flowing floodwater, for example, causing deaths due to drowning. The physical impact of heavy rains and extreme winds due to climate change can cause injuries and deaths from collapsing structures or flying debris, electrocution from damaged power lines, vehicle accidents, and other mechanisms (Greenough et al., 2001). Indirect effects such as contaminated water, compromised infrastructure, and slope instability may create risk of injury, and increase the chance of infectious disease (Ahern et al., 2005; Confalonieri et al., 2007).

Epidemiological evidence on the aggregate health impacts of storms and floods on cities is scarce, largely because a disproportionate share of the burden falls on developing countries due to vulnerability related to poor planning and prevention preparedness (Guha-Sapir and Below, 2002). Anecdotally, however, the impacts of floods and storms can be spectacularly large – for example, a two-week rainfall in 1999 resulted in extensive flooding in Caracas, Venezuela, killing an estimated 30,000 people (Jonkman, 2005). In July 2005, severe flooding in Maharashtra and Mumbai, India was caused by extremely heavy rainfall, up to 100 cm in one day; the flood killed at least 1,000 people and caused approximately US$3.5 billion in damage (meteogreece, 2005). In August 2005, when the federal flood protection system in New Orleans failed during Hurricane Katrina, 80 percent of the city was flooded, leading to at least 1,836 deaths and resulting in economic damage of greater than US$100 billion (Grossi and Muir-Wood, 2006). Globally, flooding has been estimated to have caused just over 200,000 deaths in the 30-year period ending in 2003 (Guha-Sapir et al., 2004).

7.3.3 Drought

According to the IPCC Fourth Assessment Report (2007), rising greenhouse gas concentrations are likely to increase the risk of drought in many regions. An increase in mean temperatures of 2–3 °C could result in a decline in yields from rain-fed agriculture by as much as 50 percent in North Africa and other developing country regions (IPCC, 2007). In contrast to the other impacts discussed in this section, droughts usually exert indirect pressure on the health of city dwellers. The primary causal pathway linking drought to urban health is availability of food. A decrease in regional food production reduces availability and raises prices in cities to the degree that the poor are unable to buy sufficient food. (Drought also results in water shortages; this is discussed in Section 7.3.4 below.) This reduces food consumption among the poor, and can lead to malnutrition and even to starvation in extreme cases. The UN's Food and Agriculture Organization recently estimated that at this time more than 1 billion people in the world – one in six – are undernourished because of high food prices, low incomes, and unemployment caused by the global economic downturn around the world (FAO, 2009).

A recent analysis of the impact of extreme weather (heat waves, droughts, and heavy rains) on food production in 16 low- and middle-income countries, concluded that the urban poor would be most strongly affected by reduced supply and increased price of basic grains (Ahmed et al., 2009). The urban poor in Bangladesh, Mexico, and Zambia were particularly vulnerable, according to the study.

Drought may also be an important factor in rural-to-urban migration in some parts of the world (Barrios et al., 2006), which has its own set of public health ramifications. The resulting overcrowding, lack of potable water, food, and housing can increase the spread of infectious disease, for example meningitis and human immunodeficiency virus (HIV) (Molesworth et al., 2003; del Ninno and Lundberg, 2005). Drought also affects water quality and quantity, increasing microbial and chemical load.

7.3.4 Water quantity and quality

Over and above the acute effects of flooding, climate change may reduce access to safe drinking water in cities – some of which already face shortages due to poor water distribution infrastructure and high levels of consumption. First, reductions in rainfall in some parts of the world may reduce seasonal stream flows and diminish available surface water. Surface temperatures tend to fluctuate more rapidly with reduced volumes of water. Also, rising temperatures will increase demand for water and for the services of the urban water supply system (Volschan, 2008) and the extraction of greater quantities of water from superficial and subterranean springs (Ashley and Cashman, 2006).

Rising temperatures can also threaten water systems that capture, store, and transfer water to cities. Increased evaporation rates can reduce surface water levels and fresh water availability. Cities such as Delhi, which already transports water as much as 300 kilometers to supply 15 million residents, may have to extend their supply lines, coming into competition with other users and with ecological systems (Costello et al., 2009).

Reduced surface water levels, in turn, can alter the exchange rate between groundwater and surface water, reducing the availability of fresh water supplies (Bates et al., 2008). Growing cities that turn to groundwater for their water supply may deplete nearby aquifers and create problems with subsidence, as is the case in Mexico City (Bates et al., 2008).

Cities such as Seattle, dependent on snow-melt for their water supply, are already seeing reductions in water availability (Adam et al., 2009). Some cities and towns that depend on glacial-fed water catchments are currently facing the opposite problem as glaciers melt more quickly in higher global temperatures, creating the threat of inundation (Mool et al., 2001). Over the longer term, however, these areas may face water scarcity as glaciers disappear.

At the beginning of this century, 21 percent of people in low-income countries had poor access to clean water and more than 50 percent lacked access to sanitation (Costello et al., 2009). Reduced water supplies will exacerbate this problem, particularly in informal urban settlements and slums, leading to diarrheal and other diseases.

In urban areas close to seacoasts, saline intrusion caused by sea level rise and extreme events is another factor that can affect water and food supply quality. Also, extreme rainfall events increase runoff, potentially flushing contaminants and sediments into drinking water sources (Kistemann et al., 2002). Each of these processes has the potential to affect urban dwellers, though the Walkerton, Ontario, E. coli and Campylobacter outbreak in 2000, which resulted in seven deaths and more than 2,300 cases of illness, shows that contamination of water sources from heavy rainfall and rural runoff can affect the residents of smaller towns and rural areas as well (O'Connor, 2002). City water systems that currently provide good quality water may fail under increased pathogen loading, and urban residents who rely on untreated surface water may fall ill even more frequently.

7.3.5 Air quality

Weather plays a central role in determining air pollution levels. Air flow patterns, which are likely to change as the climate warms, govern the dispersion of pollutants. High temperatures tend to accelerate secondary chemical reactions in the atmosphere, particularly those that lead to ground-level ozone. Spatial variations in emissions and air flow patterns mean that the local effects of anthropogenic changes in weather patterns defy generalization. Recent modeling has, however, illuminated some likely patterns in North America.

Knowlton *et al.* (2004) use an integrated modeling framework that combines climate and air-quality models to explore ozone concentrations in the New York metropolitan region under a range of climate scenarios. Their analysis suggests that climate change will engender a 4.5 percent increase in ozone-related mortality in the 2050s compared with the 1990s. More recent efforts to wed downscaled climate models to local atmospheric chemistry models show that air quality in Chicago, San Francisco, and several large Canadian cities is likely to suffer from increased ozone concentrations under plausible climate scenarios (Steiner *et al.*, 2006; Holloway *et al.*, 2008). In a national level analysis, Bell *et al.* (2007) developed climate scenarios and ozone models for 50 US cities. Their results show substantial variation across cities, although most project substantial increases in ozone concentrations under climate change. Cheng *et al.* (2009) conducted a similar analysis for south-central Canada, estimating that air pollution-related mortality could increase 20–30 percent by the 2050s and 30–45 percent by the 2080s, due largely to increases in ozone effects. The air pollution-related effects of temperature increases are expected to contribute much more to mortality than heat effects alone.

Climate change may also increase the frequency and severity of forest fires (Peterson and McKenzie, 2008), which can affect air quality over thousands of square kilometers.

7.3.6 Aeroallergens

Climate change will likely alter the onset of the spring pollen season in temperate zones, and some evidence suggests that the duration of the pollen season in some countries may lengthen, increasing allergic reactions (USEPA, 2008). These effects are due both to warming and to direct fertilization of plants by higher atmospheric concentrations of CO_2, and may be amplified by urban heat islands and urban CO_2 (Ziska *et al.*, 2003). Currently, the linkage between climate, aeroallergens, and health is not well understood however, and it is not known to what extent the effects may vary between cities and rural areas.

7.3.7 Vector-borne, food-borne, and water-borne diseases

According to the recent Lancet Commission on Managing the Health Effects of Climate Change, rising temperatures are expected to affect the spread and transmission of diseases carried by vectors such as mosquitoes, ticks, and mice (Costello *et al.*, 2009). Temperature affects the rates of pathogen maturation and replication in mosquitoes and insect density (Costello *et al.*, 2009). Warmer temperatures may contribute to the increased incidence of mosquito-borne diseases such as malaria, dengue, yellow fever, and West Nile. Schistosomiasis, leishmaniasis, Lyme disease, tick-borne encephalitis, hantavirus infections, and a number of other vector-borne diseases are also projected to increase as a result of climate change.

Malaria, dengue, West Nile and a number of other vector-borne illnesses are currently in a period of rapid expansion. More than 50 percent of the world's population is exposed to malaria, which is estimated to cause 300–500 million acute illnesses each year and 1.1–2.7 million deaths, of which 90 percent are estimated to occur in Africa (UN Millenium Project, 2005), and the disease is reappearing elsewhere (Reiter, 2008). There are anecdotal reports of malaria increases in some Indian cities, attributed to breeding pools of stagnant water at construction sites and to water in "tap pits" below ground level, which have been dug during times of water shortage (Nagaraj, 2009). Malaria has also surged in some South American cities (Gubler, 1998). Malaria has recently been described as "probably the most climate-sensitive vector-borne disease" (Githeko, 2007); its incidence is affected not only by temperature but by precipitation, humidity, and wind. The attribution of the recent malaria resurgence to climate change is contested by some authorities, however, who argue that other factors such as reductions in public health expenditure are more important than increased temperature in the spread of these diseases (Reiter *et al.*, 2004).

Forty million cases of dengue and several hundred thousand of dengue hemorrhagic fever occur each year (Jelinek, 2009). High levels of dengue exist in Asia. The disease is expanding rapidly in Latin America and recently appeared in North America and northern Australia. The surge in cases of dengue in regions previously considered atypical, such as the south of the United States near the Mexican border, may be related to changes in spatial distribution of the insect vector due to climate change (Barclay, 2008). Dengue is described by researchers as an "urban disease" spread mainly by the *Aedes aegypti* mosquito, which breeds in still water in containers (Lapitan *et al.*, 2009). The disease spreads in poor, densely populated areas of cities with inadequate water supply. Beebe *et al.* (2009) attribute the resurgence of dengue in Australia to the installation of large numbers of domestic water tanks as a response to recent droughts.

The first West Nile virus outbreak occurred in North America in 1999, likely as a result of an import of the vector by international air transport. Since then, the disease has spread across the United States and most of Canada, transmitted by mosquitoes that acquire the virus from infected birds. Epidemics in 2002 and 2004 were linked to locations with drought or above-average temperatures, and a more virulent mutated strain that "responds strongly to higher temperatures," suggesting that risk will increase with frequent heat waves (Karl *et al.*, 2009). The West Nile vector *Culex pipiens* (which also carries St. Louis encephalitis), is said to "thrive in city storm drains and catch basins, especially in the organically rich water that forms during drought" (Epstein and Defilippo, 2001.

It is hard to generalize about the extent of the risk that expansion of most mosquito-borne diseases will pose for urban areas, or about the extent to which climate change will contribute to this problem overall. Cities do not generally provide the best environments for reproduction of mosquitoes, though standing water in plant pots or urban detritus such as car tires can provide breeding places for these vectors (Hay *et al.*, 2002). Temperature, precipitation levels, humidity, and wind speeds, as well as the level of public health services all affect the proliferation of vectors and the spread of the diseases of concern.

The risk of food-borne illnesses is also expected to increase under climate change. *Campylobacter* is the most commonly reported gastrointestinal disease in Europe, with contaminated poultry a major source of the infection. Rapidly rising temperatures increase the colonization of broiler chicken flocks with *Campylobacter* (Kovats *et al.*, 2005). In a study of *Salmonella* infections in 10 countries in Europe, Kovats *et al.* (2004) found that illnesses were linearly associated with temperatures above 6 °C and that temperature influenced about 35 percent of all cases in six of these countries. An Australian study also found a positive association between mean temperatures in the previous month and the number of reported cases of salmonellosis in five cities (D'Souza *et al.*, 2004).

The prevalence of food-borne illnesses is undoubtedly higher in tropical low-income countries than in Europe or Australia, but the absence of public health reporting in most low-income countries makes it difficult to estimate the impact of temperature increases. The World Health Organization has undertaken a recent study to better estimate the current incidence of food-borne diseases in developing countries, but the report is not expected until 2010. It is not known whether there is a specific pattern of food-borne illness associated with urban versus rural residency, though higher temperatures in urban environments might be expected to influence the number and extent of outbreaks in cities.

Water-borne illnesses are strongly associated with heavy rainfall. Charron *et al.* (2004) chronicled a host of *E. coli*, *Campylobacter*, *Cryptosporidium*, *Toxoplasmosis*, *Giardia*, *Leptospirosis*, and non-specific gastroenteritis outbreaks in North America since 1993 that have been associated with a sustained period of rainfall or an extreme rainfall event, which contaminated water and in some cases overwhelmed water-treatment systems.

7.4 Climate change and urban health outcomes: modifying influences

As indicated in Figure 7.1, there are a number of environmental, economic, and social factors that can influence health vulnerabilities and modify the health impacts of climate change in cities around the world. These modifying influences are discussed below.

7.4.1 Local geography and environmental conditions

The environmental conditions that amplify or reduce climate-related health effects are strongly linked to local geography. A number of studies that explore the health impacts of climate change on cities, including the direct effects of heat (Luber and McGeehin, 2008); vector-borne diseases (Gage *et al.*, 2008); water-borne diseases (Patz *et al.*, 2008); and air quality (Kinney, 2008) describe the way these climate-linked health effects vary by location, emphasizing the importance of understanding spatial scales and geographic influences on health.

Some of the factors that may amplify or reduce the susceptibility of city dwellers to the health effects of climate change include:

* proximity to seacoast, especially on low-lying river deltas and coastal areas that are subsiding
* location in tornado or hurricane zones
* arid regions with limited water supply
* dependence on glacier melt or snowmass for water supply
* mid-continental location with exposure to a wide range of temperatures.

Coastal cities, especially those that inhabit low-lying deltas and cities that have grown up on the edge of large tidal rivers are especially vulnerable to floods and storms. Globally, 13 percent of the world's urban population lives in a low-elevation coastal zone (LECZ) – a coastline area with an elevation of 10 meters or less (McGranahan *et al.*, 2007). Egypt, Bangladesh, and Vietnam have 38, 46, and 55 percent of their populations living within a LECZ, the majority in densely populated urban settings. In the Netherlands, 65 percent of the urban population lives in LECZ (McGranahan *et al.*, 2007). A recent review for the Organization for Economic Cooperation and Development (OECD) examined the current and projected exposure of port cities around the world to surge-induced flood events (Nicholls *et al.*, 2008). The top 10 cities in terms of population exposure were identified as Mumbai, Guangzhou, Shanghai, Miami, Ho Chi Minh City, Kolkata, Greater New York, Osaka-Kobe, Alexandria, and New Orleans. The study identified 30 cities that contain roughly 80 percent of exposure; of these, 18 cities are located in river delta zones. Cities with the greatest exposure to extreme sea levels also tend to be those with the greatest exposure to wind damage from tropical and extra-tropical cyclones. The total urban population exposure is expected to triple by the 2070s.

Cities that are already on the path of tropical cyclones and tornados will be most at risk from increases in the intensity or extent of these storms and the related health effects. According to the United Nations Prevention Web (2009), the largest populations currently at risk are found in: Bangladesh, China, India, Japan, Korea, Madagascar, Philippines, Taiwan, and the United States. Almost 30 percent of the population of Taiwan is exposed, compared to 19 percent of the population in Philippines, 18 percent in Japan, and 15 percent in the United States.

The El Niño–Southern Oscillation (ENSO) affects temperature, precipitation, storm tracks, and the frequency of tropical cyclones in many regions of the world, especially the west coast of South America, Southeast Asia and the northwest of North America (see Chapter 3). Episodes of El Niño have been more frequent, persistent, and intense since the mid 1970s compared with the previous 100 years (WHO, 2005), but it is unclear whether this is related to anthropogenic climate change. Two very intense episodes of El Niño in 1982–3 and 1997–8 combined with land use changes caused floods, droughts, and landslides, causing many deaths and damaging infrastructure and impacting economic activity in urban areas of Asuncion, Havana, the Panama Canal watershed, and Quito (Satterthwaite *et al.*, 2007). Climate fluctuations related to ENSO are also linked to a range of diseases, including malaria, cholera, and dengue (Haines, 2008).

The effect of rising temperatures will be felt unevenly across the world. Cities in northerly latitudes and continental locations may feel the effect of increased and more extreme heat waves most, because people in these locations tend to be less acclimated to warm weather. Geographic features such as the location of nearby water bodies, topography, prevailing winds, and other factors may modify these effects (Lombardo, 1985 in Ribeiro, 2008). A recent study investigated future heat-related mortality impacts on six cities (Boston, Budapest, Dallas, Lisbon, London, and Sydney) by applying temperature projections from the UK Met Office HadCM3 climate model to temperature–mortality models (Gosling et al., 2009). The results demonstrate that mortality is more likely to increase in cities with greater temperature variability combined with health- and age-related factors rather than with the change in mean temperature alone.

The extent to which the natural local environment has been changed and/or degraded is also a major factor in amplifying or reducing the health impacts on people in cities from climate change. Deforestation, for example, can increase the impacts of heavy precipitation events. A recent assessment of natural hazards in Indonesia indicates that landslides are increasing as a result of heavy rainfalls combined with deforestation and the expansion of development into unstable hillslope areas (Marfai et al., 2008). Between 2000 and 2007, 57 landslide disasters occurred in the Central Java province. The most severe of these occurred in the town of Banjarnegara, causing 142 deaths and damaging 182 houses. The city of La Paz, Bolivia, is also prone to landslides triggered by heavy prolonged rainfall on low-income, self-built settlements on steep, unstable slopes (O'Hare and Rivas, 2005). Although more drought is predicted for this region under climate change, more intense precipitation events are also likely to occur, which will aggravate the current risks if measures are not taken to adapt.

7.4.2 Local infrastructure and urban planning

Cities depend on a wide range of infrastructure networks and systems to assure the health of their residents, as outlined in Table 7.2.

This infrastructure is meant to supply or assure the necessities of life and health to city dwellers. Transportation networks bring in food and other products, allow city dwellers to get to jobs and services and to escape from disasters; water distribution systems supply water – sometimes from very far away; sewage and garbage collection systems remove harmful wastes; energy systems provide power for lighting, cooking, food storage, heating, air conditioning, and transportation; communications systems provide warning of weather-related and other problems; emergency services help rescue residents in difficulty; and so on. When one or more of these systems is absent (as is the case in many unplanned urban slums and informal settlements around the world), works

poorly, is overwhelmed or breaks down, urban residents are more vulnerable to health problems (Hardoy and Pandiella, 2009).

The very large numbers of the urban poor who live in unplanned settlements in low- and middle-income cities have limited access to infrastructure and other services that urban residents in high-income countries take for granted. The absence of these services aggravates their vulnerability to current weather extremes and to climate change. A recent assessment of climate impacts in Kampala described some of the problems that arise in that city as a result of a very limited infrastructure system combined with heavy rainfall (Mabasi, 2009). Most residential areas and slums have no stormwater drainage systems. The natural drainage channels that exist are increasingly obstructed by construction of informal settlements as well as new commercial buildings. There is very limited solid waste collection in Kampala, which results in garbage and plants clogging drains, and leads to localized flooding even with light rainfall. Only 8 percent of Kampala houses are connected to the city's sewer system. The rest rely on pit latrines, or even dispose in plastic bags thrown into garbage pits or water channels (Kyalimpa, 2009). Uganda's national adaptation plan for action (Republic of Uganda, 2007) reports that heavy rains have already increased in intensity and frequency, resulting in a growing number of floods, mainly in the low-income settlements in low-lying areas of Kampala, which have affected almost half a million city residents. Floodwaters are usually contaminated by the overflow from latrines, septic tanks, and sewers (Mabasi, 2009). One of the health effects seen from flooding in Kampala is cholera, with 200 cases and four fatalities in 2002 (Alajo et al., 2006; Kovats and Ebi, 2006), and more than 200 cases in 2006. Other health effects of the floods were not well monitored, but it is likely that other illnesses also occurred as a result of these floods (Few and Matthies, 2006). At the time of writing – in the fall of 2009 – flood warnings are again being sounded for Kampala.

Well-built urban infrastructure and good urban planning can moderate the impacts of climate change. In the case of cities in low-income countries, urban planning that diverts the construction of informal settlements from flood-prone areas could reduce the impacts of high-intensity rain events. Developing sewage and drainage infrastructure for areas that are not currently serviced would also reduce impacts.

During the European heat wave of 2003, the cold storage systems of 25–30 percent of food-related establishments in France were found to be inadequate (Létard et al., 2004, cited in IPCC (2007a) and Bobylev, 2009). Electricity demand for air conditioning increased, but the heat undermined electricity production by raising the temperature of river water used for cooling in power plants. Six nuclear plants, dependent on river water for cooling, were shut down. A similar situation arose in July of 2009, when a summer heat wave put a third of France's nuclear power stations out of commission (Pagnamenta, 2009).

Climate change-related natural disasters can severely damage urban infrastructure which, in turn, can impede the availability of water, food, energy, transportation services, and other inputs

Table 7.2: *Urban physical infrastructure systems and their contribution to health.*

Types of urban infrastructure	Examples	Contribution to health of city dwellers
Transportation	• Roads and highways (including bridges, tunnels, culverts, street and traffic lights) • Railways, including terminals • Ports, canals and waterways, and ferries • Mass transit systems (commuter rail, subways, bus terminals) • Pedestrian walkways and bicycle paths • Airports and air navigation systems	• Transport of food, medical supplies • Access to services, including health services • Access to employment • Access for emergencies and emergency services
Water management	• Drinking water supply, including storage reservoirs, filtration, treatment, and distribution systems • Sewage collection, treatment, and disposal • Stormwater/drainage systems • Flood control systems (dikes, levees, floodgates, pumping stations, etc.)	• Provision of clean water for drinking and washing • Removal of potentially hazardous biological wastes • Control of floods
Waste management	• Waste collection • Landfills and incinerators • Hazardous waste disposal facilities	• Removal and disposal of potentially hazardous wastes
Energy	• Electrical power generation, grid, substations, etc. Natural gas storage and distribution • Petroleum storage and distribution • Steam or hot water generation and distribution for district heating systems	• Provision of energy sources for lighting, heating, cooking, refrigeration, air conditioning
Health	• Hospitals and community health-care centers • Ambulance and emergency treatment services • Public health services (vaccination, education, etc.)	• Treatment for illnesses and injuries • Prevention of illness
Communications	• Radio towers and signaling systems • Mobile phone networks • Telephone, television, and internet cable systems • Dedicated telecommunications networks (such as those used for emergency services)	• Communication of health information • Warning and advice • Emergency communications
Urban green space	• Parks and other green spaces • Water courses and wetlands • Urban trees and vegetation	• Cooling in hot weather • Natural drainage • Space for recreation
Geophysical monitoring networks	• Meteorological monitoring (and warning) systems • Tidal monitoring • Fluviometric systems (for gauging river flow)	• Advanced warning of problem conditions

relevant for the maintenance of the quality of urban life and, consequently, of health. For example, a succession of deadly midwinter ice storms that hit the eastern Canadian provinces in January 1998 demolished 130 electricity transmission towers carrying electricity from distant hydroelectric dams, felled 30,000 utility poles, and damaged millions of trees, especially in the cities of Montreal and Ottawa. More than 4.7 million Canadians were without power, in some cases for up to three weeks. The storms resulted in 28 deaths, mainly from trauma or hypothermia, and almost 1,000 serious injuries (Lecomte *et al.*, 1998). More than 100 people were treated for carbon monoxide poisoning from using generators indoors to heat their homes (Risk Management Solutions, 2008). The cities of Kingston, Montreal, Ottawa, and Smiths Falls declared states of emergency, and the army was called in to help stranded residents and to clear fallen trees and debris. Rural residents, especially dairy farmers and maple syrup producers, were also very hard hit. Climate change

is expected to bring more frequent freezing rain events to parts of Canada over the next several decades (Lemmen *et al.*, 2008).

When new or updated infrastructure systems are built, they should be planned with climate change in mind to safeguard the infrastructure investment. Sewage treatment plants should be situated to avoid flooding from intense rains, sea level rise and storm surges (see Chapter 5 on water). Combined sewage and stormwater systems, which exist in many urban centers, should be separated to avoid contaminating stormwater overflow. Water distribution systems need to be maintained to prevent the loss of large quantities of treated water through fissures in the pipes, to avoid wasting water in situations of drought and water shortage. Consideration also has to be given to the dangers of cascading failures in interdependent infrastructure such as water distribution systems that depend on electrical power. Convergent infrastructure – where electrical power lines, communications, and

[VULNERABILITY] Box 7.1   United Kingdom floods in 2007

Jennifer Penney

Clean Air Partnership

For the most part, cities in high-income countries have well-developed infrastructure, though as it is presently constructed it may not provide sufficient protection in the face of extreme weather, as evidenced by the floods that swept much of the United Kingdom in June and July of 2007. The wettest weather the UK had experienced since records began caused the "country's largest peacetime emergency since World War II" (Pitt, 2008), flooding 55,000 properties in dozens of towns and cities, stranding 7,000 people who were rescued from floodwaters by emergency crews, and causing the deaths of 13 people. Flooding of water treatment

plants and electrical substations left half a million people without drinking water or electrical power. Thousands were stranded on flooded roads, rail lines, and even schools. Several nursing homes were evacuated. Tens of thousands of people were made homeless and thousands remained so a year after the floods receded. Apart from the deaths reported by Pitt (2008), evidence of health effects from the flooding is mainly anecdotal, and includes acute stress, depression, and other mental health problems; asthma attacks resulting from exposures to mould in flood-damaged buildings; and diarrhea. Because the existing infrastructure and emergency services in the UK are well-developed, the health impacts were much less severe than they might have in the case of similar floods in poorly organized urban environments in low-income countries.

other systems are carried by the same physical infrastructure – can also be a problem (Bobylev, 2009).

7.4.3 Social and economic conditions

Social and economic conditions can also increase or decrease vulnerability to the health effects of climate change. Romero Lankao and Tribbia (2009) conducted a meta-analysis of the literature on vulnerability, adaptive capacity, and resilience in urban centers, and identified a wide range of social and economic determinants of vulnerability including:

- *individual factors* such as age, gender, ethnicity, migrant status, and pre-existing health problems (which is discussed further in Section 7.4.4 on density and demographics)
- *individual assets* such as income, employment, education, housing (quality, access)
- *collective assets* such as concentrated affluence, stability, location, services (access, quality), infrastructure (access, quality), and technology
- *institutional (political) factors* such as knowledge, policy responses, governance practices, urban planning, political power, patronage, and (not specifically mentioned by Romero Lankao and Tribbia) structural adjustment policies by international lending agencies that force national governments to reduce social spending
- *institutional (social) factors* such as social networks and community-based organizations (Romero Lankao and Tribbia, 2009).

It is beyond the scope of this chapter to discuss all of these modifying factors; however, a few of them are examined below.

7.4.3.1 Individual assets

A variety of individual assets are widely accepted as "determinants of health" and contribute to the level of vulnerability

of human populations to the health effects of climate change (PHAC, 2009). These include income, social status, education and literacy, employment and working conditions, and life skills, which are to a large extent interdependent. The poor are less well nourished, more susceptible to illness, and are less able to protect themselves and to recover from adverse events such as floods, heat waves, or droughts. Poverty limits the locations where low-income people can live – often forcing them into illegal settlements in risky locations on floodplains or unstable hillsides that have little or nothing in the way of infrastructure services (Bartlett *et al.*, 2007). In these areas they are exposed to disease vectors, including mosquitoes, resulting from of lack of liquid or solid waste management (WHO, 2005; Kjellstrom *et al.*, 2007). The poor have fewer resources and skills for building housing that is more resilient to storms and floods, or to recover when extreme weather events damage their communities, shelters, belongings, or their health (HDR, 2007).

Winchester and Szalachman (2009) recently described the vulnerability of the urban poor in Latin America and the Caribbean (LAC) to climate change. Two out of three poor people in the region live in cities. In countries such as Honduras, Nicaragua, and Paraguay, more than 50 percent of urban residents are poor. The earnings of the urban poor are low, their employment precarious, they have little access to social security, and have low levels of formal education and skills. They tend to live in overcrowded conditions (three or more people to a room) in poorly constructed housing with poor or non-existent sanitation, limited access to clean drinking water, waste in the streets, and indoor air pollution from cooking over wood. Many of the urban poor live in neighborhoods with high levels of crime and violence.

7.4.3.2 Collective assets

Collective assets include general affluence, physical infrastructure (which has already been discussed), stability, services, and technology. Affluent countries spend more on infrastructure and services that reduce vulnerability even when large disparities of income remain within the population. They are more likely to have functional health and emergency services that can intervene to reduce health impacts of extreme weather and climate change (UNEP, 2008).

Stability can also amplify or diminish the impacts of climate change on health. Unstable nations – such as Afghanistan, Democratic Republic of the Congo, Iraq, Somalia, Sudan, and Zimbabwe (Foreign Policy, 2009) – characterized by weak government with little practical control over much of their territories, absence of public services, widespread corruption and criminality, large numbers of refugees, and economic disarray, are unable to prepare for climate change or to adequately respond to extreme weather events. Several authors argue that drought and other manifestations of climate change have already contributed to the instability of countries such as Sudan (Oxfam, 2009).

7.4.3.3 Institutional factors

There are many institutional factors that can contribute to protecting urban communities from the health effects of climate change, or conversely make them more vulnerable. Cities that are institutionally well-organized and resourced are much better situated to plan climate change programs and implement them. The specific institutional factors that could reduce the vulnerability of city dwellers to climate change include:

- government stability
- public sector economic resources
- existence of infrastructure networks that provide clean water, sanitation, energy, and public transportation
- primary care and public health services with the capacity to investigate and plan for weather-related hazards
- legal authority to act (and enforcement capacity) in areas of concern, including flood plain management, land use planning, building codes, water quality standards, etc.
- political and/or executive leadership (Penney, 2007)
- government culture including "pro-poor" attitude (Satterthwaite *et al.*, 2008), orientation towards long-range planning; low levels of corruption and political patronage (Romero Lankao and Tribbia, 2009)
- access to historical and current local meteorological information, as well as to regional climate projections that can help identify weather trends, extremes and possible future conditions
- robust disaster risk management planning and emergency response systems.

See Section 7.5.1 on adaptive capacity for further discussion of these issues.

7.4.4 Density and demographics

The concentration of human populations in cities has both advantages and disadvantages with respect to the impacts of climate change. By bringing together more people, urban centers also concentrate knowledge, skills and resources, giving rise to the potential for creative and effective responses to climate change. Human populations are concentrating in cities in every region of the world (Martine, 2009). Currently, 50 percent of the global population dwells in cities, but this percentage is expected to grow above to 60 percent over the next three decades (Wilby, 2007). The world urban population is likely to increase by 3.1 billion between 2007 and 2050 (UN Population Division, 2008), as a result of general population growth, continuing rural–urban migration and the transformation of rural settlements into urban centers. Although populations are increasingly concentrating in urban areas, cities have been spreading out in recent years, with the average density of built-up areas in industrialized countries decreasing from an average of 3,545 to 2,835 persons per km^2 in the period 1990–2000, and in developing countries from 9,560 to 8,050 persons per km^2 in the same period (Dodman, 2009).

The concentration of populations in cities is widely regarded as positive for reducing greenhouse gas emissions, although a variety of other factors affect energy use and emissions from towns and cities, especially general levels of consumption. However, increasing the concentration of populations in urban areas, especially in low- and middle-income countries, increases vulnerability to the impacts of climate change. As Dodman (2009) points out, the effects of climate change over relatively small land areas can affect large numbers of people who live in dense concentrations. This is particularly worrisome when urban populations continue to grow in coastal cities that are already subject to sea level rise and cyclonic storms (McGranahan *et al.*, 2007), or arid mountain zones with water shortages (Karl *et al.*, 2009). Much of the increase in urban populations in developing countries is concentrated in informal settlements and slums (UN-HABITAT, 2003), whose lack of infrastructure is a factor in vector- and water-borne diseases, stress, and overall mortality. The spread of hard surfaces and reduction of green space in growing urban agglomerations also contributes to health effects related to heat waves and floods.

Climate change is likely to increase rural–urban and in some cases urban–urban migration, as a result of drought, flood, coastal zone inundation, and extreme weather events. Large migrations and damage to existing housing are likely to contribute to the growth of informal or squatter settlements in urban areas. Migratory processes affect the spatial distribution of endemic infectious diseases. Confalonieri (2008) described two surges in the increase of infectious disease as a result of drought-related migration: (1) an increase in the number of malaria cases in Maranhão state resulting from the return of migrants after the end of the 1982–3 El Niño-related drought; and (2) epidemics of visceral leishmaniasis, observed in some cities in the northeast of Brazil in the early 1980s and early 1990s, due to the advance of drought in this region.

In urban areas where appropriate infrastructure is in place and maintained, the concentration of population, wealth, and resources can provide the wherewithal to build climate resilience cost-effectively (Dodman, 2009). Satterthwaite *et al.* (2007)

emphasizes that making safe areas of urban land available for low-income urban groups is critical for reducing vulnerability.

Current forms of urbanization often curtail household food production and may cause environmental damages through long-distance transport (Kjellstrom *et al.*, 2007). During and after extreme events, people in cities may be cut off from food supplies or may be subject to unaffordable rises in the cost of food. Keeping production of foods close to or in urban centers can bring positive aspects for health and the environment related to urban food security and nutrition. However, it may also create some risks to human health, especially in developing countries where urban lands and irrigation waters may be contaminated by human and chemical wastes (Lock and van Veenhuizen, 2001).

In all regions, people at greatest risk of climate change are the urban poor, the elderly, children (see Box 7.2), the medically infirm, and marginalized populations. A recent study of 16 developing countries found that the urban, wage-labor-dependent poor were particularly vulnerable to climate volatility that reduces agricultural productivity and raises the price of staple foods, especially in Africa, Bangladesh, Indonesia, and Mexico (Ahmed *et al.*, 2009).

The elderly are also at increased risk, especially if they are socially isolated and living on limited incomes. They may not be able to move out of harm's way, augmenting their chances of suffering both physical stress and trauma. This was borne out in New Orleans, where 67 percent of casualties due to Hurricane Katrina were 65 years of age or older (Sharkey, 2007). Elderly people have diminished capacity to regulate body temperature, and many have underlying medical conditions that can be exacerbated by the stress of exposure to extreme heat. During the 2003 European heat wave, excess mortality was estimated at 20 percent for those aged 45–74 years, 70 percent for those aged 75–94, and 120 percent for people older than 94 (European Commission, 2008).

Children, like seniors, are considered one of the most vulnerable groups in weather-related disasters (Wilbanks *et al.*, 2007). Children make up a large proportion of the population in the countries that are most vulnerable to climate change, including the urban areas. Children under five typically make up 10–20 percent of the population in low-income countries, compared to approximately 4 percent of the population in high-income countries. Children under 18 constitute 40–50 percent of the population in the most vulnerable countries, compared to an average of 20 percent of the population in rich countries (Bartlett, 2009).

7.4.5 Health system conditions

In general, the better that public health and primary health-care systems are able to cope with current levels of disease and climate variability, the better prepared they will be for climate change (Nerlander, 2009). Health system conditions that may help reduce the impacts of climate change include universal access to primary health-care providers, laboratory services, and standardized diagnosis and reporting systems, which are crucial for national health surveillance (Nerlander, 2009).

In 1994, the US Public Health Functions Steering Committee (Frumkin *et al.*, 2008) listed ten essential services that make up a strong public health system and which should afford some protection from the health effects of climate change. These include:

1. monitoring health trends and diseases to identify community health problems
2. investigating and diagnosing health problems and health hazards in the community
3. informing and educating the public and policymakers about health issues
4. mobilizing community partnerships to identify and solve health issues
5. developing policies and plans to strengthen individual and community health efforts
6. enforcing laws and regulations that protect health and safety
7. linking people and personal health-care services
8. ensuring the competence of the health-care workforce
9. evaluating the accessibility and quality of personal and population-based health services
10. conducting new research to provide insights and solutions to health problems.

There are a number of more specific health system conditions that may amplify or reduce the health impacts of climate change. These include:

- involvement of the health sector in planning of water, sanitation, and other health-sensitive services (Nerlander, 2009)
- reliable information from meteorological monitoring systems, linked to early health warning systems to help the population protect itself from the health effects of heat, floods, storms, and other extreme weather (Costello *et al.*, 2009)
- surveillance systems that identify when climate-sensitive illnesses emerge or suddenly increase (Frumkin *et al.*, 2008)
- public health education to guide personal responses to heat, food-borne and water-borne illnesses and other climate-related illnesses (Frumkin *et al.*, 2008)
- vector control and the detection and treatment of vector-borne diseases (Nerlander, 2009).

7.5 Adaptation

In the past decade, general awareness of climate change has grown, and weather trends, extreme events, and related impacts are being associated with climate change. Many cities around the world have advocated, developed, and implemented actions to reduce greenhouse gas emissions, but until recently have not begun to consider adaptation. This is changing since weather in many cities has become noticeably more extreme. City governments are responsible for many services and infrastructure affected by climate variability and change: electricity distribution; water supply and wastewater; stormwater

---

[VULNERABILITY]   Box 7.2   Children's health and climate change

Gregorio De Simone

Instituto Oswaldo Cruz, Fiocruz

Bartlett (2008) summarized many of the vulnerabilities of children to climate change impacts:

- children are more likely to experience life-threatening diarrhea as a result of exposure to water- and sanitation-related illnesses spread by intense rainstorms and floods
- droughts that lead to food shortages are likely to lead to malnutrition, affecting the health and development of poor children
- children are particularly at risk of malaria, expected to spread as temperatures rise around the world
- asthma and other respiratory illnesses affected by air quality are on the rise, with the greatest increase among children, and asthma deaths are expected to increase 20 percent by 2016
- children have higher mortality rates than adults (except for the elderly) in heat waves, especially in cities
- the quality of care for children is reduced when families are adversely affected by weather events.

Because of immature organs and nervous systems, more rapid metabolisms, lower immunity, and lack of experience, children are especially vulnerable to the health effects of a variety of climate-related risks (Balbus and Wilson, 2000; Patz *et al.*, 2000; Bunyavanich *et al.*, 2003; Martens and McMichael, 2002; Haines and Patz, 2004). Children exposed to floods in Nepal were twice as likely to die as adults, and pre-school girls were five times more likely to die than adult women (Bartlett, 2008). Children exposed to extreme heat, vector-borne diseases, reduced food supply, scarce or contaminated water, air pollution, allergens, and weather-related disasters – all likely to be exacerbated by climate change – are also more likely to experience long-term repercussions than adults. For example, children born in Africa during drought years are significantly more likely to be malnourished or stunted (UNDP, 2007).

Some urban children are particularly at risk, especially those growing up in urban slums in low-income countries. Nowadays, hundreds of millions of children live in conditions of urban poverty, and in some slum areas 25 percent of children die before the age of five, indicating their susceptibility to disease and injury. Bartlett (2008) cites the example of Nairobi, where children in the Embakasi slum have a mortality rate of 254 per thousand for children under five, compared to 62 per thousand for children for children across the whole of Nairobi and 113 per thousand for children in Kenya's rural areas.

Climate change is an issue not just for today's children; it poses long-term risks for generations of children to come. Many of today's children will live into the 2080s, by which time climate change will be much more entrained unless actions to reduce emissions are undertaken now. (Waterston and Lenton, 2000).

---

management; roads and bridges, and public health (see relevant chapters in this book). As a consequence, a growing number of city governments are beginning to consider how to respond to extreme weather and climate change and protect the health of their residents.

The capacity to adapt to climate change varies substantially from region to region, city to city.

Factors that affect adaptive capacity include: availability of economic resources; governance; institutional capacity; sociocultural factors such as social support networks; and access to and deployment of adaptive technologies.

A key factor that will affect human health is the adequacy of economic resources available for these efforts. Individuals and families with more economic resources are able to afford emergency measures and medical services, and to stockpile supplies in preparation for a climatic event. Urban centers with greater economic resources can decrease impacts and vulnerability if they use these resources to create a more climate-resilient city. This is bolstered by new networks of towns and cities such as the Urban Leaders Adaptation Initiative in the United States and the Alliance for Resilient Cities in Canada, which meet via electronic conferencing systems to learn about recent impacts research, climate change risk assessments for municipal infrastructure, and emerging municipal adaptation programs and strategies (CCAP, 2009; Clean Air Partnership, 2009).

Adaptive capacity also depends on access to technologies, training, and skills development. Cities that have greater access to technologies can install early warning detection systems. Individual access to technologies such as computers and internet communications systems, telephones, radio, and television can decrease vulnerability by increasing access to information, weather alerts, and health advice.

As Satterthwaite and his colleagues (2007) have argued, many of the urban areas that need to adapt the most urgently to preclude major impacts have serious gaps in these building blocks of successful adaptation.

7.5.1   Reducing vulnerability to human health effects from climate change

The reduction of urban human health vulnerability can be achieved by the implementation of measures indirectly and directly related to the health sector and service. We present here the ones that can influence the reduction of health risk, being indirectly and directly linked with it.

7.5.1.1 Overview of general adaptation measures that reduce health risks

Most activities that increase urban adaptation to climate change also reduce the health impacts of climate change, even if they are not directly linked to the health sector and health services. Campbell-Lendrum and Corvalán (2007) suggest that promotion and adoption of "Healthy Cities" strategies, for example, are simultaneously beneficial for climate change mitigation, adaptation, and population health. Proactive adaptation measures – implemented before major impacts are seen – are likely to be less disruptive and costly than reactive adaptation (Dickinson, 2007). An efficient and effective emergency response by trained and well-equipped personnel can transform a catastrophic situation into one that is manageable, with minimum loss or impact to human health and well-being. Some general adaptation measures that could reduce the vulnerability of city dwellers to climate change include:

- incorporation of climate change information and projections into standards, policies, and codes such as those governing where and how to construct or update stormwater and flood control systems, transportation infrastructure and buildings, thereby increasing the safety of urban areas near coasts and rivers
- relocation of populations from floodplains and steep hillsides to safer locations to reduce deaths, diseases, and injuries from floods and landslides (especially if the relocation involves provision of soundly constructed housing, infrastructure, and transportation services)
- reducing impermeable surfaces through provision of green space, urban trees (which are likely to survive in poor neighborhoods of low-income cities only if cooking fuel is accessible and affordable), vegetation, and other means such as permeable pavements and green roofs, reducing flooding and decrease the urban heat island in cities
- extension and maintenance of essential infrastructure services such as drinking water supply and sanitation, taking account of changing climate conditions (such as a potential shortfall in water supply, or increased intensity of rainfall) to decrease illness and deaths due to water shortages or contaminated floodwaters
- energy and water conservation programs that reduce pressure on these systems in conditions of extreme heat or drought, and ensure the continuity of these critical services.

7.5.1.2 General adaptation measures for the health sector

The health sector does not directly influence many or most of the factors that directly contribute to vulnerability to climate change. However, there are a number of general adaptation measures that the health sector could undertake to improve knowledge of the health effects of climate change and motivate appropriate protective action by agencies with more direct responsibilities for the structures and services that affect the vulnerability of city dwellers to climate variability and change. Among these measures are:

- strengthening health systems in countries and cities where these are weak, ensuring universal primary care with providers who have been educated about climate-linked health problems (Campbell-Lendrum and Corvalán, 2007)
- understanding and mapping health hazards related to climate variability and change
- developing climate-based early warning systems coupled with response plans and activities and evaluation of their effectiveness (Confalonieri et al., 2007)
- analysis of meteorological data such as temperature trends and changing ecological factors such as the conditions that could support rapid reproduction of mosquitoes and other disease vectors
- tracking and assessment of diseases and health trends related to climate change
- communications on the health effects of climate change to inform the public about potential health effects and what they can do to protect themselves from weather-related illnesses and injuries
- advice to decision-makers about the likely health impacts of climate change
- collaboration with city officials responsible for the planning and design of buildings, public spaces, transportation networks, emergency management, and disaster risk reduction to ensure that health considerations are appropriately included
- communication with community leaders to encourage proactive action by individual households and communities to reduce vulnerability
- emergency medical care systems to provide trauma care and during emergency periods to help provide uninterrupted care for people with conditions requiring regular treatment (e.g., AIDS medications)
- ensuring a competent public and primary health-care workforce trained to recognize and manage health threats that may be associated with climate change
- evaluation of climate change preparedness plans, health communications and intervention strategies to ensure that these are effective in reducing risks and impacts (Ebi, 2005; Frumkin et al., 2008; Griffiths et al., 2009).

The health sector also needs to protect its own services and facilities from the impacts of climate change. This may include "climate-proofing" health centers, hospitals, nursing homes, ambulance stations, and other facilities in order to ensure that they can provide effective services during floods, storms, blackouts, and other emergencies.

The World Medical Association (2009) has strongly recommended that medical professionals be trained in climate change impacts and adaptation options and actions (see Box 7.3). Increasing the capacity of the health-care workforce and the number of health-care professionals, and promoting and encouraging health-care workers to work with other sectors may also be necessary.

7.5.2 Adaptation measures for specific health risks

In addition to general measures that reduce health risks and increase the capacity of the health service to respond to these

---

[ADAPTATION]   Box 7.3   The World Medical Association's Declaration on Health and Climate Change

In October 2009, the World Medical Association passed a Declaration on Health and Climate Change that included a large number of recommendations with respect to reducing the health impacts of climate change. Among other things, the Declaration called on physicians and national medical associations to:

- urge local, national, and international organizations focused on adaptation, mitigation, and development to involve physicians and the health-care community to ensure that unanticipated health impacts of development are minimized, while opportunities for health promotion are maximized
- build professional awareness of the importance of the environment and global climate change to personal, community, and societal health
- work to improve the ability of patients to adapt to climate change and catastrophic weather events
- work with others to educate the general public about the important effects of climate change on health and the need to both mitigate climate change and adapt to its effects
- develop concrete actionable plans/practical steps as tools for physicians to adopt in their practices; health authorities and governments should do the same for hospitals and other facilities
- advocate that governments undertake community climate change health impact assessments, widely disseminate the results, and incorporate the results into planning for mitigation and adaptation

- urge colleges and universities to develop locally appropriate and continuing medical and public health education on the clinical signs, diagnosis, and treatment of new diseases that are introduced into communities as a result of climate change, and on the management of long-term anxiety and depression that often accompany experiences of disasters
- work with others, including governments, to address the gaps in research regarding climate change and health
- report diseases that emerge in conjunction with global climate change, and participate in field investigations
- encourage governments to incorporate national medical associations and physicians into country and community emergency planning and response
- call upon governments to strengthen public health systems in order to improve the capacity of communities to adapt to climate change
- prepare physicians, physicians' offices, clinics, hospitals, and other health-care facilities for the infrastructure disruptions that accompany major emergencies, in particular by planning in advance the delivery of services during times of such disruptions.

Source: WMA Declaration of Delhi on Health and Climate Change (http://www.wma.net/en/30publications/10policies/c5/index.html)

---

risks discussed in previous sections, the following describes a number of measures designed to reduce some of the specific health risks presented in Section 7.3.

7.5.2.1 *High temperatures, heat waves, and the urban heat island*

A number of adaptation measures to reduce the urban heat island effect in cities are being studied, promoted, and implemented. Yamamoto (2004) classifies these activities into three broad categories. The first – *reduction of exhaust heat* – includes initiatives to improve the efficiency of energy-using products and air conditioning systems; optimal operation of air conditioning systems (including restraints on nighttime operations); improvements in building insulation; increasing the reflectivity of walls and roofing materials; traffic-control measures and active transportation; cooling buildings with sea or lake water; capture and use of exhaust heat from industrial plants and other point sources; use of photovoltaic and solar thermal energy.

The second category – *improvement of urban surface cover* – includes increasing the reflectivity and water-retention of paving materials; green roofs; tree planting; and provision of more open water by pond construction and other initiatives. Cities in high-income countries have recently developed a number of programs designed to expand the installation of green roofs; use of reflective roof, road, and building surfaces; and increase the urban tree canopy (Acclimatise, 2006; USEPA, 2009).

The third category – *generation of air movement through urban corridors* – involves improving the orientation of buildings and roads to take advantage of cooling wind or water pathways and the construction of large-scale parks and green spaces. Some cities have also experimented successfully with reducing heat by improving air movement. Stuttgart, for example, has designated a "wind path" that allows cool mountain air to flow into the heart of the city; and Seoul removed part of a double-decked road and restored the previously buried Cheong-Gye River, which helps channel cool sea breezes into the area around the river (Droege, 2008; Ichinose *et al.*, 2006).

Public health departments have also taken measures to address health impacts from heat waves. In conjunction with longer-term actions to reduce the urban heat island, the mandate of these groups is the more immediate task of protecting the population from heat waves. One adaptation action that has become more common in cities in high-income countries is the development of a heat-health warning system (Kalkstein *et al.*, 2008). This system will usually involve one or more of the following actions:

- public health alerts through radio and television, triggered by expected high temperatures alone, by temperature–humidity indices (such as the Humidex), synoptic analysis to identify city-specific air masses associated with high mortality, and other systems (Kovats and Ebi, 2006)
- information about the need to stay hydrated and stay cool, which may be passively disseminated (by leaflets, for example) or more actively broadcast along with the heat warnings

- advice to those who care for the elderly and other vulnerable groups (often distributed at the beginning of the summer season, and sometimes specific alerts and advice when hot days are expected)
- telephone hotlines to provide advice to concerned members of the public
- opening of air-conditioned "cooling centers"
- extended hours for public swimming pools
- activation of "buddy systems" to check on vulnerable persons in their homes or to look out for heat stress symptoms among co-workers
- outreach to the homeless; and other actions.

Heat-health warning systems that include health interventions for vulnerable populations are considered to be effective in reducing deaths (Bernard and McGeehin, 2004; Ebi et al., 2004; Koppe et al., 2004). However, few systematic studies have been done to formally evaluate the effectiveness of these specific programs for saving lives.[1]

High temperatures exacerbate poor air quality, which compounds heat-health risks. Measures to improve air quality, such as reducing emissions from vehicles, promotion of mass and active modes of transportation, and introducing incentives for clean energy sources may reduce this additional risk, as well as decrease greenhouse gas emissions (Kojima et al., 2000).

7.5.2.2 Food availability

The decline in food availability as a result of drought or floods and rising food prices impact urban poor populations heavily, reducing their intake of food and the diversity and nutritional value of foods consumed (Cohen and Garrett, 2009). Resilient cities are able to provide food to their inhabitants in the face of shock and change (Larsen and Barker-Reid, 2009). A growing number of non-governmental organizations support urban and peri-urban agriculture in parts of Africa and South America as a partial solution to this problem. Urban agriculture is promoted as a way of increasing not only food security, but the mental and emotional resilience of the population. Urban agriculture may also contribute to sustainable management of urban open spaces, including flood zones, buffer zones, steep slopes, roadsides, river banks, and water harvesting areas (Dubbeling et al., 2009). Urban agriculture also encourages the maintenance of open space in cities. Hamin and Gurran (2009) suggest that moderately dense cities with green corridors and open space throughout the city may be the most effective urban form for climate change adaptation. Food may be grown in open areas within the city or on its periphery, in backyards, school yards, on wastelands, on rooftops, in container gardens, hydroponic operations, and possibly even vertical farms (Larsen and Barker-Reid, 2009; Despommier, 2009).

A number of cities now formally support urban agriculture as a means of increasing food security and income for city dwellers, reducing greenhouse gas emissions, and/or adapting to climate change. Cities with formal policies and programs to support urban agriculture include: Accra, Kampala, and Nairobi in Africa; Bangkok, Beijing, Mumbai, and Shanghai in Asia; Brisbane and Melbourne in Australia; Brasilia and Havana in Latin America; and New York City in North America.

The risks of urban agriculture to health and the environment, if not managed properly, as well as the benefits should be considered. There are health concerns about urban agriculture, especially in those areas if wastewater is used to irrigate crops or food is grown in contaminated soils (Cole et al., 2008). Urban agriculture may also increase diseases from disease vectors attracted by agricultural activity and transmission of diseases from animal husbandry (Lock and van Veenhuizen, 2001).

7.5.2.3 Vector-borne diseases

Climate change is only one factor among many that contribute to the expansion of vector-borne diseases (Semenza and Menne, 2009). However, a wide range of vector-borne diseases are "climate-sensitive," expanding their range or living longer in warmer weather, for example, or breeding in conditions that are more frequently created by changing weather patterns. Vector-borne diseases of concern include those listed in Table 7.3.

Surveillance and mapping of these illnesses as they appear and spread, as well as the conditions that give rise to them will be an important preparatory step in adaptation. Adaptation will also require application of existing vector-control strategies, which may include (Abdallah and Burnham, 2000):

- **Environmental controls:** altering breeding sites by draining or filling places where water collects; ensuring regular disposal of refuse; maintaining clean shelters and personal hygiene

Table 7.3: *Climate-sensitive, vector-borne diseases.*

Vector	Diseases
Mosquitoes	Malaria, filariasis, dengue, yellow fever and West Nile virus
Sand flies	Leishmaniasis
Triatomines	Chagas disease
Ixodes ticks	Lyme disease, tick-borne encephalitis
Tsetse flies	African trypanosomiasis
Blackflies	Onchocerciasis
Rodents	Leptospirosis, plague, hantavirus

Source: Reid and Kovats (2009), Semenza and Menne (2009).

---

1 See Kovats and Ebi (2006) for a review of heat response measures in Europe and the United States.

- **Mechanical controls:** protective clothing, screens or bednets, traps, food covers, lids or polystyrene beads in latrines
- **Biological management:** using living organisms or products against vector larvae, such as fish that eat larvae (e.g., tilapia), bacteria that produce toxins against larvae, free-floating ferns that prevent breeding, etc.
- **Chemical controls:** including repellents, insecticides, larvicides and other pesticides.

A combination of these methods – what the World Health Organization refers to as "integrated vector management" or IVM (WHO, 2004) – will be needed for the control of most vectors. An integrated vector management strategy requires:
- knowledge of factors influencing local vector functioning, spreading of disease and rates of incidence.
- collaboration of the health sector and other sectors (such as water, solid waste, sewage disposal, stormwater, housing, construction, urban food producers, etc.)
- health communications, education and engagement with local communities (WHO, 2004).

Enhanced or new vaccination programs may be appropriate for some climate-sensitive diseases as well (Gould and Higgs, 2009; Ortíz Bultó et al., 2006).

7.5.2.4 Food-borne illnesses

Contamination of food can happen at any point at which food is grown, transported, processed, sold, stored, prepared, or eaten. Rising temperatures are expected to increase the incidence of food poisoning as microbes can multiply more quickly. Warmer weather will support the survival of flies and other pests that can contaminate food. Refrigeration failure is more likely to occur in hot weather (Menne et al., 2008). Intense rainfall and floods may also contaminate water sources from which urban residents draw water for washing or cooking food and contribute to food poisoning.

Protecting urban populations from these problems requires a strong public health response, including intensified public education alerting the population to the potential threat of increased contamination and ways to handle food and avoid food-borne diseases (D'Souza et al., 2004). Timely information to food producers and food handlers is also essential (Menne et al., 2008), as is public health inspection of places where food is commercially processed or prepared. Quality control measures for food storage and handling may need to be upgraded in light of a changing climate. General adaptation measures including those that protect electricity transmission and distribution systems and measures that reduce flooding and water contamination will also provide some protection against food-borne illnesses.

7.5.2.5 Water-borne diseases

Water-borne diseases are also likely to increase under climate change, as a result of problems associated with:

- reduced water supply – leading to reduced sanitation, personal hygiene and effluent dilution
- extreme rainfall – leading to increased pathogen loading, particularly in areas with inadequate stormwater management, aging water treatment plants or combined sewer–stormwater systems (Patz et al., 2008)
- direct effects of higher temperatures – favorable to microorganism reproduction (Semenza and Menne, 2009).

Adaptation strategies to reduce water-borne diseases include:
- improved management of water demand and maintenance of water distribution systems to help avoid critical water shortages (Menne et al., 2008)
- watershed protection such as vegetative buffers to reduce contamination of water from runoff (Patz et al., 2008)
- improved city disposal systems to capture and treat wastes
- extending and improving urban stormwater management systems incorporating expected increases in intensity of storms
- separation of combined sewage and stormwater systems and stronger regulations controlling septic systems
- "well-head alert systems" that warn water system and water supply managers when rainfall conditions approach levels of concern, similar to predictive forecasts and warning systems (Auld et al., 2004).

Public health services such as public alerts, including boil water alerts about the potential threat of increased contamination, and enhanced surveillance and monitoring programs for water-borne diseases (Ebi and Nyong, 2009) can also help protect urban populations.

7.5.3 Extreme weather events

Extreme weather events, such as extreme precipitation (intense rain, thunderstorms, heavy snowstorms, and icestorms), hurricanes, and tornadoes, can have serious impacts on the health of city dwellers through direct personal injuries; damage to housing, roads, water services and electricity distribution; disruption of transportation, employment, health services; and other mechanisms. The kinds of extreme weather expected as a result of climate change will vary from region to region and consequently adaptation responses will be different. However, effective adaptation will combine urban development strategies that incorporate knowledge of the current and future risks of extreme weather with disaster risk reduction planning and robust emergency response systems.

Recent international discussions on combining climate change adaptation and disaster risk reduction have resulted in a general framework for reducing the risks of extreme weather events (CCD, 2008; UNISDR, 2008; Mitchell and van Aalst, 2008). The Hyogo Framework for Action, agreed at the World Conference on Disaster Reduction in January 2005 sets out five broad areas for action to reduce the impacts of extreme weather (see Table 7.4).

Table 7.4: *Actions to enhance adaptation and reduce disaster risk to health from extreme weather.*

General framework	Specific areas for action
1. Make disaster risk reduction a national and local priority	• Initiate high-level policy dialogue on disaster risk reduction and adaptation to climate change • Ensure formalized multi-sectoral coordination of risk-reduction activities • Create mechanisms to engage communities
2. Identify, assess, and monitor disaster risks and enhance early warning	• Provide high-quality information about climate hazards • Conduct vulnerability assessments • Establish or strengthen early warning systems • Undertake public information programs to help people understand risks and respond to warnings • Support research programs on resilience
3. Build a culture of safety and resilience	• Disseminate best practices • Provide public information programs on local and personal actions that contribute to safety and resilience • Publicize community successes • Train the media on climate-related issues • Develop education curricula on climate adaptation and risk reduction
4. Reduce underlying risk factors	• Incorporate climate risk considerations in land use planning, water management, environmental management • Maintain and strengthen protective works such as coastal wave barriers, levees, floodways, and flood ponds • Require routine assessment and reporting of climate risks in infrastructure and building projects and other engineering works • Develop social safety nets
5. Strengthen disaster preparedness for effective response at all levels	• Revise preparedness plans and contingency plans to account for projected changes in existing hazards and prepare for new hazards • Create evacuation mechanisms and shelter facilities

Source: UNISDR (2008).

Flooding is the most frequent weather-related disaster, affects the largest number of city dwellers, and is predicted to worsen under climate change. A number of cities have developed comprehensive strategies to reduce flooding and its health and impacts. The city of Pune in India is one of these cities (see Box 7.4).

7.6 Policies and policy issues

"It is within urban centers and urban governments that so much of the battle to prevent climate change from becoming a global catastrophe will be won or lost" (Huq *et al.*, 2007).

7.6.1 Mitigation policies that benefit health

Many cities in high-income countries and in some middle-income countries have developed and are implementing programs to reduce greenhouse gas emissions. Typical strategies include energy conservation programs for buildings; expansion of combined heat and power or district heating plants; capture of methane from sewage treatment and landfill sites; renewable energy installations; innovative systems to cool downtown buildings using lake or sea water; initiatives to expand the provision and use of public transportation systems and reduce use of private vehicles; investment in active transportation through pedestrian zones, bicycle lanes and other means; and tree planting and green roof strategies.

Many of these initiatives are expected to reduce health risks and/or improve health outcomes in the short or long term. Initiatives that reduce the burning of fossil fuels for example, not only reduce carbon dioxide emissions, but also reduce the release of air contaminants and improve air quality, with a positive impact on respiratory and related diseases. Cifuentes *et al.* (2001) estimated that implementing currently available technologies to reduce fossil-fuel use could result in approximately 64,000 fewer deaths between 2000 and 2020 in Mexico City, Santiago, São Paulo, and New York alone and significantly reduced mortality in China. Embedding active transportation, walking, and cycling into urban design would improve cardiovascular health and reduce obesity (Saelens *et al.*, 2003), potentially reducing the 1.9 million deaths per year associated with physical inactivity (WHO, 2002). Studies of the benefits and monetary costs of major air pollution control efforts have concluded that benefits, at least in developed countries, far outweigh the costs (Kjellstrom *et al.*, 2007). Policies and actions to control air pollution from vehicles and industry can be important interventions for health equity as well as for mitigation of greenhouse gas emissions.

Climate change mitigation is a much lower priority for cities in low-income countries. These cities are much smaller contributors to greenhouse gas emissions overall, although the

---

[ADAPTATION]   Box 7.4   Disaster risk management in Pune, India

Jennifer Penney

Clean Air Partnership

The city of Pune in Maharashtra State, India, has a population of more than 3.5 million people and is located where the Mula and Mutha rivers meet. (UNISDR, 2009) The city suffers from floods most years (Sen *et al.*, 2003) and has had at least one severe flood every decade for the last 60 years (Bhosale, 2008). These floods have displaced and destroyed the property of families living on the river banks, disrupted transportation and communications, destroyed drainage and sewage system works, eroded the river banks, and had adverse effects on the health of Pune's residents. More than 5,000 families were adversely affected by floods in 2005 and 2006. The city anticipates more frequent floods due to climate change (UNISDR, 2009).

In 2006–7 the Pune Municipal Corporation (PMC) began preparing a comprehensive Disaster Management Plan, with the participation of several municipal departments, law enforcement agencies, the state government, and educational institutes. Participating municipal departments included the fire brigade, river improvement department, public transportation, health, public relations, education, electricity, and irrigation (Bhosale, 2008). Working together, these departments created a systematic plan to reduce flooding. They developed a detailed drainage map of the city and began by analyzing hourly rainfall intensity and how it might affect low-lying areas and sites where drainage was not functioning, identifying 39

vulnerable areas on both sides of the Mutha River. This work built on the Pune Slum Census carried out by local NGOs, who mapped slums in Pune with attention to slums that were especially vulnerable to flooding. This mapping also showed how uncoordinated municipal action in response to the flooding would negatively impact many of the slum dwellers (Sen *et al.*, 2003).

Implementation of the Disaster Management Plan involved dredging the rivers to remove silt and increase their water carrying capacity (Patil, 2008a); clearing stormwater drains; expanding drainage systems along the road network; widening flood plain areas to accommodate high water levels; lengthening bridges and enhancing soil porosity to improve infiltration (UNISDR, 2009). PMC also undertook afforestation projects in a hilly area of the city, and built small earthen check dams. The municipal government also strengthened flood monitoring and early-warning systems and protections for families. These activities have not provided complete protection from flooding; in August 2008, following two days of heavy rains, the Mula River flooded several areas and displaced almost 500 families (Patil, 2008b). However, fewer households were affected than in 2005 and 2006. In 2009 the Pune Municipal Council established a flood-control response team with 24-hour emergency numbers, and with resources to relocate residents of flooded areas.

Source: UNISDR (2009). Adaptation to Climate Change by Reducing Disaster Risks: Country Practices and Lessons. United Nations International Strategy for Disaster Reduction. Geneva.

---

consumption habits of the wealthier residents of these cities are contributing to rapid increases in emission levels (Huq *et al.*, 2007). Nevertheless, some programs in low-income countries have positive outcomes both for reducing emissions and for health. Solid fuels such as charcoal, wood, dung, and coal used for domestic purposes in low-income cities kill an estimated 1.5 million each year in developing countries. Improved stoves can cut back indoor smoke levels and fuel use considerably, which improves health, reduces costs for purchasing fuel, and reduces emissions (Kjellstrom *et al.*, 2007). A large-scale transition to cleaner-burning fossil fuels would reduce greenhouse gas emissions by a modest 1–10 percent in these countries, but reduce excess deaths by 13–38 percent (Bailis *et al.*, 2005, cited in Campbell-Lendrum and Corvalán, 2007).

7.6.2   Adaptation policies that benefit health

Mitigation strategies are designed to reduce the long-term impacts of climate change and though many such strategies may have a positive short-term impact on health, this benefit is not usually a primary goal. In contrast, many adaptation strategies are designed to reduce the impacts of extreme weather and climate change in the short- and medium-term. Most of the impacts of climate change have a health component. Hence, adaptation strategies aimed at diminishing the urban heat island effect, decreasing

floods, ensuring water or food supply, or controlling disease vectors are all expected to reduce illness, injuries, and deaths.

Many adaptation strategies coincide with the development needs of low-income cities (Satterthwaite, 2009) – the extension of water, stormwater, and sanitation infrastructure, the construction of safe and decent housing, and the provision of primary health care and public health services capable of responding to climate-related health problems.

This does not mean that all adaptation strategies will positively contribute to development or to health outcomes. If informal settlements on floodplains are razed to reduce the risk of floods without alternative housing and land arrangements made for and with the residents, for example, the affected populations may be left without shelter and worse off than before. It is important for those who promote relocation strategies as a means of reducing the impacts of floods or other climate-related problems to consider these issues.

And finally, a focus on reducing climate-related risks may not be possible in some of the most vulnerable cities that are already struggling with epidemic levels of HIV/AIDS, which is not climate-sensitive, or are trying to cope with the immediate needs of large numbers of malnourished migrants from war-torn regions.

7.6.3 Combining mitigation, adaptation, and health strategies

There are several areas where climate change mitigation and adaptation strategies overlap and reduce the health impacts of climate change especially from heat and air pollution. It may be easier to gain the support of local governments for such "win-win-win" strategies than for strategies with more limited benefits.

For example, passive methods of cooling buildings such as shading, reflective roofs, natural ventilation (convective cooling), insulation, placement of nearby trees, green roofs, and permeable pavements will all reduce heat loads on buildings, reduce the need for air conditioning, improve comfort, and reduce heat impacts for the occupants. Where electricity is produced by fossil fuels, these strategies will also reduce greenhouse gas emissions.

However, mitigation, adaptation, and health strategies do not always clearly coincide. For example, mitigation proponents tend to support high densities in cities as a means of minimizing commuter distances and vehicle emissions, and common energy schemes that reduce emissions. However, high densities also contribute to the urban heat island effect, reduce green space and trees, and increase the likelihood of urban flooding (Laukkonen et al., 2009). A coordinated local response is necessary to address these kinds of contradictions, and resolve potential conflicts (Laukkonen et al., 2009).

7.6.4 Coordinated action on climate change adaptation and health

Action to reduce the impacts of climate change on the health of city dwellers will need to occur both within and outside the traditional sphere of the health sector. The specific role of the health sector is to:

- reduce the contribution of the sector to climate change (e.g., energy conservation in health facilities)
- investigate and analyze the local health effects of current weather variability and climate change
- ensure that local staff are trained to recognize and manage emerging health threats associated with climate change
- strengthen health surveillance systems that monitor the appearance and spread of climate-sensitive illnesses
- develop and implement early warning and public outreach systems for heat waves, severe air pollution episodes, waterborne and vector-borne illnesses
- ensure that health considerations are taken into account in city-level mitigation and adaptation planning
- collaborate with other local government sectors in developing preparedness plans for events such as heat waves, wildfires, intense rainfall, floods, and windstorms. (Plans should include: neighborhood communication and response systems; shelters; transport and evacuation plans; and clinical facilities with surge capacity.

- support communities in developing the capacity to adapt through vulnerability mapping, community sanitation projects, neighborhood response systems or "buddy systems" to check on vulnerable people during extreme weather events
- ensure that the health needs of disaster victims are taken care of, including continuity of ongoing care programs (such as HIV medication and kidney dialysis), and mental health services
- make hospitals, clinics, and other health-care facilities as safe as possible from weather disasters (Frumkin and McMichael, 2998; Frumkin et al., 2008; Kovats and Akhtar, 2009).

Many of the activities that are necessary for cities to adapt to climate change are the responsibility of other sectors – city planning, energy, transportation, and water and stormwater infrastructure for example – but the health sector can play an important advocacy role in encouraging action to protect the health and lives of city dwellers from the impacts of climate change. Local public health departments have also participated in the development of citywide adaptation strategies in a number of cities internationally, including London, New York City, and Toronto (London, 2008; New York City, 2009; City of Toronto, 2008).

Public health departments can play an important role in almost all stages of citywide adaptation planning and implementation including:

- *Increasing awareness and engaging the public and stakeholders in considering the need for adaptation* – highlighting health concerns is often a good way to draw attention to the issue and to motivate action
- *Assessing city vulnerabilities to climate change* – health issues are a major aspect of vulnerability to climate change, so the participation of the health sector is critical for this stage of adaptation planning
- *Reviewing the effectiveness of existing policies and programs that seek to protect the population from climate extremes* (e.g., flood controls) – the health sector needs to be involved in reviewing the strengths and weaknesses of its own programs, and may be able to contribute information necessary for a comprehensive evaluation of the programs and services operated by other departments
- *Identifying adaptation options* – again, the health sector could contribute in identifying and analyzing adaptation options, ensuring that broad health considerations be taken into account
- Piloting adaptation strategies and incorporating adaptation into policies and programs
- *Monitoring and evaluating results* – assessing the implementation and the effectiveness of adaptation strategies from a health perspective.[2]

7.6.5 Barriers to action

While many advances have been made in understanding what must be done to protect health from climate change, there exist

---

2 See Berry, 2008 for more information on the potential role of the health sector in adaptation planning.

[ADAPTATION]   Box 7.5   Advanced Locality Management (ALM) in Mumbai, India: Fostering climate resilience through community-based approaches

Akhilesh Surjan

United Nations University, Tokyo

Located on the western coast of India, Mumbai, earlier known as Bombay, is the economic capital of India. The city has witnessed exceptional population growth (present population is 15 million including 3 million floating population) due to rapid industrialization and emergence of many business and service sector employment opportunities. The progression of settlements (including 55 percent slum dwellers) coupled with the changing lifestyle of urban folks in a rapidly globalizing city environment has resulted in widespread vulnerability in the city, which is highly prone to hydro-meteorological disasters that are projected to escalate in intensity and frequency due to climate change. Flooding and water-logging in low-lying areas is an annual occurrence in Mumbai and has become a grave concern as monsoon rains can bring the city to a halt for hours or even days. The urban populace of Mumbai is not yet prepared to deal with climate-induced catastrophes. The flood of 2005, during which 994 mm of rain fell in a single day, resulting in the deaths of more than 500 people, was the worst in the city's recorded history.

This case study is about an initiative called Advanced Locality Management (ALM), a volunteer movement with its origin in the proliferation of diseases due to filthy surroundings in certain residential areas. Concerned citizens, frustrated by the limited outreach of municipal services, formed a Street Committee in 1996 to clean up their nearby areas. Volunteers, including women and children, started cooperating in collecting garbage in the vicinity, separating household waste, and composting organic materials. This unique voluntary initiative, which began in just a few streets, was later replicated at the city level by the Municipal Corporation of Greater Mumbai (MCGM) and included in its 2006 Charter. Today, there are about 800 ALM groups covering a population of about 2 million in the city. Women manage 80 percent of the ALMs.

ALM groups contribute significantly to the separation of household waste, and recyclable waste (20 percent of total waste) is directed to identified rag pickers. These rag pickers are the local poor and are earning a livelihood through ALMs. At the neighborhood scale, about 55 percent of the organic/biodegradable waste is being composted and applied to local gardens. Construction and demolition debris is directed to low-lying areas for landfill, reducing the burden on dumping sites. ALMs prevent approximately 20 to 25 tons of garbage per day from reaching the dumps. This offers a significant contribution for curbing greenhouse gases produced by burning municipal waste. There is a significant reduction in total waste reaching disposal sites, reducing the burden on municipal systems.

Household waste not only degrades the local environment but much of this waste find its way into uncovered stormwater drains causing waterlogging and temporary flooding. As partnerships between the municipal agency and neighborhood groups, ALMs have helped people learn to care for the areas beyond their own premises and to cooperate to solve local problems. The city government supports this community-led effort, as it not only reduces the stress on municipal systems but also helps to reduce flood risk and water-borne diseases.

In addition to separating domestic garbage, clearing garbage and composting, ALMs contribute to improving water supplies and drainage, beautifying neighborhoods, maintaining roads, filling potholes, surfacing pavements and streets, and controlling pests and stray animals.

An ALM can be formed by a neighborhood or street of about 1,000 households and is registered with the local municipal ward office, which appoints an officer to respond to citizens' concerns. The ALM committee members are selected democratically and generally include a range of neighborhood representatives including women and the elderly. ALM committees play a key role in encouraging citizens to take an active role in monitoring the city administration at the ward level. At the same time, the actions of local citizens have resulted in environmental improvements and climate-related risk reduction.

The ALM movement initially spread in areas with severe environmental degradation especially in medium-income neighborhoods. Volunteers themselves contribute small amounts of money to maintain the functioning of the organizations. MCGM has ordered their officials to give priority to ALM problems and issues.

Community-owned institutions such as the ALMs are vital for local climate resilience building. ALM could help in organizing drills for disaster scenarios, and conducting training in search and rescue. By mobilizing local resources, and volunteering with other relief agencies the ALMs proved to be very helpful in the Mumbai flood in 2005.

There is active advocacy and consultation among ALM groups and municipal agencies. This encourages transparency and accountability throughout the process. Harmonizing local government–ALM relations is critical to effective functioning. Success stories in some areas can provide examples to help resolve issues in others.

The ALM process can be replicated where local issues are dealt with at the municipal level in a decentralized way. This initiative has greatly benefited community-based waste management, greening through urban gardening, etc. These efforts contributed to reducing flood and health risk in these localities. Further, collective community action is building climate resilience by promotion of recycling, water reuse, rainwater harvesting, reduced dumping/burning of waste, and so on. ALMs stand as an important example of a community-based approach for fostering climate resilience.

Source: Surjan, A. and R. Shaw (2009). Enhancing disaster resilience through local environment management: Case of Mumbai, India. Disaster Prevention and Management, 18(4), 418–433.

many barriers to action. However, there are barriers to action that are specific to cities and to city governments that deserve more discussion.

The existing urban form is a major barrier. Many cities have expanded into flood zones and up hillsides, making them more vulnerable to floods and landslides. Cities have paved over and destroyed natural drainage systems, which also makes them more vulnerable to floods. Hard surfaces that are the hallmark of city development create urban heat islands. Altering this form will take a considerable investment and many decades (Huq *et al.*, 2007).

Although research about the health effects of climate change has increased considerably in the last few years, and international organizations such as the World Health Organization and the World Medical Association have been devoting growing attention to the issue, there remains a relatively weak understanding of health impacts, especially among city officials. There needs to be more attention to analyzing the health impacts of current extreme weather on cities and assessing how future climate change could alter these impacts for better or worse. City leaders and the public need a clearer understanding of the importance of and potential for protective action.

Uncertainty about the timing and extent of impacts plays a role in the inertia of cities. Those responsible for infrastructure engineering make decisions based on past weather conditions and are reluctant (or are constrained by legal standards) to change this practice and incorporate future projections. Apparent conflicts in the predictions – for more droughts *and* for more floods in many regions, for example – also serve to confound decision-makers. The lack of regional climate projections in most areas of the world also creates a barrier to planning for the conditions of the future. Organizations such as the Public Infrastructure Engineering Vulnerability Committee (PIEVC) of Engineers Canada have been making headway in that country by developing and testing a risk assessment framework to evaluate the resilience of a variety of types of city infrastructure to climate change. Using this system, and with the involvement of city officials in each community where risk assessments are done, PIEVC has evaluated water and sewage treatment systems, bridges, roads, buildings, and other forms of infrastructure. This tool and others like it may encourage city officials in high-income countries to adapt infrastructure to the expected impacts of climate change.

There are relatively few visionary city leaders who are willing to devote political capital to strong mitigation and adaptation programs. Although public pressure for mitigation programs has been growing in high-income countries, there is relatively little pressure on political figures to develop adaptation programs, partly because urban residents do not generally connect damaging weather events to climate change. Political figures who have the next election in mind rather than long-term impacts and planning want to spend scarce financial resources on more immediate and more visible projects that will win votes.

Some adaptation measures are so complex or expensive that they can only be undertaken with the support of many different departments of the city government, or at senior levels of government. Coordination across departmental silos is notoriously difficult and the coordination across different levels of government even more so. There are examples of urban centers where interdepartmental coordination on climate change is well organized: King County, Washington, is a North American leader in this area. The UK government has also provided support and pressure for action at the city level by financing the United Kingdom Climate Impacts Programme and instituting a system of National Indicators for Local Authorities, which includes indicators on both mitigation and adaptation.

Financial resources remain a major barrier. Cities generally have relatively low capacity to raise funds, relying primarily on property taxes, and financial support from senior levels of government for expensive projects such as infrastructure has waned. Similarly the availability of international development funds for urban infrastructure projects for cities in developing countries has been reduced in recent decades.

Despite all these barriers, there are a growing number of cities that have taken up the challenge of working to reduce the health and other impacts of climate change. Networks of cities have begun to share their strategies and experiences and encourage their members to prepare for climate change. The Alliance for Resilient Cities in Canada, and the Urban Leaders Adaptation Initiative in the United States are among the leaders in this activity. Increasingly, international institutions such as WHO, UN-HABITAT, and the World Bank are supporting the preliminary efforts of cities to understand the risks of climate change and to mitigate and adapt.

7.7 Research and data needs

In recent years, research that analyzes the relationship between climate change and health has surged (WHO, 2009). Intergovernmental programs, national governments, health sector organizations, and academic institutions and national governments have initiated many new studies related to the vulnerability of populations to climate change. There has also been an increase in research on climate change and cities. The Cities and Climate Change Symposium, hosted by the World Bank in Marseilles in June 2009, made a substantial contribution to this literature. With the exception of the extensive literature on the urban heat island effect, there has not been a great deal of research that specifically investigates health effects of climate change on cities, though this is also beginning to change. There remain many uncertainties about the range and magnitude of impacts of climate change on the health of city dwellers.

The IPCC (Confalonieri *et al.*, 2007), Frumkin *et al.* (2008) and WHO (2009) have identified the need for more specific research in the following areas related to climate change and health:

- understanding the risks to health of climate change, including: identification of key health indicators to monitor; improved quantification and evaluation of current climate–health relationships especially in low- and middle-income countries; identification of vulnerable populations and life stages; emerging impacts and modeling future risks
- identifying the effectiveness (including cost-effectiveness) of health sector interventions by means of systematic reviews of the evidence base for interventions; studies of the effectiveness of informing communities and engaging them in interventions to improve climate-related health outcomes
- study of the health implications of climate-related decisions in other sectors such as energy, transportation, water and wastewater, food and agriculture
- development and testing of decision-support tools such as vulnerability and adaptation assessments and of operational predictions for weather-related hazards such as heat waves and floods; improved understanding of decision-making processes; and research on effective communications on health and climate-related issues
- improvement of methods for understanding the costs and benefits of investment in health protection including the health costs of inaction and the costs and benefits of greenhouse gas mitigation.

These recommendations for expanded and improved research are not specific to cities, though undoubtedly the knowledge derived from research in these areas would be of benefit to developing adaptation plans for urban environment. But, given the particular manifestations of climate-related health problems already evident in cities and the proportion of the world's population that lives in urban environments, a more specific program of city-centered urban research is warranted. Addressing climate change health issues with a focus on their urban manifestations – emergency preparedness, smog, infectious disease, contaminated food and water – will save lives. Research of value would include:

- improved understanding of the linkages between climate change and urban health, as well as the effects of different geographic, social, economic, environmental, and political contexts
- linkage of climate change and urban health research with existing research investigating global city health, and health equity
- empirical studies of health impacts of extreme events as they occur in cities internationally, including descriptions and assessments of the urban populations most affected; actions of the health sector and other relevant sectors during and after the events; communications and emergency response; and underlying conditions such as state of infrastructure, housing, sanitation and other factors that affected the health outcomes

- assessment of specific shortfalls in population and health data that would allow for more robust assessments of human health effects linked to extreme weather events and climate change, and development of methods, including community-based methods, to collect this data in low- and middle-income cities (e.g., Sen *et al.*, 2003)
- more city-specific climate change vulnerability assessments that include an evaluation of health risks from current and future climate
- studies of the impact of rising temperatures and heat waves on water availability and quality and related health issues in urban settings
- evaluation of the risks for health of inadequate stormwater and sewage infrastructure, and of waste collection in informal urban settlements subjected to heavy rainfall
- assessment of the potential health effects of adaptation strategies such as low-impact drainage systems, adopted in urban environments
- investigation of the urban conditions that contribute to the reproduction and expansion of specific insect or rodent vectors that carry climate-sensitive pathogens; and development, testing, and evaluation of strategies for control of these vectors in urban environments
- study of the health effects of urban mitigation strategies including strategies to reduce emissions from energy conservation, transportation, fuel switching (from solid fuels used for cooking in most low-income countries), and other interventions
- modeling of the effects of interventions to reduce the urban heat island effect in specific settings (e.g., Rosenzweig *et al.*, 2006) together with assessments of potential improvements to health outcomes as a result
- comparative evaluation of the effectiveness of existing heat alert and response systems
- research evaluating the costs and benefits of adaptation and mitigation programs expected to have impacts on health.

There are many challenges to conducting this research. In many low- and middle-income countries there is limited information. Primary care and public health services are limited, and so is information on the extent of climate-sensitive or other illnesses. Meteorological information is scant, and region-specific climate projections are not available in many parts of the world. WHO has identified several barriers in addition to these data shortages including: limited financial resources available for this kind of research; weak incentives for interdisciplinary applied research; allocation of most health research funding to technological and curative solutions rather than population health; and weak institutional partnerships between researchers and decision-makers (Campbell-Lendrum *et al.*, 2009).

To overcome barriers, WHO recommends linking researchers and decision-makers in an iterative process of consultation that involves civil society and especially representatives of vulnerable groups. WHO also recommends establishing a virtual

[ADAPTATION] Box 7.6 Toronto, Canada: Maps help to target hot weather response where it is needed most

Stephanie Gower

Healthy Public Policy Team, Toronto Public Health

In 1999, the City of Toronto implemented a heat alert and response system with the goal of reducing heat-related illness and death in Toronto. The system was the first in Canada and is considered a premiere example of climate change adaptation. Toronto calls a heat alert when a hot air mass is forecast and the likelihood excess deaths due to heat is greater than 65 percent. The alert is broadcast by local media, and advises the public to take precautions. The alert also activates Toronto's Hot Weather Response plan, which coordinates the effort of several municipal and community agencies to provide heat-related services to vulnerable populations such as socially isolated seniors, children, people with chronic and pre-existing illnesses, and people who are marginally housed or homeless. However, there is currently little information to

guide where response measures such as opening cooling centers or extending pool hours will be most likely to reach these groups.

To ensure that hot weather response services are delivered where they are most needed, Toronto Public Health (TPH) partnered with experts in geographical information systems (GIS) and epidemiology to link current knowledge about risk factors for heat-related illness with mapping data that are available for the social and physical environment in Toronto.

The project's prototype maps characterize the spatial variation of factors that are likely to increase exposure to heat in Toronto, such as surface temperature, and factors such as low income and age that may affect people's resilience to extreme heat. Two samples of these maps are shown in Box Figures 7.1 and 7.2.

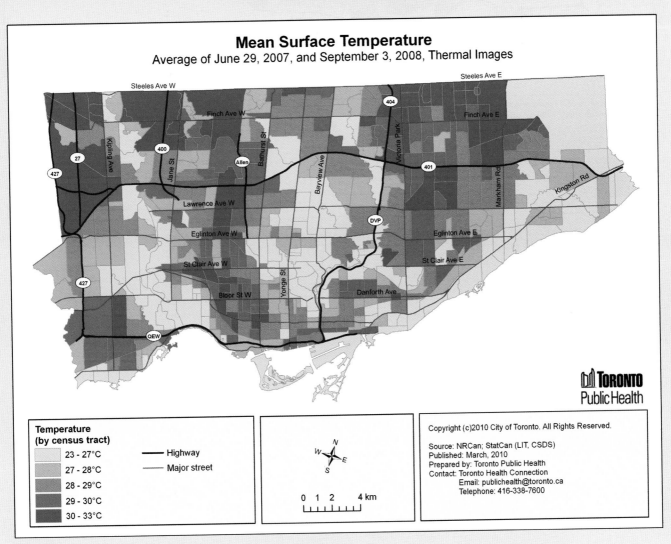

Box Figure 7.1: *Spatial variation in surface temperature (degrees C) in Toronto, Canada.*

Source: © City of Toronto 2010. Preliminary map, reprinted with the permission of Toronto Public Health.

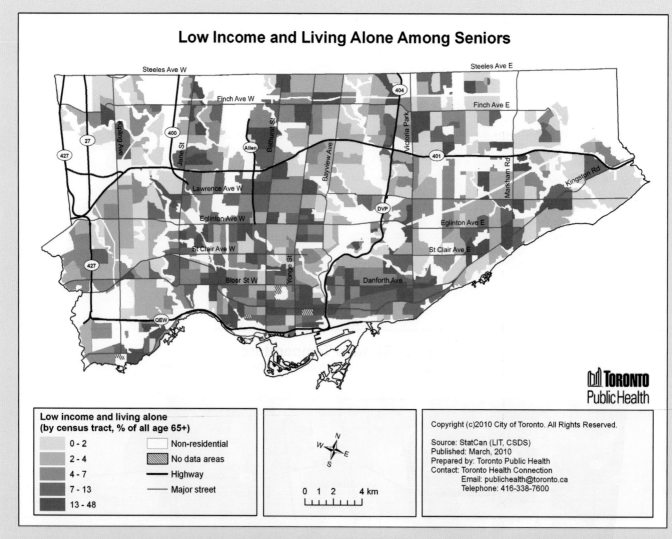

Low Income and Living Alone Among Seniors

Low income and living alone (by census tract, % of all age 65+)

- 0 - 2
- 2 - 4
- 4 - 7
- 7 - 13
- 13 - 48
- Non-residential
- No data areas
- Highway
- Major street

0 1 2 4 km

Box Figure 7.2: *Distribution of low income seniors living alone by census tract in Toronto, Canada. Low-income senior is represented as percentage of those age 65+ below the 2005 after-tax LICO (low-income cutoff).*

Source: © City of Toronto 2010. Preliminary map, reprinted with the Permission of Toronto Public Health.

To enable multiple risk factors to be considered together, the project team is now finalizing a map that integrates multiple risk factors into a summary index of overall vulnerability. To aid in short- and long-term planning, maps will also be created that overlay vulnerability to heat with access to existing hot weather facilities and services.

With support from Natural Resources, Canada's Regional Adaptation Collaboratives Program, Toronto Public Health is now testing advanced geospatial statistical techniques designed to locate pockets of vulnerable people, refining the maps to include information about population density, and determining a way to validate the approach. Before finalizing

the mapping tool and an associated map series, TPH will also gather feedback on the prototype maps from staff and volunteers who implement hot weather response across the city.

About 120 deaths per year are currently attributed to extreme heat in Toronto, and heat-related mortality could double by the 2050s and triple by the 2080s (Pengelly *et al.*, 2007). To ensure a long life for the GIS tool and associated map products, Toronto is creating guides for data maintenance and updating, and will share its methods with other interested jurisdictions. Toronto must be prepared to cope with potentially much higher health-related impacts from heat in the future.

[ADAPTATION]   Box 7.7   Pilot projects to protect Canadians from extreme heat events

Peter Berry

Health Canada

Extreme heat events pose a growing public health risk in many regions of Canada, as a result of a changing climate. The Intergovernmental Panel on Climate Change (IPCC) Fourth Assessment Report concludes that the frequency of extreme heat events is very likely to increase with climate change (IPCC, 2007a). The increase is a public health concern for Canada, which is experiencing rapid growth in the population of seniors, a group that has been identified as particularly vulnerable to the health impacts of extreme heat. Gosselin *et al.* (2008) project that in the absence of further adaptations there will be an increase of 150 excess deaths annually by 2020 in the province of Quebec, 550 excess deaths by 2050, and 1,400 by 2080.

Communities in Canada are adapting to the increase in extreme heat events by implementing heat alert and response systems (HARS), which are designed to prevent mortality and morbidity during extreme heat events by delivering timely warnings and interventions. HARS are operating in a number of Canadian cities, with established systems in communities such as Toronto and Montreal. To facilitate the development and improvement of such systems across the country, communities require information about optimal approaches to mobilize public health and emergency management officials to address these hazards implement and the most effective adaptations.

In 2007, Health Canada launched a four-year program to increase the resiliency of individuals and communities to the health impacts of extreme heat events. To accomplish this goal pilot HARS are being developed in four Canadian communities – Winnipeg (Manitoba), the Assiniboine Health Authority Region (Manitoba), Windsor (Ontario), and Fredericton (New Brunswick). The objectives of the pilot projects are to identify effective adaptations for protecting health along with protocols for the design, implementation, and evaluation of regional and municipal HARS. Communities have been selected in areas of Canada that are characterized by different climate, socio-economic, demographic, governance, and institutional conditions and challenges.

The pilot communities are developing and implementing HARS for the summer of 2011 through local participatory processes. Each community has developed an advisory committee that comprises a range of government departments and stakeholders, such as those who have a role in health care, emergency preparedness, seniors and healthy aging, social services and housing, mental health and addictions, communications, police, fire and paramedic services and utilities. These committees ensure that the HARS meet their unique community needs, in an effort to protect those most vulnerable to heat-health risks. Further, the development of measures to protect citizens is being informed by extreme heat and health vulnerability assessments, table-top (simulation) exercises, and the establishment of education and outreach strategies.

Broad collaboration among Health Canada, the pilot communities, and key stakeholders is aimed at fostering increased knowledge of heat-health risks and local adaptive capacity to support the sustainability of HARS in these communities beyond the life of the four-year program. The tools developed and learnings derived from the program will be used to inform a Heat Alert and Response System Best Practices Guidebook, which will be disseminated to public health and emergency management officials across Canada.

---

forum on climate change and health research, as well as the creation of expert panels to provide oversight, produce best practice guidance, and share tools. It would be of great value if one of these expert panels focused on climate change and health in cities.

Early research on the climate change impacts used global climate models (GCMs) in a top-down way to develop scenarios for different locations (Huq, 2008). Other studies used a more bottom-up approach to identify vulnerable places and people. Studies, now tend to combine top-down and bottom-up information to develop adaptation strategies.

The current generation of climate change impacts research is actively linked to the people whose lives will be affected, often in cities. Researchers are working with decision makers who will use their research, from project design, implementation and communication. This work includes partnerships with local governments, and non-governmental organizations (Huq, 2008).

Cities in poor countries face many constraints, including limited research capacity and lack of facilities for developing adaptation (Haines *et al.* 2004). Institutions and mechanisms are needed to promote effective interactions among researchers, policymakers, and of city groups to facilitate the appropriate incorporation of research findings into policies (Haines *et al.* 2004). The goal is to protect and improve public health in both current and future climates (Haines and Patz, 2004).

7.8 Concluding comments

Determinants of human health are forecast to worsen under climate change. Cities concentrate populations who are particularly vulnerable to the effects of climate change. When extreme weather events such as intense precipitation, cyclonic storms, or storm surges strike in cities they affect a much larger number of people than when they strike less populated regions. In both high-income and developing countries, the severe weather events associated with climate change combined with the many stresses on cities can create conditions of extreme health hazard to city residents.

Flooding is the most frequent weather-related disaster. It affects the largest number of city dwellers, and is predicted to worsen under climate change. The impacts of floods and storms can be spectacularly large in cities – leading to many thousands of deaths and resulting in economic damage in the billions.

Poorer air quality and increased ozone concentrations in metropolitan regions will engender substantial increases in ozone-related mortality. Evidence linking heat waves to mortality both from high-income countries and for the limited data available from the developing world suggests that the urban heat island effect renders city dwellers without access to cooling particularly vulnerable to heat waves. Moreover, urban air pollution related effects of temperature increases will contribute sustainably more to mortality than heat effects alone. Indirect health effects on city dwellers from climate change include reduced access to food and expanded exposure to infectious diseases.

Climate change mitigation and adaptation strategies can overlap and reduce the health impacts of climate change especially from heat and air pollution. Passive methods of cooling buildings such as shading, reflective roofs, natural ventilation (convective cooling), insulation, placement of nearby trees, green roofs and permeable pavements will all reduce heat loads on buildings, reducing the need for air conditioning, improving comfort, and reducing heat impacts for the occupants. Implementing such "win-win-win" strategies can have multiple benefits. However, mitigation, adaptation, and health strategies do not always clearly coincide and these goals need to be examined in any proposed policy.

Not all adaptation strategies will positively contribute to development or to health outcomes. If informal settlements on floodplains are razed to reduce the risk of floods without alternative housing and land arrangements made for and with the residents, for example, the affected populations may be left without shelter and worse off than before.

Barriers to effective response to climate change in cities include existing urban forms and lack of information. Altering existing urban conditions that combine with climate change to result in deaths, injuries, and illnesses in the population will be key to protecting health in cities. In low- and middle-income cities this requires a focus on meeting basic development needs – adequate housing, provision of infrastructure that supplies clean water, sanitation, and energy, education, and primary health care services. Altering the urban characteristics that worsen the impacts of extreme weather events will decrease the likelihood of events turning into health and economic disasters. Anticipating climate change over the lifetime of the urban structures can preclude investments in urban infrastructure from being jeopardized by climate change.

There is a weak understanding of the urban health impacts of climate change among officials at all levels, national, subnational, and city. The health sector can play an important advocacy role in encouraging that action is taken to protect the health and lives of city dwellers. Health officials could undertake to improve knowledge of the health effects of climate change and motivate appropriate protective action by agencies with more direct responsibilities for urban design and city services – city planning, energy, transportation, and water and storm water infrastructure. Promotion and adoption of "Healthy Cities" strategies will be simultaneously beneficial for climate change mitigation, adaptations and population health.

Many of the manifestations of climate-related health problems are already evident in cities. Provided that appropriate infrastructure is in place, urban areas with their concentration of wealth and resources provide unique opportunities to build large-scale, cost-effective resilience, thereby reducing health vulnerability to climate change. A more specific urban health-centered climate change research program is warranted to support health protection needs of the increasing proportion of the world's population residing in cities. These efforts will be successful if the efforts of local governments are supported by their national and regional governments.

Evidence of the potential impacts of climate change on urban population health is growing. The magnitude and significance of these impacts will vary according to specific modulating factors of the cities, such as population density, social, economic, political, geographic, and environmental characteristics as well as medical and infrastructure services. Local government jurisdictions over these modulating factors vary, with many local governments responsible for basic infrastructure and national governments responsible for health and other critical systems, impacts are best addressed by cooperative effort between local and senior orders of government.

There are still large uncertainties and more study is necessary. Nevertheless, urgent action is required to ensure that:

1. the connections between climate change and health in cities are made clear to public health practitioners, city planners, policymakers, and the general public
2. urban practitioners take immediate responsibility to integrate climate change into planning in all areas of public health, food systems, infrastructure, and land use planning, thereby increasing the potential to create more healthy and sustainable communities

3. adaptation strategies focus on activities that eliminate health disparities, improve neighborhood conditions, and protect those who will be most impacted by climate change

4. national governments address the health risks of climate change to avoid unnecessary costs and hardship for health care and social systems, by devoting necessary resources to developing the adaptive capacity of local governments, communities, and individuals.

Without effective adaptation strategies, climate change will increase health and other impacts in the world's cities where the majority of population resides. The costs of the health impacts due to climate change (increased illness, injury, disease, death) will further weaken stressed local social systems. Cities provide unique opportunities to marshal resources and wealth to build resilience and health protective policies and programs. The implementation of health-care adaptation measures will differ among cities, each of which has its own specific modulating influences, and different social, economic, cultural, and political realities, but concerted action for knowledge gathering and sharing can strengthen the needed efforts at the individual city level.

REFERENCES

Abdallah, S. and G. Burnham (Eds.) (2000). *Public Health Guide for Emergencies*. John Hopkins University and International Red Cross/Red Crescent Societies. Federation of Baltimore, MD.

Acclimatise (2006). *Adapting to Climate Change: Lessons for London*. London Climate Change Partnership.

Adam, J., A. Hamlet, and P. Lettenmaier (2009). Implications of global climate change for snowmelt hydrology in the twenty-first century. *Hydrological Processes*, **23**, 962–972.

Ahern, M., R. Kovats, P. Wilkinson, R. Few, and F. Matthies (2005). Global health impacts of floods: epidemiologic evidence. *Epidemiologic Reviews*, **27**, 36–46.

Ahmed, S., N. Diffenbaugh, and T. Hertel (2009). Climate volatility deepens poverty vulnerability in developing countries. *Environmental Research Letters*, **4**.

Alajo, S., J. Nakavuma, and J. Erume (2006). Cholera in endemic districts in Uganda during El Nino rains: 2002–2003. *African Health Science*, **6**, 93–97.

Alam, M. and G. Rabbani (2007). Vulnerabilities and responses to climate change for Dhaka. *Environment and Urbanization*, **19**, 81–97.

Ashley, R. and A. Cashman (2006). *Infrastructure to 2030: Telecom, Land Transport, Water and Electricity*. OECD.

Auld, H., D. MacIver, and J. Klaassen (2004). Heavy rainfall and waterborne disease outbreaks: the Walkerton example. *Journal of Toxicology and Environmental Health, Part A*, **67**, 1879–1887.

Baccini, M., A. Biggeri, G. Accetta, *et al.*(2008). Heat effects on mortality in 15 European cities. *Epidemiology*, **19**, 711–719.

Balbus, J. and M. Wilson (2000). *Human Health and Global Climate Change: A Review of Potential Impacts in the United States*, Pew Center on Global Climate Change. Accessed Feb 8, 2011, http:/lwww.~ewclimate.or~/~warming-in-depthla11re~ortslhulnanhealtl$index.cftn.

Barclay, E. (2008). Is climate change affecting dengue in the Americas? *The Lancet*, **371**, 973–974.

Barrios, S., L. Bertinelli, and E. Strobl (2006). Climatic change and rural–urban migration: the case of sub-Saharan Africa. *Journal of Urban Economics*, **60**, 357–371.

Bartlett, S. (2008). Climate change and urban children: impacts and implications for adaptation in low- and middle-income countries. *Environment and Urbanization*, **20**, 501–519.

Bartlett, S. (2009). Children: a large and vulnerable population in the context of climate change. Paper prepared for Expert Group Meeting on Population Dynamics and Climate Change, 24–25 June, 2009. Accessed Feb 8, 2011, www.unfpa.org/webdav/site/global/users/schensul/public/CCPD/papers/Bartlett%20Paper.pdf.

Bartlett, S., D. Dodman, J. Hardoy, D. Satterthwaite, and C. Tacoli (2009). Social aspects of climate change in urban areas in low- and middle-income nations. Contribution to the World Bank Fifth Urban Research Symposium *Cities and Climate Change: Responding to an Urgent Agenda*. Accessed Feb 8, 2011, www.urs2009.net/docs/papers/Satterthwaite.pdf.

Bates, B., Z. W. Kundzewicz, S. Wu, and J. P. Palutikof (2008). *Climate Change and Water*, Technical Paper of the Intergovernmental Panel on Climate Change, Geneva: IPCC Secretariat.

Beebe, N., R. Cooper, P. Mottram, and A. Sweeney (2009). Australia's dengue risk driven by human adaptation to climate change. *PLoS Neglected Tropical Diseases*, **3**, e429.

Bell, M. L., R. Goldberg, C. Hogrefe, *et al.* (2007). Climate change, ambient ozone, and health in 50 US cities. *Climatic Change*, **82**, 61–76.

Bernard, S. and M. McGeehin (2004). Municipal heatwave response plans. *American Journal of Public Health*, **94**, 1520–1521.

Berry, P. (2008). Vulnerabilities, adaptation and adaptive capacity. In J. Seguin (Ed.), *Human Health in a Changing Climate: A Canadian Assessment of Vulnerabilities and Adaptive Capacity,* Ottawa: Health Canada.

Bhosale, R. (2008). *Disaster Management and Flood Control: An Initiative by Pune Municipal Corporation*. Presentation to Building a Local Government Alliance for Disaster Risk Reduction Consultative Meeting, UNISDR, Barcelona, 22–23 May 2008.

Bicknell, J., D. Dodman and Satterthwaite, D. (2009). *Adapting Cities to Climate Change: Understanding and Addressing the Development Challenges*. London, UK: Earthscan.

Bobylev, N. (2009). Urban underground infrastructure and climate change: opportunities and threats. *Russian Academy of Sciences*. Accessed Dec 21, 2009, http://www.urs2009.net/docs/papers/ Bobylev.pdf.

Bunyavanich, S., C. Landrigan, A. McMichael, and P. Epstein (2003). The impact of climate change on child health. *Ambulatory Pediatrics*, **3**, 44–52.

Campbell-Lendrum, D. and C. Corvalán (2007). Climate change and developing-country cities: implications for environmental health and equity. *Journal of Urban Health*, **84**, doi:10 1007/s11524–007–9170-x.

Campbell-Lendrum, D., R. Bertollini, M. Neira, K. Ebi, and A. McMichael (2009). Health and climate change: a roadmap for applied research. *Lancet*, **16**, 1663–5.

CCAP [Center for Clean Air Policy] (2009). *Urban Leaders Adaptation Initiative*. Accessed Feb 8, 2011, www.ccap.org/index.php?component= resources&program=6.

CCD [Commission on Climate Change and Development] (2008). *Links Between Disaster Risk Reduction, Development and Climate Change*, UNISDR Secretariat, Geneva. Accessed Dec 21, 2009, www.ccdcommission.org/Filer/pdf/pb_disaster_risk_reduction.pdf.

CCSP (2008). *Analyses of the Effects of Global Change on Human Health and Welfare and Human Systems*, Report by the US Climate Change Science Program and the Subcommittee on Global Change Research. USEPA, Washington, DC, USA.

Charron, D., M. Thomas, D. Waltner-Toews, J. Aramini, and T. Edge (2004). Vulnerability of waterborne diseases to climate change in Canada: a review. *Journal of Toxicology and Environmental Health*, **67**, 1667–1677.

Cheng, C., M. Campbell, Q. Li, *et al.* (2009). Differential and combined impacts of extreme temperatures on human mortality in south-central Canada: Part II. *Air Quality and Atmospheric Health*, **1**, 223–235.

Cifuentes, V., N. Borja-Aburto, N. Gouveia, G. Thurston, and D. Davis (2001). Assessing the health benefits of urban air pollution reductions associated with climate change mitigation (2000–2020): Santiago, São Paulo, México City, and New York City. *Environmental Health Perspectives*, **109**, 419–426.

Clean Air Partnership (2009). *Alliance for Resilient Cities* Accessed Feb 8, 2011, www.cleanairpartnership.org/arc.

Cohen, M. and J. Garrett (2009). *The Food Price Crisis and Urban Food (In)Security*, Human Settlements Working Paper Series. Accessed Feb 8, 2011, www.iied.org/pubs/pdfs/10574IIED.pdf.

Cole, D., D. Lee-Smith, and G. Nasinyama (2008). *Healthy City Harvests: Generating Evidence to Guide Policy on Urban Agriculture*, Lima, Peru, and Kampala, Uganda: Urban Harvest and Makerere University Press.

Confalonieri, U. (2008). *Mudança Climática Global e Saúde Humana no Brasil*. Parcerias Estratégicas /Centro de Gestão e Estudos Estratégicos. Brasília: Centro de Gestão e Estudos Estratégicos: Ministério da Ciência e Tecnologia, n.27.

Confalonieri, U., B. Menne, R. Akhtar, *et al.* (2007). Human health. In *Climate Change 2007: Impacts, Adaptation and Vulnerability. Contribution of Working Group II to the Fourth Assessment Report of the Intergovernmental Panel on Climate Change*, Cambridge, UK: Cambridge University Press.

Costello, A, M. Abbas, A. Allen, and S. Ball (2009). Managing the health effects of climate change. *The Lancet*, **373**, 1693–1733.

Davis, R., P. Knappenberger., W. Novicoff, and P. Michaels (2003). Decadal changes in summer mortality in U.S. cities. *International Journal of Biometeorology*, **47**, 166–175.

Della-Marta, P., M. Haylock, J. Luterbacher, and H. Wanner (2007). Doubled length of western European summer heat waves since 1880. *Journal of Geophysical Research*, **112**.

del Ninno, C. and M. Lundberg (2005). Treading water: the long term impact of the 1998 flood on nutrition in Bangladesh. *Economic Human Biology*, **3**, 67–96.

de Sherbinin, A., A. Schiller, and A. Pulsipher (2007). The vulnerability of global cities to climate hazards. *Environment and Urbanization*, **19**, 39–64.

Despommier (2009). A farm on every floor. *New York Times*, August 23, 2009. Accessed Feb 8, 2011, www.nytimes.com/2009/08/24/opinion/24Despommier.html?_r=2.

Dickinson, T. (2007). *The Compendium of Adaptation Models for Climate Change* (1st edition), Environment Canada, Adaptation and Impacts Research Division.

Dodman, D. (2009). *Urban density and climate change: an agenda for mitigation and adaptation*. Paper prepared for Expert Group Meeting on Population Dynamics and Climate Change, London.

Droege, P. (2008). *Urban Energy Transition From Fossil Fuels to Renewable Power*. University of Newcastle, NSW, Australia and World Council for Renewable Energy: Elsevier.

D'Souza, R., N. Becker, G. Hall, and K. Moodie (2004). Does ambient temperature affect foodborne disease? *Epidemiology*, **15**(1), 86–92.

Dubbeling, M., M. C. Campbell, F. Hoekstra, and R. van Veenhuizen (2009). Building resilient cities. *Urban Agriculture Magazine*, **22**.

Ebi, K. (2005). Improving public health responses to extreme weather events. In W. Kirch, B. Menne, and R. Bertolinni (Eds.), *Extreme Weather Events and Public Health Responses*, Berlin, Germany:Springer.

Ebi, K. and A. Nyong (2009). *The Health Risks of Climate Change*, Commonwealth Health Ministers' Update. Accessed Feb 8, 2011, www.thecommonwealth.org/files/190382/FileName/2-EbiandNyong_2009.pdf.

Ebi, K., T. Teisberg, L. Kalkstein, L. Robinson, and R. Weiher (2004). Heat watch warning systems save lives: estimated costs and benefits for Philadelphia 1995–98. *Bulletin of the American Meteorological Society*, **85**, 1067–1073.

Ekelund, N. (2007). *Adapting to Climate Change in Stockholm*, City of Stockholm.

Epstein, P. and C. Defilippo (2001). West Nile virus and drought. *Global Change and Human Health*, **2**, 105–107.

European Commission (2008). *Addressing the Social Dimensions of Environmental Policy: A Study on the Linkages Between Environmental and Social Sustainability in Europe*. Accessed Feb 8, 2011, ec.europa.eu/social/BlobServlet?docId=1574&langId=en.

FAO (2009). *The State of Food Insecurity in the World 2009: Economic Crises – Impacts and Lessons Learned*. Rome: Food and Agriculture Organization of the United Nations. Few, R., H. Osbahr, L. M. Bouwer, D. Viner, and F. Sperling (2006). *Linking Climate Change Adaptation and Disaster Risk Management for Sustainable Poverty Reduction*, Report for the Vulnerability and Adaptation Resource Group, Washington.

Few, R. and F. Matthies (2006). *Flood Hazards and Health: Responding to Present and Future Risks*, London, UK: Earthscan.

Foreign Policy (2009). *Failed States Index*. Accessed Dec 21, 2009, www.foreignpolicy.com/articles/ 2009/06/22/the_2009_failed_states_index.

Frumkin, H. and A. McMichael (2008). Climate change and public health: thinking, communicating, acting. *American Journal of Preventative Medicine*, **35**, 403–410.

Frumkin, H., J. Hess, G. Luber, J. Malilay, and M. McGeehin (2008). Climate change: the public health response. *American Journal of Public Health*, **98**, 435–445.

Gage K., T. Burkot, R. Eisen, and E. Hayes (2008). Climate and vectorborne diseases. *American Journal of Preventative Medicine*, **35**, 436–50.

Gartland, L. (2008). *Urban Heat Islands: Understanding and Mitigating Heat in Urban Areas*. London, UK: EarthScan.

GFDRR (2009). *Reducing Disaster Risks for Sustainable Development*, Global Facility for Disaster Reduction and Recovery. Accessed Feb 8, 2011, http://gfdrr.org/docs/GFDRR_Brochure.pdf.

Githeko, A. (2007). Malaria, climate change, and possible impacts on populations in Africa. In *HIV, Resurgent Infections and Population Change in Africa*, International Studies in Population, Volume 6, Dordrecht, the Netherlands: Springer.

Glantz, M. H. (2008). Hurricane Katrina as a "teachable moment". *Advances in Geoscience*, **14**, 287–294. Accessed Feb 8, 2011, www.adv-geosci.net/14/287/2008/.

Gosling, S. N., G. R. McGregor, and J. A. Lowe (2009). Climate change and heat-related mortality in six cities. Part 2: climate model evaluation and projected impacts from changes in the mean and variability of temperature with climate change. *International Journal of Biometeorology*, **53**, 31–51.

Gosselin, P., D. Belanger, and B. Doyan (2008). Health impacts of climate change in Quebec. In J. Seguin (Ed.), *Human Health in a Changing Climate: A Canadian Assessment of Vulnerabilities and Adaptive Capacity*, Ottawa: Health Canada.

Gould, E., and S. Higgs (2009). Impact of climate change and other factors on emerging arbovirus diseases. *Transactions of the Royal Society of Tropical Medicine and Hygiene*, **103**, 109–121.

Greenough, G., M. McGeehin, S. Bernard, *et al.* (2001). The potential impacts of climate variability and change on health impacts of extreme weather events in the United States. *Environmental Health Perspectives*, **109**, 191–198.

Griffiths, H., M. Rao, F. Adshead, and A. Thorpe (2009). *The Health Practitioner's Guide to Climate Change: Diagnosis and Cure*. London, UK: Earthscan.

Grossi, P. and R. Muir-Wood (2006). The 1906 *San Francisco Earthquake and Fire: Perspectives on a Modern Super Catastrophe*. Newark, CA, USA: Risk Management Solutions.

Gubler, D. (1998). Resurgent vector-borne diseases as a global health problem. *Emerging Infectious Diseases*, **4**, 442–450.

Guha-Sapir, D. and R. Below (2002). *Quality and Accuracy of Disaster Data: A Comparative Analyses of 3 Global Data Sets*. Technical report, WHO Centre for Research on the Epidemiology of Disasters.

Guha-Sapir, D., D. Hargitt, and P. Hoyois (2004). *Thirty Years of Natural Disasters 1974–2003: The Numbers*. Louvain, Belgium: Presses Universitaires de Louvain.

Haines, A. (2008). Climate change and health: strengthening the evidence base for policy. *American Journal of Preventative Medicine*, **35**, 411–3.

Haines, A. and J. Patz (2004). Health effects of climate change. *JAMA*, **291**, 99–103.

Haines, A., S. Kuruvilla, and M. Borchert (2004). Bridging the implementation gap between knowledge and action for health. *Bulletin of the World Health Organization*, **82**, 724–732.

Hamin, E. and N. Gurran (2009). Urban form and climate change: balancing adaptation and mitigation in the U.S. and Australia. *Habitat International*, 33, 238–245.

Hardoy, E. and Pandiella (2009). Urban poverty and vulnerability to climate change in Latin America.In J. D. Bicknell, D. Dodman, *et al.* (Eds.), *Adapting Cities to Climate Change: Understanding and Addressing the Development Challenges*, London, UK: Earthscan.

Hay, S. D. Rogers, S. Randolph, *et al.* (2002). Hot topic or hot air? Climate change and malaria resurgence in East African highlands. *Trends in Parasitology*, 18, 530–534.

HDR (2007). *Human Development Report 2007/2008: Fighting Climate Change: Human Solidarity in a Divided World*. Human Development Report Office. Accessed Feb 8, 2011, http://hdr.undp.org/ en/reports/ global/hdr2007–2008/papers/de%20la%20fuente_alejandro _2007 .pdf.

Holloway, T., S. N. Spak, D. Barker, *et al.* (2008). Change in ozone air pollution over Chicago associated with global climate change. *Journal of Geophysical Research*, 113.

Hsieh, C., T. Aramaki, and K. Hanaki (2007). The feedback of heat rejection to air conditioning load during the night-time in subtropical climate. *Energy and Buildings*, 39, 1175–1182.

Hunter, P. R. (2003). Climate change and water-borne and vector-borne disease. *Journal of Applied Microbiology*, 94, 37–46.

Huq, S. (2008). Who are the real climate experts? *New Scientist*, No. 2649, April 2, 2008.

Huq, S., R. S. Kovats, H. Reid, and D. Satterthwaite (2007). Reducing risks to cities from disasters and climate change [Editorial]. *Environment and Urbanization*, 19, 3–15.

Ichinose, T., Y. Bai, J. Nam, and Y. Kim (2006). Mitigation of thermal stress by a large restoration of inner-city river (Cheong-Gye Stream in Seoul). *Proceedings of ICUC*, 6, 358–361.

IPCC (2007a). *Climate Change 2007: Contribution of Working Group II to the Fourth Assessment Report of the Intergovernmental Panel on Climate Change*, M. L. Parry, O. F. Canziani, J. P. Palutikof, P. J. van der Linden and C. E. Hanson (Eds), Cambridge, UK, and New York, USA: Cambridge University Press.

IPCC (2007b). Summary for Policymakers. In S. Solomon, D. Qin, M. Manning, *et al.* Eds.), *Climate Change 2007: The Physical Science Basis. Contribution of Working Group I to the Fourth Assessment Report of the Intergovernmental Panel on Climate Change*, Cambridge, UK, and New York, USA: Cambridge University Press.

Jelinek, T. (2009). Trends in the epidemiology of dengue fever and their relevance for importation into Europe. *Euro Surveillance*, 14, 19250. Accessed Dec 21, 2009, www.eurosurveillance.org/ ViewArticle. aspx?ArticleId=19250.

Jonkman, S. (2005). Global perspectives on loss of human life caused by floods. *Natural Hazards*, 34, 151–175.

Kalkstein, L., J. Greene, D. Mills, *et al.* (2008). Analog European heat waves for U.S. cities to analyze impacts on heat-related mortality. *American Meteorological Society*, 1–11.

Karl, T., J. Melillo, T. Peterson, and S. Hassol (2009). *Global Climate Change Impacts in the United States*. Cambridge, UK: Cambridge University Press.

Kinney, P. (2008). Climate change, air quality, and human health. *American Journal of Preventative Medicine*, 35, 459–67.

Kistemann, T., T. Classen, C. Koch, *et al.* (2002). Microbial load of drinking water reservoir tributaries during extreme rainfall and runoff. *Applied and Environmental Microbiology*, 68, 2188–2197.

Kjellstrom, T., S. Friel, J. Dixon, *et al.* (2007). Urban environmental health hazards and health. *Equity Journal of Urban Health*, 84, N1.

Knowlton, J., C. Rosenthal, B. Hogrefe, *et al.* (2004). Assessing ozone-related health impacts under a changing climate. *Environmental Health Perspectives*, 112, 1557.

Kojima, M., C. Brandon, and J. Shah (2000). *Improving Urban Air Quality in South Asia by Reducing Emissions from Two-Stroke Engine Vehicles*, World Bank.

Koppe, C., G. Jendritzky, R. Kovats, and B. Menne (2004). *Heat Waves: Risks and Responses* Copenhagen, Denmark: World Health Organization.Kovats, S. and R. Akhtar (2009). Climate, climate change and human health in Asian cities. *Environment and Urbanization*, 20, 165.

Kovats, R. and K. Ebi (2006). Heatwaves and public health in Europe. *European Journal of Public Health*, 16, 592–599.

Kovats, S. and S. Hajat (2008). Heat stress and public health: a critical review. *Annual Review of Public Health*, 29, 04.

Kovats, R., S. Edwards, and S. Hajat (2004). The effect of temperature on food poisoning: a time-series analysis of salmonellosis in ten European countries. *Epidemiology and Infection*, 132, 443–453.

Kovats, R., S. Edwards, D. Charron, *et al.* (2005). Climate variability and campylobacter infection: an international study. *International Journal of Biometeorology*, 49, 207–214.

Kunkel, K. E., S. A. Changnon, B. C. Reinke, and R. Arritt (1996). The July 1995 heat wave in the Midwest: a climatic perspective and critical weather factors. *Bulletin of the American Meteorological Society*, 77, 1507–1518.

Kyalimpa, J. (2009). *Looming Problems with Kampala's Water*, Inter Press Service Accessed Feb 8, 2011, http://ipsnews.net/news. asp?idnews=43592.

Laaidi, M., K. Laaidi, and J.-P. Besancenot (2006). Temperature-related mortality in France, a comparison between regions with different climates from the perspective of global warming. *International Journal of Biometeorology*, 51, 145–153.

Lapitan., J., P. Brocard, R. Atun, and C. Tantinimitkul (2009). City health system preparedness to changes in Dengue fever attributal to climate change: an exploratory case study. Presentation to the *5th Urban Research Symposium on Cities and Climate Change*, Marseille, France.

Larsen, K. and F. Barker-Reid (2009). Adapting to climate change and building urban resilience in Australia. *Urban Agriculture Magazine*, 22.

Laukkonen, J., P. K. Blanco, J. Lenhart, *et al.* (2009). Combining climate change adaptation and mitigation measures at the local level. *Habitat International*, 33, 287–292.

Lecomte, E., A. Pang, and J. Russell (1998). *Ice Storm '98*, Toronto and Boston: Institute for Catastrophic Loss Reduction and Institute for Business and Home Safety.

Lemmen, D. S., F. J. Warren, J. Lacroix, and E. Bush (2008). In *From Impacts to Adaptation: Canada in a Changing Climate 2007*, Government of Canada, Ottawa, ON, pp. 227–274.

Létard, V., H. Flandre, and S. Lepeltier (2004). *La France et les Français face à la canicule: les leçons d'une crise*. Report No. 195 (2003–2004) to the Sénat, Government of France. Cited in IPCC, 2007a. Available (in French) on www.senat.fr.

le Tertre, A., A. Lefranc, D. Eilstein, *et al.* (2006). Impact of the 2003 heatwave on all-cause mortality in 9 French cities. *Epidemiology*, 17, 75–79.

Lindsay, S. and M. Birley (1996). Climate change and malaria transmission. *Annals of Tropical Medicine and Parasitology*, 90, 573–588.

Lock, K. and R. van Veenhuizen (2001). Balancing the positive and negative health impacts. *Urban Agriculture Magazine*, 1, 3.

Lombardo, M. (1985). *Ilha de calor nas metrópoles: o exemplo de São Paulo*. São Paulo, Brazil: HUCITEC.

Luber, G. and M. McGeehin (2008). Climate change and extreme heat events. *American Journal of Preventative Medicine*, 35, 429–435.

Mabasi, T. (2009). *Assessing the Impacts, Vulnerability, Mitigation and Adaptation to Climate Change in Kampala City*, Uganda Pentecostal University, Kampala, Uganda. Accessed Dec 21, 2009, www.urs2009. net/docs/papers/Mabasi.pdf.

Marfai, L., L. King, D. Singh, *et al.* (2008). Natural hazards in Central Java Province, Indonesia: an overview. *Environmental Geography*, 56, 335–351.

Martens, P. and M. McMichael (2002). *Environmental Change, Climate, and Health: Issues and Research Methods*. Cambridge, UK: Cambridge University Press.

Martine, G. (2009). Population dynamics and policies in the context of global climate change. Paper prepared for Expert Group Meeting on Population Dynamics and Climate Change, London, June 24–25, 2009.

Mayor of London (2008). *The London Climate Change Adaptation Strategy: Draft Report*. Greater London Authority, London, UK. Accessed Dec 21, 2009, www.london.gov.uk/mayor/publications/2008/08/climate-change-adapt-strat.jsp.

McBean, G. and D. Henstra (2003). *Climate Change, Natural Hazards and Cities*, Institute for Catastrophic Loss Reduction Research Paper Series, No. 31, March 2003.

McGranahan, G., D. Balk, and B. Anderson (2007). The rising tide: assessing the risks of climate change and human settlements in low elevation coastal zones. *Environment and Urbanization*, 19, 17–37.

McMichael, A. J, D. H. Campbell-Lendrum, C. F. Corvalán, *et al.* (2003). *Climate Change and Human Health: Risks and Responses*. Geneva, Switzerland: WHO.

Meehl, G. A. and C. Tebaldi (2004). More intense, more frequent, and longer lasting heat waves in the 21st century. *Science*, **305**, 994–997.

Menne, B., F. Apfel, S. Kovats, and F. Racioppi (2008) *Protecting Health in Europe from Climate Change*. Geneva, Switzerland: WHO Press.

Mitchell, T. and M. van aalst (2008). *Convergence of disaster risk reduction And Climate Change Adaptation: A Review for DFID*. Accessed Dec 21, 2009, www.preventionweb.net/files/7853_ ConvergenceofDRandCCA1.pdf.

Molesworth, A., S. Cuevas, H. Connor, A. Morse, and M. Thomson (2003). Environmental risk and meningitis epidemics in Africa. *Emerging Infectious Diseases*, **9**, 1287–1293.

Mool, P., S. Bajracharya, and S. Joshi (2001). *Inventory of Glaciers, Glacial Lakes and Glacial Lake Outburst Floods: Monitoring and Early Warning Systems in the Hindu-Kush Himalayan Region, Nepal*.

Nagaraj, N. (2009). Malaria thrives in cities. *Times of India*. May 25, 2009.

Natural Resources Canada (2007). *Climate Change Impacts and Adaptation Division*. Accessed Feb 8, 2011, http://adaptation.nrcan.gc.ca/perspective/health_3_e.php.

Nerlander, L. (2009). *Climate Change and Health*, Commission on Climate Change and Development. Accessed Feb 8, 2011, www.ccdcommission.org/Filer/commissioners/Health.pdf.

New York City (2009). *PlaNYC*. Accessed Feb 8, 2011, www.nyc.gov/html/planyc2030/html/plan/climate.shtml.

Nicholls, R., S. Hanson, C. Herweijer, *et al.* (2008). Ranking of port cities with high exposure and vulnerability to climate extremes: exposure estimates. *OECD Working Papers*, **1**.

O'Connor (2002). *Report of the Walkerton Inquiry: The Events of May 2000 and Related Issues*. Accessed Feb 8, 2011, www.attorneygeneral.jus.gov.on.ca/english/about/pubs/walkerton/part1/.

O'Hare, G. and S. Rivas (2005). The landslide hazard and human vulnerability in La Paz City, Bolivia. *The Geographical Journal*, **171**, 239–258.

Oke, T. (1997). Urban climates and global change. In A. Perry and R Thompson (Eds.), *Applied Climatology: Principles and Practices*, London, UK: Routledge.

Ortíz Bultó, P., A. Rodríguez, A. Valencia, *et al.* (2006). Assessment of human health vulnerability to climate variability and change in Cuba. *Environmental Health Perspectives*, **114**, 1942–1949.

Oxfam (2009). *Climate Change and Security: Oxfam Fact Sheet*. Accessed Feb 8, 2011, www.oxfamamerica.org/files/climatechangeandsecurity-factsheet.pdf.

Pagnamenta, R. (2009). French add Rolls-Royce to nuclear power project. *The Times*. Accessed Feb 8, 2011, http://business. timesonline.co.uk/tol/business/industry_sectors/article6837467.ece

Patil, N. (2008a). PMC says no problem, we're ready to tackle floods. *Express India*. Accessed Feb 8, 2011, www.express india.com/latest-news/pmc-says-no-problem-were-ready-to-tackle-floods/317973/.

Patil, N. (2008b). Flood of fear engulfs helpless city. *Express India*. Accessed Feb 8, 2011, www.express india.com/latest-news/Flood-of-fear-engulfs-helpless-city/347725/

Patz, J., M. McGeehin, S. Bernard, and K. Ebi (2000). The potential health impacts of climate variability and change for the United States. *Environmental Health Perspectives*, **108**, 367–376.

Patz, J., S. Vavrus, C. Uejio, and S. McLellan (2008). Climate change and water borne disease risk in the Great Lakes region of the U.S. *American Journal of Preventative Medicine*, **35**, 451–458.

Pengelly, L., M. Campbell, C. Cheng, *et al.* (2007). Anatomy of heat waves and mortality in Toronto: Lessons for public health protection. *Canadian Journal of Public Health*, **98**, 364–368.

Penney, J. (2007). *Cities Preparing for Climate Change: A Study of Six Urban Regions*, Clean Air Partnership, Toronto.

Peterson, D. and D. McKenzie (2008). *Wildland Fire and Climate Change*, US Department of Agriculture, Forest Service, Climate Change Resource Center. Accessed Feb 8, 2011, www.fs.fed.us/ccrc/topics/wildland-fire.shtml.

PHAC (2009). *What Determines Health?* Public Health Agency of Canada. Accessed Feb 8, 2011, www. phac-aspc.gc.ca/ph-sp/determinants/index-eng.php.

Pitt, M. (2008). *Lessons Learned From the 2007 Floods: The Pitt Review*, Cabinet Office. Accessed Feb 8, 2011, www.cabinetoffice.gov.uk/thepittreview/final_report.aspx.

Reid, H. and S. Kovats (2009). Special Issue on health and climate change of TIEMPO, *Bulletin on Climate and Development*, Issue 71.

Reiter, P. (2008). Global warming and malaria: knowing the horse before hitching the cart. *Malaria Journal*, **7**, S3.

Reiter, P., C. Thomas, and P. Atkinson (2004). Global warming and malaria: a call for accuracy. *Lancet Infectious Diseases*, **4**, 323–324.

Revi, A. (2008). Climate change risk: an adaptation and mitigation agenda for Indian cities. *Environment and Urbanization*, **20**, 207–229.

Ribeiro, W. (2008). *Impactos das Mudanças Climáticas em Cidades no Brasil*. Parcerias Estratégicas /Centro de Gestão e Estudos Estratégicos. Brasília: Centro de Gestão e Estudos Estratégicos, Ministério da Ciência e Tecnologia.

Risk Management Solutions (2008). *The 1998 Ice Storm: 10-Year Retrospective*, RMS Special Report. Accessed Feb 8, 2011 www.rms.com/publications/1998_Ice_Storm_Retrospective.pdf.

Republic of Uganda (2007). Climate Change: Uganda's National Programmes of Action (2007). Accessed 20 October 2009, www.preventionweb.net/english/professional/policies/v.php?id=8578.

Robine, J., S. Cheung, S. Le Roy, H. Van Oyen, and F. Herrmann (2007).*Report on Excess Mortality in Europe During Summer 2003*. EU Community Action Programme for Public Health.

Romero Lankao, P. and J., Tribbia (2009). *Assessing Patterns of Vulnerability, Adaptive Capacity and Resilience Across Urban Centers*, National Center for Atmospheric Research, Mexico. Accessed Dec 21, 2009, www.urs2009.net/docs/papers/Romero.pdf.

Rosenzweig, C., W. Solecki and R. Solsberg (2006). *Mitigating New York City's Heat Island with Urban Forestry, Living Roofs, and Light Surfaces*, New York State Energy Research and Development Authority.

Saelens, B., J. Sallis, and L. Frank (2003). Environmental correlates of walking and cycling: findings from the transportation, urban design, and planning literatures. *Annals of Behavioral Medicine*, **25**, 80–91.

Satterthwaite, D. (2009). *Social Aspects of Climate Change in Urban Areas in Low and Middle Income Nations*. Accessed Dec 21, 2009, www.urs2009.net/docs/papers/Satterthwaite.pdf.

Satterthwaite, D., S. Huq, M. Pelling, H. Reid, and P. Romero Lankao (2007). *Adapting to Climate Change in Urban Areas: The Possibilities And Constraints in Low- and Middle-Income Nations*, Human Settlements Working Paper Series Climate Change and Cities No. 1. London, UK: IIED

Semenza, J. and B. Menne (2009). Climate change and infectious diseases in Europe. *Lancet*, **9**, 365–375.

Sen, S. N., P. K. Maiti, A. Ghosh, and R. Ghosh (2003). Status and scope of Boro rice cultivation in West Bengal. In R. K. Singh, M. Hossain, and R. Thakur (Eds.), *Boro Rice*, Los Banos: IRRI, pp. 137–156.

Sharkey, P. (2007). Survival and death in New Orleans. *Journal of Black Studies*, **37**, 482–501.

Steiner, S., R. Tonse, A. Cohen, R. Goldstein, and R. Harley (2006). Influence of future climate and emissions on regional air quality in California. *Journal of Geophysical Research*, **111**, D18303.

Surjan, A. and R. Shaw (2009). Enhancing disaster resilience through local environment management: case of Mumbai, India. *Disaster Prevention and Management*, **18**(4), 418–433.

Tan, J., Y. Zheng, G. Song, *et al.* (2007). Heat wave impacts on mortality in Shanghai, 1998 and 2003. *International Journal of Biometeorology*, **51**, 193–200.

Toronto Environment Office (2008). *Ahead of the Storm: Preparing for Climate Change*, Accessed Feb 8, 2011, www.toronto.ca/teo/pdf/ahead_of_the_storm.pdf.

UN (2005). *World Population to Increase by 2.6 Billion Over the Next 45 Years*. Press Release POP/918.

UN-HABITAT (2003). *The Challenge of Slums: Global Report on Human Settlements 2003*. Accessed Feb 8, 2011, www.unhabitat.org/pmss/listItemDetails.aspx?publicationID=1156.

UN Millenium Project (2005) *Coming to Grips with Malaria in the New Millennium*, Task Force on HIV/AIDS, Malaria, TB, and Access to Essential Medicines, Task Force on Malaria.

UN Population Division (2008). *World Urbanization Prospects: The 2007 Revision*. Accessed Feb 8, 2011, www.un.org/esa/population/publications/wup2007/2007WUP_ExecSum_web.pdf

UN Population Fund (2007). *State of the World Population 2007*. Accessed Feb 8, 2011, www.unfpa.org/swp/2007/english/introduction.html.

UNDP (2007). *Human Development Report 2007/2008, Fighting Climate Change: Human Solidarity in a Divided World*, United Nations Development Programme, New York, USA.

UNEP (2008). *Health Security through Healthy Environments*. Accessed Feb 8, 2011, www.unep.org/ health-env/pdfs/WHOLibrevilleReport.pdf.

UNISDR (2008). *Climate Change and Disaster Risk Reduction*, Briefing Note 01. Accessed Feb 8, 2011, www.preventionweb.net/files/4146_ClimateChangeDRR.pdf.

UNISDR (2009). *Adaptation to Climate Change by Reducing Disaster Risks: Country Practices and Lessons*. Geneva, Switzerland: United Nations International Strategy for Disaster Reduction.

United Nations Prevention Web (2009). www.preventionweb.net/english/.

USEPA (2008). *Global Change Research Program: Review of the Impacts of Climate Variability and Change on Aeroallergens and their Associated Effects*, Technical Report EPA/600/R-06/164F, US Environmental Protection Agency.

USEPA (2009). *Reducing Urban Heat Islands: Compendium of Strategies*, describes mitigation measures that communities can take to address the negative impacts of urban heat islands.

Volschan Jr., I. (2008). Efeitos das mudanças climáticas sobre os sistemas de abastecimento de água e de esgotamento sanitário da cidade do Rio de Janeiro: uma visão preliminar. In P. P. Gusmão, P. S. Carmo, and S. B. Vianna (Eds.), *Rio próximos 100 anos. O Aquecimento Global e a Cidade*, Rio de Janeiro, Brazil: Instituto Pereira Passos, Volume 1, pp. 199–210.

Waterston, T. and S. Lenton (2000). Public health: sustainable development, human induced global climate change, and the health of children. *Childhood*, **82**, 95–97.

WHO (2002). *The World Health Report 2002*. Geneva, Switzerland: WHO. Accessed Feb 8, 2011, www.who.int/whr/2002/en/.

WHO (2004). *Global Strategic Framework for Integrated Vector Management*. Accessed, http://whqlibdoc. who.int/hq/2004/WHO_CDS_CPE_PVC_2004_10.pdf.

WHO (2005). *El Niño and its Health Impact*, Fact sheet 192. Accessed Dec 21, 2009, www.who.int/ mediacentre/factsheets/fs192/en/index.html.

WHO (2009). *Medium-Term Strategic Plan 2008–2013*, Amended (Draft) Proposed Programme Budget 2010–2011, A62/11 Climate change and health.

Wilbanks, T. J., P. Romero Lankao, M. Bao, *et al.* (2007). Industry, settlement and society. In M. L. Parry, O. F. Canziani, J. P. Palutikof, P. J. van der Linden and C. E. Hanson (Eds.), *Climate Change 2007: Impacts, Adaptation and Vulnerability. Contributions of Working Group II to the Fourth Assessment Report of the Intergovernmental Panel on Climate Change*, Cambridge, UK: Cambridge University Press, pp. 357–390.

Wilby, R. L. (2007). A review of climate change impacts on the built environment. *Built Environment*, **33**, 1.

Winchester, L. and R. Szalachman (2009). *The Urban Poor's Vulnerability to the Impacts of Climate Change in Latin America and the Caribbean: A Policy Agenda*. Accessed Dec 21, 2009, www.urs2009.net/docs/papers/Winchester.pdf.

World Bank (2008). *Climate Resilient Cities Primer: Reducing Vulnerabilities to Climate Change Impacts and Strengthening Disaster Risk Management in East Asian Cities*. Accessed Feb 8, 2011, http://siteresources.worldbank.org/EASTASIAPACIFICEXT/Resources/climatecities_fullreport.pdf.

World Medical Association (2009). *WMA Declaration of Delhi on Health and Climate Change*, New Delhi. Accessed Feb 8, 2011, www.wma.net/en/30publications/10policies/c5/index.html.

Yamamoto, Y. (2004). Measures to mitigate urban heat islands, Chapter 6, Does ambient temperature affect foodborne disease? *Epidemiology*, **15**, 1.

Zimmerman, R., C. Restrepo, B. Nagorsky, and A. Culpen (2007). Vulnerability of the elderly during natural hazard events. In *Proceedings of the Hazards and Disasters Research Meeting*, Boulder, CO, USA. Accessed Dec 21, 2009, www.colorado.edu/hazards/work/hdrm_ proceedings.pdf.

Ziska, D., D. Gebhard, S. Frenz, *et al.* (2003). Cities as harbingers of climate change: common ragweed, urbanization, and public health. *The Journal of Allergy and Clinical Immunology*, **111**, 290–295.

Part IV

Cross-cutting issues

8

The role of urban land in climate change

Coordinating Lead Author:

Hilda Blanco (Los Angeles)

Lead Authors:

Patricia McCarney (Toronto), Susan Parnell (Cape Town), Marco Schmidt (Berlin), Karen C. Seto (New Haven)

This chapter should be cited as:

Blanco, H., P. McCarney, S. Parnell, M. Schmidt, K. C. Seto, 2011: The role of urban land in climate change. *Climate Change and Cities: First Assessment Report of the Urban Climate Change Research Network*, C. Rosenzweig, W. D. Solecki, S. A. Hammer, S. Mehrotra, Eds., Cambridge University Press, Cambridge, UK, 217–248.

8.1 Introduction

Recent IPCC reports have addressed the issue of urban land under the topic of industry, settlement, and society (IPCC, 2007). Since reviews of human settlements from the perspective of climate change have been primarily focused on climate change mitigation, topics of land cover and use, urbanization, land planning and management, land markets, property rights, and fiscal and legal issues, which will be key to responding to impacts of climate change, have not received extensive coverage. We argue in this chapter that it is important to focus on urban land as a sector or as the overarching framework in order to recognize the challenges of government coordination and integration necessary to address climate change. In incorporating urban land in climate change adaptation and mitigation efforts one would be able to include a fundamental set of strategies, such as policies concerning land conversion, land tenure, and urban land markets that have not been fully addressed.

This chapter provides an introduction to the role of urban land in climate change, discusses the potential for urban planning and management to address climate change challenges, and reviews current planning efforts focused on climate change. It is organized into several sections. This introductory section develops several key concepts, such as recent trends in urbanization, and discusses their relation to urban land and climate change. The second section focuses on urban form, impacts on ecosystems, including the urban heat island effect, and discusses the vulnerability of informal and slum settlements to climate change. The third section provides an introduction to the urban land management system, including legal aspects, the urban planning system, and urban land regulation. This is followed by a brief review of climate change risk. Two sections then focus on how urban planning can address mitigation and adaptation challenges, by reviewing current efforts and next steps. The two final sections identify policy issues and research needs requiring attention to fully enable the use of urban planning and management to mitigate and adapt to climate change.

8.1.1 Defining urban land

A recognized, global definition of urban land or urban areas has not been established. The United Nations, when it reports on urbanization or cities, uses countries' self-reports, with varying definitions. Urbanized land is typically defined as land in state-recognized cities (municipalities or local authorities), as land in agglomerations with threshold populations of from 1,000 to 5,000 persons, and in some countries in terms of density per unit area, the number of persons per square kilometer ranging from 386 in the United States, 1,500 in the People's Republic of China to 4,000 in Japan (UN Population Division, 2008). Metropolitan

areas, a concept that is defined by an integrated labor market and travel patterns rather than density, include at least one central city, and other urban areas, as well as surrounding rural land. In this report, we most often will use the term cities and urban areas interchangeably to refer to areas with urban densities, and metropolitan areas to refer to areas that incorporate fringe rural lands. Definitional issues are important not only because they determine the extent and rate of urbanization, but also because they are integral to the conceptualization of issues and problems. For example, the phenomenon of increasing metropolitanization involves population, land, and political jurisdictions or governance. A key aspect of metropolitanization is the conversion of rural lands to urban uses. Expansion of urban areas increases energy needs for travel within the area, and involves the substitution of natural vegetation by impervious surfaces, which destroys carbon sinks, and intensifies flooding and heat island effects, among other local environmental risks.

Beyond metropolitanization, two other interconnected urbanizing trends complicate the study of cities and climate change, the growth of megacities (cities over 10 million), and the convergence of metropolitan areas into mega-scale urban *megalopolis* regions (Gottmann, 1961; Gottmann and Harper, 1990), and more recently identified as *megapolitan areas* (Lang and Dhavale, 2005; Lang and Knox 2009)[1] and *mega-city-regions* (Hall, 2009). Gottmann (1961) used the merging of the metropolitan areas of Boston, New York, Philadelphia, Baltimore, and Washington, DC, or the Boswash urban corridor as a first example. Although Gottmann was at first ambivalent about whether the term referred to a physical or morphological merging of metropolitan areas, this type of conurbation of metropolitan areas, more commonly referred to as mega-city-regions, is now understood as a functional, rather than a physical, or administrative concept.[2,3] Urban space in these urban regions is defined by the "space of flows" (Castells, 1989), of people, goods, or information on a regular basis. The term mega-city-regions was first applied to East Asia regions, such as the Pearl River Delta in southern China, which include several metropolitan areas, typically with populations of ten million or more. Mega-city-regions are currently found throughout most parts of the world. Some examples include the Greater La Plata–Buenos Aires metropolitan region in Argentina, Mumbai–Pune mega-region in India, the Suez–Cairo–Alexandria urban region, and the Randstad in the west of the Netherlands.[4] The mega-city-region of central Mexico, with close to 30 million people, includes the metropolitan areas of Mexico City, Puebla, Cuernavaca, Toluca, and Pachuca, and 173 municipalities in five states.

The increase and expansion of these mega-city-regions will aggravate urban land management challenges across the world. These regions cross local and provincial political boundaries where the jurisdiction for urban land management typically lies,

---

1 Studies in the USA have identified ten megapolitan areas that house 197 million people (Lang and Knox, 2009).
2 The term mega-city-region was first applied to urbanized regions of Eastern Asia including the Pearl River Delta, the Yangtze River Delta, the Tokaido (Tokyo-Osaka) corridor, and Greater Jakarta (Lin and Ma, 1994; McGee and Robinson, 1995; Hall, 2009).
3 In the US census, Consolidated Metropolitan Statistical Areas (CMSA), composed of several Metropolitan Statistical Areas (MSAs) are similar to the concept of mega-city-regions.
4 See Hall and Pain (2006) for a study of European examples.

and thus make planning, management, and public finance more difficult to coordinate across these vast regions.

8.1.2 Cities: their natural setting, urban form, and built environment

Urban places by definition are human-dominated and constructed landscapes; however, their natural characteristics, and the management of the ecosystem services on which cities depend, are essential to their definition (Spirn, 1985; Hough, 1995; McHarg, 1995; Alberti, 2008). Furthermore, the formation of cities, their location and characteristics are heavily influenced by the economic and technological context of a people. These, in turn, are dependent on environmental features. Agricultural economies and the early cities that owed their formation to agricultural surpluses relied on water sources for urban uses, crops, and for goods transportation. For example, the alluvial basin between the Tigris and Euphrates rivers in contemporary Iraq and the Nile Delta are often cited as some of the earliest known urban concentrations dating back to at least 5000 BC (Benevolo, 1980; Kotkin, 2005) As trade increased, many major cities, such as Guangzhou with a population of over 200,000 by AD 1200, developed near and along coastlines (Ma, 1971). In a similar way, early industrialization relying on steam power gave economic advantage to coastal cities and cities along navigable waterways. Waves of European colonization of the Americas, India, Asia, Africa, and Australia, and increasing intercontinental trade from the late 1400s through the 1800s, during an era where intercontinental travel was primarily accomplished by sea, led to the establishment and increasing prominence of coastal cities in these continents, such as Boston, New York, Veracruz, Havana, Rio de Janeiro, Buenos Aires, Cairo, Cape Town, Bombay, Madras, Calcutta, Goa, Singapore, Macao, Jakarta, Adelaide, Melbourne, and Sydney (Southall, 1998).

While economic and technological factors have favored the location of cities in certain environmental settings, the geomorphology of a city and the soil conditions also affect a city's vulnerability to natural hazards, such as flooding and landslides, and can constrain urban expansion. Surface and groundwater features of an urban area are often sources of water supply and partly determine a city's drainage options. Urban coastal areas are faced with erosion and storm surges, as well as salt water intrusion into creeks and rivers. Cities in arid and semi-arid areas are subject to flooding, and special sediment problems, including desertification. The proximity or extension of urban areas into rural or undeveloped lands can increase vulnerability to wildfires. As discussed, cities are often located on or close to prime agricultural land, and urban expansion destroys these resources. As cities expand to create metropolitan regions, forested areas, which serve as carbon sinks, are degraded and destroyed. These ecological aspects of urban land use and change can constrain or expand options cities have to mitigate and adapt to climate change.

Although the terms urban form and built environment are often used interchangeably, here we use the term urban form to refer to city-scale or macro-scale patterns, and built environment to refer to micro-scale or structural aspects of cities. A city's urban form, the overall characteristics of a city's existing built environment, e.g., dispersed versus compact settlement form, the extent and pattern of open spaces and impervious surface, and the relationship of its density to destinations and transportation corridors, interacts with natural and other urban characteristics to constrain transportation options, energy use, drainage, and future urbanization. The urban form also can affect the vulnerability of a city to climate change impacts. Because of its potential impact on energy and building materials, the management of urban form is a critical area of intervention for the promotion of climate mitigation and adaptation (see Box 8.1; Boarnet and Crane, 2001; Giuliano and Narayan, 2003; Sorensen and Hess, 2007; Ewing *et al.*, 2008; Grazi and van den Bergh, 2008).

The built environment or structural aspects of cities, streets, buildings, and infrastructure systems contribute significantly to the emission of greenhouse gases, and can also amplify climate change impacts. The structure, orientation, and condition of buildings and streetscapes can increase the need for cooling and heating buildings, which are associated with the level of energy use and can account for a significant proportion of greenhouse gas emissions in a city. The extent of streetscape and the impervious surface of structures can intensify flooding and are direct determinants of the heat island effect. Conventional wastewater and drainage systems impede natural processes of evapotranspiration and can amplify flooding and drought effects.

A city's natural setting, its urban form and built environments are relatively fixed or static factors, but they are subject to future modification through urban planning and management. For example, managing the size and shape of the overall urban form through urban land use planning may provide more significant opportunities for mitigation and adaptation to climate change, and settlement patterns can be modified through redevelopment or the imposition of urban boundaries or by restricting development through land use controls.

8.1.3 Market and public good aspects of land

Urban areas are most often the economic engines of nations, containing major industrial and commercial enterprises, government centers, as well as major residential settlements. Today, in modern market economies, land for these urban uses is a market commodity, often representing a quarter or more of the total value of urban properties (O'Sullivan, 2006). In developed countries, government heavily regulates urban land, less so in developing countries where there are more diffuse and often competing systems of land regulation in place, including traditional authorities and private land barons, as well as the state. The regulation of land and its uses in developed countries typically includes institutional processes for planning, subdivision of undeveloped land, zoning, and building codes for private and public development. Public regulation of urban land in developed countries is balanced by private property rights that facilitate land and real estate markets. In developing countries, where the bulk of urbanization

219

[ADAPTATION]   Box 8.1   Report from Rome, Italy: Discovery of crossroads

Maria Paola Sutto and Richard Plunz

Urban Design Lab, Earth Institute at Columbia University

The Rome Forum was the first in a series of forums initiated by the Urban Securities Project of the Earth Institute's Urban Design Lab at Columbia University.[5] The Rome Forum, held in February 2008, aimed to investigate ways in which cities are called to expand their vision in responding to climate change. Thirty researchers from across North America and Europe participated in the forum, including engineers, anthropologists, business and media studies experts, philosophers, political scientists, urban designers, and climate scientists.

The Forum reached consensus on the following points:

1. On adaptation, cities should enhance the value of their neighborhood residential identity and their social networks and not just geographical or political boundaries. At this scale is easier to fast-track implementation of adaptation decisions at the city scale.
2. A crucial consideration is incorporating health impacts of climate change into the cycle of decision-making (co-factoring short, medium, and long-term implications), while stressing the importance of achieving social, economic, and environmental benefits.
3. A common language for intervention is needed related to goals, methods, and terminology. Urban institutions that until now have worked in isolated modes should not perceive a more integrated approach as an undue interference and loss of power within their specialized realms.
4. Environmental injustice is a key concern to avoid uneven physical dislocation and social turmoil. To achieve this goal, local knowledge as well as scientific findings should be considered.
5. Given the projected social dimensions of climate change, personal responses are important since there can be a sense of dislocation that accompanies such transformations. Social adaptation may be best addressed through "understanding by doing," where the process of taking action, either personal or collective, is continuously assessed and modified following results and discoveries.
6. At the local urban scale, the aim is to proceed from public awareness to personal action. On the city government level, political leadership is needed to clearly set forth decisions with adequate implementation and monitoring. The re-emergence of the "city-state" will be a crucial complement to national and global climate change initiatives.
7. From a communication point of view, climate change challenges are well understood when cast in a context that touches the individual directly, indicating either new economic possibilities, or threatening public well-being such as increased illnesses due to temperature change or decreased food availability due to drought/flooding.
8. Effective public communication is crucial to insure understanding and changes in attitudes. Valid communication is delivered through both statements and congruent actions. Lack of communication transmits uncertainty, and is likely to contribute to public skepticism of climate change.
9. The climate change challenges associated with the economic, social, and political systems of cities are embedded in the urban infrastructure: energy, transportation, water, waste, and food.
10. The process required to establish an appropriate assessment and response system for evaluating climate risks needs to be streamlined. Adaptation is still not fully understood by the private sector as a mainstream risk issue. More information on climate impacts is needed in the public media.
11. Overall, it is important to recognize the complexity of interactions among the natural, technological, and human components in designing for urban-scale climate change responses. We call for accelerating the process of transformation of society into one more respectful of the Earth's planetary processes and resources.

The Rome Forum proceedings are available as: R. Plunz and M. P. Sutto (Eds.) (2010), *Urban Climate Change Crossroads*. Farnham, UK: Ashgate Publishing Ltd.

---

in the first half of the twenty-first century is forecast to occur, the management of urban land is more problematic, since much urbanization occurs informally, without the protection of property rights or of adequate urban infrastructures and services, and without effective institutions for land management. Nations fall along a broad continuum that marks the extent of reliance on private property versus public management of urban land. Over the past two or three decades, there has been a shift to stronger private property rights and market orientation. Market-oriented approaches to the management of urban land are likely to make large-scale interventions in the interests of climate change mitigation or adaptation more complex, and likely more expensive.

8.1.4 Urban land and infrastructure

In the developed world, urban land is land supplied with urban infrastructure and other urban services (Kelly, 2004). Without appropriate transportation, water systems, waste disposal, and energy supplies, land cannot be properly developed to sustain urban densities with high levels of public health and wellbeing. Public health, administration, police and fire services are also crucial to ensure the safety and security of urban populations. Urban infrastructure and services in developed countries have been typically supplied by the public sector and comprehensive community planning involves both land use and infrastructure

---

5 The Rome Forum was a collaboration between the Urban Design Lab (UDL), the Urban Climate Change Research Network (UCCRN), the Fondazione Adriano Olivetti, and the Camera di Commercio Ambiente e Territorio of Rome.

planning. In developing countries, newly urbanizing areas typically lack urban infrastructure and services, both at the household level, e.g., electricity, running water, as well as at the neighborhood level, sidewalks, garbage disposal, parks. One of the most difficult aspects of climate mitigation for cities of the developing world is the lack of connection between land use planning and infrastructure provision. Thus, in the informal sections of cities (which in Africa account for as much as 80 percent of settlements) it is very difficult to use the land use management system to regulate building bulk or household infrastructure standards and make the requisite improvements required to accommodate increasing extremes associated with climate change (Parnell *et al.*, 2009).

The connections between the management of urban land and infrastructure systems can be complex. For example, as discussed in Chapter 5, land subsidence in Mexico City is due to water extraction from underground aquifers at rates greater than the aquifers can replenish themselves. In urban areas, impervious or sealed surfaces reduce the rates of aquifer replenishment. The extent of impervious surface in an urban area, in turn, can be reduced through urban planning strategies to increase greenspace, or through requirements for more permeable surfaces for certain types of uses, such as driveways or urban alleyways. Or, as discussed in Chapter 6 and below, automobile travel is responsible for a significant portion of greenhouse gas emissions in cities of developed countries. Reducing automobile use by providing adequate public transit is a major strategy to lower greenhouse gas emissions. Public transit, in turn, requires sufficient densities to make it viable. This illustrates the interactions and interdependencies between land development and infrastructure systems.

8.2 Urban form, patterns, and impacts

8.2.1 Urban form

The way in which a city expands in physical space – whether new urban development is contiguous to existing urban areas, or whether it is leapfrog development, or whether the city is compact or dispersed, shaped like a circle or an amoeba – has significant impact on energy use, resource consumption, and the ability of a region to adapt to or mitigate against climate change (Ewing *et al.*, 2008).[6] Many of these impacts are exacerbated when new growth takes the form of *urban sprawl* (Alberti, 2005), a term that refers to low-density, dispersed or even decentralized forms of urban expansion (Ewing *et al.*, 2002; Bruegmann, 2005; Flint, 2006). Compact growth can lead to efficient use of resources, whereas expansive development can strain infrastructure and natural resource availability.

Cities that developed prior to the widespread ownership of private automobiles are usually compact (Jackson, 1987). In contrast, cities that developed during the age of the car tend to be more expansive, and along road transportation corridors (Warner, 1978). For example, in car-dependent regions such as the Silicon Valley, USA, transportation infrastructure has led to expansive patterns of urban growth. In contrast, in Bangalore, India, where car ownership is relatively low, urbanization is not highway-oriented and is more compact (Reilly *et al.*, 2009). In turn, expansive patterns of urban development have led to high levels of dependence on private vehicles, which also drives an increase in fossil fuel demand (Newman and Kenworthy, 1999). See Box 8.2 for a discussion of the relation between urban form and travel behavior. These different patterns of urban form, dispersed versus compact, also have a different effect on the extent of impervious surface, and how the urban area interacts with the local and regional environment. Studies of evolving urban form suggest that cities around the world may be following similar patterns of urban growth, although different local and national policies undergird these patterns (Kenworthy, 2003; Marcotullio, 2003; Leichenko and Solecki, 2005; Seto and Fragkias, 2005). For example, cities across China follow similar physical growth trends and urban growth is largely fragmented across vast megaregions and often driven by foreign investment (Seto *et al.*, 2002; Zheng *et al.*, 2009). Consequently, managing large-scale patterns of urbanization can be a key strategy to mitigate climate change. See also the transportation chapter (Chapter 6) for a discussion of urban transportation and land use.

In addition to macroscale features of urban form, the siting of buildings, and how they use energy and water have important implications for climate change mitigation and adaptation. Many of these features have been identified and incorporated by green building rating systems, such as the US Green Building Council's rating system, Leadership in Energy and Environmental Design (LEED) in the United States and the UK Building Research Establishment's Environmental Assessment Method (BREEAM). See Chapter 7 for a discussion of building energy use.

8.2.2 Environmental impacts of urbanization

Urban expansion has profound environmental impacts that extend beyond city boundaries including: changes to microclimate, conversion of natural ecosystems, loss of agricultural land, fragmentation of natural habitats, contamination of air, soil and water, increased water use and runoff, reduced biodiversity, and introduction of non-native species (Rees, 1992; Pickett, *et al.*, 1997 El Araby, 2002; Alberti, 2005; Potere and Schneider, 2007). In addition, as urban areas expand, they change regional energy budgets, and generate greater demands for natural resources. Interactions between and among urban land use, policies, and earth system functions cannot be decoupled (Liu *et al.*, 2007; Alberti, 2008).

---

6   When compared to more compact settlements, impacts of climate change, such as heat waves, droughts, wildfires, and more intense storms, will be more difficult and costly to respond to under urban sprawl conditions. In general, more dispersed urban settlements require larger infrastructure networks and service areas, e.g., public health or emergency services, which is likely to increase the costs of responding to climate impacts.

[MITIGATION]   Box 8.2   Links between urban form, travel patterns, energy use, and greenhouse gases

Hilda Blanco

University of Southern California

Urban form is a critical factor in explaining patterns of automobile dependence and transportation energy use (Cervero, 1998; Newman and Kenworthy 1989, 1999; Kenworthy and Laube, 1999; Kenworthy, 2003; UN-HABITAT, 2008). According to these studies, the most significant urban form factor affecting travel behavior is urban density. A study of 84 cities around the world (Kenworthy, 2003) demonstrates this linkage. The study shows that higher car and energy use cities, which are the highest greenhouse gas producers, are low in population density, and conversely, that high-density cities have lower car use and lower greenhouse gas emissions due to transportation. The Kenworthy (2003) study also indicates a strong correlation between wealth and density, with lower income cities more than double the density of wealthier cities (109 versus 52 persons/ha). The strongest correlation between urban density and automobile use was among cities in high-income countries, with per capita passenger distance traveled by automobile increasing as densities decrease. This implies that as the wealth of cities in emerging economies and developing countries rises, automobile ownership and usage is likely to increase unless competitive public transit alternatives become available. Consequently, cities which aim to manage traffic congestion, minimize car and energy use, and lower greenhouse gas emissions from transport should address the issue of urban form. To accomplish this, experts recommend a policy of increasing density and mixing appropriate land uses, such as residential and retail, around areas of high public transit accessibility. The centralization of jobs in central business districts and satellite sub-centers built around transit nodes is also recommended, as well as the use of urban containment strategies, such as greenbelts or urban growth boundaries (Kenworthy, 2003). These recommendations are supported by a comprehensive recent study conducted by the US National Research Council (2009). The study examined the empirical evidence on the effects of compact development on motorized travel, energy use, and CO_2 emissions. It found that, "doubling residential density across a metropolitan area might lower household VMT

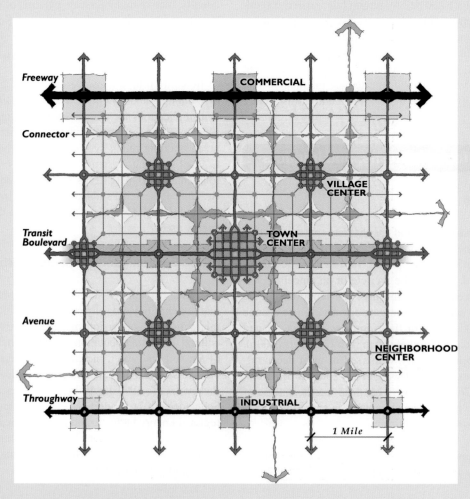

Box Figure 8.1: *Schema of transit-oriented development in an urban network.*

Source: Calthorpe, (1993).

[vehicle miles traveled] by about 5 to 12 percent, and perhaps as much as 25 percent, if coupled with higher employment concentrations, significant public transit improvements, mixed uses, and other supportive demand management measures". Density increases, transit improvements, and mixed uses are major tenets of the smart growth movement described in Box 8.3.

In the United States, a major strategy advocated to shape new development and reduce automobile use through transit villages or transit-oriented development (TODs) highlights the role of density, diversity, and design (Bernick and Cervero, 1996). Calthorpe (1993), who initiated the TOD

concept (Carlton, 2007) argues that reduction in automobile use and increase in public transit should be pursued as a regional urban form strategy that links TODs. See Box Figure 8.1 for a schematic of such an urban network that incorporates a system of higher density cluster developments with transit links at the center of the clusters. While transit-oriented compact development is the focus of the strategy, the TOD agenda is part of a broader sustainability agenda that also emphasizes a pedestrian orientation, a mix of housing types, preserving environmentally sensitive areas, creating high quality public spaces, as well as fostering development within existing neighborhoods (Dittmar and Ohland, 2004).

Cities have always relied on resources in their immediate and distant areas for their development. The location of the earliest human settlements was in fertile places. Empirical studies show that contemporary patterns of urbanization are also taking place on prime agricultural lands across the continents (Imhoff *et al.*, 1997; Seto *et al.*, 2000; Nizeyaimana *et al.*, 2001; Döös, 2002). Increasingly, urbanization in one place is driving land cover and other environmental changes over longer distances. For example, the developments of the shrimp aquaculture industry in Vietnam and Thailand have been driven largely by urban demand elsewhere (Huitric *et al.*, 2002, Lebel *et al.*, 2002; Seto and Fragkias, 2007). Urban expansion is also likely to take place in the most biologically sensitive regions. A recent study shows that 8 percent of terrestrial vertebrate species on the IUCN Red List are in peril because of urban expansion (McDonald *et al.*, 2008). While we are beginning to understand the local impacts, we have little

information on the cumulative effects of urban growth on global environmental processes (Lambin *et al.*, 2001). Urban areas are the primary source regions of many anthropogenic emissions, for instance, yet no global model of climate or biogeochemistry to date adequately represents urbanization.

8.2.2.1 Built environment and its effect on local climate

One major influence of urban spatial development and land conversion to urban uses on the local climate is the shift in the flux between the sensible and latent heat and related increases in thermal radiation. Compared to forest landscape, urban areas experience a radical change in radiation and hydrology. On a global scale, the evapotranspiration of water represents the biggest proportion of radiation conversion. The diagram in Figure 8.1 shows how global radiation is converted on the

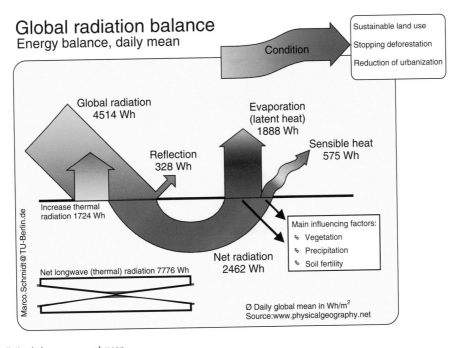

Figure 8.1: *Global daily radiation balance as annual mean.*

Source: Schmidt et al. (2007). Energy data based on www.physicalgeography.net.

[MITIGATION]   Box 8.3   Conceptualizing the generation of greenhouse gases in urban areas

Hilda Blanco

University of Southern California

Many urban activities generate greenhouse gases, but measuring their emissions at the urban scale is a new enterprise. Wackernagel and Rees (1996) originally developed the concept of the ecological footprint to provide a simple measure of the extent of land and water resources used by cities. The carbon footprint, developed out of the concept of the ecological footprint, has become a popular measure of the extent of greenhouse gas emissions generated by a person, organization, city, or nation in equivalent CO_2 emissions (Wiedmann and Minx, 2007). Connected to this concept is the earlier urban metabolism concept,

developed in a pioneering article by Wolman (1965), to analyze the energy, materials, and water inputs and outputs of a hypothetical city of one million. More recently, urban metabolism is defined as "the sum total of technical and socioeconomic processes that occur in cities, resulting in growth, production of energy and elimination of waste" (Kennedy *et al.*, 2007). Urban metabolism can be illustrated through the use of flow or Sankey diagrams. These input–output diagrams are useful for conveying existing energy flows from different sources and for different uses, as well as for developing alternative energy futures. The energy flows are represented by arrows, and the width of the arrows represents the magnitude of the flow (Schmidt 2008a, b; Suzuki *et al.*, 2009). See Box Figure 8.3 for an illustration of a Sankey or meta diagram of energy flows in Jinze Town,

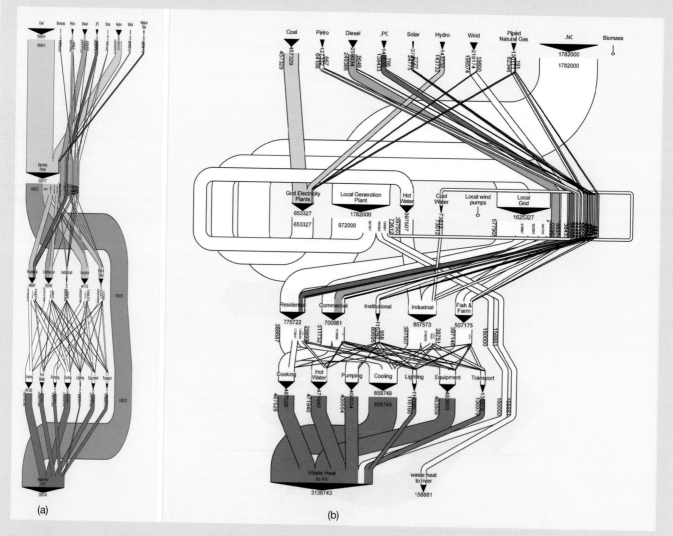

(a)

(b)

Box Figure 8.2: *(a) Meta Diagram of Jinze Town, Shanghai, Current Energy System; Box Figure 8.2(b) Meta Diagram for Jinze Town, Shanghai, Advanced System. This Meta Diagram provides a scenario for an advanced system that helps to reduce emissions and costs and increase local jobs and energy security. The advanced system represents a substantial change: for example, a local electricity generation facility is powered by liquefied natural gas, and provides a majority of electricity needs as well as hot and cool water for industry (cascading).*

Source: Suzuki et al. (2010), p. 128. Author elaboration for the diagrams by S. Moffat, with approximate data provided by Prof. Jinsheng Li, Tongji University, Shanghai.

Shanghai. The urban metabolism concept can also be used to illustrate the interrelation between urban land and infrastructures. Urban land contains the shelter for urban populations and economic activities as well as for the urban infrastructures that are major conduits for the inputs and outputs of urban metabolism. Transportation infrastructures are conduits for energy flows; water infrastructures bring in, distribute, and dispose of the waste and stormwater (compared to other materials processed through cities, water has the greatest volume). But energy flows through most urban services, e.g., water supply systems, often require electricity for pumping water. Waste materials are handled by solid waste disposal systems. Energy is brought into the city through a system of underground pipes, e.g., natural gas, or by trucks, e.g., gasoline, or above or underground through the electricity grid. Comprehensive urban planning determines the location of these infrastructures, and the densities permitted by zoning determines the sizing of much of these infrastructures, for example, the capacity of streets, or the sizing of water and wastewater pipes.

surface of the Earth into all its components. The figure depicts a mean energy flux on one square meter per day. Because of the increase in surface temperatures, 38 percent of the incoming solar radiation is directly converted to thermal radiation and 7.3 percent is reflected. The net radiation can either be converted into sensible heat (575 watt hours per square meter per day (Wh/(m^{2}d)) or consumed by evaporation, its conversion into latent heat. Representing 1888 Wh/(m^{2}d), the energy conversion by evaporation is the most important component of all, even larger than the thermal radiation converted from the incoming shortwave radiation. Additionally, evaporation influences the longwave thermal radiation due to the change in surface temperatures (Figure 8.1) (Schmidt *et al.*, 2007; Schmidt, 2009).

Urban areas modify these proportions dramatically. Instead of evaporation, the solar radiation is mainly converted to heat, reflection and longwave emissions. Figure 8.2 shows the global radiation balance of a black asphalt roof as an illustration of urban radiation changes. Instead of evaporation, most of the net radiation is converted to sensible heat instead of evaporation. Increased surface temperatures also increase the thermal radiation. Urbanization results in a large change of the small water cycle of precipitation, evaporation, and condensation. Additionally, hard materials or impervious surfaces in urban areas increase the heat capacity when compared to vegetated surfaces (Schmidt *et al.* 2007; Schmidt 2009).

Research over the last two decades has generated significant understanding of the relationship between urban areas and microclimate (Voogt and Oke, 2003; Shepherd, 2005; Souch and Grimmond, 2006; Kanda, 2007). The urban heat island effect appears stronger during the night than the day (Banta *et al.*, 1998). This effect is thought to be generated by the interaction among building geometry, land use, and urban materials (Oke, 1976; Arnfield, 2003). In addition to the heat island effect, recent studies show that urban surface characteristics affect precipitation (Shepherd, 2005: Jin and Shepherd, 2008), although the precise mechanism is unclear, and the effect seems to vary by region, with increases in rainfall reported for Tokyo, New York, and Phoenix (Bornstein and LeRoy, 1990; Fujibe, 2003; Shepherd, 2006) and decreases in winter rainfall in Europe (Trusilova *et al.*, 2008) and China (Kaufman *et al.*, 2007).

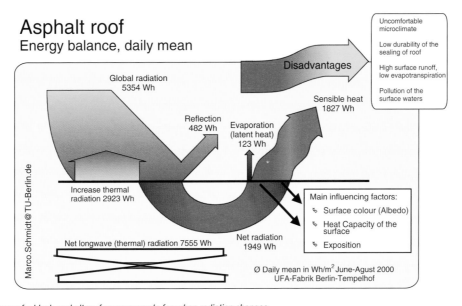

Figure 8.2: *Radiation balance of a black asphalt roof as an example for urban radiation changes.*

Source: Schmidt et al., (2007).

To adapt to the overheating of cities in summer, air conditioners are used, which additionally increase the urban heat island effect. Air conditioners use electricity or gas to "pump" heat from one place to another. This releases twice as much heat into the streets, depending on the performance of these appliances. In addition, electricity has a low energy efficiency of about 30 percent, so it is likely that two-thirds of the primary energy used to air condition buildings in cities will be released at the power plant. A major effective strategy to create more comfortable air temperatures in cities and to improve the microclimate around buildings is to green (i.e., vegetate) the surfaces, thereby "consuming" solar radiation by evapotranspiration. Vegetative coverings on buildings do present additional challenges, such as weight, but offer other opportunities as well, such as water retention, microhabitat, and aesthetic value. Climate change will likely aggravate the urban heat island effect in cities, and greening urban places will be an important climate change adaptation strategy.

8.2.3 Slums and vulnerable populations

Poor populations are more susceptible to natural disasters, and much of that susceptibility stems from the conditions of the built environment in which they live. Poverty in cities of the developing world is most often concentrated in slums or informal settlements. Although slums and informal settlements are terms often used interchangeably, informal settlements are characterized by lack of formal tenure; that is, they are settlements where people or squatters have occupied land without legal ownership or land registered in their name (Durand-Lasserve, 2006). The UN measures the degree of deprivation in slums by determining how many of five factors a household lacks: access to improved water, to sanitation, to durable housing, to sufficient living area, and to security of tenure. In 2005, according to UN-HABITAT (2008a), 36.5 percent or 810 million of the world's urban population lived in slums. Cities in Sub-Saharan Africa had the highest proportion of people living in slums, 62.2 percent, while more than half of the world's slum population lived in Asia. Most urban growth is projected to occur in cities in developing countries, and half of the urban growth "between 2001 and 2030, i.e., 1 billion people, will take place in urban slums" (UN-HABITAT, 2008a). This projection more than doubles the urban slum population of 2005 and would increase to 41 percent the worldwide urban population living in slums.

Slums and informal settlements are much more vulnerable to climate risks than the formal city. Thunderstorms highly affect these settlements due to inadequate drainage, almost complete impervious surfaces, and in many cases due to their vulnerable locations in floodplains or steep slopes. For example, in the favela Santa Marta in Rio de Janeiro, dozens of buildings were destroyed during a heavy thunderstorm in 2003. See Figures 8.3–8.5. Waste clogged the runoff areas until everything was swept away.

Figure 8.3: *Favela Santa Marta, Rio de Janeiro.*
Photo by Marco Schmidt.

Figure 8.4: *Favela Santa Marta, Rio de Janeiro.*
Photo by Marco Schmidt.

Heat waves especially affect these areas. Many slums have extremely high density. Their lack of vegetation, high area of impervious surfaces, and lack of or inadequate drainage systems differ from the rest of the city. Air conditioners are lacking, and older people have limited ability to evacuate in case of emergencies.[7] Unavailable or weak water supplies worsen the situation in case of disasters. In Rio de Janeiro alone, about 1.3 millon people live in approximately 750 *favelas*. Newcomers from rural areas have been gradually building on public land inside the city for the last 100 years.

The built environment of poorer urban households is usually at higher risk due to weaker structures, less safe city locations

---

7   As the devastating European heat wave of 2003, causing over 35,000 excess deaths (Kosatsky, 2005) demonstrated, heat waves are not only problems facing slums in developing countries. See Chapter 7 for a discussion of this.

Figure 8.5: *Informal settlements mainly established on steeply sloped public land in Rio de Janeiro.*

Photo by Marco Schmidt.

and building sites, and weaker resilience of infrastructure to withstand damage. These factors influenced the disproportionate impact of Hurricane Katrina on the poor, sick, and disabled in New Orleans, which illustrates the persistent importance of these risk factors even in highly developed countries (Briggs and Keys, 2009). Similarly, the relation between urban health and climate change risks is particularly heightened under conditions of urban poverty in cities. When basic infrastructure is inadequate, existing conditions of poor sanitation and drainage and impure drinking water are further stressed during extreme weather events and flooding, leading to the transmission of infectious diseases, which puts poor urban households at higher risk. The higher densities in urban slums add to their vulnerability. When disasters occur, because many developing countries lack the health facilities to deal with large numbers of injured patients, there are higher death tolls than in countries better equipped for disaster. The January 2010 earthquake centered near the Haitian capital Port au Prince is a tragic example of this phenomenon.

8.3   The urban land management system

Although the multiple factors that drive urban patterns may make urbanization seem like an inevitable and uncontrollable process, citizens, cities, and other levels of government have the capacity to plan and manage urbanization to meet the challenges of climate change. Climate change risks will depend on the adaptive capacities of cities and their publics. The adaptive capacity of cities fundamentally depends on urban land management sys-

tems. The urban land management system, the overall system through which decisions concerning land use are made in urban areas, is composed of interacting subsystems: the legal framework, which defines property rights and government powers over land; the planning system, which develops plans and regulations for urban development; the administrative system, which manages urban services and infrastructures; land markets, which enable the exchange or sale of land; and the fiscal system, which levies taxes and provides revenues for government services. These systems interact in various ways to reduce or amplify adaptive capacity. For example, a planning system that works in isolation from urban service delivery agencies is likely to produce plans that are not implemented, or a fiscal system with insufficient revenue sources is not likely to be able to implement plans.

8.3.1   The legal/political framework

The legal and political framework for land management systems is crucial for understanding the potential for cities to mitigate and adapt to climate change. Typically, this framework is made up of cultural values pertaining to land, legal rights to property ownership, the roles of government in securing and regulating such rights, as well as in planning and managing cities and urban areas. Property rights to land are vital because housing and economic activities require stable land tenure. Rights to urban land/property and public powers to manage urban land are complementary. A strong urban property rights regime without strong public powers would leave owners with the obligation to defend their property rights against the harmful actions of others, and subject to environmental spillover effects. Strong public powers to regulate land without well-defined property rights would leave citizens at the mercy of local or national authorities. In cities with large urban poor populations, security of tenure is generally acknowledged as the critical first step in the social and spatial integration of slums and low-income settlements. It is now clear, however, that property rights, such as the rights to develop, use, transfer, obtain financial benefits, etc. and security of tenure, that is, security of ownership or lease or dwelling, are separate and sometimes conflicting concepts, especially in developing countries (Payne, 2004) For example, programs to provide legal titles to property have been implemented in many developing countries to increase tenure security, but such programs have yielded equivocal results for the urban poor (Sjaastad and Cousins, 2008).[8]

Today, in most developed countries, cities have wide powers to plan and regulate lands within their borders. Over the past few decades, a shift towards decentralization has provided local governments in more countries responsibility over the management of their land. Based on various arguments,

---

8   Instead, a growing group of experts argue that a more effective pro-poor regularization of informal settlements and slums should include the following: moratoria on forced evictions and relocations; offers of priority relocation to safe sites for residents who live in settlements subject environmental risks; entitle all other extra-legal settlements to other forms of secure/intermediate rights but not necessarily full titles, e.g., communal tenure options (to avoid high increases in land prices); audit planning and building regulations to reduce costs and time required to develop legal shelter options, etc. (Payne, 2004).

including that local governments can be more sensitive to local preferences in the provision of public goods (Tiebout, 1956; Oates, 1972) and more accountable to citizens, 75 countries since the 1980s have ceded powers over land use or other public services to localities (Ingram and Hong, 2008). France in the 1980s (Booth, 1998) and the Netherlands in the 1990s (Blanco, 1999) are just two examples. In addition, the European Union has provided strong incentives for Central European and Baltic states to decentralize land management functions. The shift to decentralize land use and fiscal powers in developing countries, accompanied by increasing globalization and privatization, have been influenced by the Millennium Development Goals (United Nations, 2008), which rely on local governments for delivering basic services to the poor, and by international aid agencies, such as the International Monetary Fund, the UN Development Programme and the World Bank, which have promoted decentralization policies (Ingram and Hong, 2008). Decentralization of land management combined with metropolitanization and the development of mega-city-regions adds to the fragmentation of government, and is an important legal/political aspect influencing a government's ability to manage urbanization.

Public powers to manage urban growth include land use and environmental planning and regulation, as well as public infrastructure planning and management. In developed nations, there is wide variation in legal and property rights systems, as well as in the formal government system responsible for urban land planning and management. The constitutions of many countries limit governmental action in relation to land and property. For example, the US Constitution protects property rights, and thus land use regulations that affect property rights can be challenged on constitutional grounds. Since the late 1980s, a politically conservative property rights movement has used such grounds to challenge land use planning (Jacobs, 1998; Jacobs and Paulsen, 2009). In Finland and Portugal, landowners are granted constitutional rights to build on their land, and these countries have difficulties in managing urban growth. In the UK, which does not have a codified constitution, no such limits to land use planning exist. The constitutions of Italy, the Netherlands, and Spain provide that all citizens have a right to a decent home, and such constitutional provisions can offer justification for urban planning and management (Cullingworth and Nadin, 2006).

China has a distinctive land tenure system. Under China's land title system, only the state and collectives can own land, and there is no private land ownership, although individuals and corporate entities can own property above the land. Under China's 1982 Constitution, all urban land is owned by the state, but there is a system for granting, leasing, or allocating long-term (40–70 years) land use rights. Collectives own much rural land, and thus, in order to control urbanization of agricultural land, the national government has had to increasingly regulate the conversion of land uses under collective ownership. According to Article 10 of the Constitution of the People's Republic of China, all land belongs to the state, and individual farmers have no property rights. In 1988, Article 10 was amended to allow the transfer of land use rights (Sharkawy et al., 1995). Ownership of the land still remains in the hands of the state, but land use rights are available by negotiation, bid, or auction. One result of the amendment has been that farmers and collectives can rent out their land to foreign and local ventures, and large areas of farmland have been converted to urban uses. The land reform of 2008 will significantly change the land tenure system in China. Under the new system, farmers are allowed to lease or transfer land use rights (Yardley, 2008). Some argue that this new system will lead to even more rapid rates of urban expansion, with associated increases in greenhouse gas emissions.

In many parts of the developing world, colonial systems of town planning continue to operate, but relatively high and unaffordable standards have not been expanded to the settlements of indigenous people. Instead, the colonial system, e.g., in South Africa (Toulmin, 2008), continued for a small elite core, while informal markets and unregulated expansion of towns allows urban development by wealthy landlords, many of whom provide little or no service infrastructure and minimal formal security of tenure to the urban poor. With urban growth, cities also spilled into areas of land managed under traditional practices where common property rights prevailed, creating a complex web of land use regulation and property rights. For example, over the past 50 years, Latin America underwent rapid urbanization and expansion of informal developments and slums without adequate infrastructure and urban services. This is typically seen as a failure of urban planning, but legal systems have played a large role in the expansion of informal settlements in Latin America. In the case of Brazil and Colombia, for example, scholars argue that a key factor in the expansion of informal settlements was inadequate legal codes that protected private property rights while failing to recognize the public interest in land development regulations (Fernandes, 2003, 2007; Fernandes and Maldonado Copello, 2009).[9] As previously discussed, the characteristics of informal settlements and slums will make them most vulnerable to climate change impacts.

Often underlying constitutional provisions are the historical and cultural attitudes of nations towards land. Although there are likely many variations, the increasingly dominant attitude

---

9 A promising development in Brazil and Colombia is the emergence of a legal reform movement in the 1980s promoting a new legal paradigm, Urban Law, which promotes urban democracy. The proponents of Urban Law argue that civil codes contain principles that can support appropriate state intervention, emphasize the concept of the social function of property, and an integration of law and management that incorporates urban democracy and decentralization of decision-making processes. Establishing this new legal approach, Colombia enacted constitutional changes in 1991, as well as a new law (No. 388 in 1997), and Colombia's Constitutional Court has upheld this approach in its rulings. In Brazil, the 1988 Federal Constitution proposed, and the 2001 City Statute consolidated the Urban Law approach. Recent studies indicate that high court judicial decisions in Brazil have incorporated this new interpretation in 50 percent of their decisions (Fernandes, 2003, 2007; Fernandes and Maldonado Copello, 2009). This new legal approach can result in greater security of tenure for the poor, more effective urban planning, development, and infrastructure financing mechanisms, all crucial for effective climate change adaptation efforts in developing countries.

towards land in the world today is to view land as a private good, where owners are entitled to use land in its most profitable way. Still, there is cultural variation, for example, in the UK, land is conceived as a special type of property, "to be preserved and husbanded" (Cullingworth and Nadin, 2006). Land does pass the market economic criteria of a private good; it is both an excludable good (we can exclude others from the use of a land parcel), and a rivalrous good (if I put a house on this parcel, it prevents others from putting something else on it). In addition to its social function, environmental science increasingly makes clear another aspect of the special nature of land. Its use determines the health of the ecosystems on which we and other life on Earth depend. Land has living and systemic aspects that the dominant concept of property rights does not capture. Unfortunately, we have yet to adopt a modern alternative concept of land rights and obligations that adequately captures the ecological aspects of land. Instead, the increased acceptance of market economics around the world over the past three decades has diminished the variation in cultural attitudes towards land.[10] Developed and developing nations have been adopting more market-oriented property rights systems either in the absence of effective urban planning and management institutions, as in the case of many developing countries, or that weaken their traditional public powers, as in the case of many European countries, such as Denmark. In general, this trend has likely decreased the ability of governments to address climate mitigation challenges and adaptation impacts, since addressing climate challenges will require government regulations in the public interest that will restrict private property rights.

8.3.2 The urban planning system, its history and institutions

Planning is the steering function in government, and land use is a, if not the, primary concept in urban planning (Krueckeberg, 1995). A major objective of UN Agenda 21[11] explicitly recognizes the importance of comprehensive urban planning and management to achieve sustainable urban areas. Urban planning as a government function emerged in some parts of the developed world, e.g., in the UK and the USA, out of sanitary and housing reform movements in the early twentieth century. These movements were prompted by public health concerns related to infectious diseases, such as cholera, typhoid, scarlet fever, and tuberculosis, that swept through industrial cities at the turn of the twentieth century where people were housed in crowded tenements without access to light or ventilation, and lacked sanitary water, wastewater, and solid waste disposal systems. (Peterson, 2003; Lubove, 1967). The broad rationale of urban planning and regulation is to ensure public health and safety. For example, the strict residential districts in urban zoning, developed in the context of the industrial city, were designed to protect residential uses from noxious industrial uses. The modern profession of

urban planning developed as a field of studies linked to architecture and engineering faculties, beginning in the 1930s in the USA. In many parts of the world, urban planning is a specialization in architecture, engineering, or public administration schools, very often offered at the post-graduate level. Urban planning practices are now well diffused throughout the world (Ward, 1999), although there are major gaps in such practices in many countries. Several scholars (Harvey, 1989; Sassen, 1991) have argued that planning diffusion is a key aspect of economic globalization, especially given the global character of property development.[12]

Most developed countries have planning institutions that engage in long-term, comprehensive planning, zoning, and other land use regulations (Nivola, 1999). Long-range comprehensive planning refers to a type of plan that typically has a 20-year horizon, and includes citywide plans for the land uses and needed infrastructure and public facilities for a projected population and economy. Typically, local legislative bodies are responsible for approving land use plans, and thus plans have the status of law. The administering of regulations, or permits for land use actions, is carried out by an administrative department, usually under the local executive's supervision. In the USA, urban planning and the administration of land regulations are conducted by several major institutions: lay planning commissions and local legislatures, who make decisions on plans and changes to regulations; and professional departments of planning, which prepare the plans and regulations, administer them, and are part of a city's bureaucracy. Planning commissions are made up of lay citizen volunteers instead of experts or professionals, and have quasi-legislative functions subject to approval by the elected city council (Sanders and Getzels, 1987). Planning commissions, a government reform initiative from the early twentieth century, were meant to buffer planning decisions from the political process. These lay commissions, with their public hearing requirements, laid the groundwork for a more participatory approach to urban planning, which is now prevalent in the USA. This is in contrast to other departments in city government, which are governed in a more typical bureaucratic mode, lacking strong participatory processes (Blanco, 1994). The greater uncertainty associated with climate change adaptation planning will require significant new information, public education, and strong community participation to build consensus on local adaptation strategies.

8.3.3 Types of plans and their limitations

The planning system is a key element in determining a city's adaptive capacity. Urban planning is typically conducted at the city level. Local plans in developed countries typically take the form of comprehensive long-range plans, which are primarily driven by demographic and economic trends, and thus

---

10 Although several legal scholars are addressing the issue of how ecology affects the traditional concept of private property rights (Freyfogle, 2003: Goldstein, 2004).
11 UN Agenda 21, Chapter 7, sets out the objective of promoting sustainable human settlement development, including programs to ensure adequate shelter, improve urban management, and sustainable land use planning and management, as well as the provision of urban infrastructure, sustainable energy and transport systems, etc. (UNCED, 1992).
12 But diffusion need not occur in a monolithic fashion as through the authoritarian imposition of colonialism, and planning institutions have also been the subject of selective borrowing and more synthetic innovations (Ward, 1999).

[MITIGATION/ADAPTATION]   Box 8.4   Urban growth management and smart growth in the United States

Hilda Blanco

University of Southern California

Urban growth management by aiming to manage land conversion from rural to urban uses and by influencing urban form can be a key strategy for reducing greenhouse gas emissions and for adapting to climate change.

The state-wide growth management movement that emerged in the USA in the 1970s, motivated by the rise of the environmental movement, is a promising effort to provide planning mandates at the provincial/state level for coordinating inter-jurisdictional planning, and appropriate tools to accomplish this. Several of these programs require consistency between plans and regulations, and explicitly address and link land use, infrastructure, and environmental objectives. In the 1990s, reacting to continuing sprawl despite planning efforts, a new urban planning movement, smart growth, advocates several measures: (a) limiting expansion of new development through urban growth boundaries[13] or utility districts; (b) increasing residential densities in existing and new growth areas[14]; (c) promoting more mixed use and pedestrian amenities to minimize car use; (d) charging infrastructure impact fees to consumers instead of having the community in general pay through property taxes; (e) emphasizing public transit to reduce the use of private vehicles; and (e) revitalizing older existing neighborhoods (APA, 1998; Burchell *et al.*, 2000; Downs, 2001). About a dozen states in the USA have adopted state-wide measures that incorporate some of these features. The programs of the states of Oregon and Washington are perceived to be the most successful in reducing sprawl, although some studies indicate mixed results (Song and Knaap, 2004; Bae, 2007). These two state programs require planning efforts by their counties and cities to link land use, infrastructure, and environmental concerns, and incorporate urban growth boundaries as a major strategy (DeGrove and Miness, 1992; Weitz, 1999).

More recently, the Lincoln Institute of Land Policy conducted an evaluation of state-wide smart growth programs by comparing four states (Florida, Maryland, New Jersey, and Oregon) to four states without such programs (Colorado, Indiana, Texas, and Virginia). It evaluated programs on various parameters of smart growth, including growth patterns, natural resources and environmental quality, transportation, and affordable housing. The study concluded that smart growth states only marginally outperformed the states without such programs, although only one of the smart growth states, Oregon, performed best on each objective (Ingram *et al.*, 2009). Consistently, studies indicate that the most successful of these state-wide growth management efforts in the USA is in the Portland metropolitan region, which is the only elected metropolitan government in the USA (established in 1992) with metropolitan land use powers, infrastructure management, and some fiscal powers. These results suggest that effective metropolitan planning for climate change is likely to require new political arrangements.

are growth-accommodating. In some countries, however, the attitude towards trends is different, in the UK, for example, trends can be modified in more socially desirable directions (Cullingworth and Nadin, 2006). These local long-range plans incorporate various elements, including, traditionally: land use, housing, transportation, public facilities and services, natural resources or environmental protection, open space and recreation. Some cities also include economic development and urban design elements, and coastal cities include shoreline management elements. In developed countries, the connection of these long-term plans to implementation steps is not straightforward, and does not proceed in a linear fashion, since local planning departments are separate units within a multiplicity of local government agencies, including transportation, housing, water and sewer, etc. Such agencies also plan for their services, and coordination between land use planning and other related local government departments is often lacking. Reaction to this lack of coordination among city agencies in the USA led to urban growth management efforts after the 1970s that link land use plans to infrastructure planning, especially to transportation planning. But the lack of coordination among these agencies is still prevalent today, and is a major obstacle to effective growth management.

8.3.3.1   *Metropolitanization requires metropolitan planning*

As cities and their populations spread out beyond their historical borders into metropolitan areas, jurisdictional boundaries, political demarcations, and traditional governing structures and institutions of planning and land use management are becoming outdated. Increasingly, these functions demand more integrated planning, infrastructure investment, service delivery, and policy decisions than these fragmented jurisdictions can provide. Effective planning and management of these inter-municipal territories requires an integrated metropolitan approach that transcends the traditional municipal boundaries. The fragmented local government structure of metropolitan areas facilitates the conversion of agricultural, forested, or otherwise undeveloped land to urban uses. These expanding urban areas also have fiscal deficiencies and weaknesses, face heightened

---

13 Urban growth boundaries are planning lines drawn around an urban area to separate it from the rural area. These boundaries are growth accommodating: in Oregon, for example, the areas within the growth boundary are drawn to accommodate growth projections for the next 20 years, and can be adjusted to accommodate growing demand (Easley, 1992; Nelson and Dawkins, 2004).

14 Studies indicate that higher densities than typical suburban densities are more cost-effective from the perspective of infrastructure and public facilities provision, and contribute to more effective public transit (Burchell *et al.*, 2005), but we have yet to investigate whether, from the perspective of climate change, there is a point of diminishing returns or diseconomies as density increases.

challenges of metropolitan transportation, as well as deficiencies in critical physical and social infrastructures (Rusk, 1995; Ladd, 1998; Norris, 2001; Orfield, 2002; Carruthers, 2003). In addition to these challenges, cities face a context of competition with other cities at a national, continental, and often, at a global level.

The challenges of equitable development among different groups in these vast urban territories highlight the need for major improvements in the provision of public services such as health, decent shelter, education, water, and sanitation. As urban poverty worsens, especially in developing countries, poor populations have been spreading outwards, making the peripheries of these metropolitan areas the poorest and most heavily under-serviced settlements. Concerns related to an increasingly divided urban society, together with inequalities and poverty that stretch across large metropolitan areas, emphasize the need for metropolitan planning and governance frameworks that address these imbalances (McCarney, 2010).

Although decentralization could be an effective administrative reform strategy, decentralization efforts have sometimes further entangled the institutional problems facing fragmented metropolitan areas. In India, for example, under the new national strategy of the 1990s, the 74th Amendment of the Constitution (1992) conferred constitutional status on urban local bodies, and transferred to them responsibility for urban development. But whether peri-urban areas, where poor rural migrants to the cities locate, can be managed by the municipal corporations is left up to the states to determine. Thus, while some states, such as Tamil Nadu have been responsive, other states have not been (Shaw, 2005; Dupont, 2007). In Bangalore, with a population of over 7 million in 2007, the land use planning authority for the region is vested in two state agencies: the delivery of urban services is the responsibility of the elected metropolitan government, while several other independent agencies are in charge of water supply, police, transportation, etc. (Sudhira *et al.*, 2007).

8.3.4   Urban land regulation

Zoning, especially in developed countries, is a system of land classification widely used for development purposes: areas are zoned for residential, commercial, industrial, or agricultural, and each of these broad classifications can be further subdivided.[15] For example, residential districts are often subdivided into low-, medium-, and high-density districts. Zoning is composed of a map that divides the city's land into zoning districts, and the zoning code or local law, which sets out permitted uses by district and other regulations, e.g., the permitted height of buildings (Lerable, 1995). Zoning codes have been traditionally exclusive;

for example, the zoning code would not permit a high-rise apartment building in a low-density residential area. Zoning also determines the percentage of a parcel that can be built upon, as well as the height of structures, and thus can regulate solar access. Through building coverage and landscaping requirements, zoning can influence the extent of impervious surfaces, a very important determinant of drainage and the heat island effect. Zoning is empowered to restrict property rights to ensure the public's health and welfare. Traditionally, this type of zoning, which is geared to prevent the nuisances of high density or commercial or industrial activities in single family residential districts, is currently controversial since, combined with widespread automobile ownership, it segregates activities in urban areas, and increases automobile use. In the USA, for example, there is a growing movement of planners and architects that argue for mixing land uses, that is, for new zoning districts that permit several compatible uses, such as low- and medium-density residential and neighborhood commercial uses. Such mixed use districts, advocates argue, would provide more livable neighborhoods, as in historic towns and cities, where residents would be able to walk or bike instead of driving for neighborhood services, and where there would be a greater mix of social classes (APA, 1998; Grant, 2002; Talen, 2006). The mixture of land uses is an important element of a compact city strategy that could reduce the number or length of automobile trips, and thus energy use and associated greenhouse gas emissions.

Subdivision regulations (Freilich and Schultz, 1995) are major instruments for converting rural land into urban, and have been instrumental in determining the dispersed suburban pattern in the USA and other countries. Subdivision regulations are crucial for climate change, since, for example, street widths and orientation can influence ventilation patterns for a city, important for mitigating the heat island effect, and open space and drainage options will be important to adapt to increased precipitation events in many cities.

Building standards are also a major form of land regulation, administered by urban planning departments, which require adequate design, construction processes, and building materials, such as minimum room size, or heating and cooling standards to meet fire and safety codes. Very few cities develop and maintain their own building codes, New York City and Chicago being exceptions in the USA. Instead, building codes are developed either by government agencies, in which case such codes are typically mandatory for states/provinces and their cities, or by national or international associations of professionals[16] organized to develop such model codes. These latter codes are then adopted and can be amended by states and cities. Important for climate change, building codes include energy standards and regulations on heating and cooling. The updating of national and international standards is ongoing, and more energy-efficient building standards have been developed. The updating

---

15 For a comprehensive review of the state of urban planning and related tools in developed and developing countries, see the UN-HABITAT *Global Report on Human Settlements 2009: Planning Sustainable Cities* (2009).

16 Building codes are increasingly becoming consolidated and worldwide. The International Code Council, for instance, develops building codes that have become the dominant standards in the USA and are also widely adopted internationally. The heating and cooling standards of the American Society of Heating, Refrigeration, and Air-Conditioning Engineers (ASHRAE) are also widely adopted in the USA and around the world.

of building codes by cities is haphazard even in developed nations, such as the USA, where some cities retain older versions of codes, and some lack certain codes altogether, e.g., the state of Mississippi lacks a commercial building energy code, its residential energy code dates to 1992, and although cities and counties have the ability to adopt such codes, the largest city, Jackson, Mississippi, still lacked such a code.[17] The issue with building codes, however, is as much enforcement of codes as adoption of newer codes. The fatal collapse of so many structures in the 2008 Sichuan earthquake in China, for example, was not due to China lacking modern seismic codes. Instead, engineering experts have noted that the Chinese government had adopted a strong seismic design code after a devastating earthquake in 1976, but that there has been uneven application of it and lack of enforcement (Bryner, 2008). Thus, both the adoption and enforcement of energy-efficient building codes will be crucial for reducing building energy use and the emission of greenhouse gases.

Land use planning and regulations require up-to-date records on land ownership and transfers as well as on the dimensions of land parcels and structures. Local governments maintain such records in cadastral and/or land registration offices (Bogaerts and Zevenbergen, 2001). Cadastral systems provide legal protection regarding land ownership, facilitate transactions in land and real estate markets, and are necessary for levying local property taxes or fees on urban land and development. Land registration agencies are the first agencies that need to be established to develop a land management system in countries where no formal documentation exists of who owns or has rights to use land, as in many countries in Sub-Saharan Africa (Toulmin, 2008). In order for urban areas to inventory their emission of greenhouse gases and to prepare for climate change impacts, they require up-to-date data on ownership patterns and the location and condition of buildings that land registration offices maintain.

8.3.5 Development markets and their influence on urban growth

Most urban development is privately produced and exchanged through urban land and real estate markets. Real estate markets include land and the structures or development built on land. Land and real estate markets and groups who benefit from them all depend on urban growth. The role of urban land and development markets remains powerful even in countries, such as China, that have recently established such markets. In China, urban land development has largely been driven by a growing private sector and formal planning has had a relatively minor role compared to large-scale investments by overseas interests (Ding, 2007). As in most markets, developers as capitalists are seeking the highest profit. This leads to several development dynamics that have relevance to urban growth patterns. Developers seek the cheapest land for their projects, thus leading to the conversion of agricultural or otherwise undeveloped land in the periphery of existing urban areas, since these areas are cheaper than urban land (O'Sullivan, 2006). In addition, under the fragmented governance structure of metropolitan areas, if a city enacts land use regulations that marginally increase development costs, such as a higher energy-efficiency code, or if a city requires that new developments maintain a certain level of service in transportation or other infrastructure services, developers can often take their projects to localities without such regulations. Also, major developers have undue influence on local politicians. They are often major sources of local campaign financing contributions. Responding to climate change may require effective metropolitan governance that reduces the opportunities for developers to game urban planning and management systems, and rely instead on alternative coalitions to the urban growth machine (Logan and Molotch, 1987; Jonas and Wilson, 1999) to counter current urbanization trends.

8.3.6 Fiscal aspects of urban land: property taxes and city budgets

An important component of the urban land management system is the local fiscal system, the sources of revenues from which local governments draw their funds. Most local governments receive transfer of funds from national and provincial levels of government, and with increasing decentralization of government throughout the world more local governments have their own sources of revenues. In developed countries, cities typically have several sources of local taxes or fees that are used to provide local urban services, from street maintenance to water supply, public health, and safety services. For example, the property tax in the USA is the major source of revenues for local governments (Raphaelson, 2004). Property tax, the tax on the value of real estate property including land and structures, in effect provides funding for general urban services, and thus intimately links land development with the provision of local infrastructure and services. Local reliance on the property tax is not without problems. It generates inequities among cities as communities with, for example, highly valued properties (or *ratables* as they are referred to in the fiscal literature), either high-income residents with highly valued houses or large and highly valued shopping malls, generate more local tax revenues than communities with a lower valued tax base. Thus, communities with a higher proportion of ratables are able to fund better levels of urban services. In this way, reliance on the property tax can lead to fiscal or exclusionary zoning where communities, in order to attract high ratables, restrict their zoning to large, single family or other highly valued development, thereby excluding lower-income residents (White, 1975; Ladd, 1998). Fiscal zoning can aggravate affordable housing shortages, and generate greater regional road congestion and further urban sprawl as lower-income workers are forced to move farther and

---

17 However, the American Recovery and Renewal Act of 2009 (Stimulus Bill) made State Energy program grants available on condition that states adopt energy building standards for both residential and commercial buildings that meet energy standards of 2007 within 8 years of enactment of the Act (Barbour, 2009). By 2017, this will make states such as Mississippi 10 years at most behind the most recent energy standards. The passage of the American Clean Energy and Security Act (ACES) of 2009 (US House of Representatives Bill 2454 accessed at www.opencongress.org/bill/111-h2454/show) would mark a more significant milestone. The Bill requires that US DOE work with residential and commercial code developers to ensure that model codes are developed by 2010 and 2016 that will improve energy efficiency by 30 percent and by 50 percent respectively.

[ADAPTATION]   Box 8.5   Decentralization of fiscal authority in Ghanaian cities

Hilda Blanco

University of Southern California

Reliable own-source revenues will make it more feasible for local governments to respond in the future to the increased costs of coping with climate change impacts without complete reliance on their central governments.

Many African countries have devolved various government powers to local governments over the past few decades, including fiscal responsibilities. Although in most African countries, local governments still rely heavily on central–local government transfers for local budgets, localities seek ways to develop their own-source revenues. Local property tax is widely perceived as an appropriate local tax, since the land and structures are fixed and easier to monitor, and since the resource taxed and the benefits received are both local (Raphaelson, 2004). In addition, in developed countries such as the US and OECD countries, for example, property tax is a major source of own source revenues for local governments.

Decentralization began in Ghana in 1988 when the central government approved the local government law (Provincial National Defense Council Law 207) through which local government was reviewed and reorganized into district assemblies, which, as of 2008, number 166.[19] Ghana is a country on the western coast of Africa, with a population of over 21 million people, with a per capita GDP of US$682, and an urbanization rate of 36 percent. Its capital city is Accra, with a population of about 3 million. As a result of Ghana's decentralization policies, Ghanaian localities have the power to levy urban property taxes. Unlike the USA, in Ghana, property tax is levied on improvements (buildings and structures) only and not on vacant land, except in Accra's metropolitan area where, beginning in January 2008, differential property rates are being imposed on undeveloped land in different parts of the metropolitan area. The property tax on improvements is imposed on a depreciated replacement cost of the improvement. The Act on Rating (Act 462) of the Local Government Act of 1993 set the maximum tax assessment value of an owner-occupied property at 50 percent of replacement cost, and set a ceiling of 75 percent of replacement cost for other non-owner-occupied improvements. The authority for setting the actual rate, the administration of the tax, and the entitlement to the revenues from the property tax are left to the local

authorities (Jibao, 2009). This type of property tax system, where the base of the tax, or the ratable, is calculated based on the value of the property depends on the capacity of localities to assess the value of properties on a regular basis, since property values vary depending on land market conditions. This is a problem in Ghana and many other developing countries, which requires authorized evaluators or assessors of property values. Although Ghana did have over 200 registered assessors in 2007, this was not sufficient to cope with the evaluation of properties. The lack of evaluators needed in a system of taxation where the value of the property needs to be evaluated periodically means that the system is "neither practical, nor sustainable". In addition, in order to fulfill the goal of private sector participation, the Local Government Act of 1993 requires that local governments hire private companies to collect local taxes. In Accra, for example, private entities are hired to collect the property tax. The collection of taxes is fraught with many problems; for example, many houses lack house numbers, and cannot be identified to serve tax bills.

The Accra Metro Area is Ghana's largest metropolitan area with the largest industrial center. With an estimated tax base of approximately 39 trillion Cedis for about 120,000 properties, the property tax should have generated 229 billion Cedis in 2006. Instead it generated 15 billion Cedis, for a 6.71 percent tax yield. In contrast, Tema is a coastal city 25 km east of Accra with a population of about 600,000. Tema has undergone a rapid transformation from a fishing village to an industrial center, handling 70 percent of all shipments to Ghana and land-locked countries in western Africa. Unlike Accra, Tema in 2006 collected more property tax revenues than it projected, and the property tax contributes 36.7 percent of the city's own source revenues. The different experiences with the property tax in Accra and Tema are attributed to the greater percentage of industrial and commercial property in Tema. Such properties are seen as being more "tax compliant". Also, industrial and commercial properties can be taxed at higher rates.

The comparison of these two cities in Ghana indicating better tax yields under the property tax with commercial and industrial ratables could, in time, set up a dynamic comparable to the rateables chase found in US localities. Today, however, as the examples of Ghana's cities illustrate, a major problem is the lack of sufficient capacity to administer the property tax systems. This hinders the ability of local governments to supply local services including climate change adaptation.

---

farther away from places where they work.[18] On the other hand, the absence of robust fiscal systems in many cities of the developing world is likely a major reason there is inadequate capacity

to plan or enforce land use regulations at the local scale (Razin, 2000), which could be effective in mitigating climate change or adaptation planning.

---

18 An alternative to the property tax, which taxes both land and structures, is a land tax, originally proposed by Henry ([1879] 1992), where only land and not structures would be assessed a tax. Split-rate taxation is a variant of this, where land and improvements are taxed at different rates, with lower rates imposed on improvements. A land tax would be more efficient than the property tax, not distorting investment choices, and counteracting fiscal zoning. The land tax is also seen as reducing speculation on land, promoting economic development, as well as compact development (Dye and England, 2009).

19 Much of the information contained in this profile is a summary of S. S. Jibao's monograph on the property tax system in Anglophone West Africa (2009) prepared for the Lincoln Institute of Land Policy.

8.3.7 Relation to urban infrastructure systems

Land uses and urban infrastructures and public services are intimately interconnected. Typically, in developed countries, infrastructure or capital facilities planning is based on land use plans and zoning. Land use plans and regulations, for example, project a total number of new dwelling units, a certain increase of commercial or industrial space at the build out of the zoning districts. The total number and type of dwelling units or floor area become the inputs for infrastructure multipliers that are used to project traffic generated, water usage, wastewater generated, parks and open space needed, waste generated, etc.[20] These projections are, in turn, converted to new transportation or other infrastructure improvements needed to serve new development. The costs of these improvements can then be incorporated into capital facilities plans or capital improvement programs that identify the facilities required and the local and intergovernmental funding sources available to construct the facilities (Bowyer, 1993). The process just outlined, however, is an ideal process, given the sectoral divisions in local government. As discussed in Section 8.3.3, local departments of transportation, public works, parks and recreation, etc., that operate infrastructure facilities often have their own planning units with different professional practices and cultures. These departments have been slower in making the shift from their traditional role of supplying new infrastructure to managing the demand for urban growth. Without strong executive leadership or state or national mandates for integrated urban planning and management, this fragmentation of planning and operational functions makes it difficult to manage growth, and increases the climate change challenge for cities.

8.4 Climate change risk

From the perspective of urban land, climate change risk depends on the character, magnitude, rate, and variation of climate change impacts to which urban land and populations are exposed, on sensitivity factors, and on a city's adaptive capacity (IPCC, 2007). "Sensitivity refers to the degree to which a system is affected by climate change impacts. Adaptive capacity is seen as a function of behavior, resources, and technologies" (IPCC, 2007, p.720).

Chapter 3 discusses in greater detail the climate change risks and major impacts for cities: temperature increases, sea level rise, precipitation changes, and extreme weather events. We should note here, however, that from the perspective of urban land, sea level rise poses an especially significant risk, since low-elevation coastal zones (LECZ), areas less than 10 meters above sea level, take up 2 percent of the world's area, but contain 10 percent of the world's population and 13 percent of the urban population (McGranahan et al., 2007). Over 3,000 cities are located in the LECZ and more than one-half of these cities are located in Asia

(UN-HABITAT, 2008a). In Australia, 86 percent of the population lives along the coastline (Norman, 2009). LECZ cities risk inundation and destructive storm surges. Sea level rise thus poses the risk of land loss as well as loss of buildings and the infrastructure on such lands.

To determine the extent and character of the climate change hazard for specific cities requires downscaling the global projection models to take into account regional/local climate, hydrology, and other relevant factors. This is particularly important for determining the adequacy and cost-effectiveness of climate adaptation strategies. Downscaling involves data and modeling capacity often unavailable for cities in developing countries. Several recent attempts have been made to provide cities in developing countries with easily accessible methodologies to determine local risk levels, For example, the World Bank report on resilient cities (Prasad et al., 2009) outlines a process to enable cities to take the first steps in determining their susceptibility and resilience to climate change. Building on the similarities between natural hazard and climate change risks, the report sets out a simple method to establish whether a city is a "hot spot" or has high climate change risk, based on a city typology and risk characterization matrix that probes for the following: disaster-prone locations, e.g., coastal areas, plains; size; governance structure; built environment; political and economic impact of disasters; threat of natural hazards; disaster response system; vulnerability to climate change in various sectors, etc.

As discussed in Section 8.1.2, the land sensitivity factors related to climate change in cities include the natural setting of a city, its urban form, and its built environment. Although poverty is the fundamental cause of slums or informal settlements, urban form and built environment factors in urban slums amplify the vulnerability of slum populations to climate change risks.

While the natural setting, the existing settlement pattern, and the character of the built environment are relatively stable or fixed, although subject to modification, the urban planning and management system represents a major human or adaptive capacity to respond to changing conditions, including making changes in urban settlement patterns. Urban planning and management, however, are not the only important adaptive factors. Other factors include public attitudes, political leadership, and financial resources to implement certain strategies, such as engineering or retrofitting strategies. The more vulnerable the natural setting, or the urban form, or the built environment, *ceteris paribus*, the more vulnerable is a city to climate change: the weaker the capacity to change these patterns to more resilient ones, the greater the risk to a city. A city with strong institutions of planning, regulation, and management is more likely to have the capacity to prepare local mitigation or adaptation plans that take into account local conditions and engage the local populations in these efforts. Such plans are likely to be more effective in reducing climate change risk.

---

20 Professional associations typically provide these infrastructure multipliers. For example, in the case of traffic multipliers, the Institute of Transportation Engineers publishes trip generation and parking generation handbooks that assist transportation planners to estimate the traffic generated by specific land uses (ITE, 2008).

[VULNERABILITY/ADAPTATION]   Box 8.6   Vulnerability and resilience in a rapidly growing coastal city: Florianópolis, Brazil

Sandra Baptista

CIESIN, Earth Institute at Columbia University

Rapidly expanding coastal cities in low- and middle-income countries face the linked development challenges of managing accelerated urban growth, reducing social inequity and exclusion, maintaining ecosystem services, preparing for climate hazards, and adapting to climate change. Florianópolis on the coast of southern Brazil is one such city. As observed across Latin America, in Greater Florianópolis low-income households, businesses, and communities are most likely to occupy poorly built structures and informal settlements located in high-risk areas (e.g., steep slopes or low-lying lands), which often lack adequate access to basic infrastructure and services. These conditions leave poor populations highly vulnerable to climate-related hazards including high-speed winds, storm surges, heavy rainfall, floods, landslides, and post-disaster outbreaks of water-borne and vector-borne infectious diseases.

Florianópolis is the state capital of Santa Catarina. Greater Florianópolis – which extends into Santa Catarina Island and the mainland municipalities of São José, Biguaçu, and Palhoça – has emerged as a dynamic globalizing coastal city. It is one of southern Brazil's largest urban agglomerations. While in 1970 Greater Florianópolis had about 217,000 residents, it is now home to roughly 800,000 people (IBGE, 1974, 2007). The region's landscapes, ecological diversity, and urban amenities attract hundreds of thousands of tourists each year. Ecosystems include dense broadleaf forest, mangroves, coastal sandy plains, lagoons, estuaries, sandy beaches, dunes, and rocky headlands. Over the past four decades the region has experienced both ecosystem degradation and ecosystem recovery as interacting demographic, socioeconomic, land use, and institutional changes have occurred (Baptista, 2008a, b). Today, over one-third of Santa Catarina's 6 million people live near the coast. In 2000, the 37 municipalities situated directly on the coast had nearly 2 million people inhabiting about 10,000 km^2, an increase of about 1.2 million residents since 1970 (IBGE, 1974, 2001).

Since the mid twentieth century, rainfall measurements in southern Brazil have shown an increasing precipitation trend (e.g., Haylock *et al.*, 2006). Based on daily air temperatures from 27 meteorological stations in southern Brazil, Marengo and Camargo (2008) identified warming trends for the period 1960–2002. Scientists predict continued precipitation increase, positive trends for warm nights and negative trends for cold nights (Vincent *et al.*, 2005), and higher frequency and intensity of rainfall events resulting in flooding and landslides (Marengo, 2008). These changes will have consequences for ecosystems, agriculture, urban development, and human health. For example, warmer and wetter environmental conditions are likely to contribute to outbreaks of infectious diseases such as malaria and leptospirosis (Kupek *et al.*, 2000; Confalonieri, 2003).

Major infrastructure projects enabled the expansion of public services, tertiary sector growth, and real estate development in the region, thus spurring the institutionalization of modern public planning for urban development. In 1955, the first Master Plan for Florianópolis was approved and a state development plan was legislated to initiate major public works projects, including electric energy sector development (Caballero, 2002; Reis, 2002). By the late 1960s and early 1970s, the Florianópolis Metropolitan Area Development Plan prioritized the construction of a high-speed road network to economically integrate Santa Catarina Island with the mainland and promote urban expansion and tourism development (Amora, 1996; Reis, 2002). This plan led to a bayside landfill project, which added 6 square kilometers to the island, initially supporting construction of two new bridges connecting the island to the mainland and later accommodating the city's central bus terminal. During the 1970s, several new federal and state government agencies were established in the Central City District, and the jobs created stimulated new residential and commercial development.

The municipal urban planning agency IPUF (Instituto de Planejamento Urbano de Florianópolis) was established in 1977 (Adams, 2004). In 1981, the Tourism Development Plan for the Florianópolis Urban Agglomeration was proposed as part of a larger federal economic development program targeting medium-sized cities (Amora, 1996). IPUF was also involved in the preparation of Master Plans for neighboring municipalities towards metropolitan integration (Sugai, 2002). Since the 1980s – in response to observed land use trends, reevaluations by the municipal legislature and federal legislative changes – IPUF has revised the Central City District plan and has prepared plans for other submunicipal units. Furthermore, Brazil's 2001 City Statute (Federal Law No. 10257/01) institutes new tools to regularize informal settlements in urban areas and mandates municipalities to democratize local decision-making processes for urban planning, legislation, and management, including the formulation and approval of municipal master plans (Fernandes, 2007).

Since the 1980s, as impoverished urban communities have expanded in Brazil, middle- and upper-income households have responded with self-segregating behaviors and actions intended to minimize their exposure to crime, violence, and environmental hazards. As elite entrepreneurs and middle-class migrants claim emergent territories and economic opportunities in Brazil's newly expanding metropolises, they reproduce the exclusionary patterns of land use that are now characteristic of the older metropolises, while maintaining connections to national and international centers of informational, financial, and political power. The case of Florianópolis illustrates these socioeconomic dynamics and the ways in which they contribute to disparities in human well-being, including disparities in social vulnerabilities and resiliencies to climate risk. Ongoing efforts in Florianópolis, and other Brazilian cities, to develop direct democracy and formulate participatory municipal master plans have the potential to: provide important institutional mechanisms for inclusive governance and sustainable development planning, facilitate successful adaptation to climate variability and climate change, and foster more equitable outcomes.

235

8.5 How urban planning can address the mitigation of greenhouse gas emissions

Planning and management tools can help to address the critical link between emissions and urban form, particularly in terms of transportation, more compact development, building energy consumption, and conversion of land for urban use. For example, official plans, development guidelines, development permits, densification plans, infill development, reuse of buildings, transit planning and pricing, building codes, and a number of other planning tools can help to address greenhouse gas emissions in cities as climate change mitigation strategies. Effective planning, however, requires spatial data that link greenhouse gas emissions with their sources.

8.5.1 What cities are currently doing

Many cities have begun to incorporate strategies to reduce greenhouse gas emissions in their urban plans. Most mitigation plans focus on energy efficiency in buildings and the transportation sector. More difficult to address are ways to change land use patterns to reduce automobile use and increase the use of public transit and non-motorized travel. Most European urban areas have a compact core and effective public transit systems, but land conversion to urban uses and resulting sprawl and automobile use are increasing problems (Kasanko *et al.*, 2006). Several European cities in countries with strong land use powers have traditions of greenbelts and other urban growth containment measures, but their effectiveness is mixed, and in times of strong real estate markets, they are often weakened.[21] For example, the future of London's greenbelt policy, dating back to the 1940s and 1950s, was recently in question, given a strong real estate market before the 2008 recession, and the government's plans for 3 million new homes by 2020 (Brogden, 2007). In addition, the effect of greenbelts is often mixed. London's greenbelt had the effect of scattering growth beyond the greenbelt (Bruegmann, 2005). Seoul's greenbelt, inspired by London's, has been more strictly enforced for the past 30 years, but Bae and Jun (2003) argue that Seoul's metropolitan area is larger as a result of the greenbelt than it would have been without it, and this has generated a severe jobs–housing imbalance. Seoul's greenbelt policy was also recently revised, and by some reports, weakened (Bengston and Youn, 2006). In the USA, as indicated in Box 8.4, the smart growth movement has advocated compact development, mixed land uses, and public transit, all strategies with great promise to reduce automobile use and thereby energy use. But although a number of studies indicate that growth management has had some successes (Nelson, 1999; Wassmer, 2006; Yin and Sun, 2007), others point to more mixed results (Kline, 2000; Carruthers, 2002; Song and Knaap, 2004; Bae, 2007; Ingram *et al.*, 2009). A milestone in this movement and in mitigation planning is the passage in 2008 of California's Sustainable Communities and Climate Protection Act or SB 375, the first legislation in the USA to link transportation and land use planning with climate change. Major new requirements of the law are: that regional targets for greenhouse gas emission reductions be tied to land use plans; that regional planning agencies create plans, Sustainable Community Strategies, to meet such targets, even if in conflict with local plans; and, that the funding of regional transportation agencies be consistent with the regional plans. Promising as it is, SB 375 still leaves local governments in charge of their land use plans and regulations, even though it creates more incentives for regional coordination (State of California, 2008; Higgins, 2008).

8.5.2 Next steps

Leadership at the national and international levels is needed to provide either incentives or mandates to turn individual and voluntary local programs into widespread practices. Building energy efficiencies can be achieved through the adoption of more energy efficient building codes by national governments. Once adopted, national governments through national policy or incentives should require their timely adoption by provincial (sub-national, state) and local governments. Smart growth policies to reduce sprawl and automobile travel are more discussed than carried out in practice in the USA, because there are major impediments to their implementation. Two major obstacles requiring national leadership are: consumer demand for low-density living patterns, and local home rule. Local governments have the legal power over land use, but in order to implement anti-sprawl policies, such as urban growth boundaries, which should be metropolitan in scope, such policies require state-level support (Downs, 2005).

8.6 How urban planning can address the impacts of climate change

Adaptation planning fits naturally the agenda of urban and regional planning, since, unlike the global-scale strategies and benefits of mitigation planning, adaptation impacts, strategies, and benefits are all local. In addition, adapting to extreme impacts of climate change is essentially a planning challenge, which is likely to call for public, community-wide planning and not just individual or autonomous planning (Blanco and Alberti, 2009). Traditionally, the major drivers of urban plans have been population and economic projections and existing land uses. Urban planning has increasingly incorporated natural hazard constraints into the factors considered in developing land use plans. Natural hazards mitigation planning has become a subfield in planning in the past two decades, closely connected to the emergency management community but focused on mitigating, through land use planning and regulations, communities' vulnerabilities to natural hazards (Godschalk *et al.*, 1999; Mileti, 1999; Schwab *et al.*, 2005). Plans have increasingly included the

---

21 Another major urban containment strategy, greenbelts are physical areas of open space, forest, or farmland surrounding a city or metropolitan area established to become permanent barriers to urban expansion.

analysis, mapping, and regulation of environmentally sensitive, or critical areas, such as floodplains, watershed areas, or erodible soils and endangered species habitats (Steiner, 2000). Planners, for example in Washington State, work with environmental agencies and scientists to prepare such analyses using "the best available science," as state regulations require (Municipal Service Research Center, 2009). In the case of adaptation planning, hazard and environmentally sensitive areas mapping can help identify urban areas sensitive to climate change impacts, such as landslide-prone areas. However, if such mapping is based on historical records of natural disasters or static conditions, it may not take into account the variability of climate change impacts (Smit et al., 2000). As discussed, such mapping should be based on downscaled climate models that incorporate local conditions. For coastal cities, for example, determining the impacts of sea level rise will require more than the mapping of low elevation areas. Shoreline erosion, wetlands displacement, saltwater intrusion, and more severe storm surge flooding, as well as changes in sedimentation patterns, are other key impacts of sea level rise, which must be taken into account. Urban planners will need to collaborate with and contribute to interdisciplinary teams of natural scientists to produce the type of model-based impacts analysis required to prepare adaptation plans. Adaptation planning will challenge the urban planning profession to develop a wider knowledge-base of earth and natural sciences, and of interdisciplinary modeling methods.

Using the best available science and methods for adaptation planning is essential, because major urban planning adaptation strategies are likely to require restricting development on privately owned land. As urban areas at high risk to climate change impacts begin to experience more frequent and intense floods, or sea level rise with storm surges, or wildfires, at a certain point, as the frequency increases, land use restrictions will be required, and they will be contentious. A second set of land management adaptation strategies, less contentious, will involve changes to building codes to ensure that buildings on coastlines are elevated, or construction designs that have the capacity to withstand higher wind speeds, or require building materials, surfaces, and colors that reflect solar radiation (Stone and Rodgers, 2001; Gartland, 2008). Another set of strategies involves increasing the number of urban trees and the greening of cities to reduce the heat island effect (Schmidt and Kohler, 2008). In addition, there is a set of urban adaptation strategies that involve changes to urban infrastructures and public health, from elevating roads in coastal areas, or raising levees, to heat wave plans, as discussed in other chapters of this volume.

8.6.1 What cities are currently doing

While hundreds of cities across the world are working on mitigation plans, few local governments have developed adaptation plans. Of these, the city of London provided an early model for climate change adaptation planning. London's efforts were aided by a national program, the UK Climate Impacts Programme, which fostered research on climate change and led to the establishment of the London Climate Change Partnership (LCCP)

(Blanco and Alberti 2009, p. 166). The Partnership, with strong political leadership from the mayor, generated regional scenarios, and identified strategy options for climate impacts, including temperature increases, flood risks, and water availability (LCCP, 2002, 2005, 2006). In Canada, the city of Toronto's adaptation plan is also noteworthy (Blanco and Alberti 2009, p. 166). In partnership with the city, Toronto's Clean Air Partnership (CAP) undertook a climate change adaptation program. CAP led stakeholder workshops, developed a scan of climate change impacts for the city, prepared a study of six major cities' climate change adaptation efforts, and identified a set of options for two areas of impact, the urban forest and heat, including heat's impacts on health and energy use (Wieditz and Penney, 2006, 2007a, 2007b; Penney and Wieditz, 2007).

In the USA, New York City's 2007 plan, PlaNYC 2030 (New York City Office of the Mayor, 2007), is a model of an integrated, strategic plan that incorporates both mitigation and adaptation strategies to climate change and addresses energy, sea level rise, and water resource issues (Blanco and Alberti 2009, p. 165). Both London and New York City's adaptation planning efforts benefited from scientific studies and regional modeling of climate change impacts. The New York City Plan relied on the Metropolitan East Coast (MEC) Assessment for the New York region (Rosenzweig and Solecki, 2001) funded by the U.S. federal program on climate change research, and the just-completed work of the New York City Panel on Climate Change (Rosenzweig and Solecki, 2010). NYC's MEC assessment was the only one of the 18 US regional assessments that was primarily focused on urban issues. Also in 2007, Washington State, King County, where Seattle is located, published a county climate plan that incorporates both mitigation and adaptation measures (King County, 2007; Swope, 2007). The University of Washington's Climate Impacts Group provided the scientific studies that supported the plan. King County's plan included a set of guidelines for incorporating mitigation and adaptation goals into county and city agencies' plans and policies. It also identified urgent adaptation needs, such as county roads within or close to floodplains that would be impacted by more intense storms. The county plan also incorporated steps to improve the county's capacity to undertake adaptation planning, for example, by educating appropriate county staff in climate change science, and by raising public awareness on this issue. Although the plan does not include specific implementation steps, its goal of incorporating climate change adaptation considerations in all relevant county plans and projects is farsighted (Blanco and Alberti 2009).

As indicated, the development of adaptation plans is a complex process. In addition to the World Bank guide discussed in Section 8.4, in partnership with the Climate Impacts Group at the University of Washington and King County, ICLEI (Snover et al., 2007) has developed a handbook on adaptation planning. While the World Bank guide is focused on helping cities organize and develop an initial scan to determine whether a city needs to prepare a full-scale adaptation plan, ICLEI's handbook details all the steps for preparing such a plan, including tips on obtaining information, sifting and judging information, selecting

---

[MITIGATION/ADAPTATION]   Box 8.7   Benefits of urban forests in Oakville, Canada

Eva Ligeti

Clean Air Partnership

The Town of Oakville is located on the northwestern shore of Lake Ontario, 40 km west of Toronto. Oakville is home to approximately 165,000 residents and has gained much acclaim in recent years due to its progressive approach to urban forestry. In 2005, Oakville adopted the Urban Forest Effects Model (UFORE), designed by the United States Department of Agriculture, Forest Service. UFORE studies aim to quantify urban forest structure and numerous urban forest effects in cities where randomly generated sample plots combined with local pollution and weather data measure the air quality benefits provided by trees, shrubs, and other types of vegetation. These benefits are then converted to their economic value, allowing the municipality in question to develop a business case for their urban forest. Their UFORE study has enabled Oakville to estimate the replacement value of their urban forest at Canadian $878 million (from 1.9 million trees), with carbon sequestration levels of 6,000 tons/year ($141,000) and energy savings of $840,000 annually. The amount of air pollution filtered by Oakville's urban forest is equivalent to all (102 percent) of the local industrial and commercial emissions of particulate matter (PM10) and 15 percent of PM2.5, as well as over two times (243 percent) the amount of sulfur dioxide plus other criteria pollutants. The study results give decision-makers the tools they need to manage, maintain, and balance green and gray infrastructures.

The UFORE study allowed Oakville's urban foresters to develop key action items. These action items include the development of a private urban forest stewardship education program and an incentive program for private large-stature trees in order to maximize filtration of criteria pollutants and greenhouse gases. To complement this, the creation of a pro-active under-planting program in those communities at risk of decreasing urban forest canopy cover due to aging trees was proposed. Other action items include a review of the 10-year capital forecast to ensure that operating costs for rigorous, standardized tree maintenance are captured, and a review of the Tree Habitat Design Guidelines to incorporate them into the town's urban design standards. Other policy recommendations include amending the town's Official Plan to recognize the municipal urban forest as a component of the municipality's "infrastructure" and investigate the potential role of zoning bylaws to reserve the land which supports trees. Building on the results of their UFORE studies, the town developed an Urban Forest Strategic Management Plan (UFSMP) 2007–2026. A key component of this plan is the creation of an interdepartmental/interagency technical Advisory Committee that will identify, through the UFSMP, a range of future potential urban forest canopy cover targets for Oakville and the development of a private stewardship incentive program for residents and local businesses through initiatives in order to support the potential urban forest canopy targets. The town is also investigating the feasibility of the town exchanging carbon credits.

Further information regarding Oakville's urban forest can be viewed at www.oakville.ca/forestry.htm.

---

scientific advisors, etc. The handbook is based on, and draws most of its examples from, the King County Climate Plan. ICLEI has started a pilot project with four cities in the USA to test out the methodology of the handbook.

8.6.2   Next steps

Adaptation planning is still a novelty. The process to conduct such planning is evolving and many cities lack the capacity to develop such plans. The World Bank's recent guide on resilient cities (Prasad *et al.*, 2009) provides guidelines for conducting a first scan of the degree of risk a city faces, and ICLEI's handbook (2007) provides guidelines for preparing community adaptation plans. Adaptation planning, however, especially for coastal cities, will require analytic and modeling work to scale down global projections to the city scale. Most cities, whether in the developed or developing world, lack this analytic capacity in their planning departments. Urban planning, as a profession, is late in recognizing its role in adaptation planning (Blanco and

Alberti, 2009).[22] In addition, national and/or state or provincial level downscaling models and the data required for such models are needed to form the basis for local downscaling modeling. For example, the state of California, among several states at the forefront of climate change modeling, through its energy commission has developed regional climate models downscaled to 12 by 12 km grids (California Climate Action Team, 2009), but, to identify the sea level rise impacts for a coastal city, further analytic work is needed. A recent study of the city of San Diego's climate change impact scenarios required additional technical studies, including wave and sea modeling to develop impacts on several low-lying coastal areas in San Diego (Messner *et al.*, 2009). In general, a key constraint for effective local planning to address climate change impacts is the lack of regional/local-scale models of climate change. This is a widespread problem (Kehew, 2009), even in developed countries. In the United States, for example, the federal government's National Oceanic and Atmospheric Administration has established eleven Regional Integrated Sciences and

---

22 The profession is moving forward to incorporate climate change in its agenda. The most recent version of the handbook on city planning, widely used in planning schools, *Local Planning* (Hack *et al.*, 2009) includes a brief article on planning and climate change (Beatley, 2009), and recent articles in various planning journals address the topic (Wheeler, 2008; Blanco and Alberti, 2009; Wheeler *et al.*, 2009). Professional institutions, such as the Lincoln Institute of Land Policy, are beginning to publish handbooks on climate change (Condon *et al.*, 2009). In the USA, the journal of the American Planning Association will be publishing a special issue in October of 2010 on planning for climate change.

[ADAPTATION/MITIGATION]   Box 8.8   Urban land use and the urban heat island phenomenon in Shanghai, China

Stephen Solecki

C. W. Post, Long Island University

Shanghai is China's largest city and one of the largest metropolitan areas in the world at 6,340 km^2 with a population of 18.9 million as of 2008 (UNEP, 2009). Rapid urbanization and expansion has produced a strong urban heat island effect. The urban heat island phenomenon has been the cause of temperatures in excess of 7 degrees Celsius warmer in Shanghai as opposed to surrounding rural areas (Hung *et al.*, 2006). The city is what is often referred to as an urban canyon, consisting of myriad narrow streets enclosed by towering buildings (EPA, 2009). As the sun penetrates down to the streets, heat becomes trapped and cannot escape. Key to reducing the urban heat island phenomenon in Shanghai is the incorporation of green practices.

MEASURING URBAN HEAT

To better understand the effects of the urban heat island, Shanghai has been monitoring urban heat intensity. Five different weather stations are set up across Shanghai and temperature readings are taken every 30 minutes at each location. Each month, temperature readings are downloaded and analyzed (Bai *et al.*, 2000). Infrared remote sensing utilizes satellites to take thermal images in order to determine precise surface temperatures. Remote sensing devices also allow the construction of a land use database for the city. Industrial, residential, green, and business areas can be distinguished. This helps experts statistically track anthropogenic heat emission levels (Bai *et al.*, 2000). Thermal infrared remote sensing has revealed that the warmest areas in Shanghai tend to be concentrated around industrial centers, such as the Pudong Steel Factory, densely populated areas, and locations that are heavily traveled (Zhang *et al.*, 2005).

MITIGATION OF THE URBAN HEAT ISLAND EFFECT THROUGH URBAN GREENERY

Shanghai has relied heavily on integrating greenery into its urban setting to reduce the heat island effect. However, the main obstacle to implementing this policy is the lack of available space. As the population of the city grows, methods of mitigation that take up little space are essential.

Increasing vegetation through an urban forestation plan has been the most successful way of reducing the urban heat island effect in Shanghai. The city's plan stresses the importance of incorporating greenbelts into downtown Shanghai within the layout of urban development (Shanghai Agriculture, 2003). By the end of 2008, the city's parks and urban green space amounted to 33,000 hectares. Urban greenery per capita has increased to 12.5 m^2 as of 2008 from 1.0 m^2 in 1990 (UNEP, 2009). The shade provided by trees and bushes reduces the heat island effect. Since the increase of urban greenery in Shanghai, the average temperature in the city has dropped by 5 percent (Saum, 2008). The strategic planting of trees around residential areas especially can cut air conditioning use, which in turn reduces electricity usage and harmful emissions that can affect the environment.

GREEN ROOFS

Green roofs have become widespread across the city. A green roof integrates vegetation into the design of a building's covering. For cities such as Shanghai, green roofs are one of the best ways to reduce urban heat. They are a form of "cool roofs" that absorb solar energy and emit more heat back into the atmosphere than traditional roofs. As an incentive, district governments in Shanghai are willing to pay for nearly half the conversion costs of creating a green roof (Olear, 2009).

FUTURE LAND USE GOALS

By 2020, Shanghai envisions 30 percent of the city being covered with greenery (ULI, 2006). The city intends to continue to increase green space by building more small parks throughout the city. Ideally, officials want residents in the city to be no less than 500 meters away from the nearest park (ULI, 2006). Parks and other green spaces will potentially be connected via a "green grid" whereby people can get from one green space to another through a direct route. By mitigating the heat island effect, Shanghai is also contributing to the global reduction of greenhouse gas emissions.

Assessments (RISA) centers, with growing capacity for regional climate change impacts modeling, but not all regions in the United States have this regional modeling capacity (Blanco and Alberti, 2009). In order to advance adaptation planning for high risk cities, consortia of universities with climate change impacts modeling capacity, relevant state/provincial government agencies, and local planning and other agencies will be needed to provide credible impacts analyses. The UCCRN provides one example of an existing science-policy partnership for cities responding to climate change.

Although most mitigation and adaptation strategies are synergistic, recent studies point to potential conflicts, e.g., the adaptation policy of restricting building on floodplains, which is likely to reduce a city's density and the mitigation strategy of compact development (McEvoy and Handley, 2006; Stone *et al.*, 2007; Hamin and Gurran, 2009). These conflicts merit further study.

In addition, little attention has been paid to worst-case land use strategies, cases where land must be abandoned to sea level rise, or where future development should not be permitted or severely limited, and where existing development should be relocated due to frequent and extreme impacts of climate change. Issues of property rights and of public compensation, of plans and policies for relocation of populations, and of financing mechanisms for such relocations should be examined.

[ADAPTATION/MITIGATION]   Box 8.9   Initiatives towards reducing the impacts of the urban heat island (UHI) in Tokyo, Japan

Ayako Iizumi and Akhilesh Surjan

United Nations University, Tokyo

Tokyo, home to roughly 13 million people (TMG, 2009a) and one of the world's most liveable yet expensive megacities, is struggling to conquer the challenges posed by the urban heat island (UHI) effect. As a result of rapid urbanization in the last few decades, more than 80 percent of the city's surface is covered by buildings or paved roads, which retain absorbed solar and waste heat (TMG, 2003). Similarly, 80 percent of Tokyo's rivers are either reclaimed or covered to carry sewerage (Science Council of Japan, 2003). Petrochemical and iron–steel complexes occupy approximately 25,000 hectares of landfill sites in the bay area of Tokyo (TBEIC, 2009). Studies reveal that annual mean temperature in central Tokyo has increased by about 3 degrees Celsius during the past 100 years, which is five times higher than the world average (TMG, 2003; Mikami, 2006). Increased consumption of anthropogenic energy and waste heat from buildings, altered urban built environment by way of higher density of buildings, and reduced green spaces are some of the major factors that cause excessive warming in certain pockets of the city. Notably, within the past three decades, energy consumption in Tokyo has increased by 85.3 percent (TMG, 2003). The number of nights hotter than 25°C doubled from 15 per year in 1980 to more than 30 in the 1990s. The number of people suffering from heat stroke and sleep disorders in summer is growing, while localized rainfall exceeding 50mm/hour is also increasing – a phenomenon reported to have close links with UHI (TMG, 2003; Mikami, 2006; Asaba, 2008). In 2001, Metropolitan Environmental Temperature and Rainfall Observation Systems (METROS) were installed at 120 locations citywide to further investigate the condition of the heat island diurnal change of temperature patterns as they vary with land and sea breeze circulation. Of the five largest Japanese cities, only Tokyo experienced 3 percent growth in population density over the last 5 years, resulting in amplification of UHI affected areas from the central city to the western and northern suburbs, which are surrounded by hilly terrain (TMG, 2003, 2009b).

INITIATIVES TOWARDS REDUCING THE IMPACTS OF UHI IN TOKYO

The UHI, already in discussion among climatologists in the 1980s, started appearing on the policy table of the Ministry of Environment (MoE) as "an urgent matter of concern" beginning in the early 2000s. In 2004, MoE issued the Outline of the Policy Framework to Mitigate Urban Heat Island Effects, suggesting that UHI counteractions should co-benefit and synchronize with other relevant plans of the city's revitalization and environmental projects pertaining to urban planning, transportation, and climate change, without compromising the living standards of citizens. This policy encouraged the coordination of many urban activities at local level. The Tokyo Metropolitan Government (TMG) first incorporated UHI in its Environmental Master Plan of 2002 and stated in its subsequent revision in 2008 that it aims at: reducing anthropogenic heat; improving urban structure where the UHI is significantly observed; and organizing a better wind flow in certain selected areas in the central city, as well as preventing deterioration of the present heat environment in other areas including suburbs by 2016. This landmark decision was further reinforced by doubling the budget for UHI mitigation from 10.9 billion yen in 2008 to 22 billion yen in 2009. This aims to provide funding for introducing innovative pavements that block heat and absorb moisture, roadside tree plantings, increasing greening of roofs and walls, providing saplings, planting grasses on school playgrounds, planting trees for riverbank protection, installing devices to spray mists on the streets, and so on (TMG, 2009c). In addition, land rezoning has been initiated in Chiyoda ward, which is the city's government and business district. Office workers in Chiyoda ward swell its daytime population to 860,000, which is nearly 24 times greater than the ward's registered residents (MIAC, 2000). The plan for readjustment of streets and buildings is being gradually implemented in the ward to improve the ventilation environment by securing open space for land and sea breezes to come in and heat waves to pass through. Every stage of implementation, which includes research, planning, and construction, to provide links between parks, rivers, and streets involves multi-stakeholder participation, which is already embedded in city planning. The members of the Area Development Committee are drawn from business sectors, non-profit organizations, universities and research institutions. There is also built-in flexibility in existing UHI policies to progressively incorporate new scientific findings and technological developments.

TOWARDS AN INTEGRATED APPROACH

UHI is inevitably a byproduct of urbanization. It is rather complicated to clearly determine the causes of UHI with certainty since the topographical and meteorological conditions as well as patterns of anthropogenic energy consumption differ from city to city. The long-term vision and planning along with the concerted actions of the city of Tokyo that incorporate specific local characteristics is a significant step towards mitigating UHI. Redevelopment of selected city-pockets as well as careful improvement in other areas to make the city tolerant to UHI enhances Tokyo's global competence and competitiveness among many other developed or emerging urban megacities. All efforts are carefully examined and closely coordinated with other ongoing or planned projects related to climate change, city planning, energy efficiency, urban transportation, etc. Policymakers are closely observing to what extent current pilot initiatives for UHI contribute to UHI mitigation and adaptation. Creation of policies that ensure participation of stakeholders across sectors and scales is a challenging task in an urban setting. In addition to further scientific investigation and technological development, future steps in the UHI mitigation program in Tokyo call for summing up and sharing such individual projects and actions. This will not only help to effectively implement countermeasures across Tokyo city, including its suburbs where the UHI is becoming a serious concern, but also in other cities of the world facing similar problems.

[ADAPTATION]   Box 8.10   Climate change adaptation in Stockholm, Sweden

Lisa Enarsson

Climate Adaptation, Department of Environment, City of Stockholm

The city of Stockholm has been working on climate mitigation issues since the 1990s. The City Council adopted its first Action Programme in 1997 with the goal to reduce the greenhouse gas emissions to the level of emissions in 1990. The city has reached a second goal of reducing greenhouse gas emissions to 4 tons per capita through various projects on sustainable transportation, energy efficiency, and renewable energy. The current goal, approved by the Stockholm City Council, is to reach a level of 3 CO_2-eq tons per capita by 2015. The long-term goal is to be fossil-fuel free by 2050.

Undoubtedly, climate change is occurring and affecting Stockholm. A hundred years from now, Stockholmers are projected to experience the following climate change impacts:

- average temperature will rise 4 to 6 degrees Celsius
- winters will be mild and the very cold winter days below $-10\,°C$ are expected to disappear
- more precipitation – more intense rain but less snow
- risk of flooding will increase, but dry summers and heat waves may imply low water levels as well
- spring flooding will come earlier or not occur at all
- growing season will be extended by one or two months and the conditions for the natural environment in Stockholm will change
- the average water level in the Baltic Sea might rise by 0.5 meters or more.

In 2000 Stockholm experienced high water levels in Lake Mälaren. Future flooding and high water levels will cause problems. With increased burdens on the sewage system, leakage from contaminated soils and increased landslide risk will occur. A more humid climate will increase the risk of moisture damage and mold forming in buildings. The heat waves might also increase risks to health.

Stockholm's many green areas will mitigate the effects of projected climate impacts, strengthening the ecosystems. Strong ecosystems can help us cope with the changing climate by absorbing great amounts of rain, evening out water flow, providing shade, and diminishing the city's urban heat island effect. The greenery will also filter contaminants and produce oxygen.

Stockholm Royal Seaport is a new development district, mainly redeveloped on existing industrial land with new, tough environmental requirements on buildings, technical installations, and transportation. A major goal is to be fossil-fuel free by 2030. By 2020 the inhabitants are to use less than 1.5 tons CO_2-eq per capita, and the third goal is focused on adaptation. The planned adaptation is to build up the ground so that all buildings are placed 2.5 meters above the average sea level. The building materials used will be required to resist high humidity. Other requirements call for greenery on roofs, walls, yards, and parks to give shade and coolness, and absorb rain and CO_2. The rainwater from roofs and yards will be directed to the wetland zone to minimize the burden on the sewage system. Great measures will also be taken to stabilize the ecosystem. Part of the Royal Seaport is a very good habitat for species connected to oak trees. Oak is extremely valuable to Stockholm's ecosystem. The planned buildings will become a barrier to the spread of different species of oak. To counter this, the city has planned to develop a corridor with oaks and other plants to help the oak-dependent species to find their way from one habitat to the other. Stockholm hopes the species will even use the corridor as a habitat in the future. The area is also a healthy habitat for frogs. To ensure the frog habitat, the seaport redevelopment will include the construction of two ponds, a tunnel for frogs, and a wetland zone. Both the oak corridor and the frog habitat measures in the seaport redevelopment will strengthen biodiversity and the overall ecosystem.

Another development project in Stockholm is the renovation of Slussen – a floodgate between Lake Mälaren and the Baltic Sea. Here the City of Stockholm is taking adaptive measures to double its capacity so that it can better manage the risk of flooding to meet climate change impacts. The goals of this renovation are to reduce the flooding risk and the risk of low water levels during dry periods. The Slussen reconstruction will also prevent saltwater leakage from the Baltic Sea. This is important because Lake Mälaren is a drinking water reservoir. The renovation will also facilitate natural water level variation to support the ecosystem along the beaches.

A Steering Committee for Climate Adaptation was formed in 2008 by the City of Stockholm. It is led by the City Executive Office and includes members of agencies for social welfare, the fire brigade, urban development, and the environment and health administrations. The City of Stockholm is going to develop a local Climate and Vulnerability Analysis with local scenarios for extreme rain events, flooding, and heat waves, and apply the scenarios to different parts of Stockholm. The local Climate and Vulnerability Analysis is going to be included in Stockholm's Risk and Vulnerability Analysis.

This will provide the city with the information needed to assemble a plan for further adaptations throughout the entire city. For instance, Stockholm hopes to:

- look over the city's plan for tree planting
- devise new guidelines on the size of pipelines for sewage water
- come up with new ways to strengthen Stockholm's ecosystem by deciding how much greenery the urban city should have in order to cope with future climate changes.

More information on Stockholm can be found at: www.stockholm.se/klimat.

8.7   Policy issues

Urban planning and management systems constitute key adaptive institutions to mitigate greenhouse gas emissions in cities and to adapt to unavoidable impacts of climate change. However, as this review has made clear, these institutions face several impediments or constraints. The following is a list of the urban policy issues raised by this review in need of further attention grouped by topic:

Governance

- *Vertical coordination among levels of government*: What are effective mechanisms for ensuring coordination and support among national, provincial, and local governments on urban growth and climate change agendas?
- *Horizontal coordination at the metropolitan scale*: What are effective governance structures for managing urban growth in metropolitan areas? The current spatial/political mismatch between metropolitan areas and their fragmented local jurisdictions permit and encourage sprawl development, an urban pattern with high greenhouse gas emissions.
- *Horizontal coordination at the local scale*: How can cities better structure and integrate their local government agencies to deal with the systemic nature of urban growth and climate change risk? Mainstreaming and climate change teams are institutional mechanisms that have been identified as effective mechanisms for carrying out effective climate change agendas. But can an integrated climate change agenda be implemented if the underlying structure remains sectoral and fragmented?
- *Inadequate legal frameworks*: How can we better conceptualize our system of land property to capture the ecological and the social aspects of land?

Planning capacity

- *Climate change modeling at regional/local scales*: There is a lack of capacity, or of adequate capacity to model climate change impacts (e.g., sea level rise, changes in precipitation) at the local and regional scales, the level which is required for local level climate change adaptation planning, even in developed countries.
- *Planning capacity for developing countries*: How can the planning capacity of cities in developing countries be expanded to effectively address their climate change risk? For cities with high climate change risks, effective urban planning and management agencies will be crucial, and yet cities in developing countries lack such institutions and/or lack sufficient pools of professional planners and administrators.

Planning and land markets

- *Role of land markets in climate change*: How can cities change the current incentive system for the "growth machine" partners and engage such partners in climate change mitigation and adaptation agendas?

Planning strategies

- *Vulnerable populations*: Slum populations in developing countries are at highest risk from climate change. What are effective strategies to upgrade and reduce slum settlements? For developing countries, this is a particularly difficult problem, given the large percentages of urban slum populations and the insufficient resources of these cities.
- *Urban growth management*: What are the new and most promising urban growth management strategies? How effective are the existing strategies, such as urban growth boundaries, farmland preservation, or requirements for adequate infrastructure?
- *Planning strategies for existing development*: In general, urban planning is future-oriented, many urban planning climate strategies can make a difference for future development, which raises the issue of low-cost ways for cities to address existing settlement patterns. What mechanisms are available for reducing risks for *existing* settlements, especially in developing countries?
- *Worst-case scenarios land use strategies*: How do we prepare, policy-wise, for worst-case land use scenarios, i.e., the abandonment of developed urban areas due to climate change risks?
- *Conflicting strategies*: Do some adaptation and mitigation strategies conflict? Based on what criteria should we make choices among conflicting strategies?
- *Financing planning strategies*: Both for developed and developing countries, what mechanisms are available for financing strategies to reduce climate change risk?

8.8   Research needs

In addition to the need to provide downscaling modeling capacity for metropolitan areas, research on several urban land topics is currently needed: urban growth modeling; urban growth policy issues; and measurement of greenhouse gas emissions in cities.

8.8.1   Forecasting urban growth

Current urban growth models focus primarily on western and developed cities in the industrialized world where socio-economic data are usually readily and widely available. Yet, since most of the urban growth in the future will occur in the developing world, it is important to develop models that are applicable in these data-poor contexts. In less developed and developing cities, socio-economic data can be non-existent, incomplete, inaccurate, unreliable, or all the above. Forecasts of urbanization utilizing such data have been severely criticized (Cohen, 2004).

There exist two basic types of land use change models: regression-based models and transition-based models. Regression-based models aim to establish functional relationships between land use change based on historical land use data and a set of predictor variables that are used to explain the locations of future land use change on the landscape. These models use different methods to build prediction functions, including linear/logistic regression (Theobald and Hobbs, 1998; Fragkias and Seto, 2007; Arai and Akiyama, 2003) and artificial neural networks (Pijanowski *et al.*, 2002). One advantage of regression models is that they can explicitly express the effects of each predictor (spatial variables) on future land use change (Pijanowski *et al.*, 2002). Spatial transition models are characterized by transition rules, neighborhoods, and decision-makers – often called agents – within a cell. They differ from regression models in that a group of simple rules that express land use change patterns and neighborhood effects (e.g., spatial adjacency) can be incorporated to drive the prediction of land use change. Cellular automata (CA) and agent-based (AB) models have become increasingly popular tools for modeling urban growth (Batty and Xie, 1994; Clarke *et al.*, 1997; White and Engelen, 1997; Clarke and Gaydos, 1998). Many transition models have been developed over the last 20 years, but they retain a number of limitations. First, many models have significant data input requirements, limiting their utility in developing countries where data are usually sparse (Fragkias and Seto, 2007). Second, CA-based modeling techniques are still far from being mature. Despite their flexibility, there are many limitations to CA models (Torrens and O'Sullivan, 2001) and more research is needed for such models to capture the richness of urban systems.

8.8.2 Urban growth policy research

Similarly to other public administration functions, urban planning research is not well funded by traditional national research foundations. In the USA, for example, there are no urban planning or urban management research funding categories in the National Research Foundation, although NSF has over the past decade or so funded a number of interdisciplinary projects on urban land cover/land use. In particular, there is lack of funding for conducting adequate research on many of the complex policy topics identified in the section above, such as the effectiveness of urban growth strategies, or effective models of metropolitan organization.

8.8.3 Standard measurement of climate change data at the city scale

Although climate change is monitored at global and national levels according to an adopted set of measures globally agreed upon, similar statistics are rarely collected at the city level, and devising indicators on climate change at the city level is proving difficult. Several research challenges remain: a credible and globally standardized measurement for how cities impact climate change; city mitigation targets based on sound research; data on climate change impacts by category of risk and degree of vulnerability at the city scale. The World Bank's Global City Indicators Program (undated) is making progress in this area. The program has various indicators that can assist cities with climate change mitigation strategies, including for example indicators on modal shifts from road transport to rail and public transport and non-motorized transport. Indicators on cities and greenhouse gas emissions are under development that can help to create a standard and globally recognized index on cities and greenhouse gases. The program also has a pilot program that compiles indicators by cities according to their political boundaries and then aggregates these data across the area municipalities to obtain metropolitan measurements.

REFERENCES

Adams, B. (2004). Challenges and trends: the trajectory of urban preservation in Florianópolis, Santa Catarina Island, Brazil. *City & Time*, 1, 1. www.ct.ceci-br.org/novo/revista/viewissue.php?id=1.

Alberti, M. (2005). The effects of urban patterns on ecosystem function. *International Regional Science Review*, **28**(2), 168–192.

Alberti, M. (2008). *Advances in Urban Ecology: Integrating Humans and Ecological Processes in Urban Ecosystems*, New York: Springer.

Amora, A. (1996). O lugar do público no Campeche. Master's thesis, Federal University of Santa Catarina, Florianópolis, Brazil.

APA (1998). *The Principles of Smart Development. PAS Report 479.* Chicago, IL, USA: American Planning Association.

Arai T. and T. Akiyama (2003). Empirical analysis for estimating land use transition potential functions: case study in the Tokyo metropolitan region. *Computer, Environment and Urban Systems*, **28**, 65–84.

Arnfield, A. J. (2003). Two decades of urban climate research: a review of turbulence, exchanges of energy and water, and the urban heat island. *International Journal of Climatology*, **23**, 1–26.

Asaba, M. (2008). How to deal with localized heavy rainfall and flood in urban areas in an era of global warming (Ondanka jidai no shuchu-gou toshi-kouzui ni dou sonaeru ka). *Yobou jihou*, **234**, 16–21 [in Japanese].

Bae, C.-H. C. (2007). Containing sprawl. In *Incentives, Regulations, Plans: The roles of States and Nation-States in Smart Growth Planning*, G. J. Knaap (Ed.)], Cheltenham, UK: Edward Elgar Publishing, pp. 36–55.

Bai, Y., T. Ichinose, and K. Ohta (2000). *Observations and Modeling of Urban Heat Islands in Shanghai, China.* Tohoku University of Community Service and Science, Japan.Bae, C. and M. J. Jun (2003). Counterfactual planning: what if there had been no greenbelt in Seoul? *Journal of Planning Education and Research*, **22**, 374–383.

Banta, R. M., C. J. Senff, A. B. White, *et al.* (1998). Daytime buildup and nighttime transport of urban ozone in the boundary layer during a stagnation episode. *Journal of Geophysical Research: Atmospheres*, **103**(17), 22,519–22,544.

Baptista, S. (2008a). Metropolitanization and forest recovery in southern Brazil: a multiscale analysis of the Florianópolis city-region, Santa Catarina State, 1970 to 2005. *Ecology and Society*, **13**, 5. www.ecologyandsociety.org/vol13/iss2/art5/.

Baptista, S. (2008b). Forest recovery and just sustainability in the Florianópolis city-region. Ph.D. Dissertation, Rutgers, The State University of New Jersey, New Brunswick, United States.

Barbour (2009). Letter to Sect. Steven Chu on State Energy Program Assurances, dated March 23, 2009. Website of the US Department of Energy. Accessed June 20, 2009, www.energy.gov/media/Barbour_Mississippi.pdf.

Batty, M. and Y. Xie (1994). From cells to cities. *Environment and Planning B: Planning and Design*, **21**, S31–S38.

Beatley, T. (2009). Planning for global climate change. In G. Hack, E. L. Birch, P. H. Sedway, and, M. J. Silver (Eds.), *Local Planning: Contemporary Principles and Practice*, Chicago, IL, USA: International City Management Association, pp. 350–355.

Benevolo, L. (1980). *The History of the City*. Cambridge, MA, USA: MIT Press.

Bengston, D. N. and Y. C. Youn (2006). Urban containment policies and the protection of natural areas: the case of Seoul's greenbelt. *Ecology and Society*, **11**(1), 3, www.ecologyandsociety.org/vol11/iss1/art3/.

Bernick, M. and R. Cervero (1996). *Transit Villages in the 21st Century*. New York, USA: McGraw – Hill.

Blanco, H. (1994). *How to Think About Social Problems: American Pragmatism and the Idea of Planning.*, Westport, CT, USA: Greenwood Press.

Blanco, H. (1999). A United States perspective on the Dutch government's approach seeking greater cohesion in environmental and spatial policy. In D. Miller and G. DeRoo (Eds.), *Integrating City Planning and Environmental Improvement*, Aldershot, UK: Ashgate, pp. 51–58.

Blanco, H. and M. Alberti (2009). Building capacity to adapt to climate change through planning. In H. Blanco and M. Alberti (Eds.), *Hot, Congested, Crowded and Diverse: Emerging Research Agendas in Planning, Progress in Planning*, 71, 153–205.

Boarnet, M. G. and R. Crane (2001). *Travel by Design: The Influence of Urban Form on Travel*, New York: Oxford University Press.

Bogaerts, T. and J. Zevenbergen (2001). Cadastral systems: alternatives. *Computers, Environment and Urban Systems*, 25, 325–337.

Booth, P. (1998). Decentralization and land use planning in France: a 15 year review. *Policy and Politics*, 26, 89–105.

Bornstein, R. and M. LeRoy (1990). Urban barrier effects on convective and frontal thunderstorms. In Preprint Volume, *AMS Conference on Mesoscale Processes*, Boulder, CO, 25–29 January.

Bowyer, R. A. (1993). *Capital Improvements Programs: Linking Budgeting and Planning*, PAS Report 442, Chicago, IL, USA: American Planning Association.

Briggs, X. de S. and B. J. Keys (2009). Has exposure to poor neighborhoods changed in America? Race, risk and housing locations in two decades. *Urban Studies*, 46, 429–458.

Brogden, R. (2007). Rethinking greenbelt and the urban fringe. Business Section, *Farmers Weekly* (UK), 147(15), December 10.

Bruegmann, R. (2005). *Sprawl: A Compact History*. Chicago, IL, USA: University of Chicago Press.

Bryner, J. (2008). Why the China Quake was so devastating. *Live Science: Environment*. Webpage of *Imaginova*, May 15, 2008. Accessed July 11, 2009, www.livescience.com/environment/080515-quake-buildings.html.

Burchell, R. W., D. Listokin, and C. C. Galley (2000). Smart growth: more than a ghost of urban policy past, less than a bold new horizon. *Housing Policy Debate*, 11, 821–879.

Burchell, R. W., A. Downs, B. McCann, and S. Mukherji (2005). *Sprawl Costs: Economic Impacts of Unchecked Development*, Washington, DC, USA: Island Press.

Caballero, Á. (2002). Adaptação organizacional estratégica e evolução financeira no setor elétrico: o caso das Centrais Elétricas de Santa Catarina – CELESC. Master's thesis, Federal University of Santa Catarina, Florianópolis, Brazil.California

Climate Action Team (2009).: *Draft Biennial Climate Action Team Report March 2009*, CAT-1000–2009-003-D. Accessed from the California Climate Change Portal, Climate Action Team, June 2, 2009, www.climatechange.ca.gov/climate_action_team/reports/index.html.

Calthorpe, P. (undated). *The Urban Network: A New Framework for Growth*. Available from the Calthorpe Associates webpage June 10, 2010: www.calthorpe.com/files/Urban%20Network%20Paper.pdf.

Calthorpe, P. (1993). *The Next American Metropolis: Ecology, Community and the American Dream*. Princeton, NJ, USA: Princeton University Press.

Carlton, I. (2007). *Histories of Transit-Oriented Development: Perspectives on the Development of the TOD Concept*, IURD Working Paper 2009–02, Institute of Urban and Regional Development, University of California Berkeley. http://escholarship.org/uc/item/7wm9t8r6.

Carruthers, J. (2002). Evaluating the effectiveness of regulatory growth management programs: An analytic framework. *Journal of Planning Education and Research*, 21, 391–405.

Carruthers, J. I. (2003). Growth at the fringe: the influence of political fragmentation in United States metropolitan areas. *Papers in Regional Science*, 82, 475–499.

Castells, M. (1989). *The Informational City: Information Technology, Economic Restructuring and the Urban-Regional Process*. Oxford, UK: Basil Blackwell.

Cervero, R. (1998). *The Transit Metropolis: A Global Inquiry*. Washington, DC, USA: Island Press.

Clarke, K., L. Gaydos (1998). Loose-coupling a cellular automaton model and GIS: long-term urban growth prediction for San Francisco and Washington/Baltimore. *International Journal of Geographical Information Science*, 12, 699–714.

Clarke, K., S. Hoppen, and L. Gaydos (1997). A self-modifying cellular automation model of historical urbanization in the San Francisco Bay area. *Environment and Planning B: Planning and Design*, 24, 247–261.

Cohen, B. (2004). Urban growth in developing countries: a review of current trends and a caution regarding existing forecasts. *World Development*, 32(1), 23–51.

Condon, P. M., D. Cavens, and N. Miller (2009). *Urban Planning Tools for Climate Change Mitigation (Policy Focus Report)*. Cambridge, MA, USA: Lincoln Institute of Land Policy.

Confalonieri, U. (2003). Variabilidade climática, vulnerabilidade social e saúde no Brasil. *Terra Livre*, 19–20, 193–204.

Cullingworth, J. W. and V. Nadin (2006). *Town and Country Planning in the UK*, 14th edition, New York: Routledge.

DeGrove, J. and D. Miness (1992). *The New Frontier for Land Policy: Planning and Growth Management in the States*, Cambridge, MA, USA: Lincoln Institute of Land Policy.

Ding, C. , (2007). Policy and praxis of land acquisition in China. In Y. Song and C. Ding (Eds.), *Urbanization in China*, Cambridge, MA, USA: Lincoln Institute of Land Policy.

Dittmar, H. and G. Ohland (2004). *The New Transit Town: Best Practices in Transit-Oriented Development*. Washington, DC, USA: Island Press.

Döös, B. R. (2002). Population growth and loss of arable land. *Global Environmental Change: Human and Policy Dimensions*, 12, 303–311.

Downs, A. (2001). What does 'Smart Growth' really mean? *Planning*, 67, 20–25.

Downs, A. (2005). Smart growth: why we discuss it more than we do it . *Journal of the American Planning Association*, 71, 367–378.

Dupont, V. (2007). Exploring the peri-urban: a specific and non-neutral space. *Cities*, 24, 89–94.

Durand-Lasserve, A. (2006). Informal settlements and the Millennium Development Goals: Global Policy debates on property ownership and security of tenure. *Global Urban Development*, 2, 1–15.

Dye, R. F. and R. W. England (Eds.) (2009). *Land Value Taxation: Theory, Evidence, and Practice*. Cambridge, MA, USA: Lincoln Institute of Land Policy.

Easley, V. G. (1992). *Staying Inside the Lines: Urban Growth Boundaries*, PAS Report 440, Chicago, IL, USA: American Planning Association.

EPA (2009). *Urban heat island basics. Reducing Urban Heat Islands: Compendium of Strategies*. Accessed August 20, 2009, www.epa.gov/hiri/resources/pdf/BasicsCompendium.pdf.

El Araby, M. (2002). Urban growth and environmental degradation. *Cities*, 19, 389–400.

Ewing, R., R. Pendall, and D. Chen (2002). *Measuring Sprawl and Its Impact*, Smart Growth America. Accessed July 11, 2009, www.smartgrowthamerica.org/sprawlindex/MeasuringSprawl.PDF.

Ewing, R., K. Bartholomew, S. Winkelman, J. Walters, and D. Chen (2008). *Growing Cooler: The Evidence on Urban Development and Climate Change*, Washington, DC, USA: Urban Land Institute.

Fernandes, E. (2003). Del Codigo Civil al Estatuto de la Ciudad: algunas notas sobre la trayectoria del Derecho Urbanistico en Brasil. *EURE* (Santiago), 29, doi: 10.4067/S0250–71612003008700005.

Fernandes, E. (2007). Implementing the urban reform agenda in Brazil. *Environment and Urbanization*, 19, 177–189.

Fernandes, E. and M. M. Maldonado Copello (2009). Law and land policy in shifting paradigms and possibilities for action. *Land Lines*, July 2009, 14–19, Cambridge, MA, USA: Lincoln Institute of Land Policy.

Flint, A. (2006). *This Land: The Battle Over Sprawl and the Future of America*. Baltimore, MD, USA: Johns Hopkins University Press.

Freyfogle, E. (2003). *The Land We Share: Private Property and the Common Good*, Washington, DC, USA: Island Press.

Fujibe, F. (2003). Long-term surface wind changes in the Tokyo Metropolitan Area in the afternoon of sunny days in the warm season. *Journal of the Meteorological Society of Japan*, 81, 141–149.

Fragkias, M. and K. Seto (2007). Modeling urban growth in data-sparse environments: a new approach. *Environment and Planning B: Planning and Design*, **34**, 858–883.

Freilich, R. H. and M. M. Schultz (1995). *Model Subdivision Regulations: Planning and Law*, Chicago, IL, USA: American Planning Association.

Gartland, L. (2008). *Heat Island: Understanding and Mitigating Heat Islands*, London, UK: Earthscan.

Giuliano, G. and D. Narayan (2003). Another look at travel patterns and urban form: the U.S. and Great Britain. *Urban Studies*, **40**, 2295–2312.

Global City Indicators Program(undated). Accessed July 14, 2009 at www.cityindicators.org/.

Godschalk, D., T. Beatley, P. Berke, D. Brower, and E. J. Kaiser (1999). *Natural Hazard Mitigation: Recasting Disaster Policy and Planning*, Washington, DC, USA: Island Press.

Goldstein, R. J. (2004). *Ecology and Environmental Ethics: Green Wood in the Bundle of Sticks*, Aldershot, UK: Ashgate Publishing.

Gottmann, J. (1961). *Megalopolis: The Urbanized Northeastern Seaboard of the United States*, New York, USA: The Twentieth Century Fund.

Gottmann, J. and R. A. Harper (1990). *Megalopolis Revisited: The Urban Writings of Jean Gottmann*, Baltimore, MD: Johns Hopkins University Press.

Grant, J. (2002). Mixed use in theory and practice: Canadian experience in implementing a planning principle. *Journal of the American Planning Association*, **68**, 71–84.

Grazi, F. and J. Van Den Bergh (2008). Spatial organization, transport, and climate change: comparing instruments of spatial planning and policy. *Ecological Economics*, **67**, 630–639.

Hack, G., E. L. Birch, P. H. Sedway, and M. J. Silver (Eds.) (2009). *Local Planning: Contemporary Principles and Practice*. Chicago, IL, USA: International City Management Association.

Hall, P. (2009). Looking backward, looking forward: the city region of the mid-21st century. *Regional Studies*, **43**, 803–817.

Hall, P. and K. Pain (2006). *The Polycentric Metropolis: Learning from Mega-City Regions in Europe*, London, UK: Earthscan.

Hamin, E. and N. Gurran (2009). Urban form and climate change: balancing adaptation and mitigation in the US and Australia. *Habitat International*, **33**, 238–245.

Harvey, D. (1989). *The Condition of Postmodernity: An Enquiry Into the Origins of Cultural Change*. Oxford, UK: Blackwell.

Haylock, M. R., T. C. Peterson, L. M. Alves, *et al.* (2006). Trends in total and extreme South American rainfall in 1960–2000 and links with sea surface temperature. *Journal of Climate*, **19**(8), 1490–1512.

Henry, G. [1879] (1992). *Progress and Poverty*. Cambridge, MA, USA: Lincoln Institute of Land Policy.

Higgins, B. (2008). *Technical Overview of SB 375 (v.1.1)*, September 19, 2008, Sacramento, CA, USA: League of California Cities. Accessed July 1, 2009 at www.calapa.org/attachments/wysiwyg/5360/SB375TechOV.pdf.

Hough, M. (1995). *Cities and Natural Process*. New York, USA: Routledge.

Huitric, M., C. Folke, and N. Kaustky (2002). Development and government policies of the shrimp farming industry in Thailand. *Ecological Economics*, **40**, 441–455.

Hung, T., D. Uchihama, S. Ochi, and Y. Yasuoka (2006). Assessment with satellite data of the urban heat islands effects in Asian mega cities. *International Journal of Applied Earth Observation and Geo-Information*, **8**, 1, 34–48.

IBGE (Instituto Brasileiro de Geografia e Estatística) (1974). *Censo Demográfico 1970: Santa Catarina*. Rio de Janeiro, Brazil.

IBGE (Instituto Brasileiro de Geografia e Estatística) (2001). *Censo Demográfico 2000: Santa Catarina*. Rio de Janeiro, Brazil. www.ibge.gov.br/home/estatistica/populacao/default_censo_2000.shtm.

IBGE (Instituto Brasileiro de Geografia e Estatística) (2007). *Contagem da população*. Rio de Janeiro, Brazil. www.ibge.gov.br/home/estatistica/populacao/contagem2007/.

Imhoff, M. L., W. T. Lawrence, D. C. Stutzer, and C. D. Elvidge (1997). Using nighttime DMSP/OLS images of city lights to estimate the impact of urban land use on soil resources in the US. *Remote Sensing of Environment*, **59**, 105–117.

Ingram, G. K. and Y.-H. Hong (2008). The nexus of fiscal decentralization and land policies. In G. K. Ingram and Y.-H. Hong (Eds.), *Fiscal Decentralization and Land Policies*, Cambridge, MA, USA: *Lincoln Institute of Land Policy*, pp. 3–16.

Ingram, G. K., A. Carbonell, Y.-H. Hong, and A. Flint (2009). *Smart Growth Policies. An Evaluation of Programs and Outcomes*. Cambridge, MA, USA: Lincoln Institute of Land Policy.

IPCC (2007). *Climate Change 2007: Impacts, Adaptation, and Vulnerability. Contribution of Working Group II to the Fourth Assessment Report of the Intergovernmental Panel on Climate Change*, Cambridge, UK: Cambridge University Press. ITE (2008). *Trip Generation*, (8th edition), Washington, DC, USA: Institute of Transportation Engineers.

Jackson, K. (1987). *The Crabgrass Frontier: The Suburbanization of the United States*, New York, USA: Oxford University Press.

Jacobs, H. M. (Ed.) (1998). *Who Owns America? Social Conflict Over Property Rights*, Madison, WI, USA: University of Wisconsin Press.

Jacobs, H. M. and M. Paulsen (2009). Property rights: the neglected theme of 20th century American planning. *Journal of the American Planning Association*, **75**, 134–143.

Jibao, S. S. (2009). *Property Taxation in Anglophone West Africa: Regional Overview*, Lincoln Institute of Land Policy Working Paper, Cambridge, MA, USA: Lincoln Institute.

Jin M. L. and J. M. Shepherd (2008). Aerosol relationships to warm season clouds and rainfall at monthly scales over east China: urban land versus ocean. *Journal of Geophysical Research: Atmospheres*, **113**, D24S90.

Jonas, A. E. G. and D. Wilson (Eds.) (1999). *The Urban Growth Machine: Critical Perspectives Two Decades Later*. Albany, NY, USA: State University of New York Press.

Kanda, M. (2007). Progress in urban meteorology. *Journal of the Meteorological Society of Japan*, **85**, 363–383.

Kasanko, M., J. I. Barredo, C. Lavalle, *et al.* (2006). Are European Cities becoming dispersed? A comparative analysis of 15 European urban areas. *Landscape and Urban Planning*, **77**, 111–130.

Kaufmann, R. K., K. C. Seto, A. Schneider, *et al.* (2007). Climate response to rapid urban growth: evidence of a human-induced precipitation deficit. *Journal of Climate*, **20**, 2299.

Kehew, R. (2009). Projecting globally, planning locally: a progress report from four cities in developing countries. In the World Meteorological Organization's *Climate Sense*, Leicester, UK: Tudor Rose, pp. 181–184.

Kelly, E. D. (2004). *Managing Community Growth* (2nd edition), Westport, CT, USA: Praeger Publishers.

Kennedy, C., J. Cuddihy, and J. Engel-Yan (2007). The changing metabolism of cities. *Journal of Industrial Ecology*, **11**(2), 43–59.

Kenworthy, J. R. (2003). Transport energy use and greenhouse gases in urban passenger transport systems: a study of 84 global cities. *Third Conference of the Regional Government Network for Sustainable Development*, Notre Dame University, Fremantle, Western Australia, September 2003. http//cst.uwinnipeg.ca/documents/Transport_Greenhouse.pdf.

Kenworthy, J. R., F. B. Laube, *et al.* (1999). *An International Sourcebook of Automobile Dependence in Cities, 1960–1990*. Boulder, CO, USA: University Press of Colorado.

King County, Washington (2007). *King County Climate Plan*. Accessed June 24, 2009, www.metrokc.gov/exec/news/2007/pdf/climateplan.pdf.

Kline, J. D. (2000). Comparing states with and without growth management. Analysis based on indicators with policy implications. Comment. *Land Use Policy*, **17**, 349–355.

Kosatsky, T. (2005). The 2003 European heat waves. *Euro Surveillance*, **10**, 148–149.

Kotkin, J. (2005). *The City: A Global History*. New York, USA: Modern Library.

Krueckeberg, D. A. (1995). The difficult character of property: to whom do things belong. *Journal of the American Planning Association*, **61**, 301–309.

Kupek, E., M. Faversani, and J. Philippi (2000). The relationship between rainfall and human leptospirosis in Florianópolis, Brazil, 1991–1996. *Brazilian Journal of Infectious Diseases*, **4**, 131–134.

Ladd, H. (1998). *Local Government Tax and Land Use Policies in the United States: Understanding the Links*. Northampton, MA, USA: Edward Elgar Publishing.

Lambin, E. F., B. L. Turner, H. J. Geist, *et al.* (2001). The causes of land-use and land-cover change: moving beyond the myths. *Global Environmental Change: Human and Policy Dimensions*, **11**, 261–269.

Lang, R. E. and D. Dhavale (2005). *Beyond Megalopolis: Exploring America's New 'Megapolitan' Geography*, Alexandria, VA, USA: Metropolitan Institute at Virginia Tech.

Lang, R. and P. K. Knox (2009). The new metropolis: rethinking Megalopolis. *Regional Studies*, **43**, 789–802.

LCCP [London Climate Change Partnership] (2002). *London's Warming: The Impacts of Climate Change on London*, Technical Report, London, UK: Greater London Authority.

LCCP (2005). *Aims and Objectives*, London, UK: Greater London Authority.

LCCP (2006). *Adapting to Climate Change: Lessons for London*, London, UK: Greater London Authority.

Lebel, L., N. H. Tri, A. Saengnoree, *et al.* (2002). Industrial transformation and shrimp aquaculture in Thailand and Vietnam: pathways to ecological, social, and economic sustainability? *Ambio*, **31**, 311–323.

Leichenko, R. M. and W. D. Solecki (2005). Exporting the American dream: the globalization of suburban consumption landscapes. *Regional Studies*, **39**(2), 241–253.

Lerable, C. A. (1995). *Preparing a Conventional Zoning Ordinance*, PAS Report 460 Chicago, IL, USA: American Planning Association.

Li Jingsheng (Ed.) (2006). *Bridge of Jinze, Bridging to the Web*, Bridging to the Future Project. Accessed June 17, 2010, www.bridgingtothefuture.org/sites/default/files/China%20Bridging%20to%20the%20Future%20presentation.pdf.

Lin, G. C. S. and L. J. C. Ma (1994). The role of towns in Chinese regional development: the case of Guangdong Province. *International Regional Science Review*, **1**, 75–97.

Liu, J., T. Dietz, S. R. Carpenter (2007). Coupled human and natural systems. *Ambio*, **36**(8), 639–648.

Logan, J. A. and H. L. Molotch (1987). *Urban Fortunes: The Political Economy of Place*, Berkeley, CA, USA: University of California Press.

Lubove, R. (1967). *The Urban Community: Housing and Planning in the Progressive Era*, Elizabeth, NJ, USA: Prentice-Hall.

Ma, L. J. C. (1971). *Commercial Development and Urban Change in Sung China*, Michigan Geographical Society.

Marcotullio, P. J. (2003). Globalisation, urban form and environmental conditions in Asia-Pacific cities. *Urban Studies* **40**(2), 219–247.

Marengo, J. (2008). Água e mudanças climáticas. *Estudos Avançados*, **22**, 83–96.

Marengo, J. and C. Camargo (2008). Surface air temperature trends in southern Brazil for 1960–2002. *International Journal of Climatology*, **28**, 893–904.

McCarney, P. (2010). Conclusions: Governance and Challenges in Peri-Urban Areas. In M. Kurian and P. McCarney (eds.), *Peri-urban Water and Sanitation Services: Policy, Planning and Method*. Dordrecht, Germany: Springer.

McDonald, R. I., P. Kareiva, R. T. T. Forman (2008). The implications of current and future urbanization for global protected areas and biodiversity conservation. *Biological Conservation*, **141**(6), 1695–1703.

McEvoy, D. and L. Handley (2006). Adaptation and mitigation in urban areas: synergies and conflicts. *Municipal Engineer*, **159**, 185–191.

McGee, T. G. and I. Robinson (Eds.) (1995). *The Mega-Urban Regions of Southeast Asia*. Vancouver, BC, Canada: University of British Columbia Press.

McGranahan, G., D. Balk, B. Anderson (2007). The rising tide: assessing the risks of climate change and human settlements in low elevation coastal zones. *Environment and Urbanization*, **19**, 17–37.

McHarg, I. (1995). *Design with Nature*. New York, USA: Wiley.

Messner, S. S. C. Miranda, K. Green, *et al.* (2009). *Climate Change Related Impacts in the San Diego Region by 2050. Final Report*. Report CEC-500-2009-027-F, Sacramento, CA, USA: California Energy Commission.

MIAC (The Ministry of Internal Affairs and Communications) (2000). *2000 Census*. Available at the Statistics Bureau, Director General for Policy Planning and Statistical Research and Training Institute, MIAC, www.stat.go.jp/data/kokusei/2000/jutsu1/00/01.htm. Accessed January 8, 2009 [in Japanese].

Mikami, T. (2006). Recent progress in urban heat island studies: Focusing on the case studies in Tokyo. (Toshi heat island kenkyu no saishin dokoh – Tokyo no jirei wo chushin ni). E-journal *GWO*, **1**(2), 79–88 [in Japanese].

Mileti, D. (1999). *Disaster by Design: A Reassessment of Natural Hazards in the U.S.* Joseph Henry Press.

Municipal Service Research Center (2009). *Critical Areas*, Municipal Service Research Center, Seattle, WA, USA. Updated 4/09. Accessed July 14, 2009 at www.mrsc.org/Subjects/Environment/criticalpg.aspx.

Nelson, A. C. (1999). Comparing states with and without growth management. Analysis based on indicators with policy implications. *Land Use Policy*, **16**, 121–127.

Nelson, C. and C. Dawkins (2004). *Urban Containment in the United States: History, Models, and Techniques for Regional and Metropolitan Growth Management*, PAS Report 520, Chicago, IL, USA: American Planning Association.

Newman, P. W. G. and J. R. Kenworthy (1989). *Cities and Automobile Dependence: An International Sourcebook*. Aldershot, UK: Gower.

Newman, P. and J. Kenworthy (1999). *Sustainability and Cities: Overcoming Automobile Dependence*. Washington, DC, USA: Island Press.

New York City Office of the Mayor (2007). *PlaNYC. A Greener, Greater New York*. Accessed January 12, 2009, www.nyc.gov/html/planyc2030/html/downloads/download.shtml.

Nivola, P. (1999). *Laws of the Landscape: How Policies Shape Cities in Europe and America*, Washington, DC, USA: Brookings Institution Press.

Nizeyaimana, E., G. W. Petersen, M. L. Imhoff, *et al.* (2001). Assessing the impact of land conversion to urban use on soils with different productivity levels in the USA. *Soil Science Society of America Journal*, **65**, 391–402.

Norman, B. (2009). Principles for an intergovernmental agreement for coastal planning and climate change in Australia. *Habitat International*, **33**, 293–299.

Norris, D. F. (2001). Whither metropolitan governance. *Urban Affairs Review*, **36**, 532–550.

Oates, W. O. (1972). *Fiscal Federalism*. New York, USA: Harcourt Brace, Jovanovich.

Oke, T. R. (1976). The distinction between canopy and boundary-layer heat islands. *Atmosphere*, **14**, 268–277.

Olear G. (2009). A look at green roofs. *The Cooperator: The Co-op and Condo Monthly*, **29**, 4. Accessed August 15, 2009, http://cooperator.com/articles/1905/1/A-Look-at-Green-Roofs/Page1.html.

Orfield, M. (2002). *Metropolitics: The New Suburban Reality*. Washington, DC, USA: Brookings Institution Press.

O'Sullivan, A. (2006). *Urban Economics* (6th edition), New York, USA: McGraw-Hill.

Parnell, S., E. Pieterse, and V. Watson (2009). Planning for cities in the global south: an Afrocam research agenda for sustainable urban settlements. In H. Blanco and M. Alberti (Eds.), *Shaken, Shrinking, Hot, Impoverished and Informal: Emerging Research Agendas in Planning, Progress in Planning*, **72**, 233–241.

Payne, G. (2004). Land tenure and property rights: an introduction. *Habitat International*, **28**(2), 167–179.

Penney, J. and I. Wieditz (2007). *Cities Preparing for Climate Change: A Study of Six Urban Regions*, Toronto, ON, Canada: Clean Air Partnership.

Peterson, J. A. (2003). *The Birth of City Planning in the United States, 1840-1917*, Baltimore, MD, USA, and London, UK: Johns Hopkins University Press.

Pickett, S., Burch, W., Dalton, S., Foresman, T., Grove, M., & Rowntree, R. (1997). A conceptual framework for the study of human ecosystems in urban areas. *Urban Ecosystems*, **1**, 186–199.

Pickett, S. T. A., W. C. Zipperer, R. V. Pouyat, A. Flores, and R. Pirani (1997). Adopting a modern ecological view of the metropolitan landscape: the case of a greenspace system for the New York City region. *Landscape and Urban Planning*, **39**, 295–308.

Pijanowski, B., D. Brown, B. Shellito, and G. Manik (2002). Using neural networks and GIS to forecast land-use changes: a land transformation model. *Computer, Environment and Urban Systems*, **26**, 553–575.

Potere, D. and A. Schneider (2007). A critical look at representations of urban areas in global maps. *GeoJournal*. **69**, 55–80.

Prasad, N., F. Ranghieri, F. Shah, *et al.* (2009). *Climate Resilient Cities*, Washington, DC, USA: World Bank.

Raphaelson, A. H. (2004). The property tax. In J. R. Aronson and E. Schwartz (Eds.), *Management Policies in Local Government Finance*, Washington, DC, USA: International Management Association, pp. 257–288.

Razin, E. (2000). The impact of local government organization on development and disparities: a comparative perspective. *Environment and Planning C: Government and Policy*, **18**, 17–31.

Rees, W. E. (1992). Ecological footprint and appropriated carrying capacity: What urban economics leaves out. *Environment and Urbanization*, **4**, 121–130.

Reilly, M. K., M. P. O'Mara, and K. C. Seto (2009). From Bangalore to the bay area: comparing transportation and activity accessibility as drivers of urban growth. *Landscape and Urban Planning*, **92**, 24–33.

Reis, A. (2002). Permanências e transformações no espaço costeiro: formas e processo de crescimento urbano-turístico na Ilha de Santa Catarina. Ph.D. dissertation, University of São Paulo, São Paulo, Brazil.

Rozenzweig, C. and W. D. Solecki (Eds.) (2001). *Climate Change and a Global City: The Potential Consequences of Climate Variability and Change: Metro East Coast*, report for the US Global Change Research Program, national assessment of the potential consequences of climate variability and change for the United States, New York, USA: Earth Institute.

Rosenzweig, C. and W. D. Solecki (Eds.) (2010). *Climate Change Adaptation in NYC: Building a Risk Management Response*, New York City Panel on Climate Change 2010 Report. *Annals of the New York Academy of Sciences*, **1196**, 1–354.

Rusk, D. (1995). *Cities Without Suburbs*, Princeton, NJ, USA: Woodrow Wilson Center Press.

Sanders, W. and J. Getzels (1987). *The Planning Commission: Its Composition and Function*, Planning Advisory Service Reports #400, Chicago, IL, USA: American Planning Association.

Sassen, S. (1991). *The Global City: New York, London, Tokyo*. Princeton, NJ, USA: Princeton University Press.

Saum, C. (2008). Beijing and Shanghai. *Land Lines*, **20**(4), 2–6.

Schmidt, M. (2008a). The Sankey diagram in energy and material flow management, Part II: Methodology and current applications. *Journal of Industrial Ecology*, **12**(1), 82–94.

Schmidt, M. (2008b). The Sankey diagram in energy and material flow management, Part III: Methodology and Current Applications. *Journal of Industrial Ecology*, **12**(2), 173–185.

Schmidt, M. (2009). Global climate change: the wrong parameter. Rio 9-World Climate and Energy Event, 17–19 March 2009, Rio de Janeiro, Brazil. http://www.ludiaavoda.sk/dokumenty/rio2009.pdf

Schmidt, M. and M. Koehler (2008). The energy performance of green roofs and vertical gardens. In *Proceedings World Green Roof Congress*, session "Performance for Climate Change Mitigation", 17–18 September 2008, London, UK.

Schmidt, M., B. Reichmann, and C. Steffan (2007). Rainwater harvesting and evaporation for stormwater management and energy conservation. *Proceedings 2nd International Congress on Environmental Planning and Management*, 5–10 August 2007, TU Berlin. In *Landschaftsentwicklung und Umweltforschung*, pp. 221–224. www.urbanenvirongress.tu-berlin.de/.

Schwab, J. C., P. L. Gori, and S. Jeer (Eds.) (2005). *Landslide Hazards and Planning*, PAS Report 533/534, Chicago, IL, USA: American Planning Association.

Science Council of Japan (2003). *Opinions Concerning Urban Heat Island Phenomenon from Perspectives of Architecture and Urban Environment Studies (Heat island gensho no kaimei ni atatte kenchiku & toshi kankyo-gaku karano teigen)*, Science Council of Japan. Available at www.scj.go.jp/ja/info/kohyo/18pdf/1801.pdf. Accessed January 8, 2009 [in Japanese].

Seto, K. C. and M. Fragkias (2005). Quantifying spatiotemporal patterns of urban land-use change in four cities of China with time series landscape metrics. *Landscape Ecology*, **20**, 871–888.

Seto, K. C. and M. Fragkias (2007). Mangrove conversion and aquaculture development in Vietnam: a remote sensing-based approach for evaluating the Ramsar Convention on Wetlands. *Global Environmental Change*, **17**, 486–500.

Seto, K. C., R. K. Kaufmann, and C. E. Woodcock (2000). Landsat reveals China's farmland reserves, but they're vanishing fast. *Nature*, **406**, 121.

Seto, K. C., C. E. Woodcock, C. Song, *et al.* (2002). Monitoring land-use change in the Pearl River Delta using Landsat TM. *International Journal of Remote Sensing*, **23**, 1985–2004.

Shanghai Agriculture (2003). *Shanghai's Plan for Urban Forest Reveals*. Accessed August 16, 2009, http://en.shac.gov.cn/jdxw/200311/t20031117_91023.htm

.Sharkawy, M. A., X. Q. Chen, and F. Pretorius (1995). Spatial trends of urban development in China. *Journal of Real Estate Literature*, **3**(1), 47–59.

Shaw, A. (2005). Peri-urban interface of India cities. Growth, governance and local initiatives. *Economic and Political Weekly*, **40**, 129–136.

Shepherd, J. M. (2005). A review of current investigations of urban-induced rainfall and recommendations for the future. *Earth Interactions*, **9**, 1–27.

Shepherd, J. M. (2006). Evidence of urban-induced precipitation variability in arid climate regimes. *Journal of Arid Environments*, 10.1016/j.jaridenv.2006.03.022.

Sjaastad, E. and B. Cousins (2008). Formalisation of land rights in the South: An overview. *Land Use Policy*, **26**, 1–9.

Smit, B., I. Burton, R. J. T. Klein and J. Wandel (2000). An anatomy of adaptation to climate change and variability. *Climatic Change*, **45**, 223–251.

Snover, A. K., A. C. Whitely Binder, J. Kay, *et al.* (2007). *Preparing for Climate Change: A Guidebook for Local, Regional, and State Governments*, Climate Impacts Group, University of Washington, King County, WA, and ICLEI. Accessed at ICLEI website on August 11, 2009, www.iclei.org/fileadmin/user_upload/documents/Global/Progams/CCP/0709climateGUIDEweb.pdf.

Song, Y. and G.-J. Knaap (2004). Measuring urban form: is Portland winning the war on sprawl? *Journal of the American Planning Association*, **70**, 210–225.

Sorensen, A. and P. Hess (2007). *Metropolitan Form, Density, Transportation*. Toronto, ON, Canada: Neptis Foundation.

Souch, C. and S. Grimmond (2006). Applied climatology: urban climate. *Progress in Physical Geography*, 270–279.

Southall, A. (1998). *The City in Time and Space*. Cambridge, UK: Cambridge University Press.

Spirn, A.W. (1985). *The Granite Garden*. New York, USA: Basic Books.

State of California (2009). Chapter 728. S.B. 375 Chaptered 9/30/2008. Accessed at the website of Find California Statutes on July 1, 2009: www.leginfo.ca.gov/cgi-bin/statquery.

Steiner, F. (2000). *The Living Landscape: An Ecological Approach to Landscape Planning* (2nd edition), New York, USA: McGraw-Hill.

Stone, B. and M. Rodgers (2001). Urban form and thermal efficiency: How the design of cities influences the urban heat island effect. *Journal of the American Planning Association*, **67**, 186–198.

Stone, B. *et al.* (2007). Is compact growth good for air quality? *Journal of the American Planning* Association, **73**, 404–418.

Sudhira, H. S., T. V. Ramachandra, and M. H. Bala Subrahmanya (2007). City Profile : Bangalore. *Cities*, **24**, 379–390.

Sugai, M. I. (2002). Segregação silenciosa. Investimentos e distribuição sócio-espacial na Área Conurbada de Florianópolis. Ph.D. thesis, São Paulo, FAU/USP.

Suzuki, H., A. Datsur, S. Moffat, N. Yabuki, and H. Maruyama (2010). *Eco2Cities: Ecological Cities as Economic Cities*. Washington, DC: World Bank.

Swope, C. (2007). Local warming. *Governing*, December.

Talen, E. (2006). Design that enables diversity: the complications of a planning ideal. *Journal of Planning Literature*, **20**, 233–249.

TBEIC (Tokyo Bay Environmental Information Center) (2009). Available at www.tbeic.go.jp/kankyo/mizugiwa.asp. Accessed January 8, 2009[in Japanese].

Theobald, D., and N. Hobbs (1998). Forecasting rural land-use change: a comparison of regression and spatial transition-based models. *Geographical and Environmental Modeling*, **2**, 65–82.

Tiebout, C. (1956). A pure theory of local expenditure. *Journal of Political Economy*, **64**, 416–424.

TMG (Tokyo Metropolitan Government) (2003). *Framework for the Urban Heat Island Countermeasures: Toward Realization of Eco-city Tokyo (Heat island taisaku torikumi houshin – kankyo toshi Tokyo no jitsugen ni mukete)*, Tokyo: TMG. Available at www2.kankyo.metro.tokyo.jp/heat/heathoushin/heathousin.pdf [in Japanese]. Accessed January 8, 2009.

TMG (2008). *Tokyo Environmental Master Plan (revised)*, Tokyo, Japan: TMG. Available at www.kankyo.metro.tokyo.jp/kouhou/english/2008/foreword/foreword.html [in Japanese]. Accessed January 8, 2009.

TMG (2009a). *Outline of Estimated Population in Tokyo Metropolis as of December 1, 2009. (Tokyoto no Jinkou, Suikei)*, Tokyo, Japan: TMG. Available at www.toukei.metro.tokyo.jp/jsuikei/2009/js09ca0000.xls [in Japanese]. Accessed January 8, 2009.

TMG (2009b). *Population in Tokyo: Results from Census and its Analysis in Toky. (Tokyo to no Jinko, Tokyo-to ni okeru kokusei chosa kekka no suii to sono kaisetsu)*, Tokyo, Japan: TMG. Available at www.toukei.metro.tokyo.jp/tjinko/2009/to09tf01.pdf [in Japanese]. Accessed January 8, 2009.

TMG (2009c). *A List of Countermeasures Planned in Fiscal 2009. (21nendo Tokushi niyoru Heat island tasaku jigyou yotei ichran)*, Tokyo, Japan: TMG. Available at: www2.kankyo.metro.tokyo.jp/heat/summerpress/090805heat21.pdf [in Japanese]. Accessed January 8, 2009.

Torrens P. and D. O'Sullivan (2001). Cellular automata and urban simulation: where do we go from here? *Environment and Planning B: Planning and Design*, **28**, 163–168.

Toulmin, C. (2008). Securing land and property rights in sub-Saharan Africa: the role of institutions. *Land Use Policy*, **26**, 10–19.

Trusilova, K., M. Jung, G. Churkina, *et al.* (2008). Urbanization impacts on the climate in Europe: numerical experiments by the PSU–NCAR Mesoscale Model (MM5). *Journal of Applied Meteorology and Climatology*, **47**, 1442–1455.

ULI [Urban Land Institute] (2006). *Shanghai Urbanisation and EXPO 2010 Forum*, Washington D.C., USA.

UNCED [United Nations Conference on Environment and Development] (1992). *Earth Summit, Agenda 21: The UN Programme of Action from Rio*. New York, USA: United Nations. Accessed on January 12, 2009, www.un.org/esa/dsd/agenda21/index.shtml.

UNEP (2009). *UNEP Environmental Assessment Expo 2010*, Shanghai, China.

UN-HABITAT (2008a). *The State of the World's Cities 2008/2009: Harmonious Cities*. London, UK: Earthscan.UN Population Division (2008). *World Urbanization Prospects: The 2007 Revision*. Database. Glossary. Population Division Accessed 25 June 2009, www.un.org/esa/population/publications/wup2007/2007WUP_Highlights_web.pdf.

United Nations (2008). *The Millennium Development Goals Report 2008*. New York, USA: United Nations.

US National Research Council (2009). *Driving and the Built Environment: The Effects of Compact Development on Motorized Travel, Energy Use, and CO2 Emissions*, Special Report 298, Committee for the Study on the Relationships Among Development Patterns, Vehicle Miles Traveled, and Energy Consumption, Washington, DC, USA: National Research Council.

Vincent, L., T. Peterson, V. Barros, *et al.* (2005). Observed trends in indices of daily temperature extremes in South America 1960–2000. *Journal of Climate*, **18**, 5011–5023.

Voogt, J. A. and T. R. Oke (2003). Thermal remote sensing of urban climates, *Remote Sensing of Environment*, **86**, 370–384.

Wackernagel, M. and W. Rees (1996). *Our Ecological Footprint*. New York, USA: New Society Press.

Ward, S. V. (1999). The international diffusion of planning: a review and a Canadian case study. *International Planning Studies*, **4**, 53–77.

Warner, S. (1978). *Streetcar Suburbs: The Process of Growth in Boston* (2nd edition), Cambridge, MA, USA: Harvard University Press.

Wassmer, R. W. (2006). The influence of local urban containment policies and statewide growth management on the size of united state urban areas. *Journal of Regional Science*, **46**, 25–65.

Weitz, J. (1999). *Sprawl Busting*, Chicago, IL, USA: American Planning Association.

Wheeler, S. (2008). State and municipal climate change plans. The first generation. *Journal of the American Planning Association*, **74**, 481–496.

Wheeler, S., J. Randolph, and J. London (2009) Planning and climate change: an emerging research agenda. *Progress in Planning*. In press.

White, M. (1975). Fiscal zoning in fragmented metropolitan areas. In E. Mills and W. Oates (Eds.), *Fiscal Zoning and Land Use Controls*, Lexington, MA, USA: Lexington Books, pp. 31–100.

White, R. and G. Engelen (1997). Cellular automata as the basis of integrated dynamic regional modeling. *Environment and Planning B: Planning and Design*, **24**, 235–246.

Wieditz, I. and J. Penney (2006). *A Scan of Climate Change Impacts on Toronto*, Toronto, ON, Canada: Clean Air Partnership.

Wieditz, I. and J. Penney (2007a). *Climate Change Adaptation Options for Toronto's Urban Forest*, Toronto, ON, Canada: Clean Air Partnership.

Wieditz, I. and J. Penney (2007b). *Time to Tackle Toronto's Warming. Climate Change Adaptation Options to Deal with Heat in Toronto*, Toronto, ON, Canada: Clean Air Partnership.

Wiedmann, T. and J. Minx (2007). *A Definition of "Carbon Footprint"*, ISAUK Research Report 07–01, Durhau: Centre for Integrated Sustainability Analysis, ISAUK Research & Consulting. Accessed February 21, 2008, www.isa-research.co.uk/docs/ISAUK Report 07–01 carbon footprint.pdf.

Wolman, A. (1965). The metabolism of cities. *Scientific American*, **213**(3), 179–190.

Yardley, J. (2008). China enacts major land-use reform. World, *New York Times*. Accessed September 15, 2009 at www.nytimes.com/2008/10/20/world/asia/20china.html?_r=1&ref=world.

Yin, M. and J. Sun (2007). The impacts of state growth management programs on urban sprawl in the 1990s. *Urban Affairs*, **29**, 149–179.

Zhang, X., M. Liu, and F. Meng (2005). Research on the changing characteristics of the thermal field in Shanghai based on the multiple-temporal Landsat TM & ETM data. *Urban Dimensions of Environmental Change*, 79–86.

Zheng, Y., T. Chen, J. Cai, and S. Liu (2009). Regional concentration and region-based urban transition: China's mega-urban region formation in the 1990s. *Urban Geography*, **30**, 312–333.

9

Cities and climate change

The challenges for governance

Coordinating Lead Author:

Patricia McCarney (Toronto)

Lead Authors:

Hilda Blanco (Los Angeles), JoAnn Carmin (Cambridge), Michelle Colley (Montreal)

This chapter should be cited as:

McCarney, P., H. Blanco, J. Carmin, M. Colley, 2011: Cities and climate change. *Climate Change and Cities: First Assessment Report of the Urban Climate Change Research Network*, C. Rosenzweig, W. D. Solecki, S. A. Hammer, S. Mehrotra, Eds., Cambridge University Press, Cambridge, UK, 249–269.

9.1 Introduction

Formulating and implementing effective climate action in cities poses a core set of challenges for city governance. This chapter addresses the need for an empowered governance of cities worldwide, if cities and nations are to successfully confront the challenges of climate change. City governments are constrained on a number of fronts when it comes to formulating and implementing climate action. Many city governments are weakened due to only limited power over and responsibility for key public services, including planning, housing, roads and transportation systems, water, land use, drainage, waste management, and building standards resources (McCarney, 2009). In many of the poorest cities of Asia, Africa, and Latin America, informal areas of the city do not have basic services such as waste collection, piped water, storm and surface drains, and sanitation systems. While all cities and their inhabitants are at risk, the poorest cities and the most vulnerable populations are most likely to bear the greatest burden of the storms, flooding, heat waves, and other impacts anticipated to emerge from global climate change. City governments often lack powers (with respect to higher orders of government – state and national) to raise the revenues required to finance infrastructure investments and address the climate change agenda. When governance capacity is weak and constrained, cities are limited in their abilities to take programmatic action on climate change mitigation and adaptation.

9.2 Cities and climate change: six core governance challenges

New governance challenges for cities are arising as a result of new risks and vulnerabilities associated with climate change. Six core governance challenges are identified here. The body of literature on local climate governance has been growing in recent years, and a brief review of this literature is presented in Box 9.1. While each city faces unique challenges in addressing climate risks, and must find solutions that are adapted to its own context, cities globally depend on effective and long-term solutions that are based on an empowered city governance approach and acknowledge the respective contributions of a broad group of actors that cross jurisdictional and administrative boundaries (McCarney, 2006). The challenge to overcome fragmentation in urban governance is central to moving forward on the climate change agenda globally. Urban areas (and metropolitan areas, made up of more than one urban area) are defined by each country; there is no consistent definition for "urban" or what is a "municipality" throughout the world. Urban data suffer from limitations in terms of reliability and comparability due to definitional issues related to jurisdictional boundaries (McCarney, 2010). The many aspects of risk and vulnerability in cities require more integrated approaches that combine established policies related to urban governance, and management while adding new policy leverage, powers, and responsibilities to the local level of government.

9.2.1 Empowered local governance: political and fiscal

The planning and management functions in cities are more effective when local government is recognized as a legitimate partner in the governance structure of a country, and when financial powers to raise revenues and responsibilities to deliver services are commensurate with urban growth and expansion (McCarney, 2006). Cities are discussing this urban agenda with provincial and national governments around the world. As global entities dedicated to the climate change agenda increasingly point to the significant role of cities in both contributing to and mitigating climate change, the voice of cities in formulating the agenda and their role in taking action locally is also gaining momentum. As a result, the significance of well-governed, well-managed and well-financed city governments to address climate change becomes pivotal in global terms.

When cities are empowered and recognized as significant sites of governance in national and global contexts, they will gain the power to pass legislation related to greenhouse gas emissions; to encourage citizen participation, and to engage with related governmental agencies and local corporate organizations on climate change mitigation and adaptation. These conditions can set in motion a series of activities, including the ability to (McCarney, 2006):

- Build more inclusive institutions in cities for achieving environmental objectives
- To plan and implement transportation systems that support access by all citizens and rational choices on where to live and work that are in keeping with a climate change agenda for the city;
- Ensure strong and robust local economic development patterns that build sustainable economic opportunity for all citizens;
- Address land tenure and land rights in the city and thereby adopt a pro-poor set of policies governing access to and use of land in the city under changing climate conditions;
- Amend building codes and zoning bylaws and adopt flexible yet greener standards governing safer construction of housing, buildings, and infrastructure that are more resilient to climate change risks;
- Develop creative financing tools for mobilizing investments that help to overcome climate risks derived from a lack of basic infrastructure and environmental amenities for all, especially the poorest urban residents in cities.

Recently, efforts to improve urban governance have focused on the essential first step of devolution of power, authority, and resources from the central and sub-national to the municipal level (McCarney, 2006). Governed by the principle of subsidiarity, decentralization processes aim to ensure that decisions are taken, and services delivered, at the sphere of government closest to the people while remaining consistent with the nature of the decisions and services involved. Empowering cities to govern effectively is key to urban reform in countries throughout both the developed and developing worlds. The urban agenda in Canada is focussed on empowering cities with new sources of revenue and new powers to govern

[MITIGATION/ADAPTATION]   Box 9.1   Local climate governance: Barriers and motivators to formulating more targeted policies

Till Jenssen and Maike Sippel

University of Stuttgart, Institute of Energy Economics and Rational Energy Use

This box provides a systematic literature review of local climate governance (Sippel and Jenssen, 2009). A large part of the literature focuses on mitigation and cities in industrialized countries. The analysis also includes recent material on adaptation and cities in developing or industrializing countries. The review categorizes five barriers to formulating more targeted policies for local climate governance.

1. **Costs of climate policies** Costs are a crucial factor in explaining the lack of widespread citywide climate protection activities (Harrison and McIntosh Sundstrom, 2007). Some mitigation activities (such as collective energy management) are taken by local authorities because they pay off quickly (Alber and Kern, 2008). Even in the case of these "no-regret" measures, local authorities often decide against activities that have high up-front investment costs and long payback periods (Ürge-Vorsatz et al., 2007). In particular, this applies to cities with meagre financial resources, which are often situated in developing countries (Kern et al., 2005; Rezessy et al., 2006; Bai, 2007). However, many cities have not implemented measures that do not require much funding, either (Dhakal and Betsill, 2007). Concerning adaptation measures, costs are important too. However, since adaptation is a more urgent issue as many cities already face climate change impacts, adaptation activities benefit the city directly. Therefore, the willingness to finance adaptation activities is probably higher than the willingness to finance mitigation. Cities in developing countries that are most vulnerable to the impacts of climate change must often start adaptation policy from a situation of "infrastructure backlog" (Bulkeley et al., 2009).

2. **Lack of cooperation** Climate policymaking requires cooperation on different levels. First, effective implementation of both adaptation and mitigation policies needs cooperation between local stakeholders, such as authorities, businesses, and citizens (Jones et al., 2000). Second, formal city boundaries are often too small to adequately address issues such as transport (mitigation) or water management (adaptation). Nevertheless, regional cooperation on climate policies that reaches beyond city boundaries is scarce (Wilbanks and Kates, 1999). Third, cooperation between cities on the one hand and the international and national levels of climate governance on the other hand is crucial. Because of the "Tragedy of the Commons," such cooperation seems especially important for mitigation policies (Lutsey and Sperling, 2008). Policy at global and national levels may inspire local policies, enable local authorities, fund local activities, or govern local policies by authority. National policies that do not address the city level specifically, such as a national carbon tax, feed-in tariffs or energy market regulation, also significantly influence local climate poli-

cies (Fleming and Webber, 2004; Bulkeley and Kern, 2006; Schreurs, 2008; Schröder and Bulkeley, 2009).

3. **Lack of leadership and political support** Frequently, political and administrative leadership is cited as a precondition for successful local climate policymaking. Motivation and commitment of decision-makers and administrative staff members can make a difference, e.g., by putting climate protection on the agenda and convincing council members, by linking climate policies with other local issues that generate co-benefits and securing project-funding, and by overcoming fragmentation and building consensus (Betsill, 2001; Carmin et al., 2009; McCarney, 2009). Especially in the initial phase and when taking controversial decisions, political support is absolutely essential for successful climate governance. However, political support and leadership may not be sufficient where financial constraints are too high (both costs of climate policies and lack of financial and human resources) (Kern et al., 2005).

4. **Limited monitoring and evaluation of policies** The evaluation of mitigation policies requires emission inventories. Yet there is as yet no generally agreed methodology at the city level (Kern et al., 2005). Methodologies differ, depending on whether they are production or consumption based and on which sectors they include (Dodman, 2009). In addition, access to emission data is often difficult, and even more so in cities of developing and industrializing countries (Bulkeley et al., 2009; Dhakal, 2004, 2009). The combination of these two factors makes emission inventories time-consuming and costly, and the evaluation of mitigation policies difficult.

5. **Tragedy of the Commons** While local adaptation activities benefit a city directly, the benefits of local mitigation activities are non-excludable. Because the greenhouse effect of emissions occurs globally and independently of their place of origin, local mitigation measures lead to very small benefits for everyone. Therefore local climate protection underlies the logic of the "Tragedy of the Commons" (Kousky and Schneider, 2003). Some urban stakeholders argue that their cities cannot tackle climate change effectively on their own but only in a joint effort with all other local governments. Following economic rationality, implementing measures and spending local resources is not sensible if others do not take measures as well. In the absence of an adequate global climate governance framework, this results in a lack of mitigation measures (Droege, 2002; Fleming and Webber, 2004).

With regard to overcoming these barriers, the literature review found that the three key motivations for local mitigation policies are cost savings, improvement of air quality, and reduction of vulnerability (Sippel and Jenssen, 2009). In adaptation, key motivations are reduction in vulnerability, a design for smart growth, internal pressure, and improvement of a city's reputation.

Source: Sippel, M. and Jensen, T. (2009).

effectively. This is a priority goal of city mayors in their interactions with provincial and federal governments. The Federation of Canadian Municipalities and provincial associations have taken up this agenda and established an effective voice for ensuring its success.

Globally, discussions on how to enhance urban governance have identified the need for central and provincial levels of government to to be engaged in the cities agenda, and to foster the important role of cities in promoting development and civic engagement (McCarney, 2006). The national government in Brazil, for example, enacted a "City Statute" in 2001 giving municipalities the legal power to better plan urban development, to democratize local decision-making and to encourage more inclusive cities (Fernandes, 2001). The significance of national government interactions with cities, as well as by provinces/states in urban development, has been emphasized in many decentralization strategies. A new federalism is emerging that positions cities as critical partners in governing and fiscal relationship. This is increasingly recognized as a pivotal policy platform for both global actions on climate change and local responsibility for mitigating climate change and building climate resilient cities (McCarney 2006; ICLEI, 2010).

Although progress is being made, city leaders are not usually at the table when international protocols and agreements on climate change are discussed by member states and when states decide on whether to sign and support these international agreements. The vulnerability of cities to climate change risks is largely underestimated in these negotiations. Without established data and standardized city indicators on climate change, it is more challenging for cities to enter into these global discussions. With increasing urban vulnerability being recognized however, estimated simply by the fact of the increasing dominance of city dwellers worldwide and the increasing visibility of climate change vulnerabilities in cities, it has become more pressing for city governments to be considered as new sites of governance in global negotiations on climate change and in decision-making related to risk assessments. In this context, cities are increasingly joining international cooperative networks such as C40, the International Council for Local Environmental Initiatives (ICLEI), Cities for Climate Protection (CCP), and the US Conference of Mayors Climate Protection Agreement. These growing climate networks for cities are detailed in Box 9.2.

[MITIGATION/ADAPTATION] Box 9.2 Why do cities participate in global climate networks?

Taedong Lee

City University of Hong Kong

Cities and local governments that have no binding obligations to reduce greenhouse gas emissions for international treaties such as the Kyoto Protocol are nevertheless trying to tackle global climate change. In addition, these cities and municipal governments have formed cooperative networks to work together on climate change.

Which factors are associated with cities' participation in global climate change networks, even when their national governments do not or need not (i.e., non-Annex 1 countries) comply with the Kyoto Protocol? Global cities – hubs of international, economic, and policy interactions – are more likely to commit to international networks for addressing global climate change issues because they are centers of ideas and policy diffusion and because they have economic interests regarding climate change.

Whether a city joins a global climate change network voluntarily is an indication of its willingness and commitment to learn about and collaborate on climate change responses. Given that climate change is not a problem that a single local government can solve by itself, learning and collaborating through networks allows a city government to expand its capacity to handle climate change issues.

The logic of a city's participation in international environmental governance in climate change networks is related to the role cities increasingly play in the global economic context. First, global cities play a central role as hubs in the diffusion of ideas and the flow of people. A global city with a cosmopolitan identity and status as a center of diffusion provides a strong incentive

for a city to actively participate in global environmental issues. Second, international socialization takes place in a global city with numerous interactions via conferences and contacts. Conferences in global cities institutionalize socialization among policymakers, scientists, and interest groups. Third, a global city is financially motivated to cope with climate change issues.

Aside from the degree of cities' integration with globalization, vulnerability to climate change also plays a role. Given the unintended consequences and risks of climate change, it is critical to answer the question of "conflict of accountability" over why and how particular definitions of risk and responsibility are controlled and legitimated. It is important to note that the risks of climate change are unevenly distributed across geography, social classes, and demography.

In addition to cities' own attributes, the characteristics of nation states in which cities are located influence city level policies, as cities are under the hierarchy of a nation state. Attributes of the country may affect the behavior of cities' international activities. In particular, regime types are thought to be the primary factor influencing countries' participation in international treaties and better environmental outcomes.

Global cities create networks of interdependence that span international boundaries and thus encourage collective action on climate change. Global cities are more likely to commit themselves to the global environmental networks like the C40 (Cities Climate Leadership Group) and CCP (Cities for Climate Protection). The position of the city in the global economy and international transport system is crucial to socialization and the diffusion of ideas on global environmental responsibility.

Source: Lee (2010).

National governments increasingly are confronting new and emerging global agendas on climate change. Because many of these agendas find expression in cities, global commitments negotiated by national governments must be rooted in dialogue at the city level to ensure that local authorities are part of the decision-making and, as importantly, integral parts of mitigation, adaptation, and implementation processes. Stronger intergovernmental relations, local representation processes, subnational institutions, and financing mechanisms to support subnational government forms are critical needs for policymakers and leaders in all levels of government, as well as areas of focus for researchers, planners, and international agencies concerned with climate change (McCarney, 2010).

Given the global estimates that energy for heating and lighting residential and commercial building generates nearly a quarter of greenhouse gas emissions globally, and that transport contributes 13.5 percent (of which 10 percent is attributed to road transport) (United Nations 2008a), we can assume a sizeable portion of this volume of emissions is generated in cities. According to the Clinton Foundation, large cities are responsible for the majority of the greenhouse gases released into our atmosphere. Greenhouse gas emissions are usually under the control or influence of local governments since a majority of these emissions are linked to urban form that affects transportation and energy consumption. For example, according to a recent calculation in Canada by the Province of British Columbia, 43 percent of its provincial greenhouse gas emissions are within the realm of responsibility and authority of local governments (Cavens *et al.*, 2008). The World Bank estimates that the transport sector alone accounts for a third or more of total greenhouse gas emissions in metropolitan areas. Therefore cities have the potential, and indeed are becoming, the key actors in global mitigation efforts. City governments can influence patterns of energy and land use through important interventions under their control, including land use planning, urban design, zoning and local bylaws including building codes and height bylaws, transport planning including transit planning road networks, master plan and subdivision controls.

A few examples of city action in the field of climate change mitigation demonstrate the extent to which governance efforts at the city level can have real influence. For instance, the Vienna (Austria) City Council adopted the city's Climate Protection Programme as a framework for its Eco-Business plan. The results have been reductions in solid waste output by 109,300 tons, toxic solid wastes by 1,325 tons, and carbon dioxide emissions by 42,765 tons. This Eco-Business plan has saved a total of 138.7 million kWh of energy and 1,325,000 cubic meters of drinking water. The Eco-Business plan is also now being implemented in Chennai, India, and Athens, Greece. The City of Calgary, Canada, is achieving significant electricity savings and reducing greenhouse gas emissions with the EnviroSmart Retrofit Project. Most of Calgary's residential streetlights are being changed to more energy efficient flathead lenses. Streetlight wattage is being reduced from 200 W to 100 W on residential local roads and from 250 W to 150 W on collector roads (UN-HABITAT, 2008).

In the building sector, improvements to building codes and certification processes for greener buildings are being adopted by a number of cities as a means of promoting mitigation. The City of Johannesburg, South Africa has implemented measures that include retrofitting of council buildings, energy savings in water pump installations, and methane gas recovery. One set of measures already well established is the LEED certification framework that ensures a building is environmentally responsible by providing independent, third-party verification. LEED certification seeks to ensure that a building project meets the highest green building and performance measures. The average LEED certified building uses $\cong 30$ percent less energy, 30–50 percent less water and diverts up to 97 percent of its waste from landfill (US Environmental Protection Agency, 2007).

Mexico City has taken various measures to mitigate the effects of climate change by taking action in the areas of water, energy, transportation, and waste to reduce its carbon dioxide emissions. In regards to energy, Mexico City has moved to secure more sustainable housing and buildings through action to reduce energy consumption. These include establishing environmental certification systems for buildings and providing funding for new housing that integrates sustainability criteria. Various energy efficiency programs have been in place to reduce emissions, including efficient lighting in buildings, efficient street lighting, and promoting solar energy in businesses and government buildings. Regarding the water sector, Mexico City has taken action to reduce emissions from septic systems by constructing sewerage and water treatment services in areas of low methane gas. Some of the actions taken in the transportation sector include an obligatory school transportation system, which will reduce CO_2 emissions by 470,958 tons per year by ensuring that students take public transportation to school. Mexico City will also expand its transportation system and the implementation of non-motorized mobility and streetcar corridors as an effort to reduce emissions. In regards to waste management, the government plans on capturing and exploiting the bio-gas emitted from the Bordo Poniente State 4 landfill and eventually installing an electrical power plant, which will reduce emissions by 1,400,000 tons annually (Secretaria del Medio Ambiente del Distrito Federal, 2008). Additional details on the development of Mexico City's climate change mitigation efforts and also the hindrances that have affected policy change are provided in Box 9.3.

In the United States, the lack of action by the national government has resulted in a number of cities taking independent action to promote both mitigation and adaptation. As mentioned in chapter 8, King County (which includes the City of Seattle) in Washington State has shown initiative in developing a county climate plan (King County, 2007; Swope, 2007). Based on projections by the University of Washington's Climate Impacts Group for the Puget Sound, King County has developed a set of guidelines for incorporating mitigation and adaptation goals into county and city agencies. They also identified urgent adaptation needs such as those related to specific water supply pipelines or county roads within or close to floodplains. The County also is taking steps to improve its capacity to undertake adaptation planning by,

[MITIGATION]   Box 9.3   How do local authorities in Mexico City manage climate change?

Patricia Romero-Lankao

National Center for Atmospheric Research

Local authorities in Mexico City have developed a refined framing of greenhouse gas emissions and their relationships to air quality. They have designed strategies and institutional structures to target air quality, the main local concern, and to relate it to climate change; hence authorities "localized" one issue of carbon emissions by relating them to an existing local agenda. The existence of effective policy entrepreneurs and multinational networks, such as the group led by Mario Molina, Claudia Scheinbaum (Secretary of Environment of the Federal District during 2000–6), and ICLEI, played a key role in launching and shaping this agenda and in facilitating an ongoing learning process.

Nevertheless, this influence was not enough to push real and effective policy strategies and actions. Unlike the integrated and broader framing of greenhouse gases, the Local Strategy of Climate Action, and Mexico City's active participation in ICLEI and the C40 (cities climate leadership group), policy actions have remained narrow. Between 1990 and 2007, the energy sector received the biggest share of financing (between 30 and 60 percent) from three atmospheric programs, most of which was allocated to improve the quality of fuels. Standards and technologies to improve energy efficiency, reduce emissions by automobiles (e.g., catalytic converters), monitor emissions, and implement the no-driving day also received between 20 and 70 percent of the total financing.

Policymaking was constrained by diverse institutional factors. The administrative structures of governance did not align with the city's boundaries and carbon-relevant functioning. The seat of federal powers is in the Federal District, where Mexico City, the most important national economic hub, has been historically located. In its double role, Mexico City faced almost a century of contradictory processes of centralized control by the federation and institutional fragmentation of local structures and political participation. A component of the city's political reform, starting in the 1980s, focused on democracy and political rights (e.g., the legislative body of the Federal District gained considerable legislative powers), but did not change the uncoordinated and fragmented government structure of Mexico City, currently managed by various governmental entities: the Federal District with its 16 delegations, the State of Mexico with 35 conurbanized municipalities, and the federal government still exerting a strong influence on Mexico City.

Two additional components of the political reform platform, decentralization and deregulation, did not solve the issues of centralization, complexity, and fragmentation. Diverse coordinating commissions and programs have been created to address city level carbon and climate-relevant issues such as urban planning, transportation, and human settlement. The commissions have functioned as a relatively lightweight institutional instrument mobilizing relevant stakeholders to focus on key issues. Yet they did not seem to help authorities create the much-needed coordination thus far. Diverse factors explain the lack of fit, coordination, and other institutional constraints facing the city.

One factor is the disparity between the fiscal capacity of the federal government and the Federal District on the one hand and of the states and municipalities on the other. The federal government gets the lion's share of tax revenues (74.1 percent), the Federal District and Delegations receive 12.9 percent and 9.1 percent respectively, and such other entities as the State of Mexico and the municipalities get only a tiny percentage (4.5 percent). This leads to a paradox: more responsibilities are delegated – decentralized – to local authorities, but they lack the resources to undertake effective policies.

Second, authorities do not have a culture of cooperation or a common and broadly shared metropolitan vision, which may be due to the effects of both election laws and governing by diverse parties. Governors and the president are elected for a single 6-year term; municipal presidents and "*delegados*" are limited to a single three-year period, which may be a factor preventing long-term accountability in policymaking. The three tiers of government are governed by at least three different parties (National Action or PAN, Institutionalized Revolution or PRI and Democratic Revolution or PRD).

The government has lacked other features of institutional capacity (e.g., human resources, money, and power) to manage air quality and greenhouse gas emissions. Authorities lack sufficient and adequate personnel with the technical knowledge to monitor emissions and see that standards are met. On top of that, environmental efforts are constrained by institutional instability. High turnover rates among government personnel and technical staff have made training efforts less effective.

Mexico City in short has developed important efforts to curb its greenhouse gas emissions. Policy networks, political leaders, and research groups have been critical in launching a climate agenda. Nevertheless, this has not been enough to push effective policies. Policymaking has been constrained by two sets of institutional factors: the problem of fragmentation in local governance and a lack of institutional capacity.

In recent years, policy conditions have started to improve. Mexico City has become an even more effective leader in climate change, with Mayor Marcelo Ebrard, Head of the World Council of Mayors for Climate Change Action, hosting the Mexico City Cities Climate Summit before the 16th Conference of the Parties in Cancun in November, 2010. In 2008, Mexico City was the first city in Latin America to implement a Climate Action Program, which, according to the Government of Mexico City, has enabled the reduction of ~1.4 million metric tons of CO2-equivalent, ~4% of the city's emissions (Government of Mexico City, 2010). Components of the program include a zero-emission transport corridor, public bicycle system, replacement of minibuses and taxis with lower-emitting vehicles, sustainable housing, regulations to encourage the use of solar collectors in commercial and services sectors, green roofs, restoration of ecosystems outside the city, and development and use of an environmental management system. Researchers are currently evaluating these actions for their effectiveness.

Sources: Government of Mexico City (2010), Romero-Lankao (2007).

for example, entering into a collaborative agreement with the Climate Impacts Group, educating appropriate county staff in climate change science, and raising public awareness (King County, 2007). Although the plan does not include specific implementation steps, its goal of incorporating climate change adaptation considerations in all relevant county plans and projects, or mainstreaming such efforts, is indicative of proactive climate leadership and management (Blanco and Alberti, 2009).

Evidence about the potential impacts of climate change has been an influential driver of adaptation planning in cities around the world. One example where this is the case is Durban, South Africa. After learning about climate impacts projected for the global south, and conducting a vulnerability assessment, it became clear that the city and its inhabitants were at risk from climate impacts and that initiating adaptation planning was a pressing issue in addition to reducing greenhouse gas emissions. Durban is not alone in making strides in advancing adaptation as other cities globally (New York City and Quito, Ecuador, are noteworthy) are making significant progress in this arena, many without national level support for their work. Cities are progressing in planning and implementation processes at different rates. The main difference in the rate of mainstreaming appears to be the commitment of local government officials, the use of adaptation as a filter for new initiatives, the degree to which adaptation measures are linked to development goals, and the allocation of local resources to advance an adaptation agenda (Carmin *et al.*, 2009). Box 9.4 details the financial and administrative barriers to implementing adaptation strategies that city managers and local governments confront.

9.2.2 Jurisdictional boundaries: towards a metropolitan governance of climate change

In considering cities and their governance capacity to address climate change, an emerging core set of challenges reflects the complexity of city-level politics, the multiple and overlapping agency responsibilities for service sectors, and the spatial challenges associated with municipal jurisdictional boundaries. Conceptualizing vast, and often diffuse, urban territories and their spread across existing municipal boundaries and broader jurisdictions are difficult tasks. This conceptual challenge mirrors a movement in local governance reform that is in a continuous state of flux, experiment, and re-formulation.

World trends in urbanization are causing metropolitan populations to spill beyond their city limits, rendering the traditional municipal boundaries and, by extension, the traditional governing structures and institutions outdated (McCarney and Stren, 2009). As urban areas around the world continue to expand in terms of both density and horizontal space (Angel *et al.*, 2005), there is a need to govern these large areas in a coherent fashion. Highly fragmented governance arrangements in many metropolitan areas make efficient planning, management, and urban financing for area-wide service provision a difficult and ongoing challenge (Klink, 2007; Lefèvre, 2007). Climate change

action, however, requires coherence and integration across these jurisdictions. The formulation and implementation of Paris's climate plan provides a lesson on the challenge of planning within an expanding municipal boundary (Box 9.5).

This metropolitan expansion is not just in terms of population settlement and spatial sprawl but, perhaps more importantly, in terms of their social and economic spheres of influence. The functional area of cities has extended beyond the jurisdictional boundaries. Cities often have extensive labor, real estate, financial, business, and service markets that extend over the jurisdictional territories of several municipalities and, in some cases, over more than one state or provincial boundary. In a number of cases cities have spread across international boundaries. This expansion is taking place regardless of municipal jurisdictional boundaries.

Cities are the staging sites for meeting the serious challenges of climate change. When considering climate action in these large metropolitan areas, whether in terms of measuring risks, establishing indicators, or creating mitigation or adaptation strategies, the challenges of metropolitan governance and the contexts of administrative, management, and political fragmentation are critical to confront. For example, the metropolitan area of Mexico City (18 million people) extends over the territories of municipalities of two states as well as the Federal District to include as many as 58 municipalities; the economy of Buenos Aires covers the territories of the City of Buenos Aires (3 million people) and the 32 municipalities of the Province of Buenos Aires (9 million people) (McCarney and Stren, 2008). Similarly, in Africa, Metropolitan Johannesburg (7.2 million people) encompasses Ekurhuleni (made up of the East Rand), the West Rand District Municipality (the West Rand) and the City of Johannesburg (Cameron, 2005). Abidjan (with a population of 3.5 million) has expanded to encompass 196 local government units, which include municipalities and surrounding rural areas (Stren, 2007). In Asia, the Metropolitan Manila Area in the Philippines is composed of ten cities and seven municipalities, with a total population of approximately 11 million; while Cebu City comprises seven cities and six municipalities (with a population of 1,930,096) (McCarney and Stren, 2008). The Tokyo metropolitan region, with an estimated population of 35 million, contains 365 municipal areas (Sorensen, 2001). In North America, Metropolitan Minneapolis-Saint Paul (with a population of 3,502,891) is composed of 188 cities and townships (Hamilton, 1999). Portland, Oregon, with approximately 1.5 million inhabitants, covers three counties and 24 local governments (McCarney and Stren, 2008). These examples are just a few of many metropolitan areas worldwide that are growing quickly and expanding across territories, creating new pressures on the existing governing arrangements (McCarney and Stren, 2008).

These cases are not the exception. Most of the world's largest cities are made up of more than one urban unit. However, most of our comparative statistics on cities and metropolitan areas are based on data that do not attend to these differentials in units. And because metropolitan areas are rarely legally defined entities, there may be a number of different possible

[ADAPTATION]   Box 9.4   Urban climate change adaptation: Competencies and finances from an economic perspective

Bernd Hansjürgens and Joseluis Samaniego

Helmholtz Centre for Environmental Research (UFZ) and United Nations Economic Commission for Latin America and the Caribbean (CEPAL)

Most action in urban climate policy has emphasized mitigation in cities in the global north, such as London. Yet cities are increasingly aware of the challenge of adapting to global climate change. The diffusion of adaptation into cities will rely greatly on mobilizing potentials and overcoming obstacles of a political, financial, and administrative nature. Cities' adaptation spillovers (in contrast to mitigation) are site-specific, and require adequate responsibilities and resources to be addressed.

The city manager can only act on adaptation if enabled by legislative and administrative powers for climate policy measures. In many countries (especially of the global south), some key political powers needed for urban action reside at the central level. While this can be adequate for many climate mitigation issues (e.g., defining national reduction targets, distributing them among sectors, or introducing national policy measures), adaptation issues are in most cases regional or local by nature. This means that risk management (e.g., infrastructure planning, sanitation, resource management, or measures for residents' health) is primarily a task that is under local management. Equipping city managers with the information and authority to deal with adaptation issues is therefore a vital prerequisite. This requires a division of functions between national and city levels, whereby the (vertical) allocation of adaptation functions to the city level should be the rule. National adaptation measures or shared functions between the national and the city level should rather be the exception and restricted to cases where regional externalities are predominant or nationally decided and financed infrastructure is involved (pipelines, inter-urban links).

Furthermore, an effective (horizontal) distribution of functions between the various sub-national entities is required, particularly when megacities are involved. In the case of Mexico City, for example, the distribution of functions between the National Government administration, the Federal District and the administrations of the states of Mexico, Puebla, and Hidalgo, has produced overlapping competencies resulting in weak political responsibilities. Although some coordination units between these entities exist, they are not defined by clear competencies that empower the groups to take responsibility. Adaptation to climate change is harder to achieve in such a setting, because the different interests tend to conflict and do not allow an overall planning process.

To develop adaptation strategies, empowered city managers also need financial resources and competencies for managing them on a long-term basis. There are three options for cities to receive revenues:
- cities' own sources, such as fees, charges, or taxes
- taxes or (general or special-purpose) grants allocated from higher levels of government
- international funds.

One could argue that it is sufficient for senior levels of government to allocate funds to city managers for financing public expenditure and meeting the cities' adaptation requirements. This, however, seems neither feasible nor adequate. First, national governments also suffer from scarcities that limit transfers to their sub-levels. Second, and more important, city managers need some degree of autonomy to decide on adaptation actions – some of them long term – according to stable priorities. The examples of early adaptation highlight that setting and maintaining priorities is a decisive element for adaptation strategies and their long-term investments. Setting priorities is more difficult to define and maintain if the resources come – already labelled – from an external source, i.e., the national government. In contrast, it is easier to choose cities' priorities if the benefits of certain measures are balanced against the respective costs. To achieve this, it is imperative that city managers have responsibility for and decide upon both adaptation measures (and their respective inter-temporal benefits) and financial burdens.

This poses a major challenge to cities in developing countries where own-source revenues are very low in relative terms and some of that revenue is collected by national authorities. The combination of short-lasting administrations, weak local revenue, and long-term implementation of adaptation strategies calls for:
- stronger articulation between national governments and city managers to use national transfers to enhance local fiscal capacities
- city managers to take advantage of the international markets of carbon for adaptation/mitigation infrastructure (such as improved public transportation and solid waste and wastewater management)
- the design of political mechanisms to prioritize and maintain the adaptation effort over time.

In developed countries, cities can probably best achieve their adaptation strategies when they have responsibility to raise revenues from their own sources. Grants from national governments or from abroad should therefore only be given if the city managers' revenues are not sufficient or in the case of external effects. Limiting transfers (grants) to improving revenue situations (general grants) or for regional external effects (special-purpose grants) may be mechanisms to explore.

boundaries for a commonly understood extended urban area; for example, New York City and the New York Metropolitan Region, or the City of Toronto and the Greater Toronto Area (McCarney and Stren, 2008). In all these cases, different designations will mean different political arenas for policy and

planning as well as different area measurements, service areas, and populations. Not only do inconsistent definitions pose challenges for governance, for planning, and for research, but also for performance targets, indicators, and measurements in the field of climate change.

[MITIGATION] Box 9.5   An effective climate plan (CP) for Paris, France: The need to develop a metropolitan perspective

Cynthia Ghorra-Gobin

National Scientific Research Center (CREDA), Paris

Formulating and implementing a climate plan (CP) represents one more task to be completed by cities, in addition to producing their major land use plans and transportation planning documents. Climate plan development is a challenge as it requires achieving coherence among three major planning documents. This task is made more difficult when a large portion of a city's population lives in informal neighborhoods that lack basic services. Establishing a coherent and effective CP in rich and poor countries requires the delineation of an effective territory to which the plan can be applied. This is a complex task as cities often take the form of large urbanized areas composed of a number of municipalities or local governments. A CP formulated by a central city in a city-region or a metropolitan area risks a lack of coherence if it does not apply to adjacent municipalities and others that make up the greater urbanized area, as in the case of Paris, France.

The Plan Climat de Paris (PCP), adopted in October 2007, elaborates Paris's goal to reduce greenhouse gas emissions to a quarter of 2007 levels by 2050 (following European and national norms). The PCP is organized around mitigation measures dealing with buildings' energy efficiency, eco-neighborhoods, and transportation. The building section emphasizes housing (14 percent of the Parisian housing stock) as well as the need to apply "new" national norms to all buildings. Paris owns 2.5 million square meters of municipal offices and social centers, which are a focus of the climate plan's energy efficiency goals. The PCP also refers to new eco-neighborhoods including ZAC (Zone d'Aménagenent Concerté [urban development zone]) Gare de Rungis, ZAC Boucicaut, ZAC Clichy-

Batignolles and ZAC Pajol. Regarding transportation, most city council members have agreed to the goal of halving the number of cars in Paris and promoting cycling and transit as environmentally friendly alternative transport options. Paris launched Velo Lib (a local initiative that provides convenient public bike rentals) in 2007, closed a freeway for a month of the summer in 2002 to be transformed into a pedestrian walking promenade, and has begun to invest in a new fleet of streetcars.

The PCP's objectives and measures are aligned with Paris's other major planning documents (Plan Local d'Urbanisme and Plan des Déplacements Urbains). However, Paris's climate plan is only applied to the city of Paris, which is the central city within a large urban area. The Plan Climat de Paris raises two issues: it is limited to an area of 105 km^2 and does not include additional municipalities, which make up a dense urban area of 5 million inhabitants distributed over 700 km^2 (belonging to three other "departements"); and it does not address airplane greenhouse gas emissions within its transportation strategy. Paris is a global city facing the dilemma of reducing greenhouse gas emissions without compromising the local economy. The city attracts a large number of tourists, hosts a diplomatic community as France's national capital, and includes a cosmopolitan business community. Given Paris's attractiveness in different domains, air transportation has been increasing by between 3 and 5 percent per year in recent years. According to Bilan Carbone (a greenhouse gas emissions assessment tool) (http://wwwz.ademe.fr) (Agence de l'Environment et de la Maitrise de l'Energie), air transportation represents 40 percent of the volume of greenhouse gases produced by the transportation sector. Implementing a CP implies conceptualizing an optimal territorial scale for dealing with climate change within a sustainable perspective that confronts the dual challenges of economic development and social cohesion.

As a result, a significant challenge confronting the larger metropolitan centers in addressing climate change is that associated with fragmentation of local governing institutions. In the United States for example, local home rule has led to a patchwork of local governments often in fiscal competition with each other for highly valued property that yields high tax revenues. In the United States, federal and state governments have exercised very little control over land uses. The power to regulate land use and to provide for local infrastructure comes from the states, but operates primarily at the municipal level. In addition, many urban services, such as water supply, wastewater treatment, waste disposal, fire services, etc., are often supplied by limited government special districts, with their own urban agendas. In 2002, there were over 32,000 special districts, not counting school districts in the United States (Bierhanzl and Downing, 2004). This is in contrast to over 19,000 municipalities. Thus, suburbanization and increasing use of special districts to provide urban services has led to increasing governance fragmentation in metropolitan regions.

Plagued with metropolitan fragmentation, regional transportation congestion, degradation of environmental resources, and

weak land use planning regulatory frameworks, about a dozen states, including Oregon, Florida, and Washington, have established state-wide planning programs that mandate local planning and regional or county coordination. These programs have had varying degrees of effectiveness (Nelson and Moore 1996; Burby and May, 1997; Weitz and Moore, 1998; Carruthers, 2002; Wassmer, 2006; Carlson and Dierwechter, 2007; Yin and Sun, 2007). Portland Metropolitan area is considered the most successful. This is partly due to its state-wide planning program, and to its unique metropolitan governance system (Blanco, 2007). Portland's Metro regional planning agency was established in 1992. It is still the only directly elected regional government in the United States with both service (for example, solid waste, regional parks, Convention Center, etc.) and regional planning responsibilities (DeGrove and Miness, 1992; Seltzer, 2004). It encompasses the Portland Urban Growth Boundary, and serves more than 1.3 million people in Clackamas, Multnomah, and Washington counties, and the 25 cities in the Portland, metropolitan area. Metro is governed by a council president elected by the region as a whole, and six councilors elected by district; an auditor is also elected region-wide. Metro has fiscal powers, including the ability to levy property tax, sales tax, and income tax with voter approval, and limited taxes without voter approval. The case of the city of Tampere, Finland, reveals how one city, recognizing the limits of its city's jurisdictional boundary to effectively engage in climate change mitigation, launched a climate strategy process with seven neighboring municipalities in 2007. This experience is detailed in Box 9.6 below.

Metropolitan-level structures and cooperative arrangements often break down in the absence of solid legal frameworks and constitutional support for this "tier" of governance (McCarney, 1996). As a result, metropolitan authorities often lack adequate resources for governing. The challenges of equitable development between different groups in these vast urban territories point to the need for major improvements in the provision of public services such as health care, shelter and housing, education, water supply, and sanitation. Urban poverty has also been worsening and, in many cities, it too has been spreading outwards, rendering the areas on the urban periphery of these metropolitan areas some of the poorest and most heavily underserviced settlements (McCarney, 2010). Formulating effective mitigation and adaptation strategies on climate change demands more integrated planning, delivery of services, and policies than these multiple but individually bounded cities can provide. Governing in this fragmented context of multiple jurisdictional boundaries has become much more complex since a decision made in one municipality that is part of the city affects the whole urban area (McCarney, 2010).

For cities to effectively address the challenges of climate change, coordination and overcoming the problems of fragmentation in political institutions locally is a core requirement. Urban metropolitan areas demand and consume vast amounts of energy and water and other material resources that impact climate change. Cities are both victims and perpetrators of climate change. They generate significant levels of solid waste, electricity demand, transport-related emissions, and space-heating and cooling demand. Cities and local governments are well positioned to set the enabling framework for climate change mitigation strategies, as well as taking a leadership role in addressing the challenges related to hazard management as countries adapt to climate change. However, institutional fragmentation across metropolitan areas is closely related to the escalating risks associated with climate change in cities. This reality introduces new challenges of governance; in particular, what needs to be better addressed in terms of the challenge of defining new metropolitan governance systems for managing climate change.

9.2.3 Good planning and effective urban management

Emerging climate change risks identified globally create new vulnerabilities for cities. For example, the rise in extreme weather events associated with climate change places major cities, particularly those located in coastal areas, in unstable and vulnerable conditions. Global increases in natural disasters associated with climate change have shown that the nature of disasters in

cities has become more multifaceted and so must the approach to their management. Urban health is particularly threatened under conditions of urban poverty. When basic infrastructure is inadequate, poor sanitation and drainage and impure drinking water aid in the transmission of infectious diseases, which puts poor urban households at high risk. This situation is worsened under circumstances of higher densities in urban areas. Climate change vulnerabilities thus require strategic urban management and planning practices, and higher levels of investments in infrastructure, together with better-prepared local governments.

Planning and management tools can help to address the critical link between emissions and urban form, particularly in terms of transportation and building energy consumption. For example official plans, development guidelines, development permits, densification plans, transit planning and pricing building codes, and a number of other planning tools can help to address greenhouse gas emissions in cities as climate change mitigation strategies. For effective planning, spatial data that link greenhouse gas emissions with urban form and city expansion would be valuable. Such information will strengthen locally relevant policy decisions and build support and understanding by the public (Miller *et al.*, 2008).

Access to land and housing and security of tenure are critical issues in the alleviation of urban poverty worldwide (McCarney, 2010) and also relate to climate change adaptation. In cities with large urban poor populations, security of tenure is generally acknowledged as the first step in the integration of slums and low-income settlements. When tenure is uncertain, slum improvement is politically complex, both for city planners and for residents. Any intervention on the part of government is perceived as a *de facto* recognition of legal status and any improvements by residents themselves are regarded as high-risk investments owing to the lack of property rights and the threat of eviction without compensation. Hence, in considering effective planning and management in the context of climate change, the overarching policy and legal climate regarding access of the tenure in the city is critical. Pro-poor enabling legislation and land regularization instruments are crucial components of a city's agenda on climate change.

Indicators of per capita building energy consumption, of urban transport and urban density, for example, can help to inform planners and city managers on policy at several scales. At the regional scale, for example, growth and transportation policies shape major infrastructure investments that affect residents' decisions to drive or take transit. At the city scale, comprehensive development plans can help by establishing density targets that affect transit services, energy systems, and land use. At the neighborhood scale, guidelines that promote mixed-use communities can enable people to walk or cycle to meet daily needs, and at the housing scale, they can encourage building forms and orientation that reduce heating and cooling loads (Miller *et al.*, 2008).

The planning profession and planning tools to promote safer and more resilient cities can contribute to local capacity.

[MITIGATION]   Box 9.6   Tampere, Finland's climate change mitigation strategy: Addressing inter-municipal climate strategy

Lasse Peltonen and Ruusu Tuusa

Centre for Urban and Regional Studies, School of Engineering, Aalto University

Local climate policy took its first steps in Tampere[1] in the early 1990s. At that time, a local citizens' climate initiative, Tampere 21, was organized by local environmental and students' organizations. The initiative lobbied to put climate change on the local political agenda. As a result, the city adopted an environmental strategy in 1994, including a target to reduce greenhouse gas emissions within the city limits.

However, public attention and political commitment for climate action were not strong enough to sustain the issue on the agenda over the following decade. Since the 1990s, it has become evident that climate targets require sustained political commitment, and they cannot be achieved by sectoral programs only. Climate change cannot be an issue "owned" by the municipal environmental office. Furthermore, city limits are seen as too restrictive, and climate change mitigation efforts are now best addressed at the city-region scale.

Prompted by the legislation on restructuring Finnish municipalities, Tampere launched a climate strategy process jointly with seven neighboring municipalities in 2007. The present strategy is integrated with land use, traffic, housing, and municipal service, based on a set of land use and infrastructure development scenarios with respective greenhouse gas projections. The strategy seeks to produce an action plan assigning responsibilities, resources, and timetables, based on a framework agreement among the municipalities in the region. At present, Tampere has also committed itself through the EU covenant of mayors on climate change in 2009.[2]

The local climate strategy is complemented by the ILMANKOS project, which aims to promote the concrete participation of citizens and organizations in mitigating climate change. The project is split into two parts: a public involvement campaign and a research and development project monitoring and evaluating the campaign. The project is funded by the Finnish Innovation Fund (SITRA) and the city of Tampere. The research section is conducted by the Centre for Urban and Regional Studies, School of Engineering, Aalto University.

The ILMANKOS public involvement campaign seeks "to combat climate change and promote climate democracy." The first phase of the campaign has been active since September 2008. The aim of the ILMANKOS campaign is to activate citizens and organizations in the Tampere Central region to reduce greenhouse gas emissions and to take part in developing the climate strategy.

The campaign has organized thematic workshops for local organizations and citizens on issues such as housing and food, a panel on climate change for local politicians, and a lecture series on climate change. Network building among stakeholders with different backgrounds and communication with other local climate change initiatives have been integral to the campaign. The emphasis has been on relating mitigation to everyday life practices. Also, a local climate fund has been created within the campaign, funding small-scale climate initiatives and projects of NGOs, communities, or non-profit organizations.

The research and development project has monitored the campaign through participant observation, documentation of the events, and interviews with key actors. The research project will continue into 2009–10 as a citizen-centered evaluation exercise of the Tampere climate strategy and the ILMANKOS campaign. It constitutes a form of action research, which feeds into the follow-up of the campaign and climate strategy work. The mid-term findings of the project indicate that the scope of local climate strategies has changed from a sectoral and city-centered perspective to encompass the broader region. Municipal actors are seen as central to the strategy, but there is also a clear recognition of the need for broader stakeholder involvement.

The Tampere Central Region's climate strategy and publicity campaign have proceeded on separate tracks. The strategy is expert-driven and operates at the level of inter-municipal planning and administration, while the campaign is citizen-centered and practically oriented, addressing everyday life issues. Despite the visibility of climate change in the media, there is still a clear need for public information and raising awareness of the topic. The need for basic information has been a slight surprise for the campaign project team. Intermediary groups and organizations have become recognized as crucial "gatekeepers" for climate change action. This has been a key lesson of the campaign. For instance, self-governed housing companies make key decisions affecting the climate impacts of housing, and catering companies are important gatekeepers for food-related greenhouse gas emissions.

Originally the ILMANKOS campaign was planned as a short-term public involvement campaign. However, the project team sees the need for a more long-term project to further develop the networks and initiatives launched during the campaign. The planned follow-up project will target specific intermediary groups such as inter-municipal climate envoys, "climate families," and self-governed housing companies. Another aim would be to engage groups that are marginalized in the public debate on climate change, for example immigrants. The local climate fund would continue independently.

Sources: Peltonen, L.; Roininen, J.; Ahonen, S.; Nupponen, T. and Tuusa, R. (2011). Ilmastonmuutos ja kansalaisosallistuminen. ILMANKOS-hankkeen tutkimus-ja kehittämisosion loppuraportti. [Title in English: Climate Change and Citizen Participation. Final Report of the ILMANKOS Research and Development Project]. Sitran sevityksiä 45. The Finnish Innovation fund Sitra, Helsinki. Available electronically at: http://www.sitra.fi/fi/Julkaisut/sarjat/selvityksia/selvityksia.htm

1 Tampere is a city with about 200,000 inhabitants, situated in inland Finland, some 180 kilometers north of the capital, Helsinki.
2 The covenant is a political commitment for local authorities in EU countries to show their commitment to the ambitious mitigation targets of the EU climate policy. Moreover, the covenant authorities prepare a sustainable energy action plan of detailed measures through which the local authority attempts to reach the targets and reduce its CO_2 emissions. Even if authorities endorse the covenant, its focus is not only on public authorities but also on the local private sector. It also stresses the importance of citizen participation in reaching the objectives. (See www.tampere.fi/tampereinfo/tiedotus/tiedotteet/2009/t090210e.html.)

[ADAPTATION] Box 9.7 Climate-specific governance challenges faced in managing cities: A view from Durban, South Africa

Michael Sutcliffe

eThekwini Municipality, Durban

1. **There is tension between relatively short political and budgeting cycles at the local level and the long-term vision that climate change is requiring of city managers.** For example, 50–100-year planning horizons become necessary to ensure that short-term decisions do not foreclose long-term options and responses. This is a real challenge for urban governance as decision-makers battle to weigh the long-term resilience and sustainability needs of cities appropriately against short-term and more immediate needs. Addressing this disjuncture must be a critical part of any discussion around governance.

2. **In the past, cities have been planned for a fairly constrained set of predictable futures.** Climate change science requires planning for many possible and uncertain futures, and recognizing that the levels of uncertainty associated with these futures may increase. This poses a challenge for institutions to become very flexible and responsive. Such flexibility is not inherent in most local governance or government structures.

3. **Climate change, in many cities around the world, has first been picked up and championed by environmental departments.** It has therefore become stereotyped as an "environmental" issue and not understood as the developmental challenge that it actually is. This limits the ability to ensure that it is effectively mainstreamed into planning and decision-making.

4. **Local government has had limited standing, basically as an observer, in international climate change negotiations.** One of the goals for local governments to achieve in Copenhagen was acknowledgement of local government as a key implementing agent in addressing climate change. At COP16 in Cancun, Mexico, local governments were referenced in the shared vision for long-term cooperative action and in parts of the agreement regarding adaptation and capacity building. They were specifically recognized as governmental stakeholders regarding future arrangements of the intergovernmental proceedings adopted by the Subsidiary Body on Implementation of the UNFCCC, and city-wide programs were included in Clean Development Mechanisms (CDM) studies. This helps from a governance perspective to ensure that national governments acknowledge cities as climate change players and puts local governments more effectively in line for the various international funds that are being discussed.

5. **The challenge of securing funding is particularly acute in terms of the difficulties that local government has in accessing international funds, both available and proposed.** The funding streams in terms of adaptation are particularly important to cities in the global south as they come to bear the brunt of climate change impacts. Cities therefore need access to new and significant funding for adaptation. Unfortunately, Article 4.4 of the United Nations Framework Convention on Climate Change (UNFCCC) clearly states that funding is reserved for the impacts of climate change. Since much of cities' adaptation will be linked to adaptation of existing infrastructure, UNFCCC's funding is not linked to existing adaptation goals. A funding system that links Official Development Assistance (ODA) and adaptation funding, or a realignment of adaptation funding for both existing and new risks, are possible suggestions for addressing this challenge, although they are not without complications.

6. **Residual damage is going to be a complicated challenge for local government.** There is a suggestion that up to two-thirds of the potential major losses from climate change cannot be averted, for example sea level rise, desertification, and ocean acidification, because adaptation is neither economic nor feasible (Parry *et al.*, 2009). This inevitability will place enormous pressure on local patterns of governance and government. Cities' existing systems are not necessarily ready for this challenge.

However, most cities have limited planning capacity and resource commitments targeted to plan, prepare, and implement climate change response activities. Climate change action plans are often costly. The Chicago Climate Action Plan reveals the importance of securing a range of sustainable funding sources where a total of approximately US$2.8 million was contributed by 14 sources from a variety of non-profit foundations, funds, trusts, and initiatives as well as pro bono services, Illinois and Chicago government departments, and regional councils (Parzen, 2008).

In addition, there is an information crisis that seriously undermines effective urban planning (McCarney, 2006). Monitoring and data systems are needed for good planning decisions in cities, particularly cities of the developing world. City planners in poor cities are increasingly concerned with reducing vulnerability to climate change, ensuring emergency preparedness in the event of health risks, creating environmentally friendly cities, creating safer cities by re-designing public space, upgrading slums, and investing assets for pro-poor urban strategies.

Planners working in cities with appropriate resources can draw on toolkits to help decision-makers and the public understand the types of vulnerabilities that are present. Mapping tools in particular can be used to identify infrastructure, buildings, ecosystems, and populations that are vulnerable due their proximity to waterways, wetlands, floodplains, and other potential stressors (Prasad *et al.*, 2009). When combined with the development of scenarios that account for different climatic conditions and impacts such as landslides and floods, cities can use this information to set priorities and develop adaptation plans. The City of London has detailed three climate impacts in the London Climate Change Adaptation Strategy: heat waves, floods, and droughts, each considered as having both a high risk of consequence and vulnerability, as well as increasing probability (Mayor of London, 2008). The London heat wave of 2003, during which 600 residents died, was a motivator for developing a strategy to adapt to and prepare for rising temperatures in the London area. Climate change scenarios have been developed by the Hadley Centre for Climate Prediction and Research that predict that London will see the increased frequency and intensity of extreme weather, as well as a rise in the number of "very hot" summer days. As a result of this data, then Mayor of London developed strategies to mitigate the effects of heat waves. Mayor Livingstone undertook an "urban greening program" that would utilize green spaces, street trees, and urban design to enhance the cooling potential of the city. The mayor also sought to create an "Urban Heat Island Action Area" in which new development would be used to offset the heat island effect. Facilitating access to cool buildings and developing design guidelines for developers and architects were also elements of the key action plan to manage London's response to heat waves (Mayor of London, 2008).

London is prone to tidal flooding from the sea, fluvial flooding from the River Thames, surface water flooding due to the drainage system's inability to handle heavy rainfall, flooding from sewers, and flooding as a result of rising groundwater. Using maps of London that include areas at risk of tidal and fluvial floods the Greater London Authority (GLA) determined that nearly 15 percent of the metropolitan area is at risk from flooding. As a result of this analysis, the GLA proposed a review of the London Strategic Flood Response Plan, as well as improvement of the standard of flood risk management in partnership with the Environment Agency. The urban greening program is also designed to help reduce flooding, as it will improve the permeability of the urban landscape. With rising temperatures comes the possibility of drought, and with each Londoner consuming an average of 168 liters of water per day, the GLA attempted to promote and facilitate the reduction of leakage from water mains, compulsory water metering, retrofitting of London homes, as well as the encouragement of rainwater harvesting and gray water recycling. The GLA proposed publishing a Water Strategy and a Water Action Framework that will achieve a sustainable water supply–demand balance (Mayor of London, 2008).

In wealthier cities such as London, areas identified as high risk can be zoned for zero construction or only for buildings that conform to a highly regulated and appropriate standard. While these regulatory steps might be obvious, their implementation is more difficult to achieve when high-risk zones are already occupied, and different uses, densities, and status of occupation exist (McCarney, 2006). This is particularly true in poorer cities. Poverty forces many people to settle in areas of high risk and return to hazard-prone lands that have already been struck by disasters (Satterthwaite, 2009). Decisions regarding densely populated high-risk zones are contentious and often costly.

One of the crucial ways to effectively mainstream climate adaptation is to link initiatives to development goals. However, this will inevitably lead to contradictions and the need to make tradeoffs between different priorities. Some of these tradeoffs are advancing risk reduction strategies versus affordability, promoting stricter residential building regulations for disaster

resistance and safety versus flexible standards for incremental housing development by the urban poor, and self-help community development of infrastructure versus adherence to universal standards of water and sanitation services designed to avert risks of contamination in crisis situations (McCarney, 2009). Even decisions that address pressing needs can result in inequitable outcomes. For instance, in seeking to find ways to ensure the long-term availability of water to the city in response to climatic events, Quito is faced with challenges of human rights and environmental justice. One of the most controversial projects would improve water delivery to the city by damming 31 rivers. The project has the potential to secure water for the city and suburbs. However, it also is likely to reduce the water resources on which Amazonian indigenous people who live in surrounding areas depend. Examples such as these begin to reveal the need for planning and policy methods that account for the contradictions and inequities that are inherent in measures designed to promote the development of climate-resilient cities.

9.2.4 The challenge of data and measurement: evidence-based policy formulation and monitoring

Cities worldwide are entering into renewed dialogues with state/provincial and national governments to discuss the urban agenda on climate change. Cities are also increasingly engaged in global discussions on climate change. In this context, more rigorous data-driven policy analysis by cities can mean leverage in intergovernmental relations and multi-level governance negotiations.

The vulnerability of cities to climate change is largely underestimated due to lack of standardized data and weak metrics at the city level. There is no established set of city indicators that measures the effects of climate change on cities and assesses those risks, nor is there a comprehensive set of indicators with a common, accepted methodology designed to measure the impact that cities have on climate change and the role that cities play, for example, in contributing to greenhouse gas emissions. The World Bank has defined indicators as performance measures that aggregate information into a useable form. Indicators provide a useful tool in the prospective sense for policymaking and also in the retrospective sense for assessing policy implementation. Indicators also offer assistance to policymakers by aiding in comparison, evaluation, and prediction.

One cluster of challenges relates to how best to localize measurements on climate change. First, cities are responsible for the majority of the world's greenhouse gas emissions yet there is still only a very limited set of comparable measurements of climate change at the city level. While national and global measurements have advanced, a credible and globally standardized measurement for how cities impact climate change is needed. Second, and related to this, is the challenge for cities to also establish a common standard for mitigation targets that will help to lessen cities' impact on climate change. Establishing such targets requires sound research by sector that can help cities to establish benchmarks against which to measure performance

[ADAPTATION/MITIGATION] Box 9.8   Climate action planning in Quito, Ecuador

Isabelle Anguelovski and JoAnn Carmin

Massachusetts Institute of Technology

Located in the Central Andes of South America and surrounded by glaciers, Quito is highly vulnerable to the impacts of climate change. Between 1939 and 1998, the Andean region saw an increase in average temperatures of 0.11°C per decade against a global increase of 0.06°C per decade (The Government of Ecuador, UNDP, and Ministry of Environment, 2008). One impact associated with this change in temperature is that the Antisana glacier shrank by 23 percent between 1993 and 2005 (Maisincho *et al.*, 2007). This is a critical issue since this glacier and its nearby ecosystems supply a large portion of water to the city's 2.1 million inhabitants. Climate change also threatens to destroy the páramos ecosystems that regulate the hydrological system of the city's water basins. Furthermore, climate change is expected to intensify extreme weather events and rainfall in Quito. This is likely to exacerbate landslides and mudslides, stress transportation systems and infrastructure, and endanger indigenous and migrant populations living on the hillsides and slopes (Dirección Metropolitana Ambiental y Fondo Ambiental, 2008).

CLIMATE PLANNING IN QUITO

Planning for climate change in Quito was initiated in late 2006 when the former Mayor, Paco Moncayo, and the Metropolitan Council took the lead in organizing Clima Latino, a climate change conference for the Andean Community of Nations. The October 2007 event was meant to help governments in the region identify appropriate measures for climate mitigation and adaptation (Carmin *et al.*, 2009).

Climate planning became more concrete in January 2007 when Gonzalo Ortiz, a Metropolitan Councillor, gave a presentation to his fellow council members about the need for Quito to take heed of the data on temperature and glacial changes and develop a climate strategy that addressed mitigation and adaptation. With strong support from the Metropolitan Council and the Mayor, Ortiz was empowered to create an Inter-Institutional Commission. In fall 2007, the Inter-Institutional Commission presented its draft climate strategy for Quito to municipal agencies and, a few weeks later, to participants at Clima Latino. They also initiated a metropolitan-wide public consultation process as a means for identifying public concerns and suggestions for the climate strategy. After making revisions based on residents' priorities and ideas, the Inter-Institutional Commission finalized the Quito Strategy for Climate Change (EQCC) in February 2008.

PUBLIC PARTICIPATION IN CLIMATE PLANNING

The Inter-Institutional Commission in charge of the climate strategy hired the environmental NGO ECOLEX to coordinate citizen consultation and organize four workshops across the city in November and December 2007. ECOLEX was asked to engage the local population, particularly vulnerable and historically marginalized communities, as well as key social and community development organizations.

The consultation process resulted in three central concerns being raised by participants. The first concern was the need to improve air quality in Quito. In response, the EQCC includes provisions for improving and extending the public transportation system as a means for decreasing car emissions. Second, was the need to protect homes and property on hillsides from landslides and extreme weather events. The response was to include stipulations in the EQCC for the development of early warning systems and improved emergency preparedness. Further, residents raised concerns about access to potable water, given the shrinking glaciers. This concern is reflected in the EQCC commitments to studying aquifers in Quito's nearby valleys in order to define a new strategy for using these subterranean resources and working with residents to increase efficiency in water usage.

PUBLIC PARTICIPATION IN IMPLEMENTING CLIMATE INITIATIVES

The Quito strategy gives civil society actors a central place in the implementation of climate adaptation measures. Local universities and research centers monitor climate vulnerabilities, especially around the Andean glaciers, and inform decision-makers of changes so municipal adaptation measures can be adjusted as necessary. In addition, some local NGOs received funding to train indigenous farmers to improve the management of water resources in their urban agriculture practices, diversify as well as privilege native crops, and replant native tree species in hillside areas. The NGOs also train indigenous leaders to monitor variations in rainfall and flows from local rivers so that municipal staff members receive up-to-date information on changes in water levels in Quito.

CLIMATE ACTION AS BRICOLAGE

The climate planning and implementation processes in Quito reflect a longstanding commitment that public officials have to ensuring that citizens can participate in decision-making and implementation of policies and programs. Many of the issues raised by residents reflected their concerns for health, environmental quality, security, and safety. Officials and staff addressed these concerns by linking elements of the EQCC to existing priorities for development, especially in the areas of water management, land use, and transportation. It is thus possible to envision climate action as a process of bricolage, one through which planners and public officials find creative ways to respond to the needs and concerns of local residents by linking mitigation and adaptation goals to existing municipal priorities and programs.

in moving towards these targets. Ideally, these benchmarks are established with a globally comparative methodology so that global progress can also be measured in a standard format. A third set of measurement challenges relates to cities and climate change adaptation. Research on risk and vulnerability of cities to climate change needs to inform citizens and policymakers across specified categories of risk at the city level. Data by category of risk and varying degree of vulnerability can then lead to an informed policy agenda on climate adaptation and emergency preparedness.

A second set of challenges for data and improved research on cities and climate change is associated with establishing a globally comparative, standardized set of measures through common methodologies. Climate change is often monitored at global and national levels according to an adopted set of measures globally agreed upon by states. However, similar statistics are rarely collected at the city level and devising indicators on climate change at the city level is proving difficult. Furthermore, when individual cities collect and monitor data on climate change, the information is often collected using methodologies different from other cities and is analyzed and reported on in different ways. This creates further challenges for researchers, planners, and city managers when drawing comparisons across cities globally. The lack of a standardized methodology for devising indicators on climate change at the city level greatly affects the quality of research, planning, and management.

The Global City Indicators Facility (GCIF), first initiated by the World Bank and now managed at the University of Toronto, provides indicators that can assist cities with their mitigation and adaptation efforts in climate change. The GCIF has various indicators, for example, on modal shifts from road transport to rail and public transport and non-motorized transport; waste incineration; wastewater treatment and recycling. Indicators on cities and greenhouse gas emissions are being developed to help create a standard and globally recognized index on cities and greenhouse gases. More research and development of city indicators related to climate change is required. For example, measures to assess mitigation strategies in the energy supply sector, including indicators on renewable energy resources as well as the monitoring of industry practices in cities, need further development. With regard to mitigation strategies in the buildings sector, LEED certification has been a leader in promoting environmentally friendly buildings, and means to assess improvements will help to further transform the building industry.

Indicators on adaptation strategies can help cities assess progress in addressing climate change and areas requiring improvement. With regard to infrastructure, standards and regulations that integrate climate change considerations into design are as yet underdeveloped and measures of performance are not yet identified. In addition, specific land use policies for climate adaptation have not been well addressed. In the health sector, research is required on climate change health impacts necessary for informing local health policy, such as in creating

heat-health action plans. Indicators are also needed to monitor climate-sensitive diseases. More generally, the ability of health services to cope with climate change associated health risks is under-researched. The issue of energy demand (particularly in warmer cities), is shown here to be potentially very significant, especially in economic terms, and this should also be a priority (Hunt and Watkiss, 2007). Climate change impact assessments on water scarcity in cities and how cities can best create adaptation responses warrants further research, and the design of impact measures is needed. Generally, in this evolving field of climate change adaption at the city level, much more work is needed on creating standardized methodologies for measurement of impact assessment and on the evaluation of adaptation responses including the economics of adaptation (Hunt and Watkiss, 2007).

In establishing greenhouse gas reduction targets, cities have an important role to play in helping to determine an equitable distribution of reduction targets, which will help to frame mitigation strategies on climate change. Current debates on per capita emissions between inner city residents and suburban residents, between large city residents and smaller city residents, and between wealthy cities and poorer ones raises issues of equity in sharing the burden in meeting reduction targets. However, measures are weak, and an accepted methodology for determining an equitable distribution of high-level greenhouse gas reduction targets has been established (Miller et al., 2009). While it is generally assumed that suburban residents emit significantly more carbon dioxide than inner city residents, it could thus be concluded that it would be more equitable to require suburban communities to shoulder greater burdens for reductions (Miller et al., 2009). However, credible indicators on this issue are still to be refined. For example, while some estimate that suburban dwellers produce up to three times more greenhouse gases per capita than inner city dwellers, recent data (Glaeser and Kahn, 2008) suggest that this dichotomy is not so simple. They report that while per capita emissions indeed rise as you move away from the urban core of Boston, they level off once you are more than ten miles from downtown. Another exception they have found is with respect to Los Angeles, where emissions are actually lower in suburban LA than they are in the central cities of that metropolitan area. Such issues are complicated further by considering the challenges and opportunities of high-growth versus low-growth communities, as well as questions of per capita versus total reduction targets. In Canada, for example, the Province of British Columbia plans to negotiate with local governments with the goal establishing an equitable allocation on a municipality-by-municipality basis.

Finally, a new set of indicators on climate change mitigation are also needed if policymakers are to assess the capacity in communities for greenhouse gas reductions and what costs related changes would generate – physically, socially, and economically – before they can act. Policymakers need to know, for example, how redesign, modified urban form, and rebuilding of the suburbs might overcome car dependency (Miller et al., 2009).

The use of proximity-to-target methodology that quantitatively measures city-scale performance against a core set of goals, while useful in theory for measuring the distance between a city's targets and current results, providing an empirical foundation for policy benchmarking and providing context for evaluating city performance, is as yet under-developed for climate change indicators at the city level. Nonetheless, it could serve as a powerful tool for steering policy and assessing climate planning and investments in city management.

Advances have been made in environmental performance measures and empirical approaches to assessing global sustainability. The Environmental Performance Index (EPI) developed by the Yale Center for Environmental Law and Policy at Yale University and the Center for International Earth Science Information Network at Columbia University has developed 25 indicators across six policy categories that quantitatively measure country-scale performance on a core set of environmental policy goals. Country-level data and analysis on climate change have improved in recent years, but serious gaps still exist at the city level. Quantitative city data on climate change are being developed by cities in some discreet form, often adapted from these broadly accepted national level methods. However, limited availability of discrete and time series data on cities and climate change hamper efforts to diagnose emerging risks and problems, to assess policy options in terms of both mitigation and adaption strategies, and to gauge the effectiveness of city-level programs. Moreover, globally comparative indicator-based knowledge on cities and climate change is underdeveloped. Standardized indicators on climate change that allow cities to compare themselves globally are useful not for purposes of numerical ranking of cities, but for informing policy decision-making through comparative city data that leverage policy and political strategy (McCarney, 2010).

Building and adopting indicators on climate change can promote more open and transparent governance systems in cities and foster increased citizen engagement. In a review on urban sustainability indicators, Mega and Pedersen (1998) suggest that indicators should aid in decision-making at various levels to promote local information, empowerment, and democracy. They should also contribute to making the city a more important instrument for fostering citizen participation. As with indicators of sustainability, those that focus on climate change mitigation and adaptation can ensure the availability of current information about climate performance and improve policy development and implementation.

Evidence-based policymaking is made possible by advances in information technologies. Data-driven decision-making in the government domain via quantitative performance metrics can serve to measure implementation success rates, steer investments, and refine policy choices. The Global City Indicators Facility (GCIF) provides a system for cities to use globally standardized indicators as a tool for informing policymaking through the use

of international comparisons (McCarney, 2010). For example, the Secretariat of Finance in Bogota uses indicators from the Global City Indicators Facility to track the city's investments and to compare performance relative to other cities. By using indicators and such comparisons, the Secretariat of Finance "is able to evaluate and monitor performance on their investments and to benchmark their performance in comparison to other cities."[1] Similarly São Paulo is demonstrating how governments can use indicators to enhance governance and institute evidence-based policy development city.[2] They report: "the media and civil society are often skeptical of government statistics. As an active member in this global initiative (the GCIF) supported by universities and international organizations, the government of São Paulo is hoping to regain legitimacy and public confidence in government statistics by creating more transparency on its performance in city services and on improving quality of life. The Government of São Paulo recognizes the growing importance of indicators for planning, evaluating and monitoring municipal services. For example, the GCIF indicators were used during the public participation process in preparation of the City's Master Plan, Agenda 2012. The use of indicators to assist with public policy making in São Paulo has opened more effective dialogue between civil society and the local government."

When indicators are well developed and soundly articulated, they can also influence how issues are constructed in the public realm. This is an important lesson related to cities and climate change since information can help to direct behavior in building climate action. Behavioral change can result from publicly accessible information by becoming embedded in the thought and practices, and institutions of users (Innes, 1998). Hezri and Dovers (2006) argue: "as a source of policy change, learning is dependent on the presence of appropriate information with the capacity to change society's behavior" and "community indicator programs or state-of-the environment reporting are usually aimed at influencing the social construction of the policy problem". City indicators on climate change can therefore enhance understanding of the risks associated with climate change, influence opinion and behavior, shape policy, determine priorities, and thereby impact a city's relative contribution to global climate change.

9.2.5 Addressing deeper and enduring risks and long-term vulnerabilities in cities

It is important when addressing climate change risk in cities that a broader framework of risks confronting cities be considered. Cities in the twenty-first century are facing unprecedented challenges. The world's urban population is likely to reach 4.2 billion by 2020, and the urban slum population is expected to increase to 1.4 billion by 2020, meaning one out of every three people living in cities will live in impoverished, over-crowded, and insecure living conditions (McCarney, 2006). Social cohesion, safety, security, and stability are being tested by social exclusion, inequities, and shortfalls in basic services.

---

1 Interview and case study material gathered from City of Bogota: Finance Secretary, 2009.
2 Interview and case study material gathered from City of São Paulo, 2009.

The goal of promoting urban climate resilience is to ensure that settlements are vital and viable into the future. This means that climate change mitigation and adaptation are integral to a larger program of environmental, economic, and social sustainability. From a social point of view, climate resilience reflects the ability that individuals and groups have to adapt to climatic changes and impacts. The ability to cope is related to the availability of resources, particularly financial assets, political power, social status, and personal and professional networks (Adger, 2006). Some people will have the resources to relocate, retain their livelihoods, and maintain their social networks as situations change; however, others will not have the capacity to adapt. For instance, the elderly and infirm may not have the financial or familial resources needed to relocate to new residences. Those who are socially isolated may have difficulty adjusting to the disruptions around them, and individuals who do not speak the official state language may be unable to fully grasp impending threats. While there are many vulnerable populations in urban areas, the poor are at tremendous risk from climate impacts (Carmin and Zhang, 2009).

Poorer urban households are usually more vulnerable due to weaker structures, less protected city locations and building sites, and lack of resilient infrastructure to withstand climate damages. Similarly, the relation between urban health and climate change risks is particularly heightened under conditions of urban poverty in cities. When basic infrastructure is inadequate, existing conditions of poor sanitation and drainage and impure drinking water are further stressed under conditions of extreme weather events and flooding, leading to the transmission of infectious diseases, which puts poor urban households at high risk. This situation is worsened under circumstances of higher densities in urban areas. Cities in developing countries are disproportionately affected for similar reasons of vulnerability and weak institutional support and infrastructure systems (McCarney, 2006). For example, many developing countries lack the health facilities to deal with large numbers of injured patients, resulting in higher death tolls than in countries better equipped for disasters (See chapter 7). Some disasters, which may become more frequent, can paralyze entire cities and regions and permanently destroy their social and economic assets. Leadership in the governance arena is required for the adoption of sound policy on climate resilience in cities, more effective urban management of risks, and more empowered governance at the city level.

"The world Urban Forum III found that severe consequences and threats that cities are now facing as a result of climate change, pressing shortfalls in urban water, sanitation and waste management services, inadequate housing and insecurity of shelter, and the deteriorating quality of air and water in city environments, are being experienced in a context of intense urban growth of cities that increasingly manifests deepening poverty and income inequities, socio-economic exclusion (McCarney, 2006, p.8)."

The adoption, in the year 2000, of the Millennium Development Goals (MDGs) by the UN Member States documents the commitment by the international community to support the development of the poorest regions of the world and to assist the most vulnerable. All eight of the MDGs can be directly connected to the theme of vulnerability in the world's cities. Indeed, it is the world's cities and the slums within them that are pivotal platforms for the successful achievement of each MDG (McCarney, 2006). Goal 7 – to "Ensure Environmental Sustainability" – sets out three targets: to reverse the loss of environmental resources; improve access to safe drinking water; and improve the lives of slum dwellers. Linking these three targets helps to frame the challenges cities face in addressing climate change in a context of poverty.

Reducing poverty is a core challenge for urban governance and in turn addressing the deficiencies in urban infrastructure and services and sub-standard housing of slum dwellers is central to climate change adaptation. The poor have difficulty obtaining provisions and services and often are at risk of illness and death due to their compromised health and nutritional status. These factors will be heightened as climate conditions change and the poor are exposed to greater heat and humidity, higher incidence of disaster, and changing disease vectors (Kasperson and Kasperson, 2001). Urban poor who maintain subsistence lifestyles may find that they are unable to obtain fish from local waterways or fruits and vegetables from open plots as habitats and growing conditions change (Huq et al., 2007).

The situation of poverty in cities worldwide, but in particular in the less-developed regions, must be recognized as a core conditioning factor in addressing climate change and building more climate-resilient cities. This means explicitly recognizing that climate change adaptation must in tandem reduce the vulnerability of the poor in cities. To date, there are few examples of climate adaptation policies and best practices that focus on the needs of the poor or other vulnerable urban populations. However, many development and aid agencies are recommending that climate adaptation activities be aligned with pro-poor development policies. This includes ensuring that the risks to vulnerable populations are minimized and that efforts are made to enhance their capacity for independent action.

9.2.6 The challenge of inclusive governance

Cities worldwide, whether rich or poor, confront the challenge of civic engagement and how to foster an inclusive governance process in their local political environment. Governance invokes more than just political strategy; it demands attention to differentiated social circumstances and needs within the community, to accommodate different cultural values and diversity, and to engage the private sector in the governance platform on climate change.

Social cohesion, safety, security, and stability are being tested by social exclusion, inequities and shortfalls in housing and basic services in cities worldwide. Risks associated with each of these conditions are critical factors in assessing urban risks associated with climate change. Building inclusiveness in local government models is critical to overcoming the core hindrances to social and economic development for citizens.

An inclusive city government that involves long-term residents, international migrants, the poor, marginalized groups, national minorities, and indigenous peoples is fundamental to building safe, livable and climate-resilient cities. The development of new policies and mechanisms for local governance is rooted in strong grassroots participation, citizens and community groups equipped with the understanding of democratic governance to hold local and more senior levels of government accountable, the poorest and most isolated communities' representation in the public debate. Addressing risk in cities depends on a deeper understanding of the relationship between civil society and the state and the cultural competency of local government.

Inclusiveness is a key means of deepening democracy and promoting citizen involvement and social cohesion. When citizens are effectively engaged in their city's development, engaged in everyday decisions and in longer-term planning and policy development, they develop a sense of ownership of and loyalty to the city. So too are citizens more apt to embrace an action agenda on climate change if they are given such opportunities to lay claim to that agenda. If people feel more empowered to shape their own destinies in the city while embracing and participating in forging a common agenda such as climate change, then not only is governance being strengthened but that agenda is more likely to gain political traction.

Engaging citizens in the running of their city has taken many different forms. Typical forms of participatory governance include public consultations, public hearings and meetings, appointing citizens to advisory bodies inside municipal authorities, and designing community councils with stakeholder voice at municipal council sessions. These approaches are being extended to climate mitigation and adaptation efforts. While climate initiatives require the commitment and engagement of local governments, projects and programs driven by non-governmental organizations and communities are starting to emerge as important tools for promoting climate readiness. Environmental organizations have long histories of working on issues now defined as climate mitigation related, such as alternative energy, transportation, and green design. Their ongoing work in ecosystem and natural resources management, as well as the concerns many have about environmental justice, now serve as bridges to adaptation. International development and humanitarian aid organizations traditionally work in the areas of health, human services, and disaster and conflict preparedness and response. Many of these organizations are extending their efforts in these domains to account for changing disease vectors, resource conflicts, and water and food scarcity anticipated to result from climate change (Reeve *et al.*, 2008).

In addition, local citizen groups are also serving as drivers of mitigation and adaptation planning. This has been the case, for instance, in the city of Tatabanya, which is about 50 kilometers from Budapest. The residents of this former mining and industrial town formed a Local Climate Group made up of diverse individuals. In addition to working on an integrated mitigation and adaptation strategy, they have implemented a heat and UV alert program,

organized teams to assist in the development of a local climate strategy, initiated a call for proposals on energy efficient housing, established emissions reduction targets, and implemented educational and information programs (Moravcsik and Botos, 2007).

Governments are critical actors in advancing mitigation and adaptation. However, an emerging approach being used to support government adaptation initiatives is community-based adaptation (CBA). CBA is based on the premises that vulnerability to the impacts of climate change can be altered by drawing on local capacity and that local communities have the ability to assess conditions and foster change. CBA is distinguished from other participatory and collaborative approaches because it takes climate assessment and adaptation as its primary focus (Jones and Rahman, 2007). While CBA has been attempted at limited scales and often in rural locales, it has the potential to be a valuable asset in an urban climate adaptation toolkit.

Valuable research has been undertaken on recent experiments involving citizen engagement in their city's climate change development programs. Case studies on worldwide models of urban governance provide a base for considering next best steps in addressing inclusiveness in cities as they inform a deeper awareness of the intersection between civil society and government and improve our understanding of potential new institutions and paths necessary for fostering inclusiveness, empowerment, and engagement in cities globally (McCarney, 1996).

Finally, engaging the private sector in building climate resilient cities is critical if a city's climate-ready development programs are to gain traction. The private sector plays an important role in urban development and service delivery. Many of the critical urban services that are vulnerable to climate impacts are also privately owned and operated (e.g., water, power, transportation, infrastructure, and occasionally emergency preparedness). In order to be effective, efforts to increase the resilience of these services must include the private sector. Strong cooperation between private operators and public authorities is vital in order to build sustainable cities.

Cities, as centers of commerce, are vulnerable when businesses are adversely affected by climate change. In many countries the private sector is the biggest employer and a significant contributor to national income. The resilience of these businesses is critical to the cities in which they are located. The private sector is likely to be affected by physical exposure to a changing climate, regulatory risks around emissions reduction targets, competition from better-adapted businesses, and by litigation risks or risks to reputation (Llewellyn *et al.*, 2007). Businesses are increasingly aware of the potential impacts of a changing climate on raw materials, supply chains, asset design and performance, markets, products and services, and workforce health and safety (Firth and Colley, 2006).

The insurance industry, for example, has been at the forefront of business activity in assessing climate risks and opportunities, and the sector is already developing risk management processes to minimize costs arising from events driven by climate change. Many insurance companies are actively raising the profile of climate change as a business risk rather than an environmental issue. The Association of British Insurers assessed the financial risks of climate change and warned of the risk of increasing tropical storm activity and its economic impact prior to Hurricanes Katrina and Rita. In the United States, the national insurance regulator has adopted a mandatory requirement for insurance companies to disclose the financial risks they face from climate change, as well as actions the companies are taking to respond to those risks. Other sectors likely to be affected by climate change include utilities, oil and gas, mining and metals, pharmaceuticals, building and construction, and real estate, due to their reliance on global supply chains and large fixed assets. The inter-linking of international capital markets means that businesses – and the cities in which they are located – are vulnerable to climate risks globally as well as domestically (Clarke, 2002).

Inventing new norms of practice and reforming institutional procedures in cities can effectively enhance civil society and private sector involvement and create a politics and culture of inclusiveness that is essential in framing strong local governance for effective climate action in cities.

9.3 Conclusion

In conclusion, the six core governance challenges for cities in confronting climate change can be summarized as: one, a more empowered local governance in both political and fiscal terms; two, addressing jurisdictional boundaries so as to build metropolitan governance systems to better address climate change; three, establishing more effective planning and urban management practices; four, addressing data and measurement through evidence-based policy formulation and monitoring; five, addressing deeper and enduring risks and long-term vulnerabilities in cities, especially related to poverty; and, six, building more inclusive governance. From the discussion of these core challenges, four key ingredients for successful climate action emerge. These are as follows:

1. **Effective leadership** is critical for overcoming fragmentation across departments and investment sectors when building consensus on the climate change agenda in cities. Strong leadership can overcome individualism and competition across political "turf" and build recognition that more metropolitan-wide collective action is empowering at both a national and international levels. The ability to build consensus and coordination better facilitates investments in infrastructure and amenities that make the city more resilient to climate change. Strong leadership in the affairs of metropolitan governance means not only building consensus, but also aggregating these fragmented interests in a way that builds legitimacy and accountability to stakeholders in the process.

2. **Efficient financing** is a core requirement for empowered governance in cities. Success to date with efforts to confront climate change challenges in cities has been hampered due to deficient financing tools at local levels of government. The redistribution of responsibilities between different levels of government has not always been sustained by a corresponding allocation of resources or empowerment to adopt adequate financing tools to raise these resources. If these weaknesses are common at the level of individual municipalities, then the problems of raising finance to support the broader metropolitan areas are compounded. Highly fragmented governance arrangements in many metropolitan areas makes efficient financing for area-wide climate mitigation and adaptation strategies a difficult and on-going challenge. Metropolitan authorities often lack adequate resources for governing and face difficulties in raising new sources. Without a clear, permanent, and sufficient financial mechanism it is indeed quite difficult to implement the principle of territorial solidarity in the metropolitan area in order to redress social and economic inequalities in search of more climate-resilient cities.

3. **Inclusive citizen participation.** Different models of city governance can encompass different forms and degrees of citizen participation. Participation of citizens in decision-making and in the allocation of resources is challenging when principles of transparency and democracy require that the mechanisms of participation are accessible, easy to understand, and utilize simple forms of representation. Community-based adaptation strategies on climate change, transparency in climate change data on cities, and more inclusive local government planning help to build stronger involvement of urban citizens on the climate change agenda.

4. **Jurisdictional coordination** is one of the most pressing challenges common to cities worldwide. This challenge takes two forms: multi-level jurisdictional coordination of services vertically across multiple levels of government and inter-jurisdictional coordination of services horizontally across the metropolitan area. In the case of the former, the inter-governmental relations involved in the governance of cities are often in flux, with extensive and complex decentralization processes in motion in many countries worldwide. Multiple tiers of government and various levels of state agencies are involved in the climate change agenda and vertical coordination is often weak or non-existent. In the case of the latter, existing governing institutions are often horizontally fragmented, uncoordinated, and in many cases ad hoc when it comes to climate change strategy, due to multiple jurisdictional and electoral boundaries that span the territories of vast metropolitan areas. Coordination is fundamental not only in basic sectoral areas such as land, transport, energy, emergency preparedness, and related fiscal and funding solutions, but in addressing issues of poverty and social exclusion through innovative mechanisms of inter-territorial solidarity. Land-use planning across these broad urban regions is a key criterion for effective governance in the arena of climate change strategies. Territorial and spatial strategies are central in addressing climate change risks and building effective mitigation and adaptation strategies. Land use planning in peri-urban areas and the broader hinterland of cities and transport and related infrastructure planning at urban and

regional levels that emphasize territorial or spatial strategies are key functions of metropolitan institutions. Managing transportation in large metropolitan areas is essential for the advancement of the climate change agenda and addressing greenhouse gas emission targets. Transportation investments and services, however, are often implemented, financed, managed, and regulated by different governing institutions and levels of government. Coordination of these processes relies on complex inter-governmental policy networks and organizational management.

REFERENCES

Adger, W. N. (2006). Vulnerability. *Global Environmental Change*, **16**, 268–281.

Alber, G. and K. Kern (2008). Governing Climate Change in Cities: Modes of Urban Climate Governance in Multi-level Systems. In Documentation *Competitive Cities and Climate Change Conference*, Milan, Italy, October 9–10, 2008.

Angel, S., S. C. Sheppard, and D. L. Civco (2005). *The Dynamics of Global Urban Expansion*, Transport and Urban Development Department, Washington, DC, USA: World Bank.

Bai, X. (2007). Integrating global concerns into urban management: the scale argument and the readiness argument. *Journal of Industrial Ecology*, **11**, 51–92.

Betsill, M. M. (2001). Mitigating climate change in U.S. cities: opportunities and obstacles. *Local Environment*, **6**, 393–604.

Bierhanzl, E. J. and P. B. Downing (2004). User charges and special districts. In J. R. Aronson and E. Schwartz (Eds.), *Management Policies in Local Government Finance*, Washington, DC, USA: ICMA.

Blanco, H. (2007). *State Growth Management Experience in the US and Implications for Korea*, Seoul, Korea: Korean Research Institute of Human Settlements.

Blanco, H. and M. Alberti (2009). Building capacity to adapt to climate change through planning. *Progress in Planning*, **71**(3), July 2009.

Bulkeley, H. and K. Kern (2006). Local government and the governing of climate change in Germany and the U.K. *Urban Studies*, **43**, 9522–7322.

Bulkeley, H., H. Schroeder, K. Janda, *et al.* (2009). *Cities and Climate Change: The Role of Institutions, Governance and Urban Planning*. Report prepared for the World Bank Urban Research Symposium on Climate Change, June 28–30, 2009, Marseille, France.

Burby, R. and P. J. May (1997). *Making Governments Plan: State Experiments in Managing Land Use*, Baltimore, MD, USA: Johns Hopkins University Press.

Cameron, R. (2005). Metropolitan restructuring (and more restructuring) in South Africa, *Public Administration and Development*, **25**(4).

Carlson, T. and Dierwechter, Y. (2007). Effects of urban growth boundaries on residential development in Pierce County, Washington. *Professional Geographer*, **59**(2), 209–220.

Carmin, J.-A. and Yan Zhang (2009). Achieving urban climate adaptation in Europe and Central Asia. Policy Working Paper 5088. Background paper for the World Bank report, *Managing Uncertainty: Adapting to Climate Change in ECA Countries*, Washington, DC, USA: World Bank.

Carmin, J.-A., D. Roberts, and I. Anguelovski (2009). Building climate resilient cities: early lessons from early adapters. Paper presented at 5th Urban Research Symposium, Marseilles, France.

Carruthers, J. I. (2002). The Impacts of state growth management programmes: a comparative analysis. *Urban Studies*, **39**(11), 1959–1982.

Clarke, S., *et al.* (2002). *London's Warming: The Impacts of Climate Change on London*. Technical Report.

Clinton Foundation. www.clintonfoundation.org.

DeGrove, J. and D. Miness (1992). *The New Frontier for Land Policy: Planning and Growth Management in the States*. Cambridge, MA, USA: Lincoln Institute of Land Policy.

Dhakal, S. (2004). *Urban Energy Use and Greenhouse Gas Emissions in Asian Mega-Cities: Policies for a Sustainable Future*. Urban Environmental Management Project, Institute for Global Environmental Strategies (IGES), Kangawa, Japan.

Dhakal, S. (2009). Urban energy use and carbon emissions from cities in China and policy implications. *Energy Policy*, **37**, 4208–4279.

Dhakal, S. and M. Betsill (2007). Challenges of urban and regional carbon management and the scientific response. *Local Environment*, **12**, 555–945.

Dirección Metropolitana Ambiental y Fondo Ambiental (2008). *Quito Strategy for Climate Change*. Quito: DMQ.

Dodman, D. (2009). Blaming cities for climate change? An analysis of urban greenhouse gas emissions inventories. *Environment and Urbanization*, **21**, 102–581.

Droege, P. (2002). Renewable energy and the city: urban life in an age of fossil fuel depletion and climate change. *Bulletin of Science, Technology & Society*, **22**, 78–99.

Fernandez, Edésio (2001). *New Statute Aims to Make Brazilian Cities More Inclusive*. In: HABITAT Debate, Kenya, Nairobi. Vol. 7, No. 4, p.19.

Firth, J. and M. Colley (2006). *The Adaptation Tipping Point: Are UK Businesses Climate Proof?* Oxford, UK: Acclimatise and UKCIP.

Fleming, P. D. and P. H. Webber (2004). Local and regional greenhouse gas management. *Energy Policy*, **32**, 761–771.

Glaeser, E. L. and M. Kahn (2008). *The Greenness of Cities*. Cambridge, MA, USA: Rappaport Institute and Taubman Center.

Hamilton, D. (1999). *Governing Metropolitan Areas*. New York, USA: Garland Publishing.

Harrison, K. and L. McIntosh Sundstrom (2007). The comparative politics of climate change. *Global Environmental Politics*, **7**, 1–81.

Hezri, A. A. and Dovers, S. R. (2006). Sustainability indicators, policy and governance: issues for ecological economics. *Ecological Economics*, (60)**1**, 86–99.

Hunt, A. and P. Watkiss (2007). *Literature Review on Climate Change Impacts on Urban City Centres: Initial Findings*. OECD ENV/EPOC/GSP (2007) 10. Paris. 53pp.

Huq, S., S. Kovats, H. Reid, and D. Satterthwaite (2007). Editorial: Reducing risks to cities from disasters and climate change. *Environment and Urbanization*, **19**(1), 3–15.

ICLEI (2010). *Local Government Climate Roadmap*. Local Governments for Sustainability. Accessed http://www.iclei.org/index.php.id=7694

Innes, J. E. (1998). Information in communicative planning. *Journal of the American Planning Association* **64**(1), 52–63.

Jones, R. and A. Rahman (2007). Community-based adaptation. *Tiempo: A Bulletin on Climate and Development*, **64**, 17–19.

Jones, E., M. Leach, and J. Wade (2000). Local policies for DSM: the UK's home energy conservation act. *Energy Policy*, **28**, 201–211.

Kasperson, J. X. and Kasperson, R.E. (Eds.) (2001). *Global Environmental Risk*. Tokyo, Japan: United Nations University Press.

Kern, K., S. Niederhafner, S. Rechlin, and J. Wagner (2005). *Kommunaler Klimaschutz in Deutschland: Handlungsoptionen, Entwicklung und Perspektiven*. WZB Discussion Paper SP IV 2005–101.

Kessler, E., N. Prasad, F. Ranghieri, *et al.* (2009). *Climate Resilient Cites: A Primer on Reducing Vulnerabilities to Disasters*, Washington, DC, USA: World Bank.

King County (2007). *2007 King County Climate Plan*, Seattle, WA, USA: King County Government.

Klink, J. (2007). Recent perspectives on metropolitan organization, functions and governance. In E. Rojas, J. Cuaudrado-Roura, and F. J. Guell (Eds.), *Governing the Metropolis*, Washington, DC, USA: IADB.

Kousky, C. and S. H. Schneider (2003). Global climate policy: will cities lead the way? *Climate Policy*, **3**, 359–372.

Lee, T. (2010). *Global City and Climate Change Networks*. Accessed http://www.csss.washington.edu/StudentSem/TaedongLee.pdf

Lefèvre, C. (2007). Democratic governability of metropolitan areas: international experiences and lessons for Latin American cities. In E. Rojas,

J. Cuaudrado-Roura, and F. J. Guell (Eds.), *Governing the Metropolis*, Washington, DC, USA: IADB.

Llewellyn, J., C. Chaix, and J. Giese (2007). *The Business of Climate Change: Challenges and Opportunities*. New York, USA: Lehman Brothers.

Lutsey, N. and D. Sperling (2008). America's bottom-up climate change mitigation policy. *Energy Policy*, **36**, 673–685.

Maisincho, L., *et al.* (2007). *Glaciares del Ecuador: Antisana y Carihuayrazo, Informe del año 2005*. IRD-INAMHI-EMAAP-Q.

Mayor of London (2008). *The London Climate Change Adaptation Strategy*. London, UK: Greater London Authority.

McCarney, P. L. (1996). *Cities and Governance: New Directions in Latin America, Asia and Africa*. Toronto, ON, Canada: University of Toronto Press.

McCarney, P. L. (2006) *Our Future: Sustainable Cities - Turning Ideas into Action. Background Paper*. World Urban Forum UN-HABITAT. Nairobi, Kenya. 44pp.

McCarney, P. L. (2009). City indicators on climate change: implications for policy leverage and governance. Paper prepared for the World Bank's 5th Urban Research Symposium, Marseilles, France.

McCarney, P. L. and R. E. Stren (2008). Metropolitan governance: governing in a city of cities. In *State of the World's Cities Report*, Nairobi, Kenya: UN-HABITAT.

McCarney, P. (2010). *Conclusions: Governance Challenges in Urban and Peri-urban Water and Sanitation Services: Policy, Planning and Method*. DOI 10.1007/978-90-481-9425-4_13. Springer Science Business Media B.V. pp. 277–297.

Mega, V. and J. Pedersen (1998). *Urban Sustainability Indicators*. The European Foundation for the Improvement of Living and Working Condition.

Miller, N., D. Cavens, P. Condon, and R. Kellet (2009). *Policy, Urban Form, and Tools for Measuring and Managing Greenhouse Gas Emissions: The North American Problem*. University of Colorado Law Review 977. 13 pp.

Moravcsik, A. and B. Botos (2007). Tatabanya: local participation and physical regeneration of derelict areas. Presentation given in Krakow, Poland.

Nelson, A. C. and T. Moore (1996). Assessing growth management policy implementation: case study of the United States' leading growth management state. *Land Use Policy* **13**(4), 241–259.

Parry *et al.*, (2009). *Assessing the Costs of Adaptation to Climate Change: A Review of the UNFCCC and Other Recent Estimates*, International Institute for Environment and Development and Grantham Institute for Climate Change.

Parzen, J. (2008). *Lessons Learned: Creating the Chicago Climate Action Plan*. Chicago Climate Action Plan.

Prasad, N., F. Ranghieri, F. Shah, Z. Trohanis, E. Kessler, and R. Sinha. 2009. *Climate Resilient Cities: A Primer on Reducing Vulnerabilities to Disasters*. The World Bank. Washington, DC. 186pp.

Reeve, K., I. Anguelovski, and J.-A. Carmin (2008). *Climate Change Campaigns of Transnational NGOs: Summary of Survey Results*, Cambridge, MA, USA: Department of Urban Studies and Planning.

Rezessy, S., K. Dimitrov, D. Ürge-Vorsatz, and S. Baruch (2006). Municipalities and energy efficiency in countries in transition: review of factors that determine municipal involvement in the markets for energy services and energy efficient equipment, or how to augment the role of municipalities as market players. *Energy Policy*, **34**, 223–237.

Romero-Lankao, P. (2007). *How do Local Governments in Mexico City Manage Global Warming? Local Environment,* **12**(5), 519–535.

Satterthwaite, D. (2009). Social aspects of climate change in urban areas in low- and middle- income nations. Paper prepared for the World Bank 5th Urban Research Symposium, Marseilles, France.

Schreurs, M. A. (2008). From the bottom up: local and subnational climate change politics. *The Journal of Environment and Development*, **17**, 343–355.

Schroeder, H. and H. Bulkeley (2009). Global cities and the governance of climate change: what is the role of law in cities? *Fordham Urban Law Journal*, 313–359.

Secretaria del Medio Ambiente del Distrito Federal (2008). *Mexcio City Climate Action Program 2008–2012*. Mexico City.

Seltzer, E. (2004). It's not an experiment: regional planning at Metro, 1990 to the present. In C. P. Ozawa (Ed.), *The Portland Edge: Challenges and Successes in Growing Communities*, Washington, DC, USA: Island Press.

Sippel M. and T. Jenssen (2009). *What about local governance? A review of promise and problems*. MPRA Paper No. 20987. https://mpra.ub.uni-muenchen.de/20987.

Sorensen, A. (2001). Subcentres and satellite cities: Tokyo's 20th century experience of planned polycentrism. *International Planning Studies*, **6**(1), 9–32.

Stren, R. (2007). Urban governance in developing countries: experiences and challenges. In *Governing Cities in a Global Era: Urban Innovation, Competition, and Democratic Reform*, R. Hambleton and J. Gross (Eds.), New York, USA: Palgrave Macmillan.

Swope, C. (2007). Local warming. *Governing*, December, 2007.

The Government of Ecuador, UNDP, and Ministry of Environment (2008). *Adaptation to Climate Change through an Effective Governance of Water in Ecuador*. Ministry of the Environment: Quito.

United Nations (2008a). *State of the World's Cities 2008/2009*, Nairobi, Kenya: UN-HABITAT.

United Nations (2008b). *The Millennium Development Goals Report*, New York, USA: United Nations.

United States Environmental Protection Agency. National Environmental Performance Track. Available online: www.Epa.gov

Ürge-Vorsatz, D., S. Koeppel, and S. Mirasgedis (2007). Appraisal of policy instruments for reducing buildings' CO_2 emissions. *Building Research & Information*, **35**, 774–854.

Wassmer, R. W. (2006). The influence of local urban containment policies and statewide growth management on the size of United States urban areas. *Journal of Regional Science*, **46**(1), 25–66.

Weitz, J. and T. Moore, (1998). Development inside urban growth boundaries: Oregon's empirical evidence of contiguous urban form. *Journal of the American Planning Association*, **64**(4), 424–40.

Wilbanks, T. J. and R. W. Kates (1999). Global change in local places: how scale matters. In: *Climatic Change*, **43**, 601–628.

Yin, M., and Sun. (2007). The impacts of state-growth management programs on urban sprawl in the 1990s. *Journal of Urban Affairs*, **29**(2), 149–179.

Conclusion: Moving forward

Scientists and stakeholders: Key partners in urban climate change mitigation and adaptation

This volume is focused on addressing an urgent demand on the scientific community to provide new and timely information about how climate change is already affecting and will continue to affect urban areas, and how cities are responding to the challenge. Decision-makers need to know how hot their cities will become, how hydrological regimes may change, and the most effective ways to both adapt to and mitigate climate change, among many other questions.

One way forward is the creation of a process embodied by this report through which urban researchers can, over time, provide updated information and data to city decision-makers. Such an effort provides a similar science-based foundation for cities that the Intergovernmental Panel on Climate Change (IPCC) provides for countries. The Urban Climate Change Research Network (UCCRN) is an international coalition of researchers linking scholars and policy-makers in cities of all sizes throughout the world, focusing on cutting-edge science, science-policy linkages, and local mitigation and adaptation capacity. The UCCRN brought together approximately 100 authors from more than 50 cities to create this volume. In many ways it serves as both a touchstone of the current state of urban climate change science, and as precursor of even more comprehensive, integrative and collaborative work in the future. We are not alone in undertaking this work, however. Other scientific initiatives with a similar focus on cities and climate change include the ten-year Urbanization and Global Environmental Change (UGEC) project of the Human Dimensions Programme, established in 2005. The UN-Habitat's 2011 Global Report on Human Settlement also focuses on climate change and cities.

Building on these efforts, we hope that they could coalesce into an on-going series of ARC3 assessments by urban climate change researchers from small, medium, large, and mega-cities in both developing and developed countries. The ARC3 assessments would respond to the needs of urban decision-makers for practical and timely information on both mitigation and adaptation. At the same time they would provide the critical benchmarking function that will enable cities to learn over time as climate change and climate change responses unfold.

On the policy side, we see as strong allies governmental and stakeholder organizations that seek to motivate and support city-level action. The climate-related work of ICLEI-Local Governments for Sustainability, the C40-Large Cities Climate Group, the World Mayors Council on Climate Change, United Cities and Local Governments, the World Bank, UN-Habitat, and the OECD are all playing major roles in encouraging mitigation and adaptation efforts by local governments around the world. Putting emerging climate science into practice also has experienced significant forward movement as city leaders have been willing and able to take direct action to reduce greenhouse gas emissions, protect their cities against climate change impacts, and make their cities more sustainable. Evidence of this is documented time and again in this volume.

The UCCRN welcomes readers of this volume to directly comment on the usefulness of the information and areas for potential improvement, and to define their potential interest in contributing to the next report. Widening the network of linked collaborators and stakeholders will not only strengthen future research, it will expand opportunities for bringing cutting-edge science to bear as cities – the first responders – take action on climate change challenges

Contact: www.uccrn.org

Appendix A

City case studies and topics in vulnerability, adaptation and mitigation

Appendix B

Acronyms and Abbreviations

AGBA	Aglomerado Gran Buenos Aires (Greater Buenos Aires Agglomeration)
ALM	Advanced Locality Management, Mumbai, India
AMO	Atlantic Multidecadal Oscillation
AOGCM	Atmosphere-ocean general circulation model
AR4	Fourth Assessment Report (IPCC)
ARC3	First UCCRN Assessment Report on Climate Change and Cities
ASR	Aquifer storage and recovery
ASTRA	Developing Policies and Adaptation Strategies to Climate Change in the Baltic Sea Region
BaltCICA	Climate Change: Impacts, Costs and Adaptation in the Baltic Sea Region
BCCR	Bjerknes Centre for Climate Research
BREEAM	Building Research Establishment's Environmental Assessment Method, UK
BRT	Bus rapid transport
BTC	Belgian Technical Cooperation
CABA	Ciudad Autonoma de Buenos Aires (City of Buenos Aires)
CADWR	California Department of Water Resources
CAP	Clean Air Partnership, Canada
CBA	Community-based adaptation
CCAP	Center for Clean Air Policy, Washington, DC
CCCDF	Canada Climate Change Development Fund
CCCI	Cities and Climate Change Initiative (UN-HABITAT)
CCCMA	Canadian Center for Climate Modeling and Analysis
CCME	Canadian Council of Ministers of the Environment
CCP	Cities for Climate Protection
CCSR	Columbia University Center for Climate Systems Research
CDM	Clean Development Mechanism
CER	Certified emission reduction
CH_4	Methane
CHP	Combined heat and power
CIDA	Canadian International Development Agency
CIG	Climate Impacts Group, University of Washington
CLACC	Capacity Strengthening of LDCs for Adaptation to Climate Change (IIED)
CMIP	Coupled Model Intercomparison Project
CNG	Compressed natural gas
CNRM	Centre National de Rechesches Meteorolgiques (National Weather Research Centre) global climate model, Météo-France
CO_2	Carbon dioxide
CO_2-eq	Carbon dioxide equivalent
CONAMA	Comisión Nacional del Medio Ambiente, Chile (Chilean National Commission for the Environment)
COP15	15th Conference of Parties (2009), Copenhagen (UNFCCC)
COP16	16th Conference of Parties (2010), Cancun (UNFCCC)
CORDEX	Coordinated Regional Downscaling Experiment
CSH	Code for Sustainable Homes, UK
CSIRO	Commonwealth Scientific and Industrial Research Organization, Australia
CSO	Combined sewer overflow
CUNY	City University of New York
CUSCCRR	Coalition of Urban Sustainability and Climate Change Response Research, Australia
CVCCCM	Centro Virtual de Cambio Climático (Virtual Center on Climate Change) Mexico City
DG	Distributed forms of power generation and distribution
DGA	Dirección General de Aguas (National Water Authority), Chile
DoE	Department of Energy, USA
DOHMH	Department of Health and Mental Hygiene, New York City
ECHAM5	Max Planck Institute for Meteorology global climate model (5th generation)
EEA	European Environment Agency
EIA	Energy Information Administration, USA
ENSO	El Niño–Southern Oscillation
EPA	Environmental Protection Agency, USA
ERDF	European Regional Development Fund
EU-ETS	European Emissions Trading Scheme
FAO	Food and Agriculture Organization of the United Nations
FCV	Fuel cell vehicle
FEMA	Federal Emergency Management Agency, USA
FHWA	Federal Highway Administration, USA
GCIF	Global City Indicators Facility
GCM	Global climate model
GDP	Gross domestic product
GEF	Global Environment Facility
GFDL	Geophysical Fluid Dynamics Laboratory, USA
GHG	Greenhouse gas
GHMC	Greater Hyderabad Municipal Corporation, India
GIS	Geographical Information System
GISS	NASA Goddard Institute for Space Studies
GRIHA	Green Rating for Integrated Habitat Assessment
GTK	Geological Survey of Finland
GW	Gigawatt
HadCM3	Hadley Centre for Climate Prediction global climate model
HARS	Heat alert and response systems
HFA	Hyogo Framework for Action (UNISDR)
HHW	High high-water
HUA	Hyderabad urban agglomeration
ICE	Internal combustion engine
ICLEI	International Council for Local Environmental Initiatives (now ICLEI-Local Governments for Sustainability)
IDF	Intensity-duration-frequency precipitation curve
IEA	International Energy Agency
IEC	International Electrotechnical Commission
IIED	International Institute for Environment and Development, UK
INAM	Instituto Nacional de Meteorologia (National Institute of Meteorology), Mozambique
INDEC	Instituto Nacional de Estadística y Censos (National Institute for Statistics and Census), Argentina
INGC	Instituto Nacional de Gestão de Calamidades (National Institute for Disasters Management), Mozambique
INMCM	Institute for Numerical Mathematics global climate model, Russia
IOD	Indian Ocean Dipole
IPCC	Intergovernmental Panel on Climate Change
IPSL	Institut Pierre Simon Laplace global climate model, France
IPUF	Instituto de Planejamiento Urbano de Florianópolis (Urban Planning Institute of Florianópolis), Brazil
ISET	Institute for Social and Environmental Transition, Boulder, CO, USA
ITCZ	Inter-Tropical Convergence Zone
JUCCCE	Joint US-China Collaboration on Clean Energy
KCC	Kampala City Council, Uganda
LAC	Latin America and the Caribbean
LCCP	London Climate Change Partnership
LDCs	Least developed countries

LECZ	Low-elevation coastal zone	PIUBACC	Global Climate Change Research Program, Buenos Aires University
LEED	Leadership in Energy and Environmental Design	PMC	Pune Municipal Corporation, India
LGA	Local government areas	PNA	Pacific–North American teleconnection pattern
LiDAR	Light Detection and Ranging	PPP	Public-private partnership
LMDGP	Lagos Metropolitan Development and Governance Project, Nigeria	PSA	Pacific-South American teleconnection pattern
LUKAS	Local Urban Knowledge Arenas	RCM	Regional climate model
LWC	Lagos Water Corporation, Nigeria	RFID	Radio-frequency identification
		RISA	Regional Integrated Science and Assessments (NOAA)
MDG	Millennium Development Goal	RMS	Región Metropolitana de Santiago (Santiago Metropolitan Region), Chile
METROS	Metropolitan Environmental Temperature and Rainfall Observation System		
MICOA	Ministry of Coordination of Environmental Affairs, Mozambique	SAM	Southern Annular Mode
MINVU	Ministerio de Vivienda y Urbanismo (Ministry of Housing and Urban Development), Chile	SEAREG	Sea Level Change Affecting the Spatial Development in the Baltic Sea Region Project
MIROC	Frontier Research Center for Global Change, global climate model, Japan	SOI	Southern Oscillation Index
		SPM	Suspended particulate matter
MIUB	Meteorological Institute, University of Bonn, Germany	SRES	Special Report on Emissions Scenarios, IPCC
MJO	Madden–Julian Oscillation	SUD-Net	Sustainable Urban Development Network
MMC	Maputo Municipal Council, Mozambique		
MoE	Ministry of Environment, Tokyo	TDM	Transportation demand management
MPI	Max Planck Institute for Meteorology, Germany	TERI	The Energy and Resources Institute, India
MRI	Meteorological Research Institute, Japan	TMG	Tokyo Metropolitan Government
MTA	Metropolitan Transportation Authority, New York City	TOD	Transit-oriented development
MW	Megawatt	TPH	Toronto Public Health, Canada
N$_2$O	Nitrous oxide	UCCRN	Urban Climate Change Research Network
NAO	North Atlantic Oscillation	UFSMP	Urban Forest Strategic Management Plan
NAPA	National Adaptation Program of Action (UNFCCC)	UHI	Urban heat island
NARCCAP	North American Regional Climate Change Assessment Program	UNDP	United Nations Development Programme
		UNEP	United Nations Environment Programme
NASA	National Aeronautics and Space Administration, USA	UNFCCC	United Nations Framework Convention on Climate Change
NCAR	National Center for Atmospheric Research, USA		
NGO	Non-governmental organization	UNFPA	United Nations Population Fund (formerly United Nations Fund for Population Activities)
NOAA	National Oceanic and Atmospheric Administration, USA	UN-HABITAT	United Nations Human Settlements Programme
NPCC	New York City Panel on Climate Change	UNISDR	United Nations International Strategy for Disaster Reduction
NSF	National Science Foundation		
NYCDEP	New York City Department of Environmental Protection	USACE	United States Army Corps of Engineers
		USAID	United States Agency for International Development
OECD	Organization for Economic Cooperation and Development	UWMP	Urban Water Management Plan, California
OPEC	Organization of the Petroleum Exporting Countries	VKT	Vehicle kilometers traveled
		VMT	Vehicle miles traveled
PAYD	Pay-as-you-drive		
PCM	Parallel Climate Model (NCAR)	WBCSD	World Business Council for Sustainable Development
PCMDI	Program for Climate Model Diagnosis and Intercomparison, California	WCRP	World Climate Research Program
		WHO	World Health Organization
PCP	Plan Climat de Paris (Paris Climate Plan), France	WMO	World Meteorological Organization
PDO	Pacific Decadal Oscillation	WPCP	Water pollution control plant
PHAC	Public Health Agency of Canada	WRI	World Resources Institute
PIEVC	Public Infrastructure Engineering Vulnerability Committee, Canada		

Appendix C

UCCRN Steering Group, ARC3 authors, and reviewers

Steering Group

BRESSAND, Albert
Center for Energy, Marine Transportation and Public Policy,
Columbia University
New York City, USA

CONNELL, Richenda
Acclimatise
London, UK

DROEGE, Peter
School of Architecture and the Built Environment, University of Newcastle
Newcastle, Australia

GRIMM, Alice
Universidade Federal do Paraná
Curitiba, Brazil

HUQ, Saleemul
Climate Change Group, International Institute for Environment and
Development
London, UK

LIGETI, Eva
Clean Air Partnership
Toronto, Canada

NATENZON, Claudia E.
Universidad de Buenos Aires
Buenos Aires, Argentina

OMOJOLA, Ademola
University of Lagos
Lagos, Nigeria

SANCHEZ-RODRIGUEZ, Roberto
El Colegio de la Frontera Norte
Tijuana, Mexico

SCHULZ, Niels
International Institute for Applied Systems Analysis (IIASA)
Vienna, Austria

Coordinating lead authors and lead authors

BADER, Daniel
Center for Climate Systems Research, Columbia University
New York City, USA

BARATA, Martha
Instituto Oswaldo Cruz, Fiocruz
Rio de Janeiro, Brazil

BLAKE, Reginald
CUNY, New York City College of Technology
New York City, USA/Kingston, Jamaica

BLANCO, Hilda
University of Southern California
Los Angeles, USA

CARMIN, JoAnn
Massachusetts Institute of Technology
Cambridge, USA

CECIL, L. DeWayne
US Geological Survey, Global Change Program Office
Idaho Falls, USA

COLLEY, Michelle
Acclimatise
Montreal, Canada

CONNELL, Richenda
Acclimatise
London, UK

DE SIMONE, Gregorio
Instituto Oswaldo Cruz, Fiocruz
Rio de Janeiro, Brazil

DETTINGER, Michael
US Geological Survey/Scripps Institution of Oceanography
San Diego, USA

DHAKAL, Shobhakar
Institute for Global Environmental Strategies (IGES)
Tsukuba, Japan

DICKINSON, Thea
Clean Air Partnership
Toronto, Canada

FOLORUNSHO, Regina
Nigerian Institute for Oceanography and Marine Research
Lagos, Nigeria

GAFFIN, Stuart
Center for Climate Systems Research, Columbia University
New York City, USA

GERÇEK, Haluk
Istanbul Technical University
Istanbul, Turkey

GILBRIDE, Joseph
Center for Climate Systems Research, Columbia University
New York City, USA

GONZALEZ, Richard
Urban Design Lab, Columbia University
New York City, USA

GRIMM, Alice
Universidade Federal do Paraná
Curitiba, Brazil

HAMMER, Stephen A.
Mesacosa, LLC
New York City, USA

HANSON, Randall T.
US Geological Survey
San Diego, USA

HERVE-MIGNUCCI, Morgan
CDC Climat
Paris, France

HORTON, Radley
Center for Climate Systems Research,
Columbia University
New York City, USA

HYAMS, Michael
Center for Energy, Marine Transportation and Public Policy,
Columbia University
New York City, USA

ICHINOSE, Toshiaki
Nagoya University/National Institute for Environmental Studies
Nagoya, Japan

JACK, Darby
Mailman School of Public Health, Columbia University
New York City, USA

JACOB, Klaus
Lamont-Doherty Earth Observatory, Columbia University
New York City, USA

JIONG, Shu
East China Normal University
Shanghai, China

KEIRSTEAD, James
Imperial College
London, UK

LEFEVRE, Benoit
Institute for Sustainable Development and
International Relations
Paris, France

LIGETI, Eva
Clean Air Partnership
Toronto, Canada

MAJOR, David C.
Center for Climate Systems Research,
Columbia University
New York City, USA

MCCARNEY, Patricia
University of Toronto
Toronto, Canada

MEHROTRA, Shagun
Earth Institute at Columbia University
New York City, USA/Delhi, India

MITCHELL, Jeanene
University of Washington
Seattle, USA

NATENZON, Claudia E.
Universidad de Buenos Aires
Buenos Aires, Argentina

OMOJOLA, Ademola
University of Lagos
Lagos, Nigeria

PARNELL, Susan
University of Cape Town
Cape Town, South Africa

PARSHALL, Lily
Earth Institute at Columbia University
New York City, USA

PENNEY, Jennifer
Clean Air Partnership
Toronto, Canada

RAHMAN, Mizanur
International Institute for Environment and
Development
Dhaka, Bangladesh

ROSENZWEIG, Cynthia
NASA Goddard Institute for Space Studies/
Columbia University
New York City, USA

SANCHEZ-RODRIGUEZ, Roberto
El Colegio de la Frontera Norte
Tijuana, Mexico

SCHMIDT, Marco
Technische Universität Berlin
Berlin, Germany

SCHULZ, Niels
International Institute for Applied Systems
Analysis (IIASA)
Vienna, Austria

SETO, Karen C.
Yale School of Forestry and
Environmental Studies
New Haven, USA

SOLECKI, William D.
Institute for Sustainable Cities, City University of New York
New York City, USA

SRINIVASAN, Sumeeta
Harvard University
Cambridge, USA

ZIMMERMAN, Rae
New York University
New York City, USA

Case study authors*

ANGUELOVSKI, Isabelle
Department of Urban Studies and Planning, MIT
Cambridge, USA

ARIKAN, Yunus
Cities Climate Center
ICLEI–Local Governments for Sustainability
Bonn, Germany

BAPTISTA, Sandra
Center for International Earth Science Information Network
(CIESIN), Columbia University
New York City, USA

BARTH, Bernhard
UN-HABITAT
Nairobi, Kenya

*Some case studies in the book have been contributed by lead authors.

BARTON, Jonathan
Instituto de Estudios Urbanos y Territoriales,
Pontificia Universidad Católica de Chile
Santiago, Chile

BERRY, Peter
Health Canada
Ottawa, Canada

BLACK, John
Coalition of Urban Sustainability and Climate Change
Response Research
Sydney, Australia

BUDDE, Martin
Potsdam Institute for Climate Impact Research
Potsdam, Germany

CAMERON, Chris
Wellington City Council Principal Advisor Climate Change
Wellington, New Zealand

CARBONELL, Armando
Lincoln Institute of Land Policy
Cambridge, USA

CHANDIWALA, Smita
Oxford Brookes University
Oxford, UK

CHOI, Young-Soo
Climate Change Department, Seoul Metropolitan Government
Seoul, South Korea

CONDE, Cecilia
Centro de Ciencias de la Atmosfera, UNAM
Mexico City, Mexico

DE SHERBININ, Alex
Center for International Earth Science Information Network (CIESIN),
Columbia University
New York City, USA

DODMAN, David
International Institute for Environment and Development
London, UK

DOUST, Ken
Coalition of Urban Sustainability and Climate Change
Response Research
Sydney, Australia

ENARSSON, Lisa
Department of Environment, City of Stockholm
Stockholm, Sweden

ESTRADA, Francisco
Centro de Ciencias de la Atmosfera, Universidad Nacional
Autónoma de Mexico
Mexico City, Mexico

GHORRA-GOBIN, Cynthia
National Scientific Research Center (CREDA)
Paris, France

GOWER, Stephanie
Healthy Public Policy Term, Toronto Public Health
Toronto, Canada

GUPTA, Rajat
Oxford Brookes University
Oxford, UK

HANNILA, Juhani
Technical Service Centre, City of Kokkola
Kokkola, Finland

HANSJüRGENS, Bernd
Helmholtz Centre for Environmental Research
Leipzig, Germany

HEINRICHS, Dirk
German Aerospace Center
Leipzig, Germany

HOGAN, Daniel, J.
State University of Campinas
Campinas, Brazil

IIZUMI, Ayako
United Nations University
Tokyo, Japan

JENSSEN, Till
University of Stuttgart, Institute of Energy Economics
and Rational Energy Use
Stuttgart, Germany

JUNIOR, Paulo
UN-HABITAT
Maputo, Mozambique

KIM, Kwi-Gon
Seoul National University
Seoul, South Korea

KIT, Oleksandr
Potsdam Institute for Climate Impact Research
Potsdam, Germany

KOLKER, Alexander
Tulane University, Louisiana Universities Marine Consortium
New Orleans, USA

KOOJO, Charles
Urban Research and Training Consultancy (URTC)
Kampala, Uganda

KOS, Paul
Wellington City Council Senior Strategic Advisor
Urban Development
Wellington, New Zealand

LALANDE, Christophe
UN-HABITAT
Esmeraldas, Ecuador

LEE, Taedong
City University of Hong Kong
Hong Kong, China

LUDEKE, Matthias
Potsdam Institute for Climate Impact Research
Potsdam, Germany

LWASA, Shuaib
International Potato Center
Kampala, Uganda

MABIIRIZI, Frank
National Planning Authority
Kampala, Uganda

MAMONONG, Laids
UN-HABITAT
Sorsogon, Philippines

MARTINEZ, Benjamin
Centro de Ciencias de la Atmosfera, Universidad Nacional
Autónoma de Mexico
Mexico City, Mexico

MEFFERT, Douglas J.
Tulane University
New Orleans, USA

MEYER-OHLENDORF, Lutz
Potsdam Institute for Climate Impact Research
Potsdam, Germany

MUKWAYA, Paul
Makerere University
Kampala, Uganda

NELSON, Stephen A.
Tulane University
New Orleans, USA

NJENGA, Cecilia
UN-HABITAT
Nairobi, Kenya

OBALLA, Bridget
UN-HABITAT
Maputo, Mozambique

PELTONEN, Lasse
Centre for Urban and Regional Studies, Aalto University
Helsinki, Finland

PETROVIC, Nenad
Wellington City Council
Wellington, New Zealand

PLUNZ, Richard
Urban Design Lab, Columbia University
New York City, USA

RECKIEN, Diana
Potsdam Institute for Climate Impact Research
Postdam, Germany

REUSSWIG, Fritz
Potsdam Institute for Climate Impact Research
Postdam, Germany

ROMERO-LANKAO, Patricia
National Center for Atmospheric Research
Boulder, USA

SALMANIEGO, Joseluis
United Nations Economic Commission for Latin America
and the Caribbean
Santiago, Chile

SCHMIDT-THOMÉ, Philip
Geological Survey of Finland
Espoo, Finland

SEKIMPI, Deogratious
Uganda National Association of Community and Occupational Health
Kampala, Uganda

SIPPEL, Maike
University of Stuttgart, Institute of Energy Economics and
Rational Energy Use
Stuttgart, Germany

SOLECKI, Stephen
C. W. Post, Long Island University
New York City, USA

SURJAN, Akhilesh
United Nations University
Tokyo, Japan

SUTCLIFFE, Michael
eThekwini Municipality
Durban, South Africa

SUTTO, Maria Paola
Urban Design Lab, Columbia University
New York City, USA

TUUSA, Ruusu
Centre for Urban and Regional Studies, Helsinki University of Technology
Helsinki, Finland

Reviewers[1,2]

BERRY, Peter, Health Canada
BROWN, Marilyn, Georgia Institute of Technology
CAMPBELL, Monica, Toronto Public Health
FISK, David, Imperial College, London
GLEICK, Peter, Pacific
GURJAR, Bhola Ram, Indian Institute of Technology
KELLY, Eric Damian, Ball State University
KENNEDY, Chris, University of Toronto
NIJKAMP, Peter, Vrije Universiteit, Amsterdam
KILLOUGH, Brian, NASA Langley Research Center
KING, David, Columbia University

KNAAP, Garrit, University of Maryland
MACAULAY, Steve, West Yost Associates
PERERA, L.A.S. Ranjith, Asian Institute of Technology
REICHARD, Eric G., United State Geological Survey
RIND, David, NASA Goddard Institute for Space Studies
SHEPHERD, J. Marshall, University of Georgia
SUNDAR, Sanjivi, The Energy and Resources Institute
SUTCLIFFE, Michael, eThekwini Municipality, Durban
WHARTON, Laura, King County Department of Natural
   Resources and Parks

---

[1]Constructive feedback is gratefully acknowledged from Cities Alliance, United Nations Environment Programme (UNEP), UN-HABITAT and the World Bank
[2]Earlier versions of some material were reviewed by the World Bank 5th Urban Research Symposium (URS5) Scientific Committee

Index